Recent Advances in Statistics and Probability

Recent Advances in *Statistics and Probability*

Proceedings of the 4th International
Meeting of Statistics in the Basque Country

San Sebastián, Spain, 4 - 7 August, 1992

Editors: J.P. Vilaplana and M.L. Puri

///VSP///

Utrecht, The Netherlands, 1994

VSP BV
P.O. Box 346
3700 AH Zeist
The Netherlands

First published in 1994

ISBN 90-6764-170-7

CIP-DATA KONINKLIJKE BIBLIOTHEEK, DEN HAAG

Recent

Recent advances in statistics and probability / ed. by José
Pérez Vilaplana, Madan Lal Puri. - Utrecht : VSP
ISBN 90-6764-170-7 bound
NUGI 815
Subject headings: statistics.

Printed in The Netherlands by Koninklijke Wöhrmann, Zutphen.

Contributors

NARAYANASWAMY BALAKRISHNAN, Department of Mathematics and Statistics, McMaster University, Hamilton, Ontario, Canada L8S 4K1.

RICHARD E. BARLOW, Department of Industrial Engineering and Operations Research, College of Engineering, University of California, Berkeley, CA 94720, U.S.A.

KAYE E. BASFORD, Department of Agriculture, University of Queensland, Brisbane, Australia 4072.

MARÍA JESÚS BAYARRI GARCÍA, Departamento de Estadística e Investigación Operativa, Universidad de Valencia, E-46080 Valencia, España

JAMES O. BERGER, Department of Statistics, Mathematical Sciences Building, Purdue University, West Lafayette, IN 47907, U.S.A.

CHRIS BLONDIA, Department of Computer Sciences, Faculty of Mathematics and Computer Science, University of Nijmegen, NL 6525 ED Nijmegen, The Netherlands.

DAN BRADU, Department of Statistics, University of South Africa, SA-0001 Pretoria, South Africa.

OLGA CASALS TORRES, Departamento de Arquitectura de Computadores, Facultad de Informática, Universidad Politécnica de Cataluña, Campus Norte, E-08034 Barcelona, España.

ROGER M. COOKE, Department of Mathematics and Informatics, Delft University of Technology, NL-2600 GA Delft, The Netherlands.

ENDRE CSÁKI, Hungarian Academy of Sciences, Mathematical Institute, H-1053 Budapest, Hungary.

MIKLÓS CSÖRGÖ, Department of Mathematics and Statistics, Carleton University, Ottawa, Canada K1S 5B6.

WALTER T. FEDERER, Biometric Unit, College of Agriculture and Life Sciences, Cornell University, Ithaca, NY 14853-7801, U.S.A.

JOSÉ M.F. FERNÁNDES CRAVEIRINHA, Departamento de Engenieria Electrotécnica, Faculdade de Ciéncias e Tecnología, Universidade de Coimbra, /INESC–Núcleo de Coimbra, P-3000 Coimbra, Portugal.

CARLOS M. FERNÁNDEZ-JARDÓN FERNÁNDEZ, Departamento de Econo-mía y Estadística, Facultad de Ciencias Económicas y Empresariales, Universidad de Navarra, E-31080 Pamplona, España.

ANTONIA FÖLDES, Department of Mathematics, The City University of New York, St. George Campus, Staten Island, NY 10301, U.S.A.

JORGE GARCÍA VIDAL, Departamento de Arquitectura de Computadores, Facultad de Informática, Universidad Politécnica de Cataluña, Campus Norte, E-08034 Barcelona,

España.
JAMES E. GENTLE, Department of Statistics, George Mason University, Fairfax, VA 22030, U.S.A.
DAN R. GREENWAY, Department of Agriculture, University of Queensland, Brisbane, Australia 4072
KARL GRILL, Institut für Statistik und Wahrscheinlichkeitstheorie, Technische Universität Wien, A-1040 Wien, Austria.
CHRIS C. HEYDE, Dean, School of Mathematical Sciences, Institute of Advanced Studies, Australian National University, Canberra, Australia ACT 2601
MARIE HUŠKOVÁ, Katedra Pravdepodobnosti a Matematické Statistiky, Matematicko-Fyzikální Fakulta, University Karlovy, CzR-18600 Praha 8, Czech Republic.
LAJOS HORVÁRTH, Department of Mathematics, University of Utah, Saint Lake City, UT 84112, U.S.A.
RON S. KENNETT, Department of Mathematical Sciences, State University of New York, Binghamton, NY 13902-6000, U.S.A.
CHRISTOS LANGARIS, Department of Mathematics, Probability, Statistics and Operational Research Unit, University of Ioannina, GR-45110 Ioannina, Greece.
NHU D. LE, Biometric Section, British Columbia Agency Cancer, Vancouver, British Columbia, Canada V6T 1W5.
JACEK LESKOW, StatLab Director, Department of Statistics and Applied Probability, University of California, Santa Barbara, CA 93106-31110, U.S.A.
DAN LIU, Systems and Industrial Engineering Department, College of Engineering and Mines, University of Arizona, Tucson, AZ 85745, U.S.A.
JOSÉ MANUEL MARTÍNEZ FILGUEIRA, Departamento de Economía Aplicada, Facultad de Ciencias Económicas, Campus da Zapateira, Universidad de La Coruña, E-15071 La Coruña, España.
MAX B. MENDEL, Department of Industrial Engineering and Operations Research, College of Engineering, University of California, Berkeley, CA 94720, U.S.A.
N.S.K. NAIR, Computer Center, National Institute of Immunology, Shahid Jit singh Marg, New Delhi 110067, India.
SUBASH C. NARULA, Department of Mathematics, Linköping Institute of Technology, Linköping University, S-581 83 Linköping, Sweden.
GEORG NEUHAUS, Institut fur Mathematische Stochastik, Universität Hamburg, D-2000 Hamburg 13, Germany.
MARCEL F. NEUTS, Systems and Industrial Engineering Department, College of Engineering and Mines, University of Arizona, Tucson, AZ 85745, U.S.A.
JOSÉ PAIXÃO, Departamento de Estatística, Investigação Operacional e Computadores, Faculdade de Ciéncias, Universidade de Lisboa, P1300 Lisboa, Portugal.
WOLFGANG POLASEK, Institut für Statistik und Ökonometrie, Universität Basel, CH-4051 Basel, Switzerland.
ALEJANDRO QUINTELA DEL RÍO, Departamento de Matemáticas, Facultad de Informática, Universidad de La Coruña, Campus da Zapateira, E-15071 La Coruña, España.
PAL RÉVÉSZ, Institut für Statistik und Wahrscheinlichkeitstheorie, Technische Universität Wien, A-1040 Wien, Austria.
DAVID RÍOS INSÚA, Departamento de Inteligencia Artificial, Facultad de Informatica, Universidad Politécnica de Madrid, Campus de Montegacedo, E-28660 Boadilla del Monte (Madrid), España.

JORGE LUIS ROMEU, Department of Mathematics, State University of New York, Cortland, NY 13045, U.S.A.

FABRIZIO RUGGERI, CNR-IAMI, Istituto per le Applicazioni della Matemática e dell'Informática, I-20131 Milano, Italy.

O.P.SHARMA, Department of Mathematics, Indian Institute of Technology, New Delhi 110016, India.

FABIO SPIZZICHINO, Dipartimento di Matemática "Guido Castelnuovo", Università "La Sapienza", I-00185 Roma, Italy.

VINCENT A. SPOSITO[1], Statistical Laboratory and Department of Statistics, Iowa State University, Ames, IA 50011, U.S.A.

R. SYSKI, Statistics Program, College of Computer, Mathematical and Physical Sciences, University of Maryland, College Park, MA 20742-4201, U.S.A.

TRAN HUNG THAO, Institute of Mathematics, National Center for Scientific Research, 10 000 Hanoi, Viet Nam.

LINO TRALHÃO, Departamento de Matemática, Faculdade de Ciencias e Tecnologıa, Universidade de Coimbra, /INESC Núcleo de Coimbra, P-3000 Coimbra, Portugal.

JOSÉ A. VILAR FERNÁNDEZ, Departamento de Matemáticas, Facultad de Informática, Universidad de La Coruña, Campus da Zapateira, E-15071 La Coruña, España.

JUAN M. VILAR FERNÁNDEZ, Departamento de Matemáticas, Facultad de Informática, Universidad de La Coruña, Campus da Zapateira, E-15071 La Coruña, España.

JACEK WESOLOWSKI, Politechnika Warszawska, Instytut Matematyki, PL-00661 Warszawa, Poland.

SHELEMYAHU ZACKS, Department of Mathematical Sciences, State University of New York, Binghamton, NY 13902-6000, U.S.A.

JAMES V. ZIDEK, Department of Statistics, University of British Columbia, Vancouver, British Columbia, Canada V6T 1W5.

[1]Deceased

Preface

In recent years, significant progress has been made in statistical theory. New methodologies have emerged, as an attempt to brige the gap between theoretical and applied approaches. In this volume we present some of these developments, which have already had significant impacts on modeling, design, and analysis of statistical experiments.

The chapters included cover a wide range of topics of current interest in applied as well as theoretical statistics and probability. They include some aspects of the design of experiments in which there are current development, regression methods, decision theory, nonparametric theory, simulation and computational statistics, time series, reliability, and queueing network. Also included are chapters on some aspects of probability theory which, apart from their intrinsic mathematical interest, have significant applications in statistics.

The volume contains a selection of the papers presented at the Fourth International Meeting of Statistics in the Basque Country, held in the Palace of Miramar in San Sebastián, Spain, from August 4th to 7th, 1992. This event was organized under the auspicies of the Department of Applied Mathematics and Statistics and Operations Research, and the Summer Courses of the University of the Basque Country to pay homage to the Quincentennial of the Discovery of America by Christopher Columbus.

The aim of this meeting was to provide an opportunity for personal contact among scholars from different intellectual centers of the world who could represent a comprehensive cross section of contemporary thinking on some of the areas of theoretical as well as applied statistics and probability. In selecting the papers for the volume, we were guided by the advice of the referees whose names appears in the Acknowledgements following this Preface. It is a pleasure to express our appreciation to them as well as to the authors for their help and cooperation.

The volume provides a wealth of ideas for advanced graduate students and research workers in statistics and probability; it should also be useful to statisticians in industry, agriculture, engineering, medical sciences, and other fields.

The Scientific Committee of the conference consisted of V. Barnett (U.K.), R. E. Barlow (U.S.A.), J. O. Berger (U.S.A.), B. W. Conolly (U.K.), W. T. Federer (U.S.A.), A. Iztueta (Spain), M. F. Neuts (U.S.A.) Y. V. Prohorov (Russia), M. L. Puri (U.S.A.), P. Révész (Hungary), D. Ríos (Spain), J. L. Romeu (U.S.A.), P. K. Sen (U.S.A.), J. Yurramendi (Spain), S. Zacks (U.S.A.). We take this opportunity to express our sincere thanks to the members of this committee for their overall help in making the conference a great success.

These efforts resulted in the rich and active participation of about 200 scholars from 32 countries. Special thanks are due to members of the organizing committee, specially to Mr. V. Gascon (Executive Secretary), for providing an excellent atmosphere of warmth, friendliness, and generous hospitality; our best thank to all them.

We would also like to take this opportunity to express our thanks to the Spanish Government, Ministry of Education of Spain, the University of the Basque Country, the Basque Government, the Statutory Deputation of Guipúzcoa and the Summer Courses in San Sebastián of the University of the Basque Country for their financial support, and the Palace of Miramar Consortium, the mayors of San Sebastián, and Oñate, and the British Council-Elcano Programme for their collaboration.

José Pérez Vilaplana
Madan Lal Puri

Bilbao, España
Bloomington, U.S.A.

March 1993

Acknowledgments

More than 100 papers were submitted for publication in this volume. We regret that we could not publish all of them. In selecting papers for publication in this book, we were guided by the advice of the following referees, who were an indispensable part of our editorial process. We express our deepest gratitude to them for their help.

R. Adler (Israel), R. E. Barlow (U.S.A.), S. Basu (U.S.A.), P. Bauer (Austria), J. O. Berger (U.S.A.), C. Blondia (The Netherlands), S. Bose (U.S.A.), O. J. Boxma (The Netherlands), O. Bunke (Germany), E. Carlstein (U.S.A.), O. Casals Torres (Spain), B. W. Conolly (U.K.), R. M. Cooke (The Netherlands), C. Christiansen (U.S.A.), J. Crowell (U.S.A.), E. Csáki (Hungary), M. Csörgö (Canada), K-F. Cheng (Taiwan, R.O.C.), S. C. Choi (U.S.A.), A. Das Gupta (U.S.A.), P. Deheuvels (France), A. P. Dempster (U.S.A.), D. Driscoll (U.S.A.), J. P. Faria (Portugal), W. T. Federer (U.S.A.), J. M. F. Fernándes Craveirinha (Portugal), L. Fernholz (U.S.A.), A. Földes (Hungary), J. García Vidal (Spain), Th. A. Gasser (Switzerland), F. J. Girón González-Torre (Spain), J. C. Gower (Switzerland), J. E. Grizzle (U.S.A.), P. Guttorp (U.S.A.), P. Hall (Australia), J. A. Hartigan (U.S.A.), W. Hauck (U.S.A.), C. C. Heyde (Australia), L. Horváth (U.S.A.), M. Hušková (Czech Republic), T. Irony (U.S.A.), A. J. Izenmann (U.S.A.), A. Janssen (Germany), J. B. Kadane (U.S.A.), A. F. Karr (U.S.A.), C. W. Kish (U.S.A.), J. Kuelbs (U.S.A.), P. A. Lachenbruch (U.S.A.), Ch. Langaris (Greece), N. Langberg (Israel), L. Lébart (France), D. V. Lindley (U.K.), D. M. Lucantoni (U.S.A.), A. W. Marshall (U.S.A.), J. Martín (Spain), K. J. McConway (U.K.), K. S. Meier-Hellstern (U.S.A.), I. Meilijson (Israel), M. Mendel (U.S.A.), H. Migon (Brasil), M. L. Moeschberger (U.S.A.), E. Moreno (Spain), H-G. Müller (U.S.A.), B. Natvig (Norway), M. F. Neuts (U.S.A.), C. O'Cinneide (U.S.A.), A. Öztürk (Turkey), E. Perkins (Canada), L. Piccinato (Italy), D. Piccolo (Italy), M. Pollak (U.S.A.), D. Rauschenberg (U.S.A.), R. D. Reiss (Germany), P. Révész (Austria), D. Ríos Insúa (U.S.A.), R. Y. Rubinstein (Israel), P. Shaman (U.S.A.), A. N. Shiryaev (Russia), N. D. Singpurwalla (U.S.A.), V. Solana (Spain), F. Spizzichino (Italy), M. F. Steel (The Netherlands), M. S. Pepe (U.S.A.), P. Switzer (U.S.A.), Y. Takahashi (Japan), R. J. Tibshirani (Canada), A. Ullah (U.S.A.), I. Verdinelli (Italy), H. M. Wadsworth (U.S.A.), L. A. Wasserman (U.S.A.), D. Weinen (U.S.A.), W. W. Wertz (Austria), M. West (U.S.A.), W. T. Wright (U.S.A.), C. W. Wrightson (U.S.A.), E. Yashchin (U.S.A.), K. Yoshihara (Japan), S. Zacks (U.S.A.).

We apologize for any omissions.

J.P.V.
M.L.P.

Table of Contents

Conference

"Nicolás de Arriquibar"

Recent Adv. in Stat. and Prob., pp. 3-23
J. Pérez Vilaplana and M. L. Puri (Eds)

Diagnostic Procedures for Analysis of Variance

WALTER T. FEDERER
Biometrics Unit, College of Agriculture and Life Sciences, Cornell University, Ithaca, NY 14853-7801, U.S.A.

Abstract – The analysis of variance (ANOVA) developed by Sir Ronald A. Fisher around 1920 is a partitioning of the total variance into its component parts. Several uses may be made of the results from an analysis of variance such as, e.g., calculating efficiency of stratification or blocking, estimation of variance components, making significance or hypothesis tests, and obtaining an error mean square for constructing simultaneous confidence intervals and making multiple comparisons. Each use carries with it a set of assumptions about the statistical design, the response model, and the distribution of the random elements of the model. Diagnostic procedures to detect deviation from the assumptions have been developed for several types of departures. The statistical design must be such that it is representative of the population for which inferences are being made. An appropriate response model must be validated for an investigation and not merely obtained by definition. Some response models and methods for determining an appropriate response model are presented. Many response models are based upon additivity of effects. Several tests for non-additivity are discussed. Many uses of the results in an ANOVA require homoscedascity, and several procedures for detecting variance heterogeneity are available, some of which are discussed. Most uses require independent observations but this requirement is violated in certain investigations. Four situations leading to non-independence of responses are discussed. Statistical designs and analyses for removing the effect causing non-independence are presented. Patterns, trends, and discrepant observations of residual effects need to be studied. An example of a single outlier is used to demonstrate how the interpretation of results can be changed if the outlier is ignored. In modeling responses from two-factor factorials, it is desirable to do this in a parsimonious manner and retain as few parameters as possible. Bi-plot and AMMI procedures are useful in this context. Finally, after an investigator has summarized the results of an investigation, a critical examination, a post mortem, should be made for all aspects of the investigation. In particular, the statistical design, the analyses, the results, and the interpretations should be scrutinized carefully to ascertain that nothing has gone wrong and that everything makes sense.

1. INTRODUCTION

An analysis of variance (ANOVA) is a partitioning of the total variance into its component parts. The ideas in the development of ANOVA had their beginnings when the total sum of squares from a regression analysis was partitioned into that due to regression and that due to deviations from regression and when J. Arthur Harris (1913) introduced the intraclass correlation coefficient. Sir Ronald A. Fisher formalized these ideas and called this the analysis of variance in a series of publications around 1920(1918, 1923, 1925, etc.). In the published statistical literature, an ANOVA may mean different things to different writers and users of an analysis of variance. Perhaps the most frequent use associated with an analysis of variance is the computation of an F or z statistic. Many users appear to denote this use as an ANOVA. Many include the uses being made of an ANOVA as part of the definition. For the ensuing discussion, we use the definition of an ANOVA as given in the beginning of this paragraph devoid of the use to which it is put, i.e., a partitioning of the total variance into its component parts. Rather than using the name analysis of variance, perhaps the name for this procedure should have been a partitioning of variance.

To illustrate the ideas, we make use of two simple examples – a balanced one-way array such as a completely randomized experiment design (CRED) or a stratified simple random sample survey design and a balanced two-way array such as a randomized complete block experiment design (RCBED) or a two- way factorial treatment design. For a balanced one-way array the total sum of squares for rv observations from v groups of r items each is partitioned as:

$$\sum_{i=1}^{v}\sum_{j=1}^{r} Y_{ij}^2 = \sum_{i=1}^{v}\sum_{j=1}^{r}[(Y_{ij}-\bar{y}_{i.})+(\bar{y}_{i.}-\bar{y}_{..})+\bar{y}_{..}]^2 =$$

$$= \sum_{i=1}^{v}\sum_{j=1}^{r}(Y_{ij}-\bar{y}_{i.})^2 + r\sum_{i=1}^{v}(\bar{y}_{i.}-\bar{y}_{..})^2 + rv\bar{y}_{..}^2 \qquad (1)$$

where Y_{ij} is the j-th observation in the i-th group, $\bar{y}_{i.}$ is the arithmetic mean for group i, and $\bar{y}_{..}$ is the arithmetic mean of the rv observations. The results from this algebraic partitioning of the total sum of squares may be put in what is known as an analysis of variance table as in Table 1.1.

Table 1.1.
ANOVA for a one-way classification

Source of Variation	d.f.	Sum of Squares	Mean Square (Variance)
Total	rv	$\sum_{i=1}^{v}\sum_{j=1}^{r} Y_{ij}^2$	
Correction for mean	1	$rv\bar{y}_{..}^2$	
Groups	$v-1$	$r\sum_{i=1}^{v}(\bar{y}_{i.}-\bar{y}_{..})^2 = G$	$G/(v-1)$
Within groups	$v(r-1)$	$\sum_{i=1}^{v}\sum_{j=1}^{r}(Y_{ij}-\bar{y}_{i.})^2 = W$	$W/v(r-1)$

For a balanced two-way array such as a randomized complete block experiment design with r blocks and v treatments, the total sum of squares may be partitioned as follows:

$$\sum_{i=1}^{v}\sum_{j=1}^{r} Y_{ij}^2 = \sum_{i=1}^{v}\sum_{j=1}^{r}\left[(Y_{ij}-\bar{y}_{i.}-\bar{y}_{.j}+\bar{y}_{..})+(\bar{y}_{i.}-\bar{y}_{..})+(\bar{y}_{.j}-\bar{y}_{..})+\bar{y}_{..}\right]^2 =$$

$$= \sum_{i=1}^{v}\sum_{j=1}^{r}(Y_{ij} - \overline{y}_{i.} - \overline{y}_{.j} + \overline{y}_{..})^2 + r\sum_{i=1}^{v}(\overline{y}_{i.} - \overline{y}_{..})^2 + v\sum_{j=1}^{r}(\overline{y}_{.j} - \overline{y}_{..})^2 + rv\overline{y}_{..}^2, \quad (2)$$

where $\overline{y}_{.j}$ is the arithmetic mean of the j-th block and the other symbols are as defined above with groups being synonymous with treatments. We may put these sum of squares in an ANOVA as in Table 1.2.

Table 1.2.
ANOVA for a RCBED.

Source of Variation	d.f.	Sum of Squares	Mean Square (Variance)
Total	rv	$\sum_{i=1}^{v}\sum_{j=1}^{r} Y_{ij}^2$	
Correction for mean	1	$rv\overline{y}_{..}^2$	
Blocks	$r-1$	$v\sum_{j=1}^{r}(\overline{y}_{.j} - \overline{y}_{..})^2 = B$	$B/(r-1)$
Treatments	$v-1$	$r\sum_{i=1}^{v}(\overline{y}_{i.} - \overline{y}_{..})^2 = T$	$T/(v-1)$
Remainder	$(r-1)(v-1)$	$\sum_{i=1}^{v}\sum_{j=1}^{r}(Y_{ij} - \overline{y}_{i.} - \overline{y}_{.j} + \overline{y}_{..})^2 = R$	$R/(r-1)(v-1) = E$

Note that the above results make no assumptions about a response model, about effects, about independence of effects, about additivity of effects, about homoscedasticity, or about the statistical distribution of any of the elements presented. The above is purely algebraic manipulation. Requirements for making practical use of the above will vary with the application made of the results. The first requirement is that the items listed under "Source of Variation" must have practical meaning for the experimenter who obtained these results. Once it has been established that the items have meaning, the above partitioning may be used in several ways. For the partitioning in equation (1) and where the groups are strata, the use requiring the fewest assumptions is the computation of the efficiency of a stratified survey design relative to a simple random sample. The following form of a response model is implied:

$$Y_{ij} = f(\mu, \alpha_i) + \epsilon_{ij} \quad (3)$$

where $f(\mu, \alpha_i)$ is usually taken to be $\mu + \alpha_i$ with μ being an overall mean of a population and $\mu + \alpha_i$ being the true mean for the i-th stratum, and with ϵ_{ij} being a random error effect associated with the ij-th observation. A measure of efficiency of stratification versus no stratification is:

$$\frac{\sum_{i=1}^{v}\sum_{j=1}^{r}(Y_{ij} - \overline{y}_{..})^2 \,/\, (rv-1)}{W \,/\, v(r-1)} \quad (4)$$

For a RCBED, a measure of efficiency of an RCBED relative to a CRED is:

$$\frac{B+R+R\,/\,(r-1)}{E\,(rv-1)} \tag{5}$$

For this use, it is assumed that the effects are additive in the response model and that $W\,/\,v(r-1)$ is an unbiased estimate of the within stratum variance and that $R\,/\,(v-1)(r-1) = E$ is an unbiased estimate of the variance for the RCBED. This implies that the ϵ_{ij} are independently distributed with mean zero and variance parameter σ_ϵ^2. Randomization was used to obtain the unbiasedness and independence.
For a variance component analysis, a linear model of the form

$$Y_{ij} = \mu + \alpha_i + \epsilon_{ij} \qquad \text{for a CRED} \tag{6}$$

or

$$Y_{ij} = \mu + \tau_i + \beta_j + \epsilon_{ij} \qquad \text{for a RCBED} \tag{7}$$

is assumed. In addition, it is assumed that the α_i are identically and independently distributed (I.I.D.) with mean zero and variance σ_α^2, that the τ_i are I.I.D. $(0,\sigma_\tau^2)$, the β_j are I.I.D. $(0,\sigma_\beta^2)$, and that the ϵ_{ij} are I.I.D. $(0,\sigma_\epsilon^2)$. In some cases, the various variances could be a linear combination of other variance components and this would affect the use of a variance component. Note that the form of the distribution is unspecified and the effects are random.
For many situations, the effects are fixed effects and the expected value, $E(\cdot)$, of an effect is the effect parameter and not necessarily zero. For most CREDs and RCBEDs, $E(\alpha_i) = \alpha_i$ and $E(\tau_i) = \tau_i$, and it is desired to have interval estimates of these effects and perhaps to perform significance or hypothesis tests. For this, it is necessary to specify the form of the distribution for the ϵ_{ij}, which is mostly taken to be a normal distribution, i. e., N.I.I.D. Note that it only makes experimental sense to consider the β_j as random effects even though the majority of statistical methods books imply that they are fixed effects.
In the following, we shall look at techniques designed to determine if one or more of the requirements for a particular use has been violated. *Assuming* that all the requirements for a statistical procedure are satisfied does not mean that *they are* for a particular experiment. Conclusions about results may be considerably altered if the requirements for a procedure are not satisfied.

2. SELECTION OF A RESPONSE MODEL EQUATION

It is common practice in statistical literature to state "*the linear model is* ...". This is an incorrect statement in that the best anyone can do is to say "*a linear model is* ...". Proving that a linear model is unique, or even that the model is linear, is next to impossible in the majority of cases. It can, however, be a good first approximation to an appropriate response model for an investigation. In the "*chalkboard world*" of the classroom, a linear model is obtained by definition and not from the actual situation in the investigation. However, an investigator does not have this luxury, but must make a decision about which response model to use for each investigation being conducted. A good discussion of model selection may be found in Box (1980).
Box and Cox (1964) in an excellent paper, work with a parametric family of transformations of data from Y to Y^λ such that Y^λ for a specified value of λ which is a normal

homoscedastic linear model. They present procedures for separating the contributions of normality, homoscedasticity, and additivity. The two important examples they consider are:

$$Y^{(\lambda)} = \begin{cases} (Y^\lambda - 1)/\lambda \text{ or simply } Y^\lambda & \text{for} \lambda \neq 0 \\ \log Y & \text{for} \lambda = 0, \quad Y > 0 \end{cases} \tag{8}$$

and

$$Y^{(\lambda)} = \begin{cases} \left[(Y + \lambda_2)^{\lambda_1} - 1\right] / \lambda_1 & \text{for} \lambda_1 \neq 0, \quad \lambda_2 > 0 \\ \log(Y + \lambda_2) & \text{for} \lambda_1 = 0, \quad \text{and } Y > -\lambda_2 \end{cases} \tag{9}$$

They state the procedure is simpler if a normalized transformation is used for (8) and (9), respectively, as follows which for the simple power transformation is

$$Z^{(\lambda)} = (Y^\lambda - 1) \,/\, \lambda \tilde{Y}^{\lambda - 1} \;, \tag{10}$$

and for the power transformation with a shifted location is

$$Z^{(\lambda)} = \left[(Y + \lambda_2)^{\lambda_1} - 1\right] \,/\, \lambda_1 \left[\text{gm}(Y + \lambda_2)\right]^{\lambda_1 - 1} \;, \tag{11}$$

where $\tilde{Y}$ is the geometric mean of the Ys and gm $(Y + \lambda_2)$ is the geometric mean of the $(Y + \lambda_2)$s. Let $S(Z, \lambda)$ denote the residual sum of squares in an ANOVA. $S(Z, \lambda)$ is computed for a series of values of λ, and the λ producing the minimum residual sum of squares is the value to use for the power transformation. They show how to compute an approximate $(100 - \alpha)$ per cent confidence interval for λ. It should be noted that there may be no value of which fits the data set in that the correct response model for the data is not in this family of power transformations.

Another model that is rather widely used is the one for a diallel crossing experiment in genetics. This model has been found to have uses in a variety of situations, such as, e.g., psychological and personnel rating investigations. A response model of this type was introduced by W. G. Cochran in a paper by Sprague and Tatum (1941). Griffing (1956) presents models for several diallel crossing situations. One of these is discussed in some detail in Federer (1955). The concepts of general combining ability (how a line combines on the average with the other $v - 1$ lines with which it is crossed) and specific combining ability (how a line performs with a particular line) were used in constructing the response models.

Federer (1979, 1993) has constructed a number of response models for dealing with cropping systems for multiple cropping situations with and without changes in densities of the crops in a mixture. The concepts of general mixing ability, general competing ability, bi-mixing ability, bi-competing ability, tri-mixing ability, etc. as well as the effect of a second crop's density on the yield of a given crop, have been constructed and applied to specific examples. Procedures for combining the yields of the components of a mixture system were also devised. The ideas associated with diallel crossing models and those of Martin (1980) were useful in constructing these models.

For r-row by c-column designs, several variations have appeared in the literature, but the following is mostly used by textbook writers whether it is appropriate or not for the examples selected to illustrate the computations:

$$Y_{hij} = \mu + \rho_h + \gamma_i + \tau_j + \epsilon_{hij} \;, \tag{12}$$

where Y_{hij} is the response for treatment j in the h-th row and ith column, μ is a general mean effect, ρ_h is an effect for the h-th row, γ_i is an effect for the i-th column, τ_j is an effect for the j-th treatment, and ϵ_{hij} are random error effects which are I.I.D $(0, \sigma_\epsilon^2)$. Cox (1958) proposed a response model for the situation wherein the gradients within each column (or row) differ. The form of the response model is

$$Y_{ij} = \mu + \gamma_i + \beta_i a_{ij} + \tau_j + \epsilon_{ij} \,, \tag{13}$$

where Y_{ij} is the response for treatment j in column i, β_i is the linear regression of Y_{ij} on a_{ij} to depict the linear regression in column i, a_{ij} are row constants for the j-th treatment in the i-th column, measured from a zero mean at the center of the row with $\sum_{i=1}^{c} a_{ij}^2 = \sum_{j=1}^{v} a_{ij}^2 = 1$ and $v = r$, and the other effects are defined as in (12). The γ_i are the location parameters for the linear regressions. As Cox (1958) points out, other polynomial regression coefficients may be added to (13). He illustrates this by adding quadratic after linear constants and demonstrates the computations on a numerical example. Federer and Schlottfeldt (1954) and Outhwaite and Rutherford (1955) used a model with constant regressions in each row but a more appropriate response model and analysis for their example would have been (13). It should also be noted that fitting a polynomial of degree $v - 1$ as used by Outhwaite and Rutherford (1955) is the same as using model (12).

For r-row (period) by c-column designs where the treatments and rows (periods) are added sequentially to the same sampling unit and which are known as rotation and as repeated measures change-over designs, a variety of response models may be utilized depending upon the nature of the responses. There may be direct effects of the treatment in the period in which it is applied, first-, second-, etc. period carry-over effects of the previous treatment, continuing effects of treatments, and permanent effects (see e.g., Kershner and Federer, 1981, who present designs for obtaining estimates of parameters for a variety of response models).

The population structures associated with an experiment and/or treatment design has been ignored in statistical literature except for Fisher (1935) and Federer (1976*a*, 1976*b*, 1977, 1991). The latter author discusses population structures and response models for block, row-column, split-plot, and split-block designs. Kempthorne (1952), e.g., discusses randomization models over the particular set of experimental units for a particular experiment without regard as to how the sample was obtained from the population. Assuming that any sample obtained by an experimenter is representative of the target population can be grossly incorrect and often is. More attention needs to be given to the planning stages of an investigation in order to accurately make inferences about the target population.

Several response models have been proposed to take account of various forms of non-additivity in experiments. Some of these are described in the following section.

3. TESTS FOR NON-ADDITIVITY

In a classic paper, Tukey (1949) provided a one degree of freedom test for non-additivity in a two-way array such as an RCBED. The test statistic is

$$\frac{\left[\sum_{i=1}^{v}\sum_{j=1}^{r}\left(Y_{ij} - \bar{y}_{i.} - \bar{y}_{.j} + \bar{y}_{..}\right)(\bar{y}_{i.} - \bar{y}_{..})\left(\bar{y}_{.j} - \bar{y}_{..}\right)\right]^2}{\sum_{i=1}^{v}(\bar{y}_{i.} - \bar{y}_{..})^2 \sum_{j=1}^{r}(\bar{y}_{.j} - \bar{y}_{..})^2} = \frac{\left[\sum_{i=1}^{v}\sum_{j=1}^{r}\hat{\epsilon}_{ij}(\hat{\epsilon}_{ij} - \hat{\delta}_{ij})\right]^2}{\sum_{i=1}^{v}\sum_{j=1}^{r}(\hat{\epsilon}_{ij} - \hat{\delta}_{ij})^2} \tag{14}$$

which is the sum of squares due to regression of $\hat{\epsilon}_{ij}$ on $(\hat{\epsilon}_{ij} - \hat{\delta}_{ij})$. In the above

$$\hat{\epsilon}_{ij} = Y_{ij} - \overline{y}_{i.} - \overline{y}_{.j} + \overline{y}_{..} \tag{15}$$

and

$$\hat{\delta}_{ij} = Y_{ij} - \overline{y}_{i.}\,\overline{y}_{.j} \,/\, \overline{y}_{..} \,. \tag{16}$$

For a three way classification, it is suggested that the residual for the alternate multiplicative model be computed as

$$\hat{\delta}_{hij} = Y_{hij} - \overline{y}_{h..}\,\overline{y}_{.i.}\,\overline{y}_{..j} \,/\, \overline{y}^2_{...} \,, \tag{17}$$

for a four way classification, that the residual be computed as

$$\hat{\delta}_{ghij} = Y_{ghij} - \overline{y}_{g...}\,\overline{y}_{.h..}\,\overline{y}_{..i.}\,\overline{y}_{...j} \,/\, \overline{y}^3_{...} \,, \tag{18}$$

and so forth for higher way classifications. Then, use may be made of a form similar to equation (14) to compute the sum of squares for non-additivity. Note that $\hat{\delta}_{hij}$ in (17) is different from the residual used by Tukey for a latin square design (see e.g., Snedecor and Cochran, 1980, section 15.14). The above forms consider all interactions and not only two factor interactions as suggested by Tukey.

Robson (1970) studied the family of transformations

$$E[Y_{ij}] = (\mu + \rho_i + \gamma_j)^p \tag{19}$$

and reported on a test for non-additivity for $p = 2$. The residual is computed as

$$\hat{\pi}_{ij} = Y_{ij} - (\hat{\mu} + \hat{\rho}_i + \hat{\gamma}_j)^2 \,, \tag{20}$$

where

$$(\hat{\gamma}_j - \overline{\gamma}_.) = \sqrt{(\overline{y}_{.j} - \sigma^2_\rho)} - \sum_{j=1}^{r} \sqrt{(\overline{y}_{.j} - \sigma^2_\rho)} \,/r \,, \tag{21}$$

$$(\hat{\rho}_i - \overline{\rho}_.) = \sqrt{(\overline{y}_{i.} - \sigma^2_\gamma)} - \sum_{i=1}^{v} \sqrt{(\overline{y}_{i.} - \sigma^2_\gamma)} \,/v \,, \tag{22}$$

$$\sigma^2_\rho = \sum_{i=1}^{v} (\hat{\rho}_i - \overline{\rho}_.)^2 \,/\, v \,, \tag{23}$$

$$\sigma^2_\gamma = \sum_{j=1}^{r} (\hat{\gamma}_j - \overline{\gamma}_.)^2 \,/\, r \,, \tag{24}$$

and

$$\sum_{j=1}^{r} \sqrt{(\overline{y}_{.j} - \sigma^2_\rho)} \,/r \; = \sum_{i=1}^{v} \sqrt{(\overline{y}_{i.} - \sigma^2_\gamma)} \,/v \,. \tag{25}$$

Then, $\hat{\pi}_{ij}$ is substituted for $\hat{\delta}_{ij}$ in equation (14) to obtain Robson's test for non-additivity for the above model. He notes that a similar test may be constructed for $p = -1$, the reciprocal transformation. Robson (1970) states that Tukey's test compares the additive model

$$E[Y_{ij}] = \mu + \tau_i + \beta_j \,, \tag{26}$$

against the multiplicative alternative model

$$E[Y_{ij}] = (\mu + \tau_i)(\mu + \beta_j \,/\, \mu \,, \tag{27}$$

whereas his test against the additive model is for the alternative hypothesis

$$E[Y_{ij}] = (\mu + \tau_i + \beta_j)^2 \,. \tag{28}$$

A numerical comparison of Robson's and Tukey's test along with a third one was made by Federer (1970) on a numerical example involving counts of an insect and treatments for controlling the insects in an RCBED. Despite the fact that a square root transformation might be appropriate, Tukey's test recovered the largest sum of squares.

Mandel (1961, 1971) extended Tukey's test for non-additivity by regressing the residual from (15) on the difference between the residuals in (15) and (16) for each treatment (and/or for each block) in an RCBED and then computing a sum of squares to compare the v treatment regressions. Kirton (1984) extended Mandel's procedure to include simultaneously both categories of a two way classification, and to compute an ANOVA using the absolute values of the residuals both for an additive and for a non-additive model. The procedure is presented here, and the response model when both blocks and treatments non-additivity effects are included, is

$$E[Y_{ij}] = \mu + \tau_i + \beta_j + \tau_i\beta_j/\mu + (\rho_j - 1)\tau_i + (\rho_i - 1)\beta_j + \epsilon_{ij} \,, \tag{29}$$

where ρ_i is the linear regression coefficient for treatment i, i.e.,

$$\hat{\rho}_i = \frac{\sum_{j=1}^{r} Y_{ij}(\bar{y}_{.j} - \bar{y}_{..})}{\sum_{j=1}^{r}(\bar{y}_{.j} - \bar{y}_{..})^2} \,, \tag{30}$$

ρ_j is the linear regression coefficient for block j, i.e.,

$$\hat{\rho}_j = \frac{\sum_{i=1}^{v} Y_{ij}(\bar{y}_{i.} - \bar{y}_{..})}{\sum_{i=1}^{v}(\bar{y}_{i.} - \bar{y}_{..})^2} \,, \tag{31}$$

and the remaining symbols are as defined above. The formulae for the sums of squares are given in Table 3.1. In order to determine which p to use for the transformation Y^p, we compute

$$\alpha = \frac{\sum_{j=1}^{r} \hat{\beta}_j\hat{\rho}_j}{\sum_{j=1}^{r} \hat{\beta}_j^2} = \frac{\sum_{i=1}^{v} \hat{\tau}_i\hat{\rho}_i}{\sum_{i=1}^{v} \hat{\tau}_i^2} \,. \tag{32}$$

The value p is equal to $1 - \alpha\bar{y}_{..}$. Then to compare the various blocks, use $\hat{\rho}_j - 1 - \alpha\hat{\beta}_j$, and to compare the various treatments, use $\hat{\rho}_i - 1 - \alpha\hat{\tau}_i$.

There is a considerable literature on non-additivity and references may be found in the Current Index to Statistics. Also, since a considerable amount of work by D. S. Robson and his students has not been published, it would be wise for researchers in this area to check the Annual Reports of the Biometrics Unit at Cornell University for Technical Reports and Theses on this topic. The Annual Reports have been widely distributed. Also, the same type of search should be made at Princeton University, where J. W. Tukey and his students have done considerable work in this area.

Table 3.1.
Analysis of variance to check for additivity and uniformity of levels of factors for a randomized complete block design

Source of Variation	d.f.	Sum of Squares*
Total	rv	$\sum_{i=1}^{v}\sum_{j=1}^{r} Y_{ij}^2$
Correction for mean	1	$Y_{..}^2/rv = C$
Blocks	$r-1$	$\sum_{j=1}^{r} Y_{.j}^2/v - C$
Treatments	$v-1$	$\sum_{i=1}^{v} Y_{i.}^2/r - C$
Blocks × treatments	$(r-1)(v-1)$	$\sum_{i=1}^{v}\sum_{j=1}^{r}(Y_{ij} - \overline{y}_{i.} - \overline{y}_{.j} + \overline{y}_{..})^2$
Nonadditivity	1	$TNA = \dfrac{\left[\sum_{i=1}^{v}\sum_{j=1}^{r} Y_{ij}(\overline{y}_{i.} - \overline{y}_{..})(\overline{y}_{.j} - \overline{y}_{..})\right]^2}{\sum_{i=1}^{v}(\overline{y}_{i.} - \overline{y}_{..})^2 \sum_{j=1}^{r}(\overline{y}_{.j} - \overline{y}_{..})^2}$
Blocks (deviations)	$r-2$	$\sum_{j=1}^{r}(b_j - 1)^2 \sum_{i=1}^{v}(\overline{y}_{i.} - \overline{y}_{..})^2 - TNA$
Treatments (deviations)	$v-2$	$\sum_{i=1}^{v}(b_i - 1)^2 \sum_{j=1}^{r}(\overline{y}_{.j} - \overline{y}_{..})^2 - TNA$
Remainder	$(r-2)(v-2)$	by substraction

*$Y_{i.}$ and $\overline{y}_{i.}$ = i-th treatment total and mean, respectively,

$Y_{.j}$ and $\overline{y}_{.j}$ = j-th block total and mean, respectively,

$Y_{..}$ and $\overline{y}_{..}$ = grand total and mean, respectively,

$\hat{\rho}_j = b_j \sum_{i=1}^{v} Y_{ij}(\overline{y}_{i.} - \overline{y}_{..}) / \sum_{i=1}^{v}(\overline{y}_{i.} - \overline{y}_{..})^2$,

$\hat{\rho}_i = b_i \sum_{j=1}^{r} Y_{ij}(\overline{y}_{.j} - \overline{y}_{..}) / \sum_{j=1}^{r}(\overline{y}_{.j} - \overline{y}_{..})^2$.

4. HOMOSCEDASTICITY

Bartlett (1937) presented a test for homogeneity of a set of variances for a test of the null hypothesis $H_0 : \sigma_1^2 = \sigma_2^2 = \ldots = \sigma_v^2$ against the alternative H_1 : not H_0. This test, sometimes called Bartlett's M test, is

$$M = \frac{N \ln (\sum_{j=1}^{v} f_j s_j^2 / N) - \sum_{j=1}^{v} f_j \ln s_j^2}{1 + (\sum_{j=1}^{v} 1/f_j - 1/N)/3(v-1)} , \tag{33}$$

where $N = \sum_{j=1}^{v} f_j$, s_j^2 is the j-th variance in the set, ln is the natural logarithm, and M is approximately distributed as chi square with $v - 1$ degrees of freedom.
For the following tests, let $f_j = f$ degrees of freedom for all j. Then, Cochran's W test is computed as

$$W = s_{max}^2 / \sum_{j=1}^{v} s_j^2 , \tag{34}$$

where $s_{max}^2 = \max \{s_1^2, s_2^2, \ldots, s_v^2\}$. Cochran (1941) gives upper 5% points for W when $v = 3, 4, \ldots, 10$ and $f = 1, 2, 3, 4, 5, 6, 8, 10$. Hartley (1950) proposed the following test for variance heterogeneity for a set of $v > 2$ estimated variances s_j^2 :

$$F_{max} = s_{max}^2 / s_{min}^2 \tag{35}$$

where s_{max}^2 is defined above and $s_{min}^2 = \min \{s_1^2, s_2^2, \ldots, s_v^2\}$. Tables for the F_{max} statistic may be found in Pearson and Hartley (1962). The tests W and F_{max} provide for quick tests of variance heterogeneity. All three tests, M, W, and F_{max}, are non-robust in that the actual confidence or significance levels are sensitive to the form of the underlying distributions. They depend heavily upon the normality assumption. To remedy this sensitivity to the underlying distribution, Box (1953) proposed the following procedure:

i) Divide each sample into c subsamples of size m,
ii) Compute the subsample variances $s_{jk}^2, j = 1, 2, \ldots, v, k = 1, 2, \ldots, m$,
iii) Set $X_{jk} = \ln s_{jk}^2$, and then $E[X_{jk}] = \ln \sigma_j^2$, with variance $\text{Var}[X_{jk}] = 2/(m-1) + \gamma/m$, where γ stands for kurtosis,
iv) Compute a one-way analysis of variance on the X_{jk}, and
v) Use the F statistic to test H_0.

For the Box procedure, there is no firm rule for selecting the values of c and m, and the investigator is left to rely upon his own judgement. Scheffé (1959) gives more details on this procedure.
For a two-way classification such as a RCBED or a two-way factorial, one may use Spearman's rank order correlation to test for a relationship between treatment (factor) means and residual sums of squares by computing the ranks of the $\bar{y}_{.j}$ and $\sum_{i=1}^{r} \hat{\epsilon}_{ij}^2$; then, use Spearman's rank order correlation (see Federer, 1979 ; D. S. Robson and C. L. Wood, personal communication, proved that the above test procedure is indeed distributed as Spearman's rank order correlation.)
Other procedures for comparing a set of variances and covariances are discussed in Federer (1955) and Votaw (1948). Several other test procedures for comparing a set of variances may be found in published literature. Often times when variance heterogeneity is suspected or is present, it may be possible to select a transformation of the data to reduce or eliminate the heterogeneity (See, e.g., Federer, 1955 chapter II, Snedecor and Cochran, 1980, chapter 15, and Box and Cox, 1964). Also, it may be desirable in certain situations to deal with unequal variances and use such procedures as a Behrens-Fisher method (See, e.g., Grimes and Federer, 1984).

5. INDEPENDENCE OF e.u.s.

Non-independence of responses in an investigation may take several forms. Four such forms are

i) Carry-over or residual effects from treatments in previous periods on the e.u.,
ii) Competition between treatments in adjacent e.u.s.,
iii) Competition among treatments (intercropping mixtures, mixtures of drugs, mixtures of programs, etc.) which occur in the same e.u., and
iv) Gradients from s.u. to s.u. or e.u. to e.u. within each stratum or block of an investigation.

Under (i), numerous experiment designs and statistical analyses have been developed to take into account the effects of treatments in previous periods on the response of a treatment in the present period. Some experiment designs for obtaining estimates of carry-over effects are variously known as repeated measures and double change-over or reversal designs. For three and four treatments (letters), particular designs are:

3 treatments
column

period	1	2	3	4	5	6
1	A	B	C	A	B	C
2	B	C	A	C	A	B
3	C	A	B	B	C	A

4 treatments
column

period	1	2	3	4
1	A	B	C	D
2	D	A	B	C
3	B	C	D	A
4	C	D	A	B

Note that each letter is preceded by and is followed by every other letter, but not itself, an equal number of times. These designs are balanced for residual or carry-over effects. For four treatments, an alternate experiment design would be to use a set of three orthogonal latin squares of order four to form 12 columns. This results in an ED which is balanced for carry-over effects. Many forms of EDs and various response models have appeared in published literature (See, e.g., Kershner and Federer, 1981). These designs allow estimation of the various direct and carryover effects of treatments when the same material is used in several periods.

Competition in experiments can take two forms, i.e., intra-experimental unit and inter-experimental unit competition. In many types of investigations, intra-e.u. competition is not a concern except as it may relate to the density in an e.u. In other types of investigation, such as growing a mixture of cultivars in the same area (one form of intercropping), using a mixture of various procedures, a mixture of drugs, or compounds to treat patients, using a mixture of various educational or recreational programs, etc., intra-e.u. competition effects and their estimation are a main concern of the investigation. To obtain estimates of the various effects, a major problem is to select appropriate treatment designs and response models (See Federer, 1979, 1993, and Federer and Raghavarao, 1987), whereas the selection of an appropriate ED is usually straight-forward. The appropriate treatment combinations for inclusion in a design depends upon the types of effects to be estimated and included in the selected response model.

Competition between adjacent e.u.s under (ii) will make for dependence among e.u.s in an experiment or investigation. Some treatments make be good competitors relative to their neighbors while others may be unaffected or poor competitors. Experimenters usually try to conduct experiments in such a manner as to eliminate any effect of competition by using space or border material, by discarding edge material of an e.u., by changing the size and/or shape of an e.u. to minimize or eliminate the effect of competition, or by constructing EDs and statistical analyses to estimate treatment effects free of competition effects. For the last, Kempton (1982), Besag and Kempton (1986), and Federer and Basford (1991), have constructed various EDs and statistical analyses for this purpose. Kempton (1982), e.g., has constructed experiment designs balanced for competition effects from the two opposite sides of an e.u. Sequences of treatments (letters) in the various blocks may be of the form ABA (or BAB) for two treatments, ABACBCA for three treatments, ABCDADBACBDCA for four treatments, etc. Note that each letter precedes and follows each of the other letters but not itself and that the length of the sequence increases rapidly with the number of letters v. The sequences may be shortened if it is only required that a letter be adjacent to each of the other letters. E.g., for three letters ABCA and for four letters ABCDACBD are sufficient for each letter to be adjacent to each of the other letters. For large v, the length of a sequence becomes highly impractical. For two-dimensional arrangements of e.u.s, Federer and Basford (1991) constructed EDs balanced for competition effects in both rows and columns. Three families of designs were given. For the first family, row-column latin square type designs are selected such that treatments precede and follow each other an equal number of times in both rows and columns. These have been denoted as complete latin squares but could have equally well have been denoted as double change-over designs in both rows and columns. For $v = 4$ and 6, the EDs are:

$v = 4$
column

period	1	2	3	4
1	A	D	B	C
2	D	C	A	B
3	B	A	C	D
4	C	B	D	A

$v = 6$
column

period	1	2	3	4	5	6
1	A	F	B	E	C	D
2	F	E	A	D	B	C
3	B	A	C	F	D	E
4	E	D	F	C	A	B
5	C	B	D	A	E	F
6	D	C	E	B	F	A

Designs of the above type for $v = 4$ and 6 do not allow solutions for competition effects for each of the treatments under the following response model:

$$Y_{fghijkm} = \mu + \rho_f + \gamma_g + \tau_h + \alpha_{1i} + \alpha_{2j} + \alpha_{3k} + \alpha_{4m} + \epsilon_{fghijkm} \,, \tag{36}$$

where the first four terms and the last term are defined as for equation (12) and the $\alpha_{1i}, \alpha_{2j}, \alpha_{3k}$ and α_{4m} denote the competition effects from the four adjacent e.u.s of the fgh-th e.u. When the e.u. is rectangular rather than square, the authors show how to change (36) to account for the unequal areas of adjacency. Also, they give a suggestion of how to handle edge effects of an experiment. For this first family of EDs, solutions for all competition effects are only available for all even $v \geq 8$. For $v = 4$, only one linear combination (i.e., only one non-zero eigenvalue for the competition effects matrix) of the α's has a solution; for $v = 6$, four linear combinations have solutions (four non-zero eigenvalues) (See Federer and Basford, 1991). If the balance in rows (or columns) is changed, then a solution for all α's results. A second family of EDs may be constructed by repeating the last row of the designs from the first family of designs. Then, solutions for $v = 4$ and 6 are now possible. The third family of designs may be constructed from a particular type of F-square constructed as follows for $v = 3$ treatments:

$$F(6;2,2,2) = \begin{bmatrix} 1 & 1 \\ 1 & 1 \end{bmatrix} \otimes \begin{bmatrix} A & B & C \\ B & C & A \\ C & A & B \end{bmatrix} = \begin{bmatrix} A & B & C & A & B & C \\ B & C & A & B & C & A \\ C & A & B & C & A & B \\ A & B & C & A & B & C \\ B & C & A & B & C & A \\ C & A & B & C & A & B \end{bmatrix} \quad (37)$$

where $\otimes$ denotes a Kronecker product. Instead of the first matrix being 2×2, it may be any size not necessarily square (Federer, 1993, chapter 9). Then the rows and columns of the resulting F-square are permuted in the same manner as for the first family of EDs to obtain row and column balance for competition effects. For $v = 3$, the design is:

	column					
row	1	2	3	4	5	6
1	A	C	B	B	C	A
2	C	B	A	A	B	C
3	B	A	C	C	A	B
4	B	A	C	C	A	B
5	C	B	A	A	B	C
6	A	C	B	B	C	A

The competition effect matrix for $v = 3$ treatments in the above design has only one non-zero eigenvalue but has $v - 1$ non-zero eigenvalues for $v \geq 4$, i.e., solutions are possible for all competition effects. The above EDs are balanced for competition effects in both rows and columns. Note that all treatments precede and follow each other including themselves in both rows and columns an equal number of times.

For competition experiments, one could compute a single degree of freedom sum of squares corresponding to that for the largest eigenvalue from the competition effects matrix. This will be denoted as Kempton's one degree of freedom sum of squares for competition as the idea came from his papers. This one degree of freedom sum of squares could be computed for every experiment where competition might occur, as a diagnostic statistic for competition and could be used in much the same manner as Tukey's one degree of freedom for non-additivity. These two diagnostic statistics could be included in computer software packages and used on a routine basis as aids in detecting deviations from the usual linear model when using an ANOVA.

In animal experiments wherein animals are in the same pen or are born to the same litter, competition among animals for food and space is a fact of life and must be dealt with in the design and analysis of experiments. One procedure for doing this is to estimate a component of variance due to competition. Competition may change with litter or pen size; if it can be modeled as a function of litter or pen size, then a component of variance due to competition may be estimated (see Federer and Ladipo, 1978).

Another type of non-independence occurs when there are gradients in the responses within strata or blocks (type (iv) above). A serial correlation between adjacent e.u.s is introduced. This type of situation has received much attention in the literature under the name nearest neighbor (NN) design and analysis. NN analyses are designed to remove the effect of this serial correlation from the estimates of treatment effects. The idea has been around for many years and has recently been rejuvenated (See, e.g., Wilkinson et al., 1983, Stroup and Mulitze, 1991, and the list of references in these papers.). For a NN analysis, the serial order of the e.u.s in the investigation is taken into account. A response model of this form for a RCBED is:

$$Y_{ijk} = \mu + \tau_i + \rho_j + f_k + \epsilon_{ijk} \ , \tag{38}$$

where $f_k = f_{k-1} + \sigma_f \phi_k$ for $k = 2, \ldots, n$, $f_k = \sigma_f \phi_k$ for $k = 1$ and k is the order within a block, and the other parameters are defined as in equation (7); σ_f is a constant and ϕ_k are random deviations associated with deviations from serial order within a block. For a standard RCBED, $n = v$. First order differences between adjacent e.u.s within a block are computed as:

$$Y_{ij2} - Y_{ij1} \ , \ Y_{ij3} - Y_{ij2} \ , \ Y_{ij4} - Y_{ij3} \ , \ \ldots \ , \ Y_{ijn} - Y_{ij(n-1)} \tag{39}$$

and second order differences are computed as:

$$Y_{ij2} - (Y_{ij1} + Y_{ij3})/2 \ , \ Y_{ij3} - (Y_{ij2} + Y_{ij4})/2 \ , \ \ldots \ , \ Y_{ij(n-1)} - (Y_{ij(n-2)} + Y_{ijn})/2 \tag{40}$$

where k runs serially in each block. Using first order or both first and second order differences, a NN analysis is performed. Since blocking does not control gradients within a block, nearest neighbor analyses have been found useful in analyzing results from several types of experiments.

6. STUDY OF RESIDUALS

For any response model, the residuals may be computed and investigated for patterns, trends, outliers, or other types of behavior which would affect the assumptions involved in the use of the results in an ANOVA. For example, a serial correlation of the serially ordered residuals within each of r blocks could be computed with the associated sum of squares to obtain r single degree of freedom sums of squares for detecting gradients within blocks. If the gradients or trends were the same in every block, a single degree of freedom sum of squares could be computed as a single degree of freedom diagnostic test for trend. Such a test could also be included in computer packages for routine use in testing for trend.

Major advances in studying residuals have been made by J. W. Tukey and associates over the last 40 years. The terms modern data analysis, exploratory data analysis, and study of residuals are used to include many aspects of studying residuals from a response model. Two of several books by Tukey and associates are by Hoaglin, Mosteller, and Tukey (1985, 1985). The numerical example in Table 6.1 is presented to illustrate how a study of residuals pointed to a single outlier which greatly affected the interpretation of the experimental results from an ANOVA. The largest residual in the table is 2.461 / 16 which is associated with the observation 1.035. Note that this is an impossible result in that the dry weight measurement divided by the wet weight measurement on the same material must be less than or equal to one! As is often the situation, an investigator submits the data for statistical analysis using computer software and fails to note this discrepant result. A study of the original data for possible gross errors is usually not made, especially for large data sets. After computing the residuals, it may be noted that a pattern exists in that every residual in the same row or column as the observation 1.035 is negative while the one associated with the observation is a large positive residual. This is only one type of pattern that residuals may take. For this particular example, the experimenter would have used a residual mean square that was eight times too large, thus affecting significance tests and confidence intervals. Biological, medical, nutritional, and engineering researchers often use ratios as used here. It should be noted that such ratios often tend to eliminate differences among numerator or among denominator responses in that ratios tend to be a constant proportion. For this example, the coefficient of variation after removing the effect of the outlier, was about three percent which is much smaller than for wet or dry weights analyzed separately.

Table 6.1.
Wheat yields of dry/wet grain weight from a RCBED
with four nitrogen treatments and four blocks

nitrogen applied

Block	none	early	middle	late	Total $Y_{.j}$
1	0.718	0.732	0.734	0.792	2.976
2	0.725	0.781	0.725	0.716	2.947
3	0.704	1.035	0.763	0.758	3.260
4	0.726	0.765	0.738	0.781	3.010
Total	2.873	3.313	2.960	3.047	12.193

16 Residuals $(Y_{ij} - \bar{y}_{i.} - \bar{y}_{.j} + \bar{y}_{..})$

Block	none	early	middle	late	Total	Sum of absolute values
1	0.285	−1.251	0.193	0.773	0	0.156375
2	0.513	−0.351	0.165	−0.327	0	0.084750
3	−1.075	2.461	−0.479	−0.907	0	0.307625
4	0.277	−0.859	0.121	0.461	0	0.107375
Total	0	0	0	0	0	–
Sum of absolute values	0.134375	0.307625	0.059875	0.154250	–	0.656125

ANOVA

Source of Variation	d.f.	Sum of Squares	Mean Squares
Total	16	9.381715	
Correction for mean	1	9.291828	
Block	3	.015443	.005147
Nitrogen	3	.027149	.009050
Block × Nitrogen	9	.047295	.005255
Outlier suspect	1	.042059	.042059
Remainder	8	.005236	.000655

$CV = \sqrt{.000655} \,/\, (11.920) \,/\, 16 = 3.4\%$

H.C. Kirton (1984) suggested that an ANOVA be computed using the absolute values of the residuals and adjusting the "degrees of freedom" column so that they add up to $(r-1)(v-1)$, or $[(r-1)(v-1)-k]$ if there are k missing plot values. The residuals using the computed missing plot value of 0.762 are given in Table 6.2. ANOVAs on the two sets of residuals are also presented. For the original values in Table 6.1, the "*mean squares*" for blocks and for treatments is about five times that for the remainder. When residuals are computed using the missing plot value 0.762 and computing an ANOVA on the absolute values of the residuals, the "*mean squares*" for blocks and for treatments is approximately equal to that for remainder. Of course, these ratios of "*mean squares*" do not follow an F distribution, but since F is robust to non-normality, it can be considered to be a fair approximation.

A study of the block and treatment of absolute values of the residuals in Table 6.1, indicates that the treatment early and block 3 means were much higher than the other means. This again would point to the discrepant observation 1.035.

Table 6.2.

Missing plot value 0.762 inserted for discrepant value 1.035 of Table 6.1

ANOVA on $|\hat{\epsilon}_{ij}|$ values of Table 6.1 as per H.C. Kirton

Source of Variation	"Degrees of Freedom"*	Sum of Squares	"Mean Square"
Total	$(r-1)(v-1)=9$	0.047295	0.005255
Mean	$c=9/16$	0.026906	–
Block	$c(r-1)=27/16$	0.007543	0.004470
Treatment	$c(v-1)=27/16$	0.008111	0 004807
Remainder	$c(r-1)(v-1)=81/16$	0.004735	0.000935

$^*c=(r-1)(v-1)/rv=9/16$.

Residuals times 16 using 0.762 in place of 1.035

Block	Treatment none	early	middle	late	Sum of absolute residuals
1	0.012	−0.432	−0.080	0.500	0.0640
2	0.240	0.468	−0.108	−0.600	0.0885
3	−0.256	0.004*	0.340	−0.088	0.0430
4	0.004	−0.040	−0.152	0.188	0.0240
Sum of absolute residuals	0.0320	0.0590	0.0425	0.0860	0.2195

* zero within rounding error on 0.762

ANOVA on absolute values of residuals in above Table as per H.C. Kirton

Source of Variation	"Degrees of Freedom"*	Sum of Squares	"Mean Square"
Total	$(r-1)(v-1)-1=8$	0.005236	0.000655
Mean	$c'=8/15$	0.003011	–
Block	$c'(r-1)=24/15$	0.000577	0.000361
Treatment	$c'(v-1)=24/15$	0.000416	0.000260
Remainder	$c'(rv-r-v)=64/15$	0.001232	0.000289

$^*c'=[(r-1)(v-1)-1]/(rv-1)=8/15$

As stated, there are many methods for investigating whether patterns, trends, and outliers occur in the residuals used to compute an error mean square for statistical analyses. Instead of studying this type of residual, the investigator may focus attention on the

interaction terms from a two factor factorial and use some of the same methods. It is often desirable to model factorial responses with as few parameters as possible (parsimony). The interaction terms are treated as residuals. Bradu and Gabriel (1978) present a method known as bi-plot as a diagnostic tool in searching for an appropriate model. Their general results contain several of the procedures described previously such as Tukey's one degree of freedom for non-additivity and Mandel's procedure. They discuss bi-plotting using the original observations, deviations from the overall mean, and residuals or interaction terms. Gauch (1988) made use of their ideas to develop an additive main effects and multiplicative interaction (AMMI) model for two factor studies such as genotype and environment. The method has been successfully applied to a variety of experiments in agriculture. A principal components analysis is applied to the residuals (interactions). Often only the first and perhaps the second principal components are sufficient to model the response. Using a bi-plot aids in the interpretation of the data.
An AMMI response model for a RCBED with ab treatments in a two factor factorial with a levels of factor A and b levels of factor B is:

$$Y_{ijk} = \mu + \rho_i + \alpha_j + \beta_k + \sum_{h=1}^{n} \lambda_h \gamma_{hj} \delta_{hk} + \pi_{jk} + \epsilon_{ijk} \ , \tag{41}$$

where μ, ρ_i, and ϵ_{ijk} are as defined for (7), α_j is the additive effect of the j-th level of factor A, β_k is the additive effect of the k-th level of the factor B, λ_h is the singular value for interaction principal component h, γ_{hj} and δ_{hk} are the two factor eigenvectors for principal component h, and π_{jk} is the residual left for interaction after fitting n principal components to the interaction terms $\overline{y}_{.jk} - \overline{y}_{.j.} - \overline{y}_{..k} + \overline{y}_{...} = \widehat{\alpha\beta}_{jk}$. Note that $\hat{\alpha}_j = \overline{y}_{.j.} - \overline{y}_{...}$ and $\hat{\beta}_k = \overline{y}_{..k} - \overline{y}_{...}$ and the usual principal components analysis constraints are used, i.e.,

$$\sum_{j=1}^{a} \alpha_j = \sum_{k=1}^{b} \beta_k = 0 \quad , \quad \sum_{j=1}^{a} \gamma_j^2 = \sum_{k=1}^{b} \delta_k^2 = 1 \tag{42}$$

and every eigenvector is constrained to be orthogonal to all previous eigenvectors, so that for $h \neq h'$

$$\sum_{j=1}^{a} \gamma_{hj} \gamma_{h'j} = \sum_{k=1}^{b} \delta_{hj} \delta_{h'j} = 0 \ . \tag{43}$$

The maximum number of principal components to be fitted is the minimum of $(a-1)$ and $(b-1)$. For any data set, zero to min $\{a-1, b-1\}$, principal components will be fitted in the AMMI model family, i.e., AMMI0, AMMI1, AMMI2, ..., AMMIF, $F = \min\{a-1, b-1\}$. AMMI0 corresponds to the no interaction case and AMMIF corresponds to a consideration of the means $\overline{y}_{.jk}$ and comparisons among these ab means, e.g. multiple comparisons. The eigenvalue for component h is $r\lambda_h^2$. There is a controversy in the literature about the number of degrees of freedom to assign to the sum of squares associated with each principal component. Gauch (1992) appears to have resolved this dilemma.
The steps in the AMMI model analysis for two-factor factorials are

i) Compute $\widehat{\alpha\beta}_{jk}$,
ii) Select min $\{a-1, b-1\}$ to determine which levels of factors A and B are to be used as variates,
iii) Fit a principal components analysis,
iv) Let n, usually only $h = 1$ or $h = 2$, be the number of principal components to be used as determined by having the residual interaction mean square approximately equal to the error mean square,

v) Obtain the estimated jk-th cell means as $\hat{y}_{.jk} = \hat{\mu} + \hat{\alpha}_j + \hat{\beta}_k + \sum_{h=1}^{n} \lambda_h \gamma_{hj} \delta_{hk}$, and

vi) Prepare a bi-plot of the two-way array of the values from (v) using the adjusted values $\hat{y}_{.j.}$ and $\hat{y}_{..k}$ as the abscissas for the plot and the γ_{hj} and δ_{hk} values as the ordinates for one bi-plot

(The name bi-plot is used to denote that two sets of bivariates, $\hat{y}_{.j.}$ and γ_{hj} and $\hat{y}_{..k}$ and δ_{hk}, are plotted on the same graph.). A second bi-plot of $\gamma_{1j}\sqrt{\lambda_1}$ values against $\gamma_{2j}\sqrt{\lambda_2}$ values is also used to aid in interpreting results from factorial experiments; the $\delta_{1j}\sqrt{\lambda_1}$ values are plotted against the $\delta_{2j}\sqrt{\lambda_2}$ values on the same bi-plot. Patterns in this latter bi-plot are indicative of particular types of models for the interaction terms. Plots of this nature can be most helpful in the interpretation of results from an experiment.

7. POST MORTEMS ON AN INVESTIGATION

In light of the various procedures employed in the analysis of the results of an investigation, the investigator makes a decision on the model and the particular statistical analysis to use for the data. In addition, the nature of the observations, the manner in which they were obtained, and the theoretical background for such observations should be studied carefully. From a study of numerous data sets, it has been found that nearly always there is something peculiar about the results, about some or all of the observations, or about the way in which the observations were obtained. There usually is "*something wrong*" somewhere with a data set. Thus, a healthy attitude for an approach to analyzing any data set is to critically examine all aspects of the investigation prior to performing any statistical computations. Then, after all analyses have been made and conclusions drawn, a post mortem diagnosis of the entire procedure is in order. Unless the results and conclusions are repeatable by other researchers in the field, the results of an investigation are not of much use.

Another situation in experimentation is the occurrence of unequal sample sizes. Any of several books on linear models provide statistical procedures for unbalanced data. There appears, however, to be something missing in their descriptions in that all analyses presented are conditional upon the particular sample size configuration that resulted for this experiment. If, however, it can be demonstrated that treatment response is not a function of sample patterns in an experiment, then the analyses are also unconditional. One place where these analyses have been used extensively is in animal breeding studies. Here sample size is often related to treatment response, e.g., the best bulls always get the most cows!

Examples IV-1 and IV-2 in Federer (1955) are illustrations of what can go wrong in performing a statistical analysis on a set of data. The first example was purported to be a CRED whereas it most likely was a single replicate of a RCBED with multiple sampling units in each experimental unit. This type of mistaken ED is of frequent occurrence in published literature (See Federer, 1975). The second example had random sample sizes for the three types of plant. Here sample size is most likely related to treatment in that the germination and survival rate for aberrants is lower than for the other two types.

8. DISCUSSION

Some general considerations for checking on the assumptions behind a statistical procedure are discussed. Several more could have been included. Several of the procedures described herein have as their goal the partitioning of the residual (block × treatment, for example) sum of squares into a part attributable to an effect such as non-additivity and a part due to error variation. The Tukey one degree of freedom sum of squares for non-additivity is an example of this as this sum of squares is removed from the block × treatment sum of squares to obtain the estimated error mean square. The estimated treatment effects are unaffected whether or not the blocks × treatment sum of squares is partitioned. Likewise, the interaction sum of squares for a two factor factorial may be partitioned into two parts, one part to explain the interaction and a second part representing the estimated noise or error mean square. The AMMI procedure is an example of this kind of partitioning. The factor effects are considered to be additive and unaffected by the manner in which the interaction sum of squares is partitioned.

Other procedures discussed above have as their goal the adjustment of treatment effects for other effects. When competition effects between adjacent e.u.s are present, it is desirable to adjust direct treatment effects for effects of competition of adjacent units. Direct effects of treatments are adjusted for residual effects of previous treatments in repeated measures experiments. Nearest neighbor analyses have as their goal the adjustment of treatment effects for local gradients within blocks. The two different goals described above are not incompatible, and both may be desired when analyzing the results from an experiment. The extraction of all available information from an experiment should be the ultimate target for any data analyst.

REFERENCES

[1] BARTLETT, M. S. (1937). Properties of Sufficiency and Statistical Tests. *Proceedings of the Royal Society (London), Series A*, **160**, 268-282.

[2] BESAG, J. AND R. A. KEMPTON (1986). Statistical Analysis of Field Experiments Using Neighboring Plots. *Biometrics*, **42**, 231-251.

[3] BOX, G. E. P. (1953). Non-normality and Test on Variances. *Biometrika*, **40**, 318-335.

[4] BOX, G. E. P. (1980). Sampling and Bayes' Inference in Scientific Modelling and Robustness. *Journal of the Royal Statistical Society, Series A*, **143**, 383-430.

[5] BOX, G. E. P. AND D. R. COX (1964). An Analysis of Transformations. *Journal of the Royal Statistical Society, Series B*, **26**, 211-252.

[6] BRADU, D. AND K. R. GABRIEL (1978). The Biplot as a Diagnostic Tool for Models of Two-way Tables. *Technometrics*, **20**, 47-68.

[7] COCHRAN, W. G. (1941). The Distribution of the Largest of a Set of Estimated Variances as a Fraction of Their Total. *Annals of Eugenics*, **11**, 47-52.

[8] COX, C. P. (1958). The Analysis of Latin Square Designs with Individual Curvatures in One Direction. *Journal of of the Royal Statistical Society, Series B*, **20**, 193-204.

[9] FEDERER, W.T. (1955) *Experimental Design - Theory and Application.* Macmillan Co., New York (Republished by Oxford and IBH Publishing Co., New Delhi in 1967 and in 1974).

[10] FEDERER, W. T. (1970). *An Application of Three Tests for Nonadditivity.* BU-343-M in the Technical Report Series of the Biometrics Unit, Cornell University, October.

[11] FEDERER, W.T. (1975). *The Misunderstood Split Plot.* In *Applied Statistics* [R.P. Gupta, editor], North Holland Publishing Co., Amsterdam, Oxford.

[12] FEDERER, W. T. (1976a). Sampling, Blocking, and Model Considerations for the Completely Randomized, Randomized Complete Block, and Incomplete Block Experiment Designs. *Biometrische Zeitschrift*, **18**, 7, 511-525.

[13] FEDERER, W. T. (1976b). Sampling, Blocking, and Model Considerations for r-row by c-column Experiment Designs. *Biometrische Zeitschrift*, **18**, 8, 595-607.

[14] FEDERER, W. T. (1977). Sampling, Blocking, and Model Considerations for Split Plot and Split Block Designs. *Biometrical Journal*, **19**, 3, 181-200.

[15] FEDERER, W. T. (1979). Statistical Designs and Response Models for Mixtures of Cultivars. *Agronomy Journal*, **71**, 701-706.

[16] FEDERER, W. T. (1979). *Response Equations and Data Analysis for Some Simple Experiment Designs.* BU-676-M in the Technical Report Series of the Biometrics Unit, Cornell University, June.

[17] FEDERER, W. T. (1991). *Statistics and Society. Data Collection and Interpretation.* Marcel Dekker, Inc. New York, (Second edition).

[18] FEDERER, W. T. (1993). *Statistical Design and Analysis for Intercropping Experiments.* Springer Verlag, New York.

[19] FEDERER, W. T. AND K. E. BASFORD (1991). Competing Effects Designs and Response Models for Two-dimensional Field Arrangements. *Biometrics*, **47**, 1461-1472.

[20] FEDERER, W. T. AND O. O. LADIPO (1978). A Component of Variance Due to Competition. *Australian Journal of Statistics*, **20**, 2, 99-110.

[21] FEDERER, W. T. AND D. RAGHAVARAO (1987). Response Models and Minimal Designs for Mixtures of n of m Items Useful for Intercropping and Other Investigations. *Biometrika*, **74**, 571-577.

[22] FEDERER, W. T. AND C. S. SCHLOTTFELDT (1954). The Use of Covariance to Control Gradients in Experiments. *Biometrics*, **10**, 282-290.

[23] FISHER, R. A. (1918). The Correlation between Relatives on the Supposition of Mendelian Inheritance. *Transactions of the Royal Society of Edinburg*, **LII**, 399-433.

[24] FISHER, R. A. (1925). *Statistical Methods for Research Workers.* Oliver and Boyd, Edinburgh. (First edition).

[25] FISHER, R. A. (1935). *The Design of Experiments.* Oliver and Boyd, Edinburgh, (First edition).

[26] FISHER, R, A. AND W. A. MACKENZIE (1923). Studies in Crop Variation II. The Manurial Response of Different Potato Varieties. *Journal of Agricultural Science*, **13**, 311-320.

[27] GAUCH, H. G., JR. (1988). Model Selection and Validation for Yield Trials with Interaction. *Biometrics*, **44**, 705-715.

[28] GAUCH, H. G., JR. (1992). Degrees of Freedom in the Mandel Model for Two-way Tables. (In the process of publication).

[29] GRIFFING, B. (1956). Concept of General and Specific Ability in Relation to Diallel Crossing Systems. *Australian Journal of Biological Science*, **9**, 463-493.

[30] GRIMES, B. A. AND W. T. FEDERER (1984). *Comparison of Means from Populations with Unequal Variances.* In *W. G. Cochran's Impact on Statistics* [Editors, P. S. R. S. Rao and J. Sedransk], John Wiley and Sons, Inc., New York, pp. 353-374.

[31] HARRIS, J. A. (1913). On the Calculation of Intra-class and Inter-class Coefficients of Correlation from Class Moments when the Number of Possible Combinations is Large. *Biometrika*, **9**, 446-472.

[32] HARTLEY, H. O. (1950). The Maximum F–ratio as a Short-cut Test for Heterogeneity of Variance. *Biometrika*, **37**, 308-312.

[33] HOAGLIN, D. C., F. MOSTELLER, AND J. W. TUKEY (1985). *Fundamentals of Exploratory Analyses.* John Wiley and Sons, Inc., New York, Chichester, Brisbane, Toronto, Singapore.

[34] HOAGLIN, D. C., F. MOSTELLER, AND J. W. TUKEY (1985). *Exploring Data Tables, Trends, and Shapes.* John Wiley and Sons, Inc., New York Chichester, Brisbane, Toronto, Singapore.

[35] KEMPTHORNE, O. (1952). *Design and Analysis of Experiments.* John Wiley and Sons, Inc., New York.

[36] KEMPTON, R. A. (1982). Adjustment for Competition between Varieties in Plant Breeding Trials. *Journal of Agricultural Science*, **98**, 599-611.

[37] KEMPTON, R.A. (1984). The Use of Biplots in Interpreting Variety by Environment Interactions. *Journal of Agricultural Science, Cambridge*, **3**, 123-135.

[38] KERSHNER, R. P. AND W. T. FEDERER (1981). Two-treatment Cross-over Designs for Estimating a Variety of Effects. *Journal of the American Statistical Association*, **76**, 612-619.

[39] KIRTON, H. C. (1984). *Personal Communication.*

[40] MANDEL, J. (1961). Non-additivity in Two-way Analysis of Variance. *Journal of the American Statistical Association*, **56**, 878-888.

[41] MANDEL J. (1971). A New Analysis of Variance Model for Non-additive Data. *Technometrics*, **13**, 1-18.

[42] MARTIN, K. J. (1980). A Partition of a Two-factor Interaction, with an Agricultural Example. *Applied Statistics*, **29**, 149-155.

[43] OUTHWAITE, A. D. AND A. RUTHERFORD (1955). Covariance Analysis as an Alternative to Stratification in the Control of Gradients. *Biometrics*, **11**, 431-440.

[44] PEARSON, E. S. AND H. O. HARTLEY (1962). *Biometrika Tables for Statisticians, Volume 1.* Cambridge University Press, London. (2nd edition).

[45] ROBSON, D. S. (1970). *Tests for Non-additivity Viewed as Tests of the Hypothesis of No Interaction. A Preliminary Report.* BU-334-M in the Technical Report Series of the Biometrics Unit, Cornell University, October.

[46] SCHEFFÉ, H. (1959). *The Analysis of Variance.* John Wiley and Sons, Inc. New York.

[47] SNEDECOR, G. W. AND W. G. COCHRAN (1980). *Statistical Methods.* Iowa State University Press, Ames, Iowa. (Seventh edition).

[48] SPRAGUE, G. F. AND L. A. TATUM (1941). General vs. Specific Combining Ability in Single Crosses of Corn. *Journal of the American Society of Agronomy*, **34**, 923-932.

[49] STROUP, W. W. AND D. K. MULITZE (1991). Nearest Neighbor Adjusted Best Linear Unbiased Prediction. *The American Statistician*, **45**, 3, 194-200.

[50] TUKEY, J. W. (1949). One Degree of Freedom for Non-additivity. *Biometrics*, **5**, 232-242.

[51] VOTAW, D. F., JR. (1948). Testing Compound Symmetry in a Normal Multivariate Distribution. *Annals of Mathematical Statistics*, **19**, 447-473.

[52] WILKINSON, G. N., S. R. ECKERT, T. W. HANCOCK, AND O. MAYO (1983). Nearest Neighbour (NN) Analysis of Field Experiments (with discussion). *Journal of the Royal Statistical Society, Series B*, **44**, 151-211.

Design of Experiments and Multivariate Analysis

Recent Adv. in Stat. and Prob., pp. 27-40
J. Pérez Vilaplana and M. L. Puri (Eds)

Clustering of Treatment Means Using the Mixture Method

K.E. BASFORD[1] and D.R. GREENWAY[2]
[1] Department of Agriculture, The University of Queensland, Queensland 4072, Australia
[2] Department of Agriculture, The University of Queensland, Queensland 4072, Australia

Abstract – A common statistical problem is the comparison of means obtained from experiments designed to compare several treatments. Where an analysis of variance leads to a significant F test, the mixture method of clustering can be considered as an alternative to multiple comparison procedures for the partitioning of treatment means into relatively homogeneous groups.

One limitation to the use of this technique, when clustering is carried out on the treatment means, is the existence of distinct observations corresponding to single member groups. This situation occurs frequently when summarising experimental data. A practical course of action involving the clustering of treatment means using mixture models with normal component distributions is presented, with available computer software modified to enable identification and removal of single member groups. The method is demonstrated with several real data sets, and the effect of the removal of single member groups on the estimated log likelihood at convergence is discussed. Changes to the estimated log likelihood resulting from the removal of distinct observations must be fully understood before the related difficulties of determining the most appropriate number of groups consistent with the data can be resolved.

In cases where partitioning of treatment means into non-overlapping groups is a desirable outcome, the mixture method of clustering appears to provide a useful summary of the data. Additionally, it is possible to assess the strength of the association between a mean and the group to which it is assigned by examining the estimated posterior probabilities of group membership.

1. INTRODUCTION

Many experiments are designed to compare a number of different treatments, say varieties in an agricultural science context. When analysing the data obtained from these experiments, it is often useful, following a significant F test in Analysis of Variance, to allocate the treatment means into non-overlapping groups. Tukey (1949), when discussing a hypothetical example said, " *We wish to separate the varieties into distinguishable groups as often as we can without too frequently separating varieties which should stay together*".

Considerable work has been done in this area and various multiple comparison procedures have been developed. A survey of the relevant literature can be found in O'Neill and Wetherill (1971), Miller (1981) and more recently Hochberg and Tamhane (1987).

Two of the most widely used multiple comparison procedures are the Studentized range methods, developed primarily by Tukey (1949), and the F test procedure of Gabriel (1964). Both of these procedures lead to an underscoring of those subsets of the ordered treatment means which are judged to be not significantly different from one another. However, the homogeneous subsets produced by these methods commonly overlap. Many users prefer the treatment means to be partitioned into non-overlapping distinct groups as the interpretation is then considerably simplified.

The possibility of using cluster analysis, instead of multiple comparison procedures to achieve these aims was suggested by Plackett (1971) in his discussion of the paper by O'Neill and Wetherill (1971). The technique of clustering uses the measurements on a set of elements to identify clusters or groups such that the elements are relatively homogeneous within the clusters, and heterogeneous between the clusters. There are many examples where various clustering techniques have been used to partition treatment means into groups (Scott and Knot, 1974; Jolliffe, 1975; Binder, 1978, 1981; Aitkin, 1980; Menzefricke, 1981, Carmer and Lin, 1983; Caliński and Corsten, 1985; and Basford and McLachlan, 1985*b*).

More recently, considerable emphasis has been placed on the use of mixture models for the purpose of clustering treatment means (Basford and McLachlan, 1985*b*; and McLachlan and Basford, 1988, Chapter 6).

Under this approach, the treatment means are considered as a sample from a mixture of several populations in various proportions. The aim is to identify those treatment means which have been selected from the same underlying population within the mixture. Estimates of the parameters of the probability density functions of these underlying populations can be obtained using the likelihood principle. Then the treatment means can be allocated to these populations or groups on the basis of their estimated posterior probabilities of group membership. In this way, treatment means are partitioned into a number of discrete, relatively homogeneous groups.

An advantage of this procedure is that it is possible to assess the strength of the association between any treatment mean and the group to which it is assigned by examination of the estimated posterior probabilities of group membership. This can be useful in identifying which treatment means have clear links with a particular group, and which means do not. The mixture method is model based, in that the form of the density of an observation from each of the underlying populations has to be specified. The most common approach is to take these densities to be Gaussian or normal and compute the maximum likelihood solution using the EM algorithm (Dempster, Laird and Rubin, 1977).

Basford and McLachlan (1985*b*) and McLachlan and Basford (1988, Chapter 6) used the mixture maximum likelihood approach to cluster treatment means from Randomized Complete Block Design (RCBD). They assumed that the variances in each of the underlying populations were equal. This variance was estimated by a combination of the within treatment variability, as estimated by the error mean square in analysis of variance, and the variation of the treatment means about the underlying population or group means. However, the present approach ignores this within treatment variability and considers the treatment means as raw data or individual observations.

This approach has several advantages over the previous method of clustering treatment means. Most importantly, calculations are not dependent on the experimental design used. Basford and McLachlan (1985*b*) had to model random and fixed block effects in a RCBD quite separately and showed that considerable computational problems were associated with random block effects, even for a moderate number of treatments. Providing separate mixture models to group treatment means obtained from different experimental designs (or with different assumptions) is not a particularly attractive or feasible alternative to a multiple comparison procedure. The present approach also allows consideration of unrestricted group variances, a more realistic condition when there is no a priori reason for the variances among the treatment means within the groups to be the same (or equal). Although the within treatment variability is not used in this model, some protection can be given by only clustering the treatment means when the F test in the analysis of variance is significant. In addition, the group variances (variation about the group means) would not be expected to be substantially smaller than the variance of a treatment mean from the original analysis of variance.

Unfortunately, difficulties occur with the application of the mixture method of clustering when it is assumed that the data come from a mixture of normal populations with unrestricted variances. These theoretical considerations are discussed in the next section. A more practical problem occurs when one treatment in an experiment is substantially different from all the others. When applying a clustering procedure, it would be expected that only this one treatment mean would be allocated to a particular group, with no other treatment having any probability of belonging to it. Under these conditions (single member group and unrestricted group variances), the estimate of this group's variance would be zero, and the EM algorithm would converge towards a singularity in the likelihood.

The present research examines a practical method of overcoming this problem, and illustrates it with two real data sets. The procedure involves the identification of any individual treatment mean which is sufficiently distinct to result in the convergence of the EM algorithm towards a singularity. Once identified, this observation is removed from the data set and the mixture model is fitted to the remaining observations. This procedure can be applied assuming either equal or unrestricted group variances.

Another related difficulty encountered when using the mixture method is the determination of the smallest number of groups compatible with the data. As the clustering is being carried out on treatment means, it is unlikely that there is prior knowledge of the number of groups in the underlying mixture, and so a method of determining the most appropriate number of groups is necessary. An obvious way of approaching the problem is to use the likelihood ratio test statistic to test for the smallest value of g compatible with the data. Unfortunately, minus twice the logarithm of the test statistic does not have its usual asymptotic null distribution of chi–squared with degrees of freedom equal to the difference in the number of parameters in the two hypotheses (McLachlan and Basford, 1988, Chapter 1). An alternative method developed by Akaike (1973) [see also Bozdogan and Sclove (1984), and Bozdogan (1986, 1987)], is based on the maximum estimated log likelihood, with an adjustment for the number of parameters in the model. As both methods use the log likelihood at convergence, it is necessary to understand what happens to the estimated log likelihood when observations are removed from the data set (and the group number is correspondingly reduced). Changes in the log likelihood estimates under these conditions are investigated here.

Given that it is currently uncertain how best to theoretically compare log likelihoods, the problem is compounded when single member groups are present. Thus, the appropriate number of groups to represent the data analysed here is obtained subjectively, and the results compared to those obtained by other researchers.

In the following sections, the observed treatment means are considered as raw data and the terminology refers to a treatment mean as an individual or single observation. It is these observations which are being grouped using the mixture method of clustering.

2. MIXTURE MODELS

Let $x_1, \ldots, x_n$ denote the observed values of a random sample of n p-dimensional observations ($p = 1$ for univariate data), taken from a superpopulation G, which is a mixture of an unknown finite number, say g, of populations $G_1, \ldots, G_g$, in some proportion $\pi_1, \ldots, \pi_g$ respectively, with

$$\sum_{i=1}^{g} \pi_i = 1 \quad \text{and} \quad \pi_i \geq 0 \quad (i = 1, \ldots, g). \tag{1}$$

The probability density function (p.d.f.) of an observation x in G can therefore be written in the finite mixture form

$$f(x, \phi) = \sum_{i=1}^{g} \pi_i f_i(x, \theta) \tag{2}$$

where θ denotes the vector of unknown population parameters, $\phi = (\pi', \theta')'$ is the vector of all unknown parameters, and $f_i(x, \theta)$ is the p.d.f. corresponding to G_i.
The log likelihood of ϕ is given by

$$\ln L(\phi) = \sum_{j=1}^{n} \ln \left[\sum_{i=1}^{g} \pi_i f_i(x_j, \theta) \right]. \tag{3}$$

Under the assumption of underlying normal density functions with unrestricted variances, (the normal heterogeneous mixture model),

$$x_j \sim \mathcal{N}(\mu_i, V_i) \quad \text{in} \quad G_i \quad (i = 1, \ldots, g), \tag{4}$$

the likelihood function, $L(\phi)$, is unbounded, and so the maximum likelihood estimator (MLE) of ϕ does not exist (Kiefer and Wolfowitz, 1956). However, Kiefer (1978) verified that for univariate data there is a sequence of roots of the likelihood equation,

$$\frac{\partial \ln L(\phi)}{\partial \phi} = 0 \tag{5}$$

which is consistent, asymptotically normal and efficient. With probability tending to one, these roots correspond to the local maximum of $L(\phi)$. Because with mixture models the likelihood equation (5) has multiple roots, there is the problem of which root to choose. This problem is discussed in detail in Basford and McLachlan (1985a) and McLachlan and Basford (1988, Chapter 2). With equal variances and the normality assumption (the normal homogeneous mixture model),

$$x_j \sim \mathcal{N}(\mu_i, V) \quad \text{in} \quad G_i \quad (i = 1, \ldots, g), \tag{6}$$

the MLE of ϕ does exist and is strongly consistent.
Once the likelihood solution of ϕ has been obtained, estimates of the posterior probabilities of population membership can be formed for each x_j to give a probabilistic clustering. The posterior probability that x_j belongs to G_i is given by

$$\tau_{ij} = \frac{\pi_i f_i(x_j, \theta)}{\sum_{u=1}^{g} \pi_u f_u(x_j, \theta)}. \quad (i = 1, \ldots, g;\ j = 1, \ldots, n) \tag{7}$$

and under the normality assumption this can be written as

$$\tau_{ij} = \frac{\pi_i |V_i|^{-\frac{p}{2}} \exp\left[-\frac{1}{2}(x_j - \mu_i)' V_i^{-1} (x_j - \mu_i)\right]}{\sum_{u=1}^{g} \pi_u |V_u|^{-\frac{p}{2}} \exp\left[-\frac{1}{2}(x_j - \mu_u)' V_u^{-1} (x_j - \mu_u)\right]} \tag{8}$$

where

$$\pi_i = \frac{\sum_{j=1}^{n} \tau_{ij}}{n} \qquad (i = 1, \ldots, g) \tag{9}$$

$$\mu_i = \frac{\sum_{j=1}^{n} \tau_{ij} x_j}{\sum_{j=1}^{n} \tau_{ij}} \qquad (i = 1, \ldots, g) \tag{10}$$

and

$$V_i = \frac{\sum_{j=1}^{n} \tau_{ij}(x_j - \mu_i)(x_j - \mu_i)'}{\sum_{j=1}^{n} \tau_{ij}} \qquad (i = 1, \ldots, g) \tag{11}$$

Each x_j is then assigned to the population or group to which it has the highest estimated posterior probability of belonging, i.e. to G_u if

$$\hat{\tau}_{uj} > \hat{\tau}_{ij} \qquad (i = 1, \ldots, g\ ;\ i \neq u)\,. \tag{12}$$

3. IDENTIFICATION AND REMOVAL OF DISTINCT OBSERVATIONS

When fitting a mixture model with unrestricted variances (normal heterogeneous mixture model) to univariate data, the EM algorithm converges towards a singularity in the likelihood if any group has a zero variance. This will occur if one observation is sufficiently different or "distinct" from all other observations in the data set for it to be assigned to a group to which no other observation has any probability of belonging. Under these conditions, only one observation will contribute to the calculation of the group mean (which will be equal to that particular observation) and the group variance (which will be zero). Such an observation will be referred to as being "distinct", i.e. it is sufficiently different from all the others to be considered as a single member group.

In the case of fitting a mixture model with equal variances (normal homogeneous mixture model), the common variance is calculated as

$$V_c = \frac{1}{n} \sum_{i=1}^{g} \left[\sum_{j=1}^{n} \tau_{ij}\right] V_i\,. \tag{13}$$

As this variance is used in the iterative calculation of log likelihood, singularities do not occur. In addition, the posterior probabilities are calculated as

$$\tau_{ij} = \frac{\pi_i |V_c|^{-\frac{1}{2}} \exp\left[-\frac{1}{2}(x_j - \mu_i)' V_c^{-1} (x_j - \mu_i)\right]}{\sum_{u=1}^{g} \pi_u |V_c|^{-\frac{1}{2}} \exp\left[-\frac{1}{2}(x_j - \mu_u)' V_c^{-1} (x_j - \mu_u)\right]} \tag{14}$$

using the estimate of common variance. It is therefore unlikely that only one observation will have a non zero probability of belonging to a particular group. This will be discussed further in the examples.
However, if there is a group to which only one observation has a non zero probability of belonging, the inclusion of that group, whose estimated variance is zero, in the calculation of common variance leads to an underestimate of the variance of the remaining observations.

4. EFFECT ON LIKELIHOOD ESTIMATE

As the usual method of assessing the clustering most consistent with the data is by maximising the log likelihood, it is important to consider the effect that the removal of distinct observations has on this estimate. This is particularly so when attempting to use the log likelihood to determine the appropriate number of groups compatible with the data. Comparisons might need to be made between a log likelihood which has been calculated after removal of one or more observations and one which has been calculated on the full data set.
Once an individual observation has been identified as being distinct, it and the group to which it has been assigned, can be removed from the data and the mixture model fitted to the remaining data. This means that subsequent estimates of log likelihood ($\ln L(\phi)$) are based on a reduced number of individuals (n to $n-1$ as the observation identified as being distinct no longer contributes) and groups (g to $g-1$).
When data are analysed assuming unrestricted group variances, the improvement in the contribution of each of the remaining observations to $\ln L(\phi)$ after the removal of the distinct observation is the same. The contribution of each individual x_j to $\ln L(\phi)$ is

$$\ln\left\{\sum_{i=1}^{g}\pi_i(2\pi)^{-1/2}|V_i|^{-1/2}\exp\left[-\frac{1}{2}(x_j-\mu_i)'V_i^{-1}(x_j-\mu_i)\right]\right\}. \tag{15}$$

The removal of a distinct individual and the resultant reduction in the number of groups has no effect on the estimated means and variances of the remaining groups. Thus the only part of equation (15) that will change for any remaining observation will be π_i. For example, if individual 1 belonging to group 1 was identified as distinct, the form of equation (9) after removal of this individual (and group) would be

$$\pi_i = \frac{\sum_{j=2}^{n}\tau_{ij}}{n-1} = \frac{\sum_{j=1}^{n}\tau_{ij}}{n-1} \qquad (i=2,\ldots,g) \tag{16}$$

because

$$\tau_{i1} = 0 \qquad \text{for} \qquad (i=2,\ldots,g) \tag{17}$$

Thus π_i will increase because the denominator has decreased from n to $n-1$ while the numerator has not changed. In effect, the change from equation (9) to equation (16) is equivalent to multiplication by a constant factor of $n/(n-1)$. As a result, when fitting a normal heterogeneous mixture model, the increase in the contribution of each of the remaining individual observations to the log likelihood following the removal of a distinct individual can be expressed as

$$\ln\left[\frac{n}{n-1}\right] \tag{18}$$

and the total improvement in the estimate of $\ln L(\phi)$ as

$$(n-1)\ln\left[\frac{n}{n-1}\right]. \tag{19}$$

This is approximately 1 for large n.
When equal group variances are assumed, the major effect on $\ln L(\phi)$ is the removal of the contribution to the log likelihood made by the observation identified as being distinct. The removal of a distinct individual also leads to an increase in the common variance given in equation (13), and to an increase in π_i previously discussed. Consequently, the magnitude of the improvement in the contribution of each remaining individual to $\ln L(\phi)$ differs.

5. APPLICATION

Two real data sets are chosen to illustrate the mixture maximum likelihood approach to clustering treatment means. For both of these examples, identification and removal of distinct observations is demonstrated, and the resultant effect on the likelihood estimate examined. These sets have been used by other researchers to demonstrate various techniques for clustering means. One of them was also used by McLachlan and Basford (1988, Chapter 6) to demonstrate their previously mentioned method of clustering treatment means using a mixture model with the estimate of common group variance including the within treatment variation obtained from the analysis of variance. Here the treatment means are considered as raw data or individual observations, and the results are compared to those obtained by other methods.
Each data set is analysed using an amended version of the FORTRAN program KMM (listed in the monograph by McLachlan and Basford, 1988) assuming equal as well as unrestricted group variances, referred to as the homogeneous mixture model and heterogeneous mixture model, respectively. In its original form, this program did not allow for one or more of the group variances to be equal to zero when the group variances were unrestricted. However, by including criteria for identification of distinct individuals, based on posterior probabilities of group membership and convergence of any group variance to zero, these observations can be removed from the data set prior to a singularity in the likelihood occurring.
The same technique of identification and removal of distinct means can be used when the group variances are assumed to be equal. As the EM algorithm does not converge to a singularity in this case, it is possible to compare the results obtained when an observation is removed from the data set with those obtained when it is retained.
To evaluate the effect of the removal of individual observations on estimates of the log likelihood for the heterogeneous mixture model, the program was amended to remove the distinct observation in the iteration prior to a singularity occurring, i.e. when the variance of the group to which it was assigned was very small, but not quite zero. It was therefore possible to compare the contributions of individual observations to estimates of the log likelihood in a similar way to that possible with the homogeneous mixture model.

6. ILLUSTRATIVE EXAMPLES

This section presents the analysis of each data set, together with a brief discussion of the results.

Example 1

This data set was originally presented by Duncan (1955) and has been used by a number of researchers including Scott and Knott (1974), Cox and Spjøtvoll (1982), Caliński and Corsten (1985), and McLachlan and Basford (1988, Chapter 6) to demonstrate various clustering techniques. It represents yield (in bushels per acre) of 7 varieties of barley as follows:

1	2	3	4	5	6	7
49.6	58.1	61.0	61.5	67.6	71.2	71.3

McLachlan and Basford (1988) applied a previously discussed mixture model to these data and reported the partitioning, (1) (2 – 4) (5 – 7). A partitioning into 3 groups was also suggested by Scott and Knott (1974) and Plackett in the discussion of O'Neill and Wetherill (1971). Accordingly, the data were analysed here with $g = 3$.

In this instance, the first variety was identified as being distinct with both equal and unrestricted group variances. Although not illustrated here, this is not always so, particularly in the former situation. For example, only one observation might be assigned to a particular group but other observations may have some probability of belonging to it. The estimate of the group variance would not be zero and the observation would not be identified as being distinct.

Table 1.
Individual contributions to $\ln L(\phi)$ for the variety mean yields in Example 1, when fitting a normal homogeneous mixture model with $g = 3$.

Group	Variety	Original Procedure $\ln L(\phi)$	Amended Procedure $\ln L(\phi)$
1	1	−3.2664	Removed
2	2	−3.1555	−2.9374
2	3	−2.3111	−2.2136
2	4	−2.5462	−2.4151
3	5	−3.4939	−3.2273
3	6	−2.4726	−2.3520
3	7	−2.5271	−2.3987
$\ln L(\phi)$		−19.7728	−15.5441
Common Variance (V_c)		2.2325	2.6051

In the case of equal variances, the estimated log likelihood was −15.5441 and partitioning was consistent with that reported by McLachlan and Basford (1988, Chapter 6), Scott and Knot, (1974), and Cox and Spjøtvoll (1982). The contribution of each variety to the log likelihood at convergence was determined both when variety 1 was removed and when it was retained (Table 1).

The removal of variety 1 from the data resulted in a considerable improvement in the log likelihood estimate. This improvement came from two sources. The first and major one was the removal of the contribution of the first variety to the log likelihood. The second one was the improvement in each variety's contribution to $\ln L(\phi)$which occured after the removal of the first variety, even though the estimate of common variance had increased. However, with equal group variances, the size of this improvement was not the same for each variety.

Analysis of the data with unrestricted variances gave a final estimate of $\ln L(\phi)$ of −11.562 with a final partitioning of (1) (2 – 5) (6 – 7). This partitioning was slightly different to that obtained with the assumption of equal group variances as it placed variety 5 into group 2, rather than group 3. However, this grouping was presented as a possible solution by Cox and Spjøtvoll (1982).

The estimates of individual contributions to $\ln L(\phi)$ for the two iterations prior to the group 1 variance being equal to zero were obtained for the original procedure, and compared with the contributions of each variety to the log likelihood at the iteration prior to group 1 variance being equal to zero in the amended procedure (Table 2).

In this instance, there was no substantial difference between the original and amended procedures in the estimate of $\ln L(\phi)$ at iteration $s-1$ owing to the slow convergence of the program to a singularity. However, if the contribution of variety 1 was ignored, as in the amended procedure, a higher estimate of the log likelihood was obtained.

Table 2.
Estimates of individual contributions to $\ln L(\phi)$ for the variety mean yields in Example 1, when fitting a normal heterogeneous mixture model with $g = 3$

Group	Variety	Original Procedure		Amended Procedure
		Iteration $s-2$	Iteration $s-1$	Iteration $s-1$
1	1	−3.3667	0.8630	Removed
2	2	−3.3875	−3.3674	−3.2132
2	3	−2.7837	−2.7699	−2.6157
2	4	−2.7495	−2.7369	−2.5828
2	5	−3.9879	−3.9957	−3.8415
3	6	0.1932	0.1932	0.3473
3	7	0.1927	0.1928	0.3470
$\ln L(\phi)$		−15.8890	−11.6210	−11.5590
(V_1)		2.828	$5.746E-3$	–

Iteration s = iteration in which singularity occurs.

Example 2

This example concerns the mean volume (in millilitres) of loaves baked from 17 varieties of wheat. These data were originally presented by Duncan (1965), and have subsequently been used by researchers to illustrate various clustering techniques (Joliffe, 1975; Cox and Spjøtvoll, 1982; and Caliński and Corsten, 1985). The mean volumes were as follows:

1	2	3	4	5	6	7	8	9
654	729	755	801	828	829	846	853	861

10	11	12	13	14	15	16	17
903	908	922	933	951	977	987	1030

When analysed assuming equal group variances and partitioned into 5 groups, the final partition of (1) (2 − 3) (4 − 9) (10 − 14) (15 − 17) was consistent with that reported by other researchers. The final estimate of $\ln L(\phi)$ was −98.195 and common variance estimated as 363.451. No observation was identified as being distinct because of the effect of the large common variance on the calculation of posterior probabilities. However, when the data were partitioned into 6 groups the estimate of $\ln L(\phi)$ was −89.68, a substantially better result. The smaller estimate of common variance resulted in the first variety being

identified as distinct. The individual contributions to the log likelihood are presented both when variety 1 was removed and when it was retained (Table 3).
Analysis of the data with unrestricted group variances and 5 groups gave a partitioning of (1) (2 – 4) (5 – 9) (10 – 13) (14 – 17) and a final estimate of ln $L(\phi)$ of –90.323, which

Table 3.
Individual contributions to ln $L(\phi)$ for the variety mean volumes in Example 2, when fitting a normal homogeneous mixture model with $g = 6$

Group	Variety	Original Procedure ln $L(\phi)$	Amended Procedure ln $L(\phi)$
1	1	−6.5203	Removed
2	2	−6.1645	−6.1182
2	3	−6.1517	−6.1018
3	4	−7.1878	−6.9878
3	5	−4.8671	−4.8345
3	6	−4.8361	−4.8058
3	7	−4.9131	−4.8767
3	8	−5.2757	−5.2129
3	9	−5.9243	−5.8134
4	10	−5.5913	−5.5340
4	11	−5.3102	−5.2697
4	12	−5.0443	−5.0119
4	13	−5.3550	−5.2871
5	14	−6.1152	−5.9900
5	15	−5.5768	−5.5585
5	16	−5.8530	−5.8041
6	17	−6.5043	−6.4757
ln $L(\phi)$		−97.1908	−89.6820
Common Variance (V_c)		253.747	273.333

was substantially higher than that obtained for five groups groups under the assumption of equal group variances. The first variety was identified as being distinct. While this result differed from what has been found previously, Cox and Spjøtvoll (1982) indicated that variety 4 could belong to either group 2 or 3, and likewise variety 14 to either group 4 or 5. The contributions of each variety to ln $L(\phi)$ in the iterations of the program prior to the identification of the distinct individual were obtained (Table 4).
When the data were partitioned into 6 groups, variety 17 was identified as being distinct as well as variety 1. The estimate of log likelihood based on the remaining 4 groups and 15 varieties was –82.463, with the final partitioning being (1) (2–4) (5–9) (10–13) (14–16) (17). This was also a substantially better estimate than that obtained for 5 groups. Owing to the occurrence of two singularities at different points within the program, it was not possible to compare the amended and original procedures in the iteration prior to both group 1 and group 6 variances converging to zero. However, examination of the individual contributions to the log likelihood just prior to the identification and removal of the first distinct observation showed a similar pattern to that previously found.

7. LOG LIKELIHOOD ESTIMATION

As can be seen from the examples presented, the identification and removal of distinct observations has a marked effect on the estimated log likelihood.
In the case where equal group variances were assumed, this effect could be readily identified and assessed by comparing results obtained from the original and amended procedures. In the case of unrestricted group variances, a comparison could be made between

Table 4.
Estimates of individual contributions to $\ln L(\phi)$ for the variety mean volumes in Example 2, when fitting a normal heterogeneous mixture model with $g = 5$

Group	Variety	Original Procedure Iteration $s-2$	Original Procedure Iteration $s-1$	Amended Procedure Iteration $s-1$
1	1	−6.0740	3.9575	Removed
2	2	−6.7244	−6.7167	−6.6561
2	3	−6.2105	−6.1934	−6.1328
2	4	−6.5700	−6.5799	−6.5193
3	5	−5.3460	−5.3457	−5.2850
3	6	−5.2764	−5.2758	−5.2152
3	7	−4.7884	−4.7881	−4.7274
3	8	−5.0205	−5.0206	−4.9600
3	9	−5.6072	−5.6078	−5.5471
4	10	−5.4824	−5.4764	−5.4158
4	11	−5.1334	−5.1256	−5.0650
4	12	−4.9983	−5.0025	−4.9419
4	13	−5.5995	−5.6186	−5.5580
5	14	−6.1027	−6.0909	−6.0302
5	15	−5.7701	−5.7662	−5.7056
5	16	−5.7874	−5.7873	−5.7267
5	17	−6.9010	−6.9040	−6.8433
$\ln L(\phi)$		−97.3920	−87.3420	−90.3294
(V_1)		104.598	$1.219E-7$	–

Iteration s = iteration in which singularity occurs.

the original and the amended procedures at the iteration prior to a singularity occurring provided only one mean had been identified as being distinct.
It should be noted that the data in both examples gave log likelihood estimates less than zero. Hence removal of an individual observation resulted in an increased estimate of $\ln L(\phi)$. If the log likelihood was greater than zero, and the contribution of the distinct observation was positive, the opposite would occur.
When equal group variances were assumed, the major effect on the estimate of the log likelihood occured as a result of the removal of the contribution made by the observation that the amended procedure identified as being distinct. Both data sets analysed showed that the distinct observation made a substantial contribution to the log likelihood estimate in the original procedure (Table 5). This was attributed to $\ln L(\phi)$ being calculated using the common variance, rather than the variances associated with particular groups.

In addition, it would be expected that some adjustment to $\ln L(\phi)$ would result from the reduction in number of groups (g to $g-1$), number of means (n to $n-1$), and, in the case of equal variances, the resultant increase in the estimate of common variance, V_c, as defined in equation (13).

Table 5.
Contribution to estimate of $\ln L(\phi)$ in original program of individuals identified by amended program as being distinct, (for fitting a normal homogeneous mixture model)

Example	Contribution of Distinct Individual	Final $\ln L(\phi)$ (Original Procedure)
1	−3.2664	−19.773
2	−6.5203	−97.191

As illustrated in the examples, removal of a distinct observation resulted in an increase in each of the remaining observation's contribution to the log likelihood and therefore an increase in the estimate of $\ln L(\phi)$ (Tables 6 and 7).

Table 6.
Total increase in $\ln L(\phi)$ from the observations remaining after removal of a distinct individual, in fitting a normal homogeneous mixture model

	Estimate of $\ln L(\phi)$		
Example	Original Procedure	Amended Procedure	Difference
1	−16.5064	−15.5441	0.9623
2	−90.6705	−89.6820	0.9885

For the case of unrestricted group variances, the total improvement in $\ln L(\phi)$ for the remaining observations was given by equation (19). This theoretical improvement was determined for the above examples, along with the actual values observed (Table 7). Even though the number of observations being grouped was not large in either case, the improvement was reasonably close to 1.
Although the magnitude of the improvement in the contribution of each individual to $\ln L(\phi)$ differed in the homogeneous mixture model, the total improvement in the estimate of $\ln L(\phi)$ (Table 6) was quite consistent with that obtained for the heterogeneous mixture model (Table 7).

Table 7.
Total increase in $\ln L(\phi)$ at iteration "$s-1$" from the observations remaining after the removal of the distinct individual, in fitting a normal heterogeneous mixture model

Example	Initial n	Increase	$(n-1)\ln[n/(n-1)]$
1	7	0.9250	0.9249
2 ($g=5$)	17	0.9701	0.9700
2 ($g=6$)	17	0.9704*	0.9700

* Calculated prior to first singularity only.

8. CONCLUSION

Clustering of means using the mixture maximum likelihood approach provides a useful alternative to multiple comparison techniques when the partitioning of treatment means into nonoverlapping, homogeneous groups is considered desirable. When the variances of the underlying groups are assumed to be equal, the results obtained using this method were consistent with those reported by other researchers using different clustering techniques, and also with those obtained by McLachlan and Basford (1988, Chapter 6). When the group variances were unrestricted, the groupings obtained were not always identical to the homogeneous mixture model solution, but were consistent with other groupings of the treatment means quoted by researchers whose method provided several appropriate options. Similar results were obtained from other data sets not reported here.
One difficulty still to be overcome is the determination of the smallest number of groups consistent with the data, particularly in view of the effect on the log likelihood of the removal of those treatment means identified as distinct. This study reports an initial investigation for univariate data, but more research is clearly required, particularly for multivariate data.

REFERENCES

[1] AITKIN, M. (1980). Mixture Applications of the EM Algorithm in GLIM. *Compstat 1980, Proc. Computational Statistics.* Vienna: Physica Verlag, pp. 537–541

[2] AKAIKE, H. (1973). Information Theory and an Extension of the Maximum Likelihood Principle, In *2nd International Symposium on Information Theory, [eds. B.N. Petrov and F. Csaki]*, Akademiai Kiado, Budapest, 267–281.

[3] BASFORD, K.E. AND MCLACHLAN, G.J. (1985a). Likelihood Estimation with Normal Mixture Models. *Applied Statistics*, **34**, 282–289.

[4] BASFORD, K.E. AND MCLACHLAN, G.J. (1985b). Cluster Analysis in a Randomized Complete Block Design. *Comm. Statist. Theor. Meth.*, **14**, 451–463.

[5] BINDER, D.A. (1978). Bayesian Cluster Analysis. *Biometrika*, **65**, 31–38.

[6] BINDER, D.A. (1981). Approximations to Bayesian Clustering Rules. *Biometrika*, **68**, 275–285

[7] BOZDOGAN, H. (1986). Multisample Cluster Analysis as a Alternative to Multiple Comparison Procedures. *Bulletin of Informatics and Cybernetics*, **22** , 95–130.

[8] BOZDOGAN, H. (1987). Model Selection and Akaike's Information Criterion (AIC): The General Theory and its Analytical Extensions. *Psychometrika*, **52**, 345–370.

[9] BOZDOGAN, H. AND SCLOVE, S.L. (1984). Multisample Cluster Analysis Using Akaike's Information Criterion. *Ann. Inst. Statist. Math.*, **36**, 163–180.

[10] CALIŃSKI, T. AND CORSTEN, L.C.A. (1985). Clustering Mean in ANOVA by Simultaneous Testing. *Biometrics*, **41**, 39–48.

[11] CARMER, S.G. AND LIN, W.T. (1983). Type I Error Rates for Divisive Clustering Methods for Grouping Means in Analysis of Variance. *Commun. Statist.Simula. Comput.*, **12**, 451–466.

[12] COX, D.R. AND SPJØTVOLL, E. (1982). On Partitioning Means into Groups. *Scand. J. Statist.*, **9**, 147–152.

[13] DEMPSTER, A.P., LAIRD, N.M. AND RUBIN, D.B. (1977). Maximum Likelihood from Incomplete Data via the EM Algorithm. *J.R. Statist. Soc. B.*, **39**, 1–38.

[14] DUNCAN, D.B. (1955). Multiple Range and Multiple F Tests. *Biometrics*, **11**, 1–42.

[15] DUNCAN, D.B. (1965). A Bayesian Approach to Multiple Comparisons. *Technometrics* , **7**,171–222.

[16] GABRIEL, K.R. (1964). A Procedure for Testing the Homogeneity of all Sets of Means in Analysis of Variance. *Biometrics*, **20**, 459–477.

[17] HOCHBERG, Y., AND TAMHANE A.C. (1987). *Multiple Comparison Procedures.* John Wiley and Sons, New York.

[18] JOLLIFFE, I.T. (1975). Cluster Analysis as a Multiple Comparison Method. In *Applied Statistics*, [R.P. Gupta (ed.)], 159–168. Amsterdam, North Holland.

[19] KIEFER, J. AND WOLFOWITZ J. (1956). Consistency of the Maximum Likelihood Estimates in the Presence of Infinitely Many Incidental Parameters. *Ann. Math. Statist.*, **27**, 887–906.

[20] KIEFER, N.M. (1978). Discrete Parameter Variation: Efficient Estimation of a Switching Regression Model. *Econometrica*, **46**, 427–434.

[21] MCLACHLAN, G.J. AND BASFORD, K.E. (1988). *Mixture Models: Inference and Applications to Clustering.* Marcel Dekker, Inc., New York.

[22] MENZEFRICKE, U. (1981). Bayesian Clustering of Data Sets. *Commun. Statist.Theor. Meth.*, **A10**, 65–77.

[23] MILLER, R.G. (1981). *Simultaneous Statistical Inference.* Second Edition. Springer Verlag, New York.

[24] O'NEILL, R. AND WETHERILL, G.B. (1971). The Present State of Multiple Comparison Methods (with discussion). *J.R. Statist. Soc. B.*, **33**, 218–250.

[25] PLACKETT, R.L. (1971). Contribution to the Discussion of Paper by R. O'Neill and G.B. Wetherill. *J.R. Statist. Soc B.*, **33**, 242–244.

[26] SCOTT, A.J. AND KNOTT, M. (1974). A Cluster Analysis Method for Grouping Means in the Analysis of Variance, *Biometrics*, **30**, 507–512.

[27] TUKEY J.W. (1949). Comparing Individual Means in the Analysis of Variance. *Biometrics*, **5**, 99–114.

Recent Adv. in Stat. and Prob., pp. 41-57
J. Pérez Vilaplana and M. L. Puri (Eds)

Validation of Multivariate Monte Carlo Studies*

J.L. ROMEU
Department of Mathematics, SUNY-Cortland, Cortland, NY 13045, U.S.A.

Abstract – Validating a simulation study is a complex but necessary process. All study results depend on the strength of the validation statement. In Monte Carlo simulations, validation opportunities become particularly reduced. The multidimensionality issue only increases the problem complexity.
In this paper, a three-phase validation scheme based on the multivariate generation methods adopted in the study, is presented and explained in detail. Examples of the implementation of such a scheme, in three large Monte Carlo power studies, are described in detail.

1. INTRODUCTION

Recent computing advances (e.g., evergrowing power of PC's, parallel processing) have spurred the use of Monte Carlo techniques in statistical work. From engineering applications (Romeu, 1985), to comparison of methods (Romeu, 1989), to teaching (Romeu, 1986) or methodological research (Romeu, 1988), Monte Carlo and system simulation methods have become an important working tool for the modern statistician.
No longer is the practitioner constrained by the dimension of the problems. Hence, we are increasingly seeing the development of Monte Carlo techniques in the areas of multivariate statistics. And comparisons of multivariate methods (Ozturk and Romeu, 1992), small sample studies (Romeu, 1992a) or validation of statistical theories (Romeu, 1992b) are proliferating by day.
However, Monte Carlo results are only as good as the faith we can have in the validity of such studies. And building this faith becomes increasingly difficult when dealing with multivariate Monte Carlo. For, in addition to the significant differences with system simulation that inhibit the use of specific types of validation techniques, we add the multidimensionality problem.

Table 1.
Multivariate Normality Tests Compared

– Multivariate Q_n: Cholesky version.	– Koziol's Angles Test.
– Multivariate Q_n: Sigma Inverse version.	– Koziol's Chi Square Test.
– Mardia's Skewness Test.	– Malkovich and Afifi's Test.
– Mardia's Kurtosis Test.	– Royston's Test.
– Cox and Small Test,	– Hawkins' Test.

*This research was undertaken under a Dr. Nuala McGann SUNY/UUP Award and a Supercomputer Grant from the Cornell Theory Center

Table 2.
Multivariate Statistical Alternatives

- Bivariate Morgenstern (with two parameters).
- Bivariate Kinchine (with two parameters).
- Bivariate Regression (with two parameters).
- Pearson Type II (with $m = 10, 6, 4$).
- Pearson Type VII (with $m = 10, 6, 2$).
- Mixtures of Normals (with two mixing parameters).
- Student t (with 8 degrees of freedom).
- Chi-square (with 10 degrees of freedom).
- Generalized Lambda Distribution (three versions).
- Uniform (0,1).

In this paper, we describe a *three-phase validation scheme* for multivariate Monte Carlo studies. This methodology is derived from our experiences in planning and implementing extensive studies of this type. For example, in Romeu (1990), we compared ten multivariate normality (MVN) Goodness of Fit (GOF) tests (Table 1), under twelve non-normal alternatives (Figure 1 and Table 2). In the comparison we used experimental settings with two, four and eight p-variates, four sample sizes and two covariance structures (Figure 2), for a total of 288 experimental treatments or simulation runs. We will use this and other similar experiences to illustrate the application of this validation methodology.

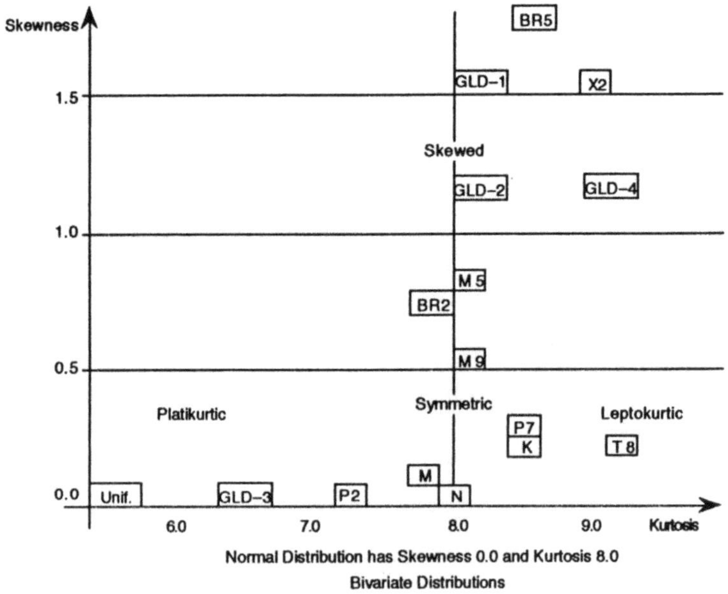

Figure 1. Schematic of the statistical alternatives design, with respect to their relation to skewness/kurtosis. [Alternatives considered are: Normal (N); Uniform (Unif.); t-distribution (T8); Chi–square (X2); Morgenstern (M); Khinchine (K); Pearson Type II (P2) and Type VII (P7); Mixtures of Normals (M5, M9) and Bivariate Regression(BR)].

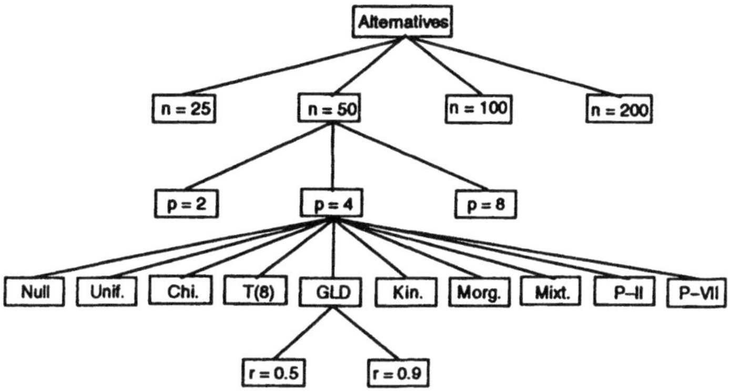

Figure 2. Schematic of the Monte Carlo study experimental design tree.

The three proposed validation phases are:

i) planning, during the design stage of the study,

ii) concurrent, as we move along the study itself and

iii) final stage, using the study results.

Carried out in such a way, validation becomes a researcher's quality control tool instead of just an activity performed to satisfy a client or a journal reviewer.

A good validation methodology prevents our learning, at the end of hundreds of runs, that our experiment somehow went wrong, and that we could have detected and corrected the problem earlier, if a carefull monitoring (validation) scheme had been implemented.

2. PLANNING STAGE

Monte Carlo studies are driven by statistical problems with intractable or messy mathematical solutions. Otherwise, the use of the Monte Carlo approach would be ill-advised. However, there is frequently an associated problem (the asymptotic version, a special case) with a well known closed form solution. It is during the initial literature search, while researching the theory behind the problem, that its associated solved version can be brought out to light. We may also find, during this initial research, that previous work exists in this general area with some reliable numerical results. And these activities will provide our first validation parameters.

For example, in Romeu (1992a) we studied and compared the small sample properties of ten MVN GOF tests, through our empirically derived small sample critical values. However, asymptotic distributions existed for some of these tests. And we used them to validate our work, by showing how the empirical critical values actually tended to the asymptotic ones, as $n \to \infty$.

We also found, during our literature search, that Mardia (1970, 1979) and Koziol (1982, 1983 and 1986) had obtained limited subsets of empirical critical values for their tests. We used these numerical results to check and validate our work in progress.

Power studies require the generation of well specified types of statistical alternatives. This activity constitutues the main challenge in a multivariate Monte Carlo study. But it also provides one of its most useful validation tools. For, by carefull investigaton of the

statistical alternatives used, their properties and their generation methods, we can find additional validation parameters with which to check our work.
In Romeu (1990) we classified the twelve statistical alternatives used in the power study into *purely skewed*, *purely kurtic* and *combined*, based on their first four moments. We also discovered that most MVN GOF tests investigated were either *skewed-prone* or *kurtic-prone*. And we classified them as such. For example, we verified how, in the bivariate skewness vs. kurtosis plane, Pearson Type II distribution yielded zero skewness and kurtosis smaller than that of the bivariate Normal (Figure 2). We realized we could use a combination of Mardia's Skewness and Kurtosis tests, applied to the generated samples, to construct another bidimensional plot (Figure 3). And that we could use these plots as validation tools. For both types of plots graphed the alternative distributions in different ways, but in compatible Skewness vs. Kurtosis planes. Generation methods were validated by verifying that both bivariate distribution classifications (the first plane representing the theoretical and the second the empirical conditions) would be consistent.

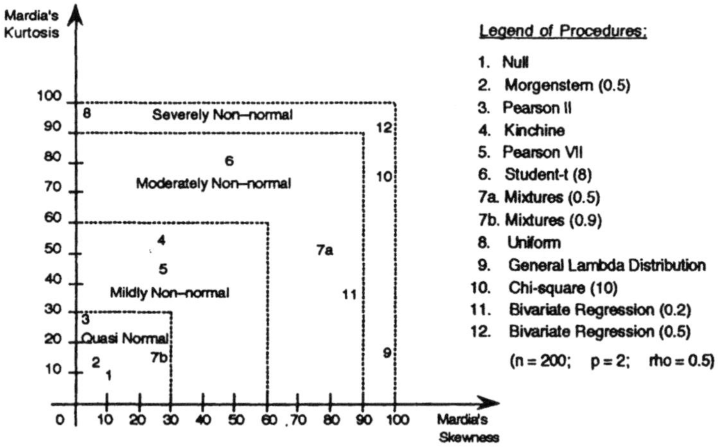

Figure 3. Achieved power of Mardia's skewness/kurtosis test, by procedures.

Up to now, we have been discussing our use of methods for generating multivariate distributions in the validation procedure. However, multivariate generation is not a trivial problem. And there exist several approaches to it. We surveyed them and organized the material into two broad groups which we call

i) indirect approaches to generating multivariate distributions and

ii) direct methods.

The indirect approaches are based on combining natural or empirical univariate distributions, given a covariance structure. But they do not place other constrains on the theoretical properties of the resulting (unknown) multivariate distribution. Such methods are easy to implement and have been widely used.
For example, Gnanadesikan (1977), obtained bivariate correlated distributions by first generating two independent random variates $Z_1, Z_2 \sim \mathcal{F}$. From them, two other variables Y_1, Y_2 are obtained by letting $Y_1 = Z_1$ and $Y_2 = \rho Z_1 + \sqrt{1-\rho^2} Z_2$ with $Corr\{Y_1, Y_2\} =$

ρ. Or, as performed by Loh (1986), following Andrews et al. (1973), by applying a transformation g to each coordinate of a bivariate normal.

Some advantages of these combinations of distributions are their simplicity and realistic flavor. Their major disadvantage consists in poor control of some parameters: skewness, kurtosis and marginal variances. And also, that the resulting multivariate distribution are unknown, except in the case that the original $Z_i \sim N(\mu, \sigma)$ and $g = I$.

An alternative is to generate the random variates from an empirical family of distributions. Shapiro and Gross (1981) list criteria that empirical families should meet: *i)* easy to select, and *ii)* to generate, and to *ii)* include as wide a variety of shapes as possible. Shapiro and Gross also classify the distributions exclusively based on their third and fourth moments, $\sqrt{\beta_1}$ and β_2. Empirical families allow us to control these moments with ease.

Three widely used univariate empirical families are (*a*) the Generalized Lambda Distribution (GLD), (*b*) Johnson's Family and (*c*) Pearson's Family. The GLD family was originally developed for Monte Carlo studies. It is based on p, a percentile of the distribution $\mathcal{F}$:

$$x_p = \mathcal{F}^{-1}(p), \quad 0 \leq p \leq 1$$

The GLD family is defined in terms of these percentile functions by:

$$x_p = R(p) = \lambda_1 + \frac{p^{\lambda_3} - (1-p)^{\lambda_4}}{\lambda_2}, \quad 0 \leq p \leq 1$$

$$f(x) = \frac{1}{R'(p)} = \frac{\lambda_2}{\lambda_3 p^{\lambda_3 - 1} + \lambda_4 p^{\lambda_4 - 1}}, \quad 0 \leq p \leq 1$$

Tables for the four lambda parameters of the GLD, for given values of $\sqrt{\beta_1}, \beta_2$, are available (Ramberg, Dudewicz, Tadikmalla and Mykytka (1979)). The GLD allows the exploration of the effect of a change in skewness, given a fixed kurtosis or viceversa, with relative ease.

Johnson's system is based on the following transformation: $z = \gamma + \eta\kappa_j(x; \xi, \lambda)$, $j = 1, 2, 3$, (Shapiro and Gross (1981)), where $z \sim N(0, 1)$, where $\gamma, \eta, \xi, \lambda$ are parameters and where $\kappa_j(x; \xi, \lambda)$, $j = 1, 2, 3$, are three functional forms, each defining one of the three subfamilies in the system. Johnson's system, partitions the β_1 vs β_2 plane into two non overlapping regions: S_U, S_B, separated by S_L, the family of the Lognormal distributions.

Pearson's families of distribution (Kendall and Stuart (1966)) are defined by the equation:

$$\frac{df}{dx} = \frac{(x-a)f}{b_0 + b_1 x + b_2 x^2}, \quad b_i \in \mathcal{R}, \quad i = 0, 1, 2$$

where f is the density of the random variable X. Pearson defines seven family types. For example, his Type II is the Beta and Type III, the Gamma distribution.

The main advantage in using empirical families of distributions consists in the larger control we have on the distribution's parameters. One serious disadvantage is their restricted domain, resulting in somewhat artificial distributions.

Since the resulting multivariate distributions obtained from such combinations of univariates are not known, we called this approach the *indirect* approach. However, we can check for the *known* covariance structure and skewness/kurtosis. In Romeu (1990) we generated combinations of GLD to obtain experimentally required skewness. We used these prespecified values as validation parameters with which to check our results.

We can, similarly, achieve a prespecified covariance structure with *mixtures* of MVN distributions. The resulting unknown multivariate distributions help assess the effect of data contamination on power. Let $\mathbf{X} \sim \mathcal{F}$, where:

$$\mathcal{F} = p_0\, MVN_p(\mu_1, \Sigma_1) + (1 - p_0) MVN_p(\mu_2, \Sigma_2)$$

$$Cov(\mathbf{X}) = p_0\Sigma_1 + (1 - p_0)\Sigma_2 + p_0(1 - p_0)(\mu_1 - \mu_2)(\mu_1 - \mu_2)'$$

There are many possible combinations formed by varying the parameters given by vector μ_i, covariance matrix Σ_i, for $i = 1, 2$, and the mixing parameter p_0. Based on the graphical study by Johnson (1987), based on bivariate mixtures, and seeking a mildly versus a severely contaminated alternative, Romeu (1990) selected $\mu_1' = (\mathbf{0}, \ldots, \mathbf{0})$ and $\mu_2' = (\mathbf{1}, \ldots, \mathbf{1})$, $\rho_1 = 0.5$ and $\rho_2 = 0.9$ and covariance matrices as:

$$\Sigma_i = \begin{pmatrix} 1 & \rho_i & \ldots & \rho_i \\ \ldots & \ldots & \ldots & \ldots \\ \rho_i & \ldots & \rho_i & 1 \end{pmatrix} \qquad i = 1, 2$$

The second approach to generating multivariate distribution, which we call *direct*, is more efficient but complex and often mathematically involved or intractable. Among the methods included in this group are

i) conditional distribution,
ii) transformation of marginals, and
iii) factorization.

The *conditional* distribution approach to generating a random vector (r.v.) $\mathbf{X}$ requires, first, the derivation of the p marginal distributions X_i. Then, of the successive conditional distributions of $X_j|X_{j-1}, \ldots, X_1$, for $j = 2, \ldots, p$. This is not always easy or feasible. For the *transformation* approach, a function $g(Y) = \mathbf{X}$ must be found such that $\mathcal{F}(g(Y)) = \mathcal{F}(\mathbf{X})$. Then, we proceed by generating, first, the easier multivariate Y. Then transforming it to $\mathbf{X}$ via g. The problem with this approach is that function g is not always available. For details, see Johnson, Wang and Ramberg (1984).

A frequent application of the above technique is in the generation of $MVN_p(\mu, \Sigma)$, from $MVN_p(0, I)$, via a Cholesky factorization A of the required covariance matrix $AA' = \Sigma$. Then, $g(Y) = AY + \mu$.

The multivariate Johnson (transformation) system (Johnson, 1987) offers the possibility of specifying many *controlled* multivariate distributions. But their derivation becomes mathematically involved and often intractable as p increases.

As in Johnson's univariate system, mentioned above, one of the four established transformations is performed on each of the p marginals. Then, the resulting joint multivariate distribution is obtained. Johnson has derived the densities of the transformed bivariate distributions. He has obtained relational functions between the original and resulting parameters and distributional moments, and has graphed, the bivariate distributions obtained with such transformations. They allow the study of specific types/levels of departures from the null, in a controlled environment. But for $p > 2$ the derivations become mathematically involved.

A comparison of the bivariate contours from Johnson's multivariate system with those obtained by mixtures of multivariate normals, appear on pages 64 to 82 and 56 to 51, respectively, in Johnson (1987). One notices how, with a convenient combination of the mixture parameters, similar statistical alternatives can be obtained. However, one ends

up with with less information, using this simpler method. We opted for this second approach, in Romeu (1990), to generate some of our skewed distributions.
The third approach, which we have called *factorization*, obtains a multivariate r.v. by multiplication of two other ones via the Elliptically Contoured (EC) distributions (Johnson (1987)). EC are defined in terms of the subclass of spherically symmetrical distributions. A p dimensional $\mathbf{X} \sim \mathcal{F}$ is spherically symmetrical if $\mathcal{F}(\mathcal{X}) = \mathcal{F}(\mathcal{P}\mathcal{X})$, for all $p \times p$ orthogonal matrices P. Geometrically speaking, spherically symmetrical distributions are invariant under rotations and include the normal, t and the symmetrical cases of the Pearson, Johnson and GLD families.
We say (and denote) $\mathbf{X} \sim EC_p(\mu, \Sigma; g)$ if its density:

$$f(x) = \kappa_p |\Sigma|^{-1/2} g\Big((\mathbf{X}_i - \mu)'\Sigma^{-1}(\mathbf{X}_i - \mu)\Big)$$

where κ_p is a normalizing constant and $g(.)$ a continous variable.
Therefore, $\mathbf{X}$ can be generated by multiplying R by $U^{(p)}$:

$$\mathbf{X} = RB\,U^{(p)} + \mu$$

where R is a positive random variable, independent of $U^{(p)}$, having the distribution of $\sqrt{(\mathbf{X}_i - \mu)'\Sigma^{-1}(\mathbf{X}_i - \mu)}$. And B is a $p \times p$ matrix such that $BB' = \Sigma$. Finally, $U^{(p)}$ is a random vector uniformly distributed on the unit hypersphere. Since $U^{(p)}$ is always the same, R^2 is the driver of the distribution of $\mathbf{X}$.
The univariate R^2 has density:

$$h(z) = \frac{\pi^{p/2}}{\Gamma(p/2)} \kappa_p \, z^{p/2-1} \, g(z)$$

where

$$z = (\mathbf{X} - \mu)'\Sigma^{-1}(\mathbf{X} - \mu)$$

For the multivariate normal (p), R^2 is the χ^2_p and $\kappa_p = (2\pi)^{-p/2}$ and g the identity. For the Pearson Type II, R^2 is $Beta(p/2, m+1)$. And for Pearson Type VII, R^2 is the univariate Pearson Type VI. This last type is generated via:

$$R^2 = Y/(1-Y), \qquad \text{where} \qquad Y \sim Beta(p/2, m - 1/2)$$

In Romeu (1990) we selected two elliptically contoured distributions: Pearson's type II and VII (Johnson (1987); Chmielewski (1981)), with parameters $m = 10, 6, 2$. Both these distributions are close to being multivariate and marginally normal, for large p.
The density function of the p-dimensional Pearson II distribution is:

$$f(x) = \frac{\Gamma(\frac{p}{2} + m + 1)}{\Gamma(m+1)\pi^{\frac{p}{2}}} |\Sigma|^{-\frac{1}{2}} \{1 - (\mathbf{X}_i - \mu)'\Sigma^{-1}(\mathbf{X}_i - \mu)\}^m$$

Its marginals are also Pearson type II distribution, with kurtosis:

$$\frac{3(m + \frac{p}{2} + 1)}{m + \frac{p}{2} + 2} \to 3 \qquad \text{as } m, p \to \infty$$

For $p = 4$ and $p = 8$, kurtosis is approximately 2.8, close to the Normal kurtosis of 3.0.

The density function for the p-dimensional Pearson type VII is:

$$f(x) = \frac{\Gamma(m)}{\Gamma(m-\frac{p}{2})}|\Sigma|^{-\frac{1}{2}}\{1+(\mathbf{X}_i-\mu)'\Sigma^{-1}(\mathbf{X}_i-\mu)\}^{-m}$$

Its marginal distribution is, again, Pearson VII. We used the property that marginals of EC distributions are also EC (in our case Pearson distributed) as another validation parameter in our Monte Carlo study.
A more recent approach to this problem is that of Rangaswamy, Weiner and Ozturk (1992). They decompose the multivariate $\mathbf{X} \sim \mathcal{F}$ using the factorization $\mathbf{X} = SZ$. Here, $Z \sim MVN_p(0,\Sigma)$ and S is a univariate r.v. driving the multivariate distribution of $\mathbf{X}$ such that:

$$f_{\mathbf{X}}(X) = (2\pi)^{-p/2}|M|^{-1/2}h_p(q); \qquad \text{where} \qquad q = \mathbf{X}'M^{-1}\mathbf{X}$$

where

$$h_p(q) = \int_0^\infty s^{-p} exp(\frac{-q}{2s^2}) f_S(s)ds \qquad \text{and} \qquad \Sigma = M\mathcal{E}(S^2)$$

One can make $M = \Sigma$ by redefining $s' = \frac{s}{\mathcal{E}(f)}$. And one can obtain a general multivariate r.v. Y, by letting:

$$Y = A\mathbf{X} + b \qquad \text{where} \qquad AA' = \Sigma;\ b \text{ a vector}$$

For example, for the case of multivariate Laplace, S is distributed as a Rayleigh.
In Romeu (1992b) a two-phase approach was used to validate such generator of a multivariate Laplace. First, with $p = 1$, let $z \sim N(0,1)$ and $w \sim exp(1)$. Making the transformation $y = \sqrt{2w}$ we obtain a Rayleigh distribution, with $\mathcal{E}(y^2) = 2$. Redefining $s = \frac{y}{\sqrt{2}} = \sqrt{w}$ we obtain a more convenient r.v. with expectation $\mathcal{E}(s) = 1$. This yields $M = \Sigma$ and the density of S is now:

$$f_S(s) = 2s\, exp(-s^2)$$

To obtain the distribution of the quadratic form $q = x'x = x^2$ Romeu finds, for $p = 1$ and following Rangaswami et al., that:

$$h_p(q) = \int_0^\infty s^{-p}\, exp(\frac{-q}{2s^2}) f_S(s)ds = \sqrt{\pi}\, exp(-\sqrt{2q})$$

And the distribution of the quadratic form q becomes:

$$f_Q(q) = \frac{\sqrt{2q}}{e} xp(-\sqrt{2q})$$

Then, by making the transformation $t = \sqrt{2q}$ Romeu obtains that $t \sim exp(1)$. Hence, for $x = sz$, with s, z and $h_p(.)$ as above with $p = 1$ we have:

$$f_X(x) = \sqrt{2\pi}\,|\Sigma|^{-1/2}\, h_p(q) = \frac{1}{\sqrt{2}} exp(-\sqrt{2}|x|)$$

Again making the transformation $t = \sqrt{2}\,|x|$ Romeu obtains a Double Exponential, which is a special case of the Laplace Distribution with parameter $(\lambda = \sqrt{2})$.

Samples of size $n = 25, 50, 100, 200$ from the univariate x factored as above ($x = sz$) were generated. And the variables x, s, q were then tested for the goodness-of-fits for, respectively, the double exponential, Rayleigh and exponential distributions. Then, in phase II we generated bivariate Laplace and, using the marginal property of EC, tested the two components of $\mathbf{X}$ and the quadratic form q. It is worth noticing that, as we increase p the resulting density functions become mathematically involved, with embedded Bessel Functions. In such cases, obtaining integrable closed forms, for the Distribution Function (CDF), manageable in a simulation program, becomes difficult.

Finally, some remarks about random number generators (RNG). In our Monte Carlo studies, we used the (IMSL) routine DRNUN, for generating the Uniform. IMSL has six different variations of its multiplicative congruential uniform generator (with or without shuffling). Each new pseudorandom uniform variate x_i, $i = 1, \ldots, n$, is generated by:

$$x_i = c_j x_{i-1} \, mod(2^{31} - 1)\,, \quad i = 1, \ldots, n$$

Shuffling uses a table of 128 uniform random variates from which the subsequent stream of uniform variates is randomly taken/replaced to prevent any possible autocorrelation in the pseudorandom variate stream. Empirical results by Nance and Overstreet suggest that shuffling does lead to an improvement (Bratley et al. (1983)) of pseudorandom generators. On the other hand, Dudewicz and Ralley (1981) show this is not always the case.

For a comprehensive discussion of pseudorandom uniform generators or pseudorandom generators at large, refer to Bratley, Fox and Schrage (1983). For a comparison of their statistical performances and for source codes, refer to Dudewicz and Ralley (1981).

As implemented in Romeu (1990, 1992), the generation of elliptically contoured distributions requires, first, the generation of the random vector $U^{(p)}$, uniformly distributed in the unit hypersphere (where the uniform distribution induces equally likely directions). There are several schemes for obtaining such distribution (Johnson (1987)) when $p = 2, 3, 4$, and some of these schemes can be extended to cases of $p > 4$. However, the most general method, which we used extensively, is to generate p *iid* univariate standard normal variates and, then, to define:

$$U_j = \frac{X_j}{\sqrt{X_1^2 + \ldots + X_p^2}}, \quad j = 1, \ldots, p$$

The above discussed multivariate statistical distributions fulfill two important functions. First, they yield the main validation parameters we are seeking. Then, they cover the essential needs of generating alternatives in a multivariate Monte Carlo study, i.e.:

1. Shape (as provided by skewness, kurtosis, contamination, etc.).
2. Covariance structure (as provided by the respective ρ in the covariance matrix (low, medium on high).
3. Sample size (small, medium and large).
4. Number of p-variates (few, moderate and large).
5. Significance level (of the test statistic nominal α) corresponding to the tail values say of ten, five and one percent points.

With these generators we can assess the effects of a distribution that is skewed, instead of symmetrical; peaked or flatter than the standard kurtosis; both, skewed and peaked or flat, or suffer from data contamination.

In the initial stage of a Monte Carlo study, the requirements for the experimental design, the alternative distributions to be used, and the particular generation methods are selected. Hence, this is the proper stage to define the validation parameters and the validation roadmap for the two subsequent stages.

3. CONCURRENT STAGE

The second validation phase starts with the computer implementation of the study and is carried out, as a control tool, throughout the entire experiment. It consists in verifying, along the different generation stages, that the validation parameters identified in Phase I are met. If one validation parameter fails to be met, a halt and thorough investigation of the causes should be undertaken. After verification and correction, the Monte Carlo study can proceed.

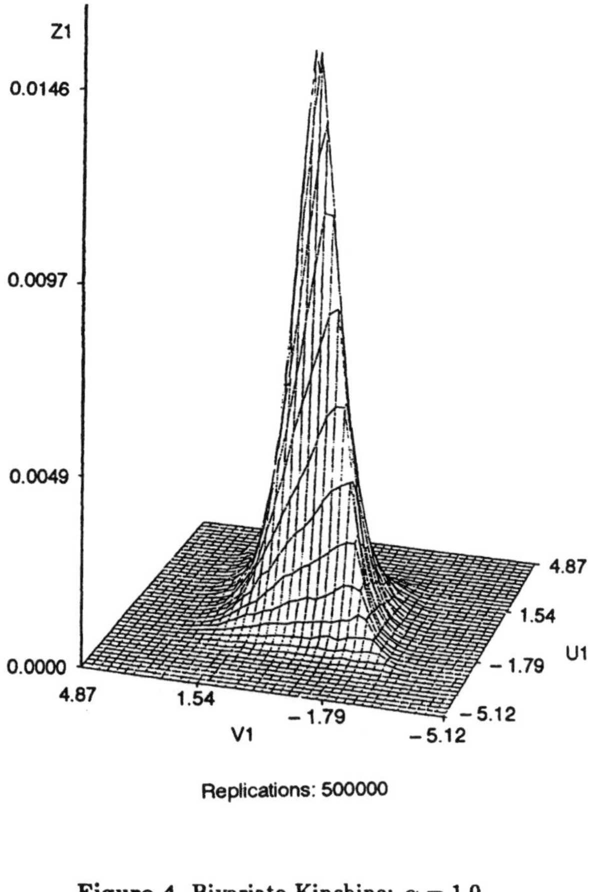

Figure 4. Bivariate Kinchine: $\alpha = 1.0$.

The first activity of this second stage concerns the verification of any and all off-the-shelf software to be used. Several unpleasant surprises have encouraged us to (i) program many of our routines (Press et al. (1986)) and to (i) never trust any canned software.
For example, to check the validity and accuracy of the IMSL uniform RNG option 2 (URN04 in Dudewicz and Ralley (1981)), Romeu (1990) first selected a better though

slower one, implementable in our machine: URN14. We ran two non-normal alternatives that depended heavily on the generation of $U(0,1)$ random variates. We verified that the results, obtained with IMSL and URN14 RNG's were very close (tables of differences are given in Romeu (1990)). Differences between empirical powers obtained from each generator, were overwhelmingly less than 0.01. Hypothesis tests on these differences were not statistically significant.

By following the validation roadmap drafted during Phase I, we can compare incoming results with those available from previous work in the literature. In Romeu (1990) we verified how our results agreed with those obtained by Mardia and Koziol, for special cases of the p-values.

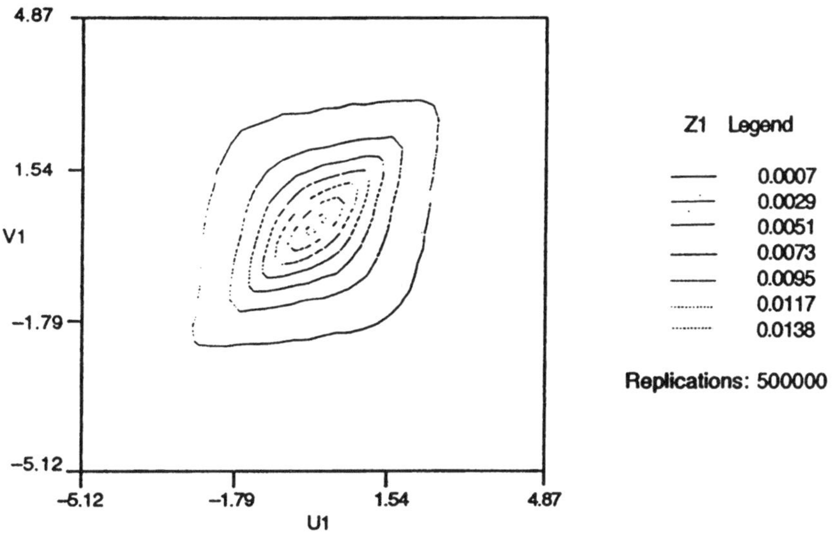

Figure 5. Bivariate Kinchine: $\alpha = 1.0$.

Another validation procedure consisted in checking the results under the null. For, under this hypothesis, the empirical powers should match the nominal significance level α. In Romeu (1990), three standard deviations about the empirical powers, obtained using $\sqrt{\frac{\alpha(1-\alpha)}{n}}$, were used to test that they were distributed about the nominal α.

Next, all distribution properties identified in Phase I were graphically checked. Bivariate measures of covariance, skewness and kurtosis were especially usefull. We plotted bivariate densities and graphed bivariate contours from the data generated from our distribution simulators (Figures 4 and 5). And we compared them with those appearing in the studies by Johnson (1987) to verify we had achieved accuracy and the effect desired in them.

We also used Mardia's Skewness and Kurtosis MVN GOF tests plot described in the previous section (Figure 3), to assess all generated distributions. And we verified whether or not, the positions of the different distributions were consistent with those in Figure 2. For, both graphs assessed, in different ways, the same Skewness/Kurtosis characteristics. Notice, in Figure 3, how severely skewed (GLD-1) and severely kurtic (Uniform) distributions, lay far out on the graph axes. And how severely skewed-kurtic distributions like

χ^2_{10} and Bivariate Regression fall far out on the bisection of the first quadrant. And how quasi normal distributions remain about (0.1, 0.1), the nominal significance level of these tests.

Sample covariance matrices and marginal distribution parameters were also checked and compared with the theoretical ones, before launching the production runs. Sensitivity analyses were performed by letting p-variate correlation coefficient ρ vary, and comparing the results.

If there were two ways of generating the same random variable, we took advantage of this for the validation process. We implemented a small comparison to check that the results were close. For example, in Romeu (1990) we simulated the Pearson Type VII distribution of a random variable $\mathbf{X}$, via the transformation $\mathbf{X} = (\sqrt{S}/\nu)^{-1}Z + \mu$, with $Z \sim MNV_p(0, \Sigma)$ and $S \sim \chi^2_\nu$. Then we generated this distribution again, via the factorization ($\mathbf{X} = RBU^{(p)} + \mu$) approach discussed in Section 2. Results were verified and the EC method was then selected for production runs.

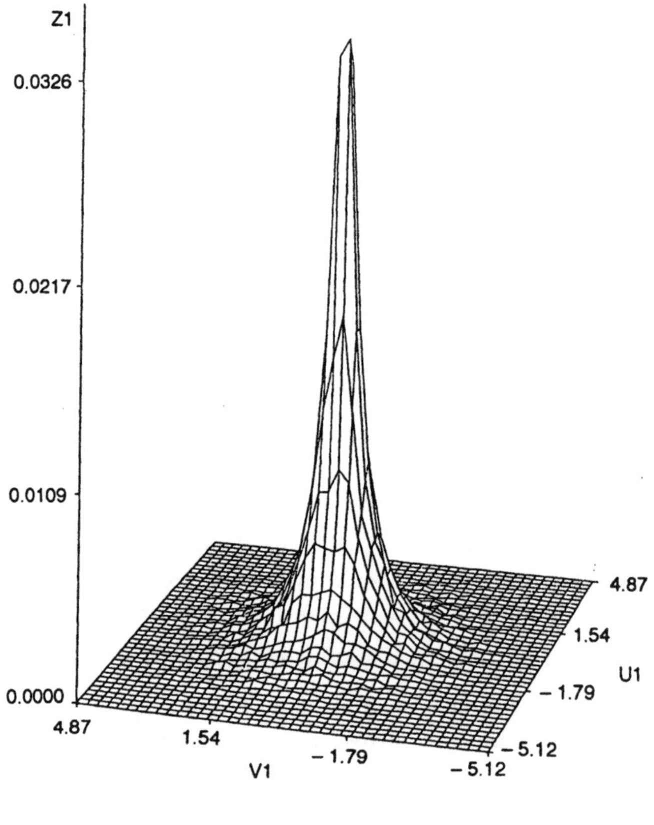

Figure 6. Bivariate Laplace.

In another type of study, dealing with the validation of a theoretical generation method, Romeu (1992b) ran simulations for $n = 25, 50, 100, 200$ as explained above. We checked the fit of model variables x, s, q. We graphed the bivariate distribution, from the generated data, as a *visual validation* approach (Figure 6). We also conducted sentitivity analysis by generating closely related, but different distributions (Laplace with other values for λ). All results were positive.

Documentation of all these validation procedures and their results should be carefully kept and included in the final report with the study findings.

4. FINAL ANALYSIS

This last validation stage is undertaken after the Monte Carlo study is completed and the final results are available. The type of validation performed in this stage is highly dependent on the objectives of the study. However, all Monte Carlo studies have in common that the subject problem is not tractable *in the study setting.* But that there usually exists another setting, where asymptotic results or particular cases do have a solution which is available for comparison.

Table 3.
Regressions of ecv on Sample Sizes

ROW	p	eta	Bo	C.V.	sigma	IoF	MVN GOF Test:
1	2	0.90	7.790	7.78	0.027	0.99	Mardia's Skew.
2	2	0.95	9.670	9.49	0.044	0.98	Mardia's Skew.
3	2	0.99	13.770	13.28	0.065	0.98	Mardia's Skew.
4	2	0.90	4.580	4.61	0.014	0.98	Cox and Small
5	2	0.95	5.990	5.99	0.032	0.94	Cox and Small
6	2	0.99	9.100	9.21	0.078	0.94	Cox and Small
7	2	0.90	0.989	*	5.500†	0.99	Malkovich-Afifi
8	2	0.95	0.989	*	5.800†	0.99	Malkovich-Afifi
9	2	0.99	0.989	*	7.500†	0.99	Malkovich-Afifi
10	2	0.90	1.630	1.65	0.012	0.99	Mardia's Kurt.(UB)
11	2	0.95	2.080	1.95	0.016	0.99	Mardia's Kurt.(UB)
12	2	0.99	2.900	2.58	0.036	0.92	Mardia's Kurt.(UB)
13	2	0.90	−1.550	−1.65	0.011	0.96	Mardia's Kurt. (LB)
14	2	0.95	−1.770	−1.95	0.018	0.95	Mardia's Kurt. (LB)
15	2	0.99	−2.180	−2.58	0.032	0.93	Mardia's Kurt. (LB)
16	2	0.90	1.060	*	5.200†	0.14‡	Hawkins
17	2	0.95	1.320	*	7.700†	0.04‡	Hawkins
18	2	0.99	1.940	*	1.500†	0.08‡	Hawkins
19	2	0.90	4.630	4.61	0.037	0.30‡	Koziol's Angles
20	2	0.95	5.990	5.99	0.060	0.11‡	Koziol's Angles
21	2	0.99	9.110	9.21	0.157	0.04‡	Koziol's Angles

* Empirical tests; no asymptotic distribution available.
‡ Critical Value was independent of sample size.
† Exponential notation; four decimal places (i.e 5.5*e-4).
NOTE: Empirical critical values (*ecv*) were regressed on sample sizes (n = 25(25)200) for each percentile (*eta*). The simple linear regression intercept (*Bo*) is compared with the method asymptotic percentile (*C.V.*) if it exists. *IoF* is the regression index of fit and *sigma* is the standard deviation of the independent term *Bo*.

For example, Romeu (1990) obtained small sample empirical critical values (*ecv*) for several MVN GOF tests, for samples $n = 25(25)200$, for $\rho = 0.5, 0.9$ and $p = 2(1)6(2)10$. Several of these tests also had asymptotic distributions. We used the asymptotic critical values as a validation parameter, by regressing the empirical ones on sample size. And then, comparing the regression independent term, β_0, with the corresponding asymptotic value.

In Table 3, we present some of these regression results. We regressed, by percentile $\eta = 0.90, 0.95, 0.99$, the ecv's on sample size n:

$$ecv_\eta = \beta_0 + \beta_1 n^{-1}$$

Results in Table 3 include, β_0, its standard deviation σ_β, the corresponding asymptotic critical value CV, and the regression index of fit R^2. As expected, as $n \to \infty$, our empirical values approached the asymptotic critical values.

Another way of validating a Monte Carlo study using the final results, in Romeu (1990), consisted of assessing the precision of the empirical critical values for $n = 200$. Following Dudewicz and van der Meulen (1984), we obtain two order statistics: $Z_{(r)}, Z_{(s)}$, such that;

$$s - 1 + 0.5 = \xi_\alpha \sqrt{Np(1-p)} + Np$$

$$r - 0.5 = -\xi_\alpha \sqrt{Np(1-p)} + Np$$

where ξ_α is the normal standard percentile evaluated at α and s, r are the corresponding positions in the ordered sample $Z_{(i)}$, $i = 1, \ldots, n$, such that:

$$\mathcal{P}\{Z_{(r)} < \xi_p < Z_{(s)}\} \geq p$$

We calculated approximate 95% conficence intervals (C.I.) for the relevant critical values. Examples of such calculations are presented in Table 4, for $p = 2, 8$, $n = 200$ and $\rho = 0.5, 0.9$. There, we show, the point estimate of the 95th percentile (*ecv*), the corresponding lower/upper bounds of its approximate 95% C.I. $(Z_{(r)}, Z_{(s)})$, and the asymptotic critical value. We can verify how these C.I. cover the asymptotic values in some MVN GOF tests that converge faster (e.g. Mardia's Skewness test) while they do not cover (for $n = 200$) the asymptotic value in those tests that converge slower (e.g. Mardia's Kurtosis, converges for $n > 200$).

Any other similar asymptotic result or special case can be exploited for performing validation at this final stage. In Romeu (1992b), the distribution of interest was the K-Distribution, which requires the manipulation of Bessel functions. Laplace, a special case of the K, with parameter $\alpha = 1$, was used. Such special cases will surface during the literature search performed in Phase I.

5. CONCLUSIONS

Validation of *multivariate Monte Carlo studies* constitutes an involved, time consuming but necessary activity. It really consists of a three-phase, continous process that starts with the initial planning of the study and concludes after the study final results are obtained.

As opposed to system simulation, Monte Carlo studies do not have real data to compare with, from an operating system. As opposed to univariate statistics, multivariate introduces yet another level of complexity to the problem. For now, the distributions obtained by transformations are much more complex. They have too many parameters and factors to control.

However, some of these same characteristics can be used as a validating tool. If we have correctly performed transformations and generation of variates, the output must exhibit certain prespecified caracteristics that we want to measure. Another problem arises because, sometimes, these characteristics are difficult to measure.

In any case, the confidence in the results of any Monte Carlo study lies in the faith we can place in it. Hence, all efforts allocated to the validation process are well justified.

Table 4.
95% Nonparametric Confidence Intervals.

ROW	p	rho	eta	CV	LB	UB	MVN GOF Test:
1	2	0.5	0.95	9.49	9.17	9.660	Mardia's Skew.
2	2	0.5	0.90	1.65	1.52	1.610	Mardia's Kurt.
3	2	0.5	0.95	5.99	6.11	6.533	Royston
4	2	0.5	0.95	5.99	5.94	6.290	Cox and Small
5	2	0.5	0.95	5.99	5.85	6.110	Koziol's Angles
6	2	0.9	0.95	9.49	9.11	9.590	Mardia's Skew.
7	2	0.9	0.90	1.65	1.48	1.570	Mardia's Kurt.
8	2	0.9	0.95	4.61	4.78	5.070	Royston
9	2	0.9	0.95	5.99	5.85	6.230	Cox and Small
10	2	0.9	0.95	5.99	5.86	6.220	Koziol's Angles
11	5	0.5	0.95	49.80	48.32	49.710	Mardia's Skew.
12	5	0.5	0.90	1.65	1.23	1.340	Mardia's Kurt.
13	5	0.5	0.95	11.07	10.83	11.410	Royston
14	5	0.5	0.95	11.07	20.05	26.990	Koziol's Angles
15	5	0.9	0.95	49.80	48.43	49.760	Mardia's Skew.
16	5	0.9	0.90	1.65	1.27	1.410	Mardia's Kurt.
17	5	0.9	0.95	5.33	8.24	8.800	Royston
18	5	0.9	0.95	11.07	18.97	22.920	Koziol's Angles
19	5	0.5	0.99	57.34	56.41	58.920	Mardia's Skew.
20	5	0.5	0.98	2.05	2.02	2.300	Mardia's Kurt.
21	8	0.5	0.95	145.98	143.32	145.610	Mardia's Skew.
22	8	0.5	0.90	1.65	1.08	1.190	Mardia's Kurt.
23	8	0.5	0.95	15.51	13.53	14.480	Royston
24	8	0.5	0.95	15.51	16.25	17.090	Koziol's Angles
25	8	0.9	0.95	145.98	142.14	144.750	Mardia's Skew.
26	8	0.9	0.90	1.65	1.06	1.200	Mardia's Kurt.
27	8	0.9	0.95	5.55	9.12	9.660	Royston
28	8	0.9	0.95	15.51	18.25	19.530	Koziol's Angles

* *Rho* is the p-variate correlation coefficient.

* *ecv*'s *LB* (confidence interval lower bounds) and *UB* (upper bounds) were empirically obtained with 10,000 replications for $p = 2$ and with 5,000 replications for $p > 2$.

* *CV* is the asymptotic critical value, for the corresponding percentile, (*eta*), of 90, 95, 98 or 99 percent for the test in question.

There is a final conclusion, that surfaces inevitably in the mind of the educator, after implementing an experience such as this. We have seen how the validation process requires extensive study and implementation of distributions, its moments, estimators, transformations, generation, hypothesis testing and experimental design, among other statistical topics.

Therefore, a *project* such as the *design, implementation and validation of* a small *multivariate Monte Carlo power study* lends itself, beautifully, for an *applied graduate statistics course.*

REFERENCES

[1] ANDERSON, T.W. (1984). *An Introduction to Multivariate Analysis.*John Wiley and Sons.

[2] ANDREWS, D.F., GNANADESIKAN, R. AND WARNER, J.L. (1973). Methods for Assessing Multivariate Normality. In: *Multivariate Analysis.* Academic Press.

[3] BRATLEY, P; FOX, B. AND SCHRAGE, L. (1983). *A Guide To Simulation.* Springer-Verlag.

[4] CAMBANIS, S.; HUANG, S. AND SIMONS, G. (1981). On the Theory of Elliptically Contoured Distributions. *J. Multivar. Annal.,* **11**, 368–385.

[5] CHMIELEWSKI, M.A. (1981). A Re-Appraisal of Tests for Normality. *Comm. Stat.- Theor. Meth.,* **a10(20)**, 2005–2014.

[6] DUDEWICZ, E.J. AND RALLEY, T.G. (1981). *The Handbook of Random Number Generation and Testing With TESTRAND Computer Code.* American Sciences Press Inc.

[7] DUDEWICZ, E.J. AND VAN DER MEULEN, E.C. (1984). On Assessing the Precision of Simulation Estimates of Percentile Points. *Amer. Jour. Math. Manag. Sci.,* **4**, 3–4, 335–343.

[8] GNANADESIKAN, R. (1977). *Methods of Statistical Data Analysis of Multivariate Observations.* John Wiley and Sons.

[9] JOHNSON, M.E. (1987). *Multivariate Statistical Simulation.* John Wiley and Sons.

[10] JOHNSON, N.L. AND KOTZ, S. (1970). *Distributions in Statistics: Continous, Univariate Distributions.* John Wiley and Sons.

[11] JOHNSON, M.E., CHIANG, W. AND RAMBERG, J. S. (1984). Generation of Continuous Multivariate Distributions for Statistical Applications. *Amer.Jour. Math.Manag.Sci.,* **4**, 225–248.

[12] JOHNSON, R.A. AND WICHERN, D.E. (1982). *Applied Multivariate Analysis.* Prentice Hall.

[13] JOHNSON M. AND RAMBERG, J.S. (1982). The Johnson Translation System in Monte Carlo Studies. *Comm. Stat.- Simula.Comput.,* **11**, 5, 521–525.

[14] KENDALL, M.G. AND STUART, A. (1966). *The Advanced Theory of Statistics.* Charles Griffin and Co.

[15] KOZIOL, J.A. (1982). A Class of Invariant Procedures for Assessing Multivariate Normality. *Biometrika,* **69**, 423–427.

[16] KOZIOL, J.A. (1983). On Assessing Multivariate Normality. *JRRS, Ser.B,* **45**, 358–361.

[17] KOZIOL, J.A. (1986). Assessing Multivariate Normality: A Compendium. *Comm.Stat.,* **15**, 2763–2783.

[18] LOH W. (1986). Testing Multivariate Normality by Simulation. *Jour. Statist. Comput. Simul.,* **26**, 243–252.

[19] MARDIA, K.V., KENT, J.T. AND BIBBY, J.M. (1979). *Multivariate Analysis.* Academic Press.

[20] MARDIA K.V. (1970). Measures of Multivariate Skewness and Kurtosis With Applications. *Biometrika,* **57**, 519–530.

[21] MARDIA, K.V. (1975). Assessment of Multinormality and the Robustness of Hotelling T Test. *Appl.Statist.*, **24**, 163–171.

[22] OZTURK, A. AND ROMEU, J.L. (1992). A New Graphical Test for Multivariate Normality. *Comm.in Statist. (Simula.)*, **21**, 1.

[23] PRESS, W.H.; FLANNERY, B.P.; TEUKOLSKY, S. A. AND VETTERLING, W.T. (1986). *Numerical Recipes: the Art of Scientific Computing.* Cambridge University Press.

[24] RAMBERG, J.S.; DUDEWICZ, E.J.; TADIKAMALLA, P.R. AND MYKYTKA, E.F. (1979). Probability Distributions and its Uses in Fitting Data.*Technometrics*, **21**, 2 , 201–214.

[25] RANGASWAMY, M.; WEINER, D. AND OZTURK, A. (1992). Computer Generation of Correlated Non Gaussian Clutter for Radar Signal Detection. *IEEE Trans.Aerosp. Electr.Sys.* (To appear).

[26] ROMEU, J.L. (1985). A Simulation Approach for the Analysis and Forecast of Software Productivity. *Journal of Computers and Industrial Engineering*, **9**, 2.

[27] ROMEU, J.L. (1986). Teaching Engineering Statistics With Simulation: A Classroom Experience. *Journal of the Institute of Statisticians*, **35**, 4.

[28] ROMEU, J.L. (1988). Another Look at the Comparison of the Non Overlapping Batch Means and Area STS Simulation Output Analyses Procedures. In: *Investigación Operativa - IO'88*, [J.P.Vilaplana (ed.)]. The Basque Country University Press, San Sebastian, Spain.

[29] ROMEU, J.L. (1989). A Small Sample Monte Carlo Study of Four System Reliability Bounds. *Journal of Computers and Industrial Engineering*, **16**, 1.

[30] ROMEU, J.L. (1990). *Development and Evaluation of a General Procedure for Assessing Multivariate Normality.* CASE Center Technical Report 9022. Syracuse University.

[31] ROMEU, J.L. (1991). A New Multivariate Normality Goodness of Fit Test With Graphical Applications. *Proceedings of the Computers and Industrial Engineering Conference.*

[32] ROMEU, J.L. (1992a). Small Sample Empirical Critical Values as a Tool for the Comparison of Multivariate Normality Goodness of Fit Tests. *Proceedings of the Conference on the Interface Between Statistics and Computer Science.*

[33] ROMEU, J.L. (1992b). *Monte Carlo Validation of a Theoretical Model for Generating Non Gaussian Radar Clutter.* Final Report. U.S. Air Force Faculty Summer Research Program. (To Appear).

[34] SHAPIRO, S. AND GROSS, A. (1981). *Statistical Modeling Techniques.* Marcel Dekker.

[35] TONG, Y.L. (1990). *The Multivariate Normal Distribution.* Springer-Verlag.

Recent Adv. in Stat. and Prob., pp. 59-66
J. Pérez Vilaplana and M. L. Puri (Eds)

Comparison of Variants of Barrodale and Roberts' L_1 Algorithm

V.A. SPOSITO[1]; J.E. GENTLE[2] and S.C. NARULA[3]
[1] Department of Statistics, Iowa State University, Ames, IA 50011, U.S.A. (deceased)
[2] George Mason University, Fairfax, VA 22030, U.S.A.
[3] School of Business, Virginia Commonwealth University, Richmond, VA 23284, U.S.A., and Department of Mathematics, Linköping Institute of Technology, S-581 83 Linköping, Sweden

Abstract – Barrodale and Roberts proposed an efficient algorithm to solve the L_1 multiple linear regression model that can be formulated as a linear programming problem with a very special structure. Various modifications of their algorithm have been proposed that take further advantage of this special structure.
We report the results of a computational study comparing various modifications of Barrodale and Roberts's computer program for solving the L_1 multiple regression.

1. INTRODUCTION

Consider the multiple linear regression model:

$$\mathbf{y} = \mathbf{X}\beta + \mathbf{e}$$

where $\mathbf{y}$ is an $n-$vector of the values of a response variable corresponding to $\mathbf{X}$, an $n \times k$ matrix of the values of regressor (or predictor) variables, β is a $k-$vector of unknown parameters, and $\mathbf{e}$ is an $n-$vector of unobservable random errors. Assume one desired to estimate β under the criterion of minimizing the sum of absolute deviations, (L_1 problem); i.e.,

$$\min_{\beta} \sum_{i=1}^{n} | y_i - \sum_{j=1}^{k} \beta_i x_{ij} |$$

The earliest reported algorithms for obtaining L_1 estimates were described by Edgeworth in the 19th century. However, due to computational difficulties, other procedures were not considered for nearly 70 years. Charnes, Cooper and Ferguson (1955) showed that the L_1 problem could be formulated as a standard linear programming problem. In particular, the L_1 problem can be expressed as:

$$\begin{aligned} \text{minimize} \quad & \mathbf{1}'(\mathbf{e}^+ + \mathbf{e}^-) \\ \text{subject to} \quad & \mathbf{X}\beta + \mathbf{I}\mathbf{e}^+ - \mathbf{I}\mathbf{e}^- = \mathbf{y} \\ & \mathbf{e}^+, \mathbf{e}^- \geq \mathbf{0} \\ & \beta \in E^k \end{aligned} \tag{1}$$

where $\mathbf{e}^+$ and $\mathbf{e}^-$ are $n \times 1$ vectors, $\mathbf{1}$ is an n–vector of ones and $\mathbf{I}$ is an $n \times n$ identity matrix.

Wagner (1959) underscored the fact that the dual problem of (1) needs only a basis of size $k \times k$, hence suggesting that simple upper bounding steps incorporated into simplex procedure would be an efficient procedure for obtaining L_1 estimates. The dual problem of (1) is:

$$\begin{aligned} \text{minimize} \quad & \mathbf{y}'\mathbf{d} \\ \text{subject to} \quad & \mathbf{X}'\mathbf{d} = \mathbf{0} \\ & -1 \le d_i \le 1 \qquad \text{for all } i \end{aligned}$$

Various simplex-based computational procedures have been developed for multiple regression problems since 1959; however, Barrodale and Roberts' (1973) modified simplex procedure or variants of this procedure has been noted to be the most efficient procedures available in the literature [see Dielman and Pfaffenberger (1984) or Narula (1987)]. In particular, Barrodale and Roberts (BR) showed how several simplex iterations could be combined into one. Armstrong, Frome and Kung (1979), (AFK), proposed a modification of BR's algorithm that uses a LU decomposition to maintain the current basis and requires less sorting. Bloomfield and Steiger (1983) modified BR's procedure further by determining a pivot that improves on a normalized steepest edge criterion proposed by Bartels, Conn and Sinclair (1978).

Dielman and Pfaffenberger (1984) have summarized several studies comparing various L_1 computer programs. They reported that for multiple linear regression, when the number of parameters, k, is less than four, a descent algorithm described by Wesolowsky (1981) appears best. For larger values of k, they recommended AFK's computer program. However, Bloomfield and Steiger (1983) , (BS), reported that their modifications of BR's procedure appears to be the best for all values of k.

One objective of our study was to compare the computational efficiency of the available computer programs based on Barrodale and Roberts' original algorithm and its variants that are considered most promising. A further objective of the study was to consider the efficiency of these computer codes when one computes and uses an L_2 estimator as a starting solution. Sklar and Armstrong (1982), in a small study, suggested that AFK's original program can be greatly improved if one computed and used an L_2 estimator $\hat{\beta}$; namely, using observations corresponding to the k smallest absolute least squares residuals as the initial basis. McCormick and Sposito (1976) have shown that by updating the $\mathbf{y}$ vector in (1) with $\mathbf{y} - \mathbf{X}\hat{\beta}$ in BR's program an optimal L_1 solution can be obtained in fewer iterations than without this modification.

The rest of the paper is organized as follows: In Section 2, we briefly review several simplex based L_1 algorithms. In Section 3, we present some preliminary considerations for the study. In Section 4, we present the results of the comparative study. We conclude the paper with a few remarks.

2. SIMPLEX-BASED L_1 ALGORITHMS

In view of various computational studies, the modifications of Barrodale and Roberts' algorithm by Amstrong, Frome and Kung (1979) and Bloomfield and Steiger (1983) are considered to be very efficient. These simplex-based algorithms exploit the following features of the L_1 problem:

a) There exists an L_1 solution that passes through at least k observations when $\mathbf{X}$ is of full rank k.
b) Each procedure considers the primal problem (1) and notes that due to the structure of the constraints an initial feasible solution always exists; i.e., if $y_i \geq 0$ for all i, let $e_i^+ = y_i$ and $e_i^- = 0$ for all i and let $\beta = \mathbf{0}$; for any $y_i < 0$, multiply this constraint by minus one and let $e_i^- = y_i$. This avoids the need of using artificial variables.
c) Each algorithm combines several simplex iterations into one iteration by determining a pivot which will yield the greatest reduction in the objective function value in (1).
d) Each procedure has two stages. In Stage I, each algorithm determines an initial extreme fit w.r.t. the β's, i.e., a solution which passes through k observations.
e) In Stage II the algorithms iterate from one extreme point to the next such that the consecutive fits differ by one zero residual.

These simplex-based algorithms differ in Stage I in how they select a new nonbasic variable to leave the basis and in the way they determine the current basic variable to leave the basis. Before describing the differences among the procedures we outline BR's computational procedure. Let (1) be denoted as

$$\mathbf{A} = \begin{bmatrix} \mathbf{X} & \mathbf{y} \\ \mathbf{I} & \underline{\mathbf{0}} \end{bmatrix} \tag{2}$$

where $(\mathbf{X} \,|\, \mathbf{y})$ is an $n \times (k+1)$ matrix, $\mathbf{I}$ is a $k \times k$ identity matrix an $\underline{\mathbf{0}}$ is a vector of zeros in $E^{\mathbf{k}}$. Here, the initial basis consists of the deviation variables.
Stage I begins with $\beta^* = \mathbf{0}$ and with the most negative directional derivative determining the next entering nonbasic vector. Moreover, only the regression coefficients, β_j's, are considered as candidates to enter the basic solution in Stage I. It consists of k steps. Each step, i, activates a new coordinate direction; hence locating a steepest edge, say δ_i. At the end of Stage I a set of independent edge direction vectors, $\delta_1, \ldots, \delta_k$, is determined. These directional derivatives are obtained from the first n rows of (2) as follows:
Let

$$Z = \{i \,|\, a_{i,p+1} = 0 \quad \text{for } i = 1, 2, \ldots, n\} \quad \text{and } N = \overline{Z}\,.$$

Then for each m $(m = 1, 2, \ldots, k)$ compute

$$\begin{aligned} g_m &= \sum_Z |\, a_{i_m} \,| \\ h_m &= \sum_N a_{i_m} \operatorname{sign}(a_{i,k+1}) \end{aligned}$$

and

$$f_m = \min(g_m - h_m, g_m + h_m)\,. \tag{3}$$

The entering vector, $\mathbf{q}$, is determined by $f_q = \min\{f_m \,|\, m \text{ nonbasic}\}$ for each step $(\leq p)$ in Stage I. This step consists of exactly the same sequence of calculations used in the standard simplex procedure to determine the entering nonbasic variable with the most negative marginal cost.
The next step in BR's algorithm is to determine the basic vector that leaves the current basis. The standard simplex procedure determines the outgoing basic vector by determining a pivot that will preserve feasibility while allowing the new incoming variable to increase form its set value of zero. BR noted that the value of the objective function can be decreased further if one modifies the standard simplex pivot rule by allowing a pivot

which will lose feasibility. They note that at this point a feasible solution can be easily obtained by multiplying any infeasible row, i, by minus one and interchanging e_i^+ and e_i^- in (1). Hence, BR essentially locates a pivot which will by-pass several simplex iterations and at the same time yield the greatest reduction in the sum of absolute deviations.
Bloomfield and Steiger (1983) noted that BR's selection sequence can be recast to correspond to a weighted median calculation for the line search of finding the minimum of (1) along a line $\beta^* + t\underline{\delta}$ with directions δ_i, $i = 1, 2, \ldots, k$. This yields a solution $t = y_r / x_{r_q}$ which is the weighted median of $(y_i / x_{i_q}$ with weights $| x_{i_q} |$ where r denotes the pivot row. Hence, the multiple pivot sequence in Barrodale and Roberts's procedure is essentially a steepest edge criterion for determining the direction to move from a given estimate. This computational step has been modified by Armstrong, Frome and Kung (1979) using a partial sort routine to obtain the weighted median a certain set of ratios in an explicit manner and also using an LU decomposition to maintain the current basis.
After the appropriate pivot row, r, has been determined using BR's algorithm, then the q−th column is divided by a_{r_q}, (x_{r_q}), and for $i \neq q$ the i−th column is updated by subtracting a_{r_i} times the new q−th column for all rows in (2). This yield the following updated structure:

$$\mathbf{A}_1 = \begin{bmatrix} \mathbf{XD}_1 & \mathbf{r} \\ \mathbf{D}_1 & -\beta_1^* \end{bmatrix} \tag{4}$$

where $\mathbf{D}_1$ is the $k \times k$ matrix of direction columns, $\mathbf{XD}_1$ is an $n \times k$ matrix of inner products $\{ \langle x_i, \underline{\delta} \rangle \}$ and $\mathbf{r}$ denotes the residuals for the first fit β_1^*.
After k iterations Stage I ends and Stage II begins. This stage involves selecting the entering vector q such $f_q = \min(f_m)$, i.e., the incoming vector is selected as that nonbasic vector having the most negative marginal cost, and the outgoing vector is again selected to maximize a decrease in the values of the sum of absolute deviations as in the Stage I.
BS's algorithm incorporates the following modifications in BR's algorithm. Instead of using a steepest edge approach they use a normalized steepest edge criterion. In particular, the entering nonbasic vector is selected by determining:

$$f_q = \max\{(-g_m - h_m, -g_m + h_m) / (-g_m) \quad \text{for } m = 1, 2, \ldots, k\}$$

in Stage I, or (3), of BR's algorithm. Furthermore, the pivot row is selected by determining the weighted median of $\{a_{i,k+1}/a_{i_q};\ i = 1, 2, \ldots, n\}$ with weights $| a_{i,k+1} |$; i.e., the weighted median of $\{y_i / x_{i_q};\ i = 1, 2, \ldots, n\}$ with weights $| y_i |$.

Table 1.
Program's storage and constants requirements

Program Name	*Author*	*Storage Required*	*Stopping Constants*
BARR	Barrodale and Roberts (1974)	$5n + p(n + p + 3)$	TOLER = .9537 E-06 BIG = .7237 E76
LINORM	Armstrong *et al.* (1979)	$6n + p(n + p + 5)$	ACU = 1.0 E-6 BIG = 1.0 E15
BLAD1	Bloomfield and Steiger (1983)	$4n + p(2n + 4)$	None

3. PRELIMINARY CONSIDERATIONS

In this study we compared the computer codes of Barrodale and Roberts (1974) (BR), Armstrong, Frome and Kung (1979) (AFK), and Bloomfield and Steiger (1983) (BS). The codes of BR and AFK were obtained from the distribution services of the journals that published these computer codes; the code of BS was obtained from the authors. All of the codes are in Fortran.

These codes have different storage requirements and values of stopping constants; these are shown in Table 1.

The test data used in our study were generated as follows: Each x_i was generated as $\mathcal{U}(0, 1)$ using IMSL's subroutine RNUM. Each y_i was formed as $\sum_{i=1}^{k} x_i$ plus an independent $\mathcal{N}(0, 9)$ random error term generated from IMSL's subroutine GGNQF. The number of observations, n, and the number of parameters, k, were varied over the intervals [50, 100, 300, 500, 1000, 2000, 3000] and [3, 4, 5, 6, 10, 15], respectively. Each computer code was executed 300 times for each combination of n and k. The study was conducted at Iowa State University using an NAS 9160 computer when the system was essentially empty. In the tables that follows, the timing results were recorded where the first number is the mean time in milliseconds over 300 random data sets and the maximum time required over this set is contained in parenthesis.

Table 2a.
Average (maximum) CPU time (milliseconds) for k = 3 and 5

n	k	BR	AFK	BS
50	3	5.19 (8.45)	5.63 (9.73)	4.55 (7.68)
	5	11.12 (16.90)	12.82 (19.71)	10.46 (20.99)
100	3	11.57 (18.94)	13.34 (21.50)	9.48 (16.38)
	5	25.45 (36.61)	28.84 (43.52)	24.25 (35.84)
300	3	49.33 (73.73)	52.76 (101.12)	35.47 (63.23)
	5	112.48 (158.46)	112.13 (163.84)	91.52 (146.18)
500	3	101.63 (173.06)	104.10 (165.89)	64.69 (101.63)
	5	244.90 (395.51)	216.43 (333.82)	161.76 (216.58)
1000	3	302.11 (425.98)	302.68 (538.62)	148.30 (217.34)
	5	659.55 (843.78)	559.30 (831.23)	402.89 (624.13)
2000	3	892.58 (1269.5)	925.29 (1624.3)	327.77 (553.73)
	5	1962.02 (2400.5)	1541.63 (2352.9)	897.90 (1444.9)
3000	3	1837.06 (2503.7)	1775.63 (3535.9)	511.39 (744.70)
	5	3953.50 (5063.7)	3178.55 (4355.1)	1418.07 (1976.6)

4. RESULTS OF THE STUDY

4.1. Original Computer Codes

We initially considered the unaltered programs of BR, AFK and BS for k = 3, 5, 10, and 15 over various values of n. Results in Table 2*a* show that for k = 3 and 5, BS's program required the least average CPU time. From the results in Table 2*b*, we observe that for k = 10 and 15, BR's program appears very efficient for values of n up to 300; for n between 300 and 1000 and p = 10 there seems to be no difference between the times for AFK and BS; for $n > 1000$ BS is preferable. For $n \geq 500$ and k = 15, AFK's code is the most efficient.

Table 2*c* gives a summary of Tables 2*a* and 2*b*; where BR– BS, for example, denotes that BR's is *slightly* more efficient than BS's code.

Table 2b.
Average (maximum) CPU time (milliseconds) for $k = 10$ and 15

n	k	BR	AFK	BS
50	10	35.28 (59.39)	41.41 (63.23)	39.73 (51.97)
	15	67.64 (87.04)	114.57 (167.94)	84.29 (116.99)
100	10	88.98 (134.14)	99.40 (157.44)	89.04 (126.98)
	15	175.41 (242.69)	242.42 (324.35)	192.98 (282.11)
300	10	390.70 (527.87)	355.46 (498.94)	352.21 (495.36)
	15	789.01 (990.70)	791.10 (1070.6)	786.90 (1053.4)
500	10	793.45 (1004.0)	652.54 (905.47)	669.10 (902.40)
	15	1590.10 (2052.6)	1386.20 (1792.3)	1580.92 (2067.4)
1000	10	2069.11 (2521.3)	1620.22 (2026.8)	1624.09 (2303.5)
	15	4275.46 (5211.9)	3142.90 (3755.3)	3732.10 (4684.8)
2000	10	5830.10 (7147.8)	4125.15 (5734.4)	3770.20 (5448.5)
	15	11172.80 (12818.2)	7874.60 (10270.9)	8824.50 (12022.0)
3000	10	10868.10 (13766.7)	7339.44 (10232.6)	5681.32 (8320.1)
	15	20079.27 (24780.3)	13476.30 (16102.4)	14083.90 (18657.8)

Based on Table 2*c*, we propose the following guidelines:

Use BS if $k \le 8 + .1\sqrt{n}$

Use BR if $k \ge 7$ with $n \le 200$

Use AFK if $k \ge 12$ with $n \ge 300$

Use BS or AFK if $k \in [8, 12]$ with $n \in (200, 1400)$.

Table 2c.
Summary of Tables 2a/2b

	k			
n	3	5	10	15
50	BS	BS	BR	BR
100	BS	BS	BR–BS	BR
300	BS	BS	BS–AFK	(no diff.)
500	BS	BS	AFK–BS	AFK
1000	BS	BS	AFK–BS	AFK
2000	BS	BS	BS	AFK
3000	BS	BS	BS	AFK

4.2. Computer Codes with an Advanced L_2 Start

It is well-known, [see McCormick and Sposito (1976) or Sposito, Smith and McCormick (1978)], that by computing and using an L_2 start one can solve L_1 regression problems more efficiently; i.e., using $\mathbf{y} - \mathbf{X}\hat{\beta}$ instead of $\mathbf{y}$ in (1). Also AFK's program can be modified to initially obtain a starting basis which corresponds to the smallest absolute L_2 residuals until m linearly independent x's are determined. Hence, in order to make a more comparable study we also modified BR and BS's codes to initially compute an L_2 solution and, then update the right-hand-side in (1). The modified codes are denoted

by AFK*, BR* and BS*. In the CPU times reported, the total computational effort was taken into consideration.
We report times for $k = 4$, 5 and 6 in Table 3*a*, and for $k = 10$ and 15 in Table 3*b*.

Table 3a.
Average (maximum) CPU time (milliseconds) for $k = 4$, 5 and 6 computing and using an L_2 start.

n	k	BR	AFK	BS
50	4	5.99 (9.98)	4.82 (9.98)	6.31 (9.73)
	5	8.61 (13.31)	6.71 (12.03)	9.40 (16.38)
	6	11.23 (20.22)	9.10 (15.87)	12.51 (21.25)
100	4	12.69 (21.76)	9.62 (17.41)	12.70 (20.22)
	5	17.92 (30.98)	13.93 (30.46)	18.47 (30.46)
	6	23.78 (36.35)	17.83 (37.89)	25.73 (42.75)
300	4	44.57 (74.24)	31.95 (57.34)	42.67 (79.36)
	5	62.13 (99.58)	47.11 (75.78)	61.29 (95.74)
	6	83.99 (122.11)	62.37 (105.73)	88.93 (138.75)
500	4	81.09 (149.25)	60.45 (111.62)	74.39 (123.39)
	5	111.03 (187.39)	85.64 (146.94)	110.02 (167.42)
	6	154.83 (248.58)	116.40 (194.82)	158.28 (231.17)
1000	4	179.19 (271.36)	126.68 (209.15)	162.97 (274.43)
	5	252.16 (490.24)	185.54 (407.30)	238.96 (374.02)
	6	350.23 (514.30)	252.47 (398.08)	336.11 (541.18)
2000	4	388.72 (597.76)	289.54 (525.31)	355.07 (589.57)
	5	553.13 (838.14)	403.75 (637.70)	507.49 (832.26)
	6	788.01 (1336.32)	566.78 (990.72)	725.73 (1073.41)

AFK* codes clearly outperforms BR* and BS* for nearly all cases considered. This advantage in AFK*'s code stems from the fact that upon computing the L_2 estimator the p smallest absolute L_2 residuals denote a starting basis associated with the β's whereas the basis associated with BR* and BS* are composed of the deviations, (e_i^+ and e_i^-). BR* outperformed AFK* (and BS*) for $k = 15$ with $n \leq 100$.

Table 3b.
Average (maximum) CPU time (milliseconds) for $k = 10$, and 15 computing and using an L_2 start.

n	k	BR	AFK	BS
50	10	24.62 (38.40)	23.06 (42.22)	30.83 (43.78)
	15	47.68 (65.54)	56.07 (102.91)	65.03 (97.79)
100	10	55.85 (78.08)	46.99 (68.35)	65.02 (94.98)
	15	106.06 (150.78)	112.69 (205.31)	142.78 (207.87)
300	10	203.18 (311.81)	155.93 (286.46)	220.84 (342.53)
	15	410.70 (594.32)	342.29 (528.90)	490.08 (649.73)
500	10	360.96 (540.67)	283.49 (424.70)	401.72 (605.44)
	15	775.41 (1176.58)	652.69 (934.40)	907.17 (1289.98)
1000	10	836.25 (1265.66)	631.09 (1060.35)	893.74 (1355.26)
	15	1745.62 (2505.47)	1401.15 (2009.60)	1965.92 (2927.87)
2000	10	1964.18 (2882.30)	1429.82 (2127.87)	2000.11 (3020.29)
	15	4009.12 (6245.12)	3119.19 (4601.34)	4448.33 (5811.97)

5. CONCLUSION

Based on our study for $k = 3$ to 5 as shown in Table 2c, BS's original program requires less computer time than AFK or BR's unaltered programs for all sample sizes considered. In general, BS is preferred if $k \leq 8 + .1\sqrt{n}$. For $k = 10$ or 15, BR's program performs quite well for $n \leq 200$; for larger sample sizes, AFK is generally preferred for $n \in (300, 1000]$, whereas BS is faster for $k = 10$ with $n \geq 2000$. General guidelines are given in Section 4. If one computes and uses an L_2 solution, then AFK* is highly recommended for all values of k and n where the total CPU time includes computing the L_2 solution and obtaining a starting basis.

REFERENCES

[1] Armstrong, R.D., Frome, E.L. and Kung, D.S. (1979). A Revised Simplex Algorithm for the Absolute Deviation Curve Fitting Problem. *Comm. Statist.*, **B8**, 175 – 190.

[2] Barrodale, I. and Roberts, F.D.K. (1973). An Improved Algorithm for Discrete L_1 Linear Approximation. *S.I.A.M. J. Numer. Anal.*, **10**, 839 – 848.

[3] Barrodale, I. and Roberts, F.D.K. (1974). Solutions of an Overdetermined System of Equations in the L_1–norm. *Commun. of the ACM*, **17**, 319 – 320.

[4] Bartels, R.H., Conn, A.R. and Sinclair, J.W. (1978). Minimization Techniques for Piecewise Differentiable Functions: The L_1 Solution of an Overdetermined Linear System. *S.I.A.M. J. Numer. Anal.*, **15**, 224 – 241.

[5] Bloomfield, P. and Steiger, W. (1983). *Least Absolute Deviations Theory – Applications and Algorithms.* Birkhauser, Boston, Mass.

[6] Charnes, A., Cooper, W.W. and Ferguson, R.O. (1955). Optimal Estimation of Executive Compensation by Linear Programming. *Management Science*, **4**, 169 – 197.

[7] Dielman, T. and Pfaffenberger, R. (1984). Computational Algorithms for Calculating Least Absolute Value and Chebyshev Estimates for Multiple Regression. *American Journal of Mathematical and Management Sciences*, **4**, 169 – 197.

[8] Edgeworth, F.Y. (1888). On a New Method of Reducing Observations Relating to Several Quantities. *Phil. Mag.*, **25**, 184 – 191.

[9] McCormick, G.F. and Sposito, V.A. (1976), Using the L_2–estimator in L_1–estimation. *S.I.A.M. J. Numer. Anal.*, **13**, 337 – 343.

[10] Narula, S.C. (1987 . The Minimum Sum of Absolute Error Regression. *J. Quality Technology*, **19**, 37 – 45.

[11] Sklar, M. and Armstrong, R.D. (1982). Least Absolute Value and Chebyshev Estimation Utilizing Least Squares Results. *J. Math. Prog.*, **24**, 346 – 352.

[12] Sposito, V.A., Smith, W.C. and McCormick, G.F. (1978). *Minimizing the Sum of Absolute Deviations.* Vandenhoeck and Rupercht, Gottingen, Germany.

[13] Wagner, H.M. (1959). Linear Programming Techniques for Regression Analysis. *J. Amer. Statist. Assoc.*, **54**, 206 – 212.

[14] Wesolowsky, G.O. (1981). A New Descent Algorithm for the Least Absolute Value Regression Problem. *Comm. Statist.*, **B10**, 479 – 491.

Recent Adv. in Stat. and Prob., pp. 67-70
J. Pérez Vilaplana and M. L. Puri (Eds)

The Kagan Classification of Multivariate Distributions and the Central Limit Theorem

J. WESOLOWSKI
Politechnika Warszawska, Instytut Matematyki, Plac Jednosci Robotniczej 1, 00661 Warszawa, Poland 00661

Abstract – A version of the Lapounov type central limit theorem for triangular array of dependent random variables – belonging row-wisely to the Kagan $\mathcal{D}_{k_n,2}$ classes – is proved.

1. INTRODUCTION

In Kagan (1988) the following classification of multivariate probability distributions has been introduced: A n-dimensional random vector X (or its distribution) belongs to the class $\mathcal{D}_{n.k}$ $(k \leq n)$ if its characteristic function has the form

$$\phi(\mathbf{t}) = \prod_{1\leq i_1<\ldots<i_k\leq n} R_{i_1\ldots i_k}(t_{i_1}, \ldots, t_{i_k}) \tag{1}$$

for any $\mathbf{t} = (t_1, \ldots, t_n) \in \mathbf{R}^n$, where $R_{i_1,\ldots i_k}$ is a continuous complex function with $R_{i_1,\ldots i_k}(0, \ldots, 0) = 1$, $1 \leq i_1 < \ldots < i_k \leq n$. If (1) holds in a neighborhood of the origin then X belongs to the class $\mathcal{D}_{n,k}(loc)$.
Many interesting properties of these classes, together with some applications and examples are given in the above mentioned paper. Its main result lies in some extensions of the celebrated Darmois–Skitovitch Theorem (on characterizing the normal law by independence of linear statistics in independent random variables) and the Cramèr Theorem (on decomposition of the normal law). The core of these extensions was a weakening of the independence assumption by applying $\mathcal{D}_{n.k}$ classes. And this is the essential feature of these classes – they are gradual generalizations of independence of components of a random vector. Observe that if $X \in \mathcal{D}_{n,1}$ then its components are independent and if $X \in \mathcal{D}_{n,1}(loc)$ then its components are quasi-independent.
The investigations have been continued in Kagan (1989) where a similar concept of (n, k)-equivalence has been studied (being an extension of the identity of distributions). In Wesolowski (1991a) a formula expressing the characteristic function of $\mathcal{D}_{n,k}$ distribution in terms of characteristic functions of its k-dimensional marginals was obtained. Its immediate consequences are some results on determining the distribution of $\mathcal{D}_{n,k}$ random vectors by k-dimensional marginal distributions. This formula was also applied in Wesolowski (1991b) where the relations between Gaussian conditional structure of the second order and $\mathcal{D}_{n,k}$ classes have been studied.
The factorization of the characteristic function given in Wesolowski (1991a) is also the core of the proof of the Lapounov type central limit theorem for $\mathcal{D}_{k_n,2}$ classes. This is the

subject of the present paper. We consider random vectors $X_n = (X_{n,1}, \ldots, X_{n,k_n})$ with k_n tending to infinity together with n.

2. CENTRAL LIMIT THEOREM OF THE LAPOUNOV TYPE

Recall that the classical Lapounov central limit theorem for triangular array has the following form:

Theorem 1 *Let $X_n = (X_{n,1}, \ldots, X_{n,k_n})$ be a random vector with independent zero-mean components, $n = 1, 2, \ldots$ If*

$$\lim_{n\to\infty} \sum_{i=1}^{k_n} E X_{n,i}^2 = \sigma^2 > 0,$$

$$\lim_{n\to\infty} \sum_{i=1}^{k_n} E \mid X_{n,i} \mid^3 = 0$$

then $S_n = X_{n,1} + \ldots + X_{n,k_n}$ converges in distribution to the normal law with the mean zero and the variance σ^2 ($S_n \xrightarrow{d} \mathcal{N}(0, \sigma^2)$) as $n \to \infty$.

Observe that in the above theorem X_n belongs to the class $\mathcal{D}_{k_n,1}$. Now we are interested in the next Kagan class i.e. we try to obtain the Lapounov type theorem replacing the assumption of independence by considering X_n from the class $\mathcal{D}_{k_n,2}$.

Theorem 2 *Let $X_n = (X_{n,1}, \ldots, X_{n,k_n}) \in \mathcal{D}_{k_n,2}$ be a zero-mean random vector, $n = 1, 2, \ldots$ If*

$$\lim_{n\to\infty} \sum_{i,j=1}^{k_n} E X_{n,i} X_{n,j} = \sigma^2 > 0, \tag{2}$$

$$\lim_{n\to\infty} k_n \sum_{i=1}^{k_n} E X_{n,i}^2 = \tau^2, \tag{3}$$

$$\lim_{n\to\infty} k_n \sum_{i=1}^{k_n} E \mid X_{n,i} \mid^3 = 0, \tag{4}$$

then $S_n \xrightarrow{d} \mathcal{N}(0, \sigma^2)$ as $n \to \infty$.

The proof of the above theorem is given in Section 3. Note that, in spite of the inclusions $\mathcal{D}_{n,1} \subset \mathcal{D}_{n,2}$, due to the relations (2) and (3), Theorem 2 can not be applied for X_n's with independent components. The formulas (2) and (3) jointly are the analogues of the first condition in the Lapounov Theorem and (4) works as the Lapounov condition.
Consider now zero-mean random vector $X_n \in \mathcal{D}_{k_n,k}$, $n = 1, 2, \ldots$, where k is an arbitrary natural number. Then by Theorem 1 in Wesolowski (1991a) in some neighborhood of the origin we have

$$E \exp(itS_n) = \prod_{r=1}^{k} \left(\prod_{1 \le i_1 < \ldots < i_r \le k_n} \phi_{n;i_1,\ldots,i_r}(t, \ldots, t) \right)^{a(k_n,k,r)}, \tag{5}$$

where $a(m, k, r) = \sum_{i=0}^{k-r} \binom{m-r}{i} (-1)^i$ and $\phi_{n;i_1,\ldots,i_r}$ is a characteristic function of the marginal $(X_{n,i_1}, \ldots, X_{n,i_r})$, $1 \le i_1 < \ldots < i_r \le k_n$.
We hope that by applying to (5) a reasoning similar to the proof given in Section 3 the following general theorem may be proved, which is formulated as a conjecture since we have not been able to prove it yet.

Conjecture 1 *Let $X_n = (X_{n,1}, \dots, X_{n,k_n}) \in \mathcal{D}_{k_n,k}$ be a zero-mean random vector, $n = 1, 2, \dots$, where k is a fixed natural number and $k_n \geq k$ for each n. If (2) and*

$$\lim_{n\to\infty} \binom{k_n}{k-1} \sum_{i=1}^{k_n} E X_{n,i}^2 = \tau^2 < \infty,$$

$$\lim_{n\to\infty} \binom{k_n}{k-1} \sum_{i=1}^{k_n} E \mid X_{n,i} \mid^3 = 0,$$

hold then $S_n \xrightarrow{d} \mathcal{N}(0, \sigma^2)$ as $n \to \infty$,

Note that Theorems 1 and 2 are immediate consequences of the above conjecture if only it is true. It suffices to take $k = 1, 2$, respectively.

3. PROOF

Consider independent random variables $Y_{n,j}$, $j = 1, 2, \dots, k_n^2$ such that

$$Y_{n,(j-1)k_n+i} \stackrel{d}{=} X_{n,j}, \qquad j, i = 1, \dots, k_n,$$

where $\stackrel{d}{=}$ denotes equation of distributions, $n = 1, 2, \dots$ Observe that

$$\sum_{i=1}^{k_n^2} E Y_{n,i}^2 = k_n \sum_{i=1}^{k_n} E X_{n,i}^2$$

and

$$\sum_{i=1}^{k_n^2} E \mid Y_{n,i} \mid^3 = k_n \sum_{i=1}^{k_n} E \mid X_{n,i} \mid^3 .$$

Consequently by (3) and (4) the assumptions of Theorem 1 are fulfilled. Hence

$$\left(\prod_{1 \leq i \leq k_n} \phi_{n;i}(t) \right)^{k_n} \to \exp\left(-\frac{t^2 \tau^2}{2} \right) \tag{6}$$

for any real t as $n \to \infty$.
Define now independent random variables $Z_{n;i,j}$, $1 \leq i < j \leq k_n$ such that

$$Z_{n;i,j} \stackrel{d}{=} X_{n,i} + X_{n,j}$$

for any $n = 1, 2, \dots$ Since

$$\sum_{1 \leq i < j \leq k_n} E (X_{n,i} + X_{n,j})^2 = (k_n - 2) \sum_{1 \leq i \leq k_n} E X_{n,i}^2 + \sum_{i,j=1}^{k_n} E X_{n,i} X_{n,j}$$

then by (2) and (3)

$$\lim_{n\to\infty} \sum_{1 \leq i < j \leq k_n} E Z_{n;i,j}^2 = \tau^2 + \sigma^2 \tag{7}$$

On the other hand

$$\begin{aligned} \sum_{1 \leq i < j \leq k_n} E \mid X_{n,i} + X_{n,j} \mid^3 &\leq 4 \sum_{1 \leq i < j \leq k_n} (E \mid X_{n,i} \mid^3 + E \mid X_{n,j} \mid^3) \\ &\leq 4(k_n - 1) \sum_{1 \leq i \leq k_n} E \mid X_{n,i} \mid^3 . \end{aligned}$$

Consequently by (4)

$$\lim_{n\to\infty} \sum_{1\le i<j\le k_n} E \mid Z_{n;i,j} \mid^3 = 0. \tag{8}$$

The formulas (7) and (8) together with Theorem 1 yield

$$\lim_{n\to\infty} \prod_{1\le i<j\le k_n} \phi_{n;i,j}(t) = \exp\left(-\frac{t^2(\tau^2+\sigma^2)}{2}\right) \tag{9}$$

for any real t.

The equations (6) and (9) imply that there is a neighborhood V of the origin such that for any $n > N$, where N is sufficiently large $\phi_{n;i}(t)$, $i = 1, 2, \ldots, k_n$ and $\phi_{n;i,j}(t,t)$, $1 \le i < j \le k_n$ are non-zero for every $t \in V$. Hence by Theorem 1 from Wesolowski (1991a) - put $k = 2$ in the formula (5) above - we obtain for any $t \in V$ and any $n > N$

$$E \exp(itS_n) = \frac{\prod_{1\le i<j\le k_n} \phi_{n;i,j}(t,t)}{\left(\prod_{1\le i\le k_n} \phi_{n;i}(t)\right)^{k_n-2}}$$

Once again applying (6) and (9) we find that for $t \in V$

$$E \exp(itS_n) \to \exp\left(-\frac{t^2\sigma^2}{2}\right). \tag{10}$$

To complete the proof observe that by the Tschebyshev inequality

$$P(\mid S_n \mid > \lambda) \le \lambda^{-2} E\, S_n^2$$

and (2) the sequence of distributions of $S_n, n = 1, 2, \ldots$ is tight. Consequently from the Prokhorov Theorem it follows that each subsequence contains a convergent subsequence. Thus we can assume that S_n converges in distribution. Hence the characteristic functions of $S_n, n = 1, 2, \ldots$ are convergent to some characteristic function ψ which, by (10), in a neighborhood of the origin is equal to the characteristic function of $\mathcal{N}(0, \sigma^2)$. It is well known that in such a case for any real t

$$\psi(t) = \exp\left(-\frac{t^2\sigma^2}{2}\right). \qquad \blacksquare$$

REFERENCES

[1] Kagan, A.M. (1988) New Classes of Dependent Random Variables and a Generalization of the Darmois–Skitovitch Theorem to the Case of Several Forms. *Teor. Veroyat. i Primen.*, **33**, 305–315 (in Russian).

[2] Kagan, A.M. (1989) Generalized Condition of the Identity of Distributions of Random Vectors in Connection with the Analytic Theory of Linear Forms in Independent Random Variables. *Teor. Veroyat. i Primen.*, **34**, 370–375 (in Russian).

[3] Wesolowski, J. (1991a) On Determination of Multivariate Distribution by its Marginals for the Kagan Class. *J. Multivar. Anal.*, **36**, 314–319.

[4] Wesolowski, J. (1991b) Gaussian Conditional Structure of the Second Order and the Kagan Classification of Multivariate Distributions. *J. Multivar. Anal.*, **39**, 79–86.

Nonparametric Estimation

Recent Adv. in Stat. and Prob., pp. 73-90
J. Pérez Vilaplana and M. L. Puri (Eds)

Relationships Between Product Moments of Order Statistics from Non-Identically Distributed Variables

N. BALAKRISHNAN
Department of Mathematics and Statistics, McMaster University, Hamilton, Ontario, Canada L8S 4K1

Abstract – In this paper, we derive several recurrence relations and identities satisfied by the product moments of order statistics in the general case when the order statistics arise from n independent and non-identically distributed variables. These generalize many well-known results on order statistics in the i.i.d. case given by David (1981) and Arnold and Balakrishnan (1989). Applications of these results to robustness studies is indicated.

1. INTRODUCTION

Several recurrence relations and identities satisfied by the single and product moments of order statistics arising from n independent and identically distributed variables are available in literature. All these results have been analyzed in great detail by Arnold and Balakrishnan (1989); see also David (1981). These results have been recently extended by Balakrishnan (1987, 1988a,b) to the case when the order statistics arise from a sample containing a single outlier; these developments have also been analyzed by Arnold and Balakrishnan (1989).

For the general case when the order statistics arise from n independent and non-identically distributed variables, Balakrishnan (1988c, 1989), Balasubramanian and Beg (1990), Bapat and Beg (1989a,b), Balasubramanian, Bapat and Beg (1991) and David (1993) derived many relations and identities satisfied by the single moments of order statistics. These results and some extensions are presented recently by Balakrishnan (1992a).

In this paper, we consider the most general case when the order statistics arise from n independent and non-identically distributed variables and first present in Section 2 some basic formulae and notations to be used. In Section 3, we establish several recurrence relations and identities satisfied by the product moments of order statistics. Reference may be made to Beg (1991) for some results already in this direction. By a systematic use of these results, we determine the maximum number of single and product moments to be evaluated for the calculation of means, variances and covariances of order statistics arising from n independent and non-identically distributed variables, assuming these quantities for order statistics arising from $n-1$ (and less) independent and non-identically distributed variables to be known. These are exactly the same bounds as derived by Joshi and Balakrishnan (1982) in the i.i.d. case. In Section 4, we derive some recurrence relations and identities satisfied by covariances of order statistics. In Section 5, we consider the special case when all the n variables are symmetrically distributed and establish some relations satisfied by the single and the product moments of order statistics in this case. By making use of these results, we determine once again the maximum number of single and product moments to be evaluated for the calculation of means, variances and covariances of order statistics arising from n independent and non-identically distributed variables, and these bounds are improved ones as compared to those determined in Section 3. Once again, these bounds are exactly the same as those obtained by Joshi (1971) (also see David, 1981; Arnold and Balakrishnan, 1989) in the i.i.d. case. Some applications of these results to robustness studies are indicated in Section 6. For this purpose, results for the special case of order statistics arising from a p-outlier model can easily be deduced from the results developed in this paper. Finally, in Section 7 we draw some conclusions and suggest some possible generalizations.

2. BASIC FORMULAE AND NOTATIONS

Let $X_1, X_2, \ldots, X_n$ be n independent absolutely continuous random variables with X_i having a cumulative distribution function $F_i(x)$ and a probability density function $f_i(x)$. Let $X_{1:n} \leq X_{2:n} \leq \ldots \leq X_{n:n}$ be the order statistics obtained by arranging the n variables in increasing order of magnitude.

Then, it can be easily shown that the density function of $X_{r:n}$ $(r = 1, 2, \ldots, n)$ is given by

$$f_{r:n}(x) = \frac{1}{(r-1)!(n-r)!} \sum_P F_{i_1}(x) \ldots F_{i_{r-1}}(x) f_{i_r}(x) \left\{1 - F_{i_{r+1}}(x)\right\} \ldots \{1 - F_{i_n}(x)\},$$

$$-\infty < x < \infty, \qquad (2.1)$$

where $\sum_P$ denotes the sum over all $n!$ permutations $(i_1, i_2, \ldots, i_n)$ of $(1, 2, \ldots, n)$. Similarly, the joint density function of $X_{r:n}$ and $X_{s:n}$ $(1 \leq r < s \leq n)$ is given by

$$f_{r,s:n}(x,y) = \frac{1}{(r-1)!(s-r-1)!(n-s)!} \sum_P F_{i_1}(x) \ldots F_{i_{r-1}}(x) f_{i_r}(x) \left\{F_{i_{r+1}}(y) - F_{i_{r+1}}(x)\right\}$$

$$\ldots \left\{F_{i_{s-1}}(y) - F_{i_{s-1}}(x)\right\} f_{i_s}(y) \left\{1 - F_{i_{s+1}}(y)\right\} \ldots \{1 - F_{i_n}(y)\},$$

$$-\infty < x < \infty, \qquad (2.2)$$

As pointed out by Vaughan and Venables (1972), the density function of $X_{r:n}$ and the joint density function of $X_{r:n}$ and $X_{s:n}$ in Eqs. (2.1) and (2.2), respectively, can be expressed in terms of permanents of some matrices. Specifically, they have shown that

$$f_{r:n}(x) = \frac{1}{(r-1)!(n-r)!} \begin{vmatrix} F_1(x) & \ldots & F_n(x) \\ f_1(x) & \ldots & f_n(x) \\ 1 - F_1(x) & \ldots & 1 - F_n(x) \end{vmatrix}^{+} \begin{matrix} r-1 \text{ rows} \\ \\ n-r \text{ rows} \end{matrix}$$

$$1 \leq r \leq n, \ -\infty < x < \infty, \qquad (2.3)$$

and

$$f_{r,s:n}(x,y) = \frac{1}{(r-1)!(s-r-1)!(n-s)!} \times$$

$$\times \begin{vmatrix} F_1(x) & \cdots & F_n(x) \\ f_1(x) & \cdots & f_n(x) \\ F_1(y)-F_1(x) & \cdots & F_n(y)-F_n(x) \\ f_1(y) & \cdots & f_n(y) \\ 1-F_1(y) & \cdots & 1-F_n(y) \end{vmatrix}^{+} \begin{array}{l} r-1 \text{ rows} \\ \\ s-r-1 \text{ rows} \\ \\ n-s \text{ rows} \end{array}$$

$$1 \le r < s \le n, \ -\infty < x < y < \infty, \tag{2.4}$$

where $\overset{++}{|A|}$ denotes the permanent of matrix A.

Let us denote the single moments $E(X_{r:n}^k)$ by $\mu_{r:n}^{(k)}$ for $1 \le r \le n$ and $k \ge 1$, and the product moments $E(X_{r:n}X_{s:n})$ by $\mu_{r,s:n}$ for $1 \le r < s \le n$. For convenience, let us denote $\mu_{r:n}^{(1)}$ by $\mu_{r:n}$ and $\mu_{r:n}^{(2)}$ by $\mu_{r,r:n}$. We shall also denote $\mathrm{Var}(X_{r:n}) = \mu_{r:n}^{(2)} - \mu_{r:n}^2$ by $\sigma_{r,r:n}$ for $1 \le r \le n$, and $\mathrm{Cov}(X_{r:n}, X_{s:n}) = \mu_{r,s:n} - \mu_{r:n}\mu_{s:n}$ by $\sigma_{r,s:n}$ for $1 \le r < s \le n$.

Further, let us denote the kth raw moment of the variable X_i by $\mu_{1:1}^{(k)}(i)$ and the variance of the variable X_i by $\sigma_{1,1:1}(i)$. We shall also denote the rth order statistic arising from $n-m$ variables (obtained from deleting the m variables $X_{i_1}, X_{i_2}, \ldots, X_{i_m}$ from the original n variables) by $X_{r:n-m}^{[i_1,\ldots,i_m]}$, its density function by $f_{r:n-m}^{[i_1,\ldots,i_m]}(x)$, the joint density function of $X_{r:n-m}^{[i_1,\ldots,i_m]}$ and $X_{s:n-m}^{[i_1,\ldots,i_m]}$ by $f_{r,s:n-m}^{[i_1,\ldots,i_m]}(x,y)$, $E(X_{r:n-m}^{k[i_1,\ldots,i_m]})$ by $\mu_{r:n-m}^{(k)[i_1,\ldots,i_m]}$, $E(X_{r:n-m}^{[i_1,\ldots,i_m]} X_{s:n-m}^{[i_1,\ldots,i_m]})$ by $\mu_{r,s:n-m}^{[i_1,\ldots,i_m]}$, $\mathrm{Var}(X_{r:n-m}^{[i_1,\ldots,i_m]})$ by $\sigma_{r,r:n-m}^{[i_1,\ldots,i_m]}$, and $\mathrm{Cov}(X_{r:n-m}^{[i_1,\ldots,i_m]}, X_{s:n-m}^{[i_1,\ldots,i_m]})$ by $\sigma_{r,s:n-m}^{[i_1,\ldots,i_m]}$. Finally, we define the following quantities which are necessary for the development of results presented in the following sections:

$$\bar{\mu}_{r:n-1}^{(k)} = \frac{1}{n}\sum_{i=1}^{n} \mu_{r:n-1}^{(k)[i]}, \tag{2.5}$$

$$S_{r:n-m}^{(k)} = \sum_{1 \le i_1 < \ldots < i_m \le n} \mu_{r:n-m}^{(k)[i_1,\ldots,i_m]}, \tag{2.6}$$

$$S_{r,s:n-m} = \sum_{1 \le i_1 < \ldots < i_m \le n} \mu_{r,s:n-m}^{[i_1,\ldots,i_m]}, \tag{2.7}$$

and

$$T_{r:n-m,\ s:m} = \sum_{1 \le i_1 < \ldots < i_m \le n} \mu_{r:n-m}^{[i_1,\ldots,i_m]} \mu_{s:m}^{[i_{m+1},\ldots,i_n]}. \tag{2.8}$$

3. RELATIONSHIPS FOR PRODUCT MOMENTS

In this section we derive several recurrence relations and identities satisfied by the product moments $\mu_{r,s:n}$. Then, by making use of these results we determine the maximum number of single and product moments to be evaluated for the calculation of means, variances and covariances of order statistics arising from n independent and non-identically distributed variables, assuming these quantities to be known for order statistics arising from $n-1$ (and less) variables. These results generalize several well-known results on product moments of order statistics for the i.i.d. case presented in detail by Govindarajulu (1963), Joshi

and Balakrishnan (1982), David (1981) and Arnold and Balakrishnan (1989). Refer also to the work of Beg (1991).
For convenience, let us denote

$$\begin{aligned}\mathbf{R}_1 &= \{(x,y): -\infty < x \le y < \infty\},\\ \mathbf{R}_2 &= \{(x,y): -\infty < y < x < \infty\},\end{aligned}$$

and

$$\mathbf{R}^2 = \mathbf{R}_1 \cup \mathbf{R}_2 = \{(x,y): -\infty < x < \infty,\ -\infty < y < \infty\}.$$

Result 1: For $n \ge 2$,

$$\sum_{r=1}^{n}\sum_{s=1}^{n}\mu_{r,s:n} = \sum_{i=1}^{n}\sigma_{1,1:1}(i) + \left(\sum_{i=1}^{n}\mu_{1:1}(i)\right)^2 = \sum_{i=1}^{n}\sigma_{1,1:1}(i) + S_{1:1}^2. \tag{3.1}$$

PROOF: By starting with the identity

$$\sum_{r=1}^{n}\sum_{s=1}^{n}X_{r:n}X_{s:n} = \sum_{r=1}^{n}\sum_{s=1}^{n}X_rX_s = \sum_{i=1}^{n}X_i^2 + \sum_{1\le i\ne j\le n}X_iX_j$$

and taking expectation on both sides, we obtain

$$\sum_{r=1}^{n}\sum_{s=1}^{n}\mu_{r,s:n} = \sum_{i=1}^{n}\mu_{1:1}^{(2)}(i) + \sum_{1\le i\ne j\le n}\mu_{1:1}(i)\mu_{1:1}(j) =$$

$$= \sum_{i=1}^{n}\mu_{1:1}^{(2)}(i) + \sum_{i=1}^{n}\mu_{1:1}(i)\left\{\sum_{j=1}^{n}\mu_{1:1}(j) - \mu_{1:1}(i)\right\} = \sum_{i=1}^{n}\left\{\mu_{1:1}^{(2)}(i) - \mu_{1:1}^2(i)\right\} + \left(\sum_{i=1}^{n}\mu_{1:1}(i)\right)^2$$

which gives the identity in Eq. (3.1).
Result 2: For $n \ge 2$,

$$\sum_{r=1}^{n-1}\sum_{s=r+1}^{n}\mu_{r,s:n} = \frac{1}{2}\left(S_{1:1}^2 - \sum_{i=1}^{n}\mu_{1:1}^2(i)\right). \tag{3.2}$$

PROOF: Since

$$\sum_{r=1}^{n}\sum_{s=1}^{n}\mu_{r,s:n} = \sum_{r=1}^{n}\mu_{r:n}^{(2)} + 2\sum_{r=1}^{n-1}\sum_{s=r+1}^{n}\mu_{r,s:n}$$

and that

$$\sum_{r=1}^{n}\mu_{r:n}^{(2)} = S_{1:1}^{(2)},$$

we immediately obtain from Result 1 that

$$\sum_{r=1}^{n-1}\sum_{s=r+1}^{n}\mu_{r,s:n} = \frac{1}{2}\left\{\sum_{r=1}^{n}\sum_{s=1}^{n}\mu_{r,s:n} - S_{1:1}^{(2)}\right\} = \frac{1}{2}\left\{\left(\sum_{i=1}^{n}\mu_{1:1}(i)\right)^2 - \sum_{i=1}^{n}\mu_{1:1}^2(i)\right\}$$

which gives the identity in Eq. (3.2).
Result 3: For $2 \le r < s \le n$,

$$(r-1)\mu_{r,s:n} + (s-r)\mu_{r-1,s:n} + (n-s+1)\mu_{r-1,s-1:n} = \sum_{i=1}^{n}\mu_{r-1,s-1:n-1}^{[i]} = S_{r-1,s-1:n-1}. \tag{3.3}$$

PROOF: By considering the expression of $(r-1)f_{r,s:n}(x,y)$ from Eq. (2.4) and expanding the permanent by its first row, we get

$$(r-1)f_{r,s:n}(x,y) = \sum_{i=1}^{n} F_i(x) f^{[i]}_{r-1,s-1:n-1}(x,y). \tag{3.4}$$

Next, by considering the expression of $(s-r)f_{r-1,s:n}(x,y)$ from Eq. (2.4) and expanding the permanent by its rth row, we get

$$(s-r)f_{r-1,s:n}(x,y) = \sum_{i=1}^{n} \left(F_i(y) - F_i(x)\right) f^{[i]}_{r-1,s-1:n-1}(x,y). \tag{3.5}$$

Finally, by considering the expression of $(n-s+1)f_{r-1,s-1:n}(x,y)$ from Eq. (2.4) and expanding the permanent by its last row, we get

$$(n-s+1)f_{r-1,s-1:n}(x,y) = \sum_{i=1}^{n} \left(1 - F_i(y)\right) f^{[i]}_{r-1,s-1:n-1}(x,y). \tag{3.6}$$

Upon adding Eqs. (3.4), (3.5) and (3.6), we get

$$(r-1)f_{r,s:n}(x,y) + (s-r)f_{r-1,s:n}(x,y) + (n-s+1)f_{r-1,s-1:n}(x,y) = \\ = \sum_{i=1}^{n} f^{[i]}_{r-1,s-1:n-1}(x,y). \tag{3.7}$$

The recurrence relation in Eq. (3.3) follows immediately upon multiplying both sides of (3.7) by xy and then integrating over the region $\mathbf{R}_1$.

Remark 1: It is easy to note that Result 3 will enable one to calculate all the product moments $\mu_{r,s:n}$ $(1 \le r < s \le n)$ with the knowledge of $n-1$ suitably chosen moments like, for example, the immediate upper-diagonal product moments $\mu_{r,r+1:n}$ $(1 \le r \le n-1)$. This bound can be improved as given in Theorem 2. For the i.i.d. case, Result 3 was proved by Govindarajulu (1963), in the special case of normal by Teichroew (1956), and in its general form by Balakrishnan (1988c) and Bapat and Beg (1989a).

Result 4: For $1 \le r < s \le n$,

$$\mu_{r,s:n} = \sum_{i=r}^{s-1} \sum_{j=n-s+i+1}^{n} (-1)^{j+n-r-s+1} \binom{i-1}{r-1} \binom{j-i-1}{n-s} S_{i,i+1:j}, \tag{3.8}$$

$$\mu_{r,s:n} = \sum_{i=s-r}^{s-1} \sum_{j=n-s+i+1}^{n} (-1)^{n-j-r+1} \binom{i-1}{s-r-1} \binom{j-i-1}{n-s} S_{1,i+1:j}, \tag{3.9}$$

and

$$\mu_{r,s:n} = \sum_{i=s-r}^{n-r} \sum_{j=r+i}^{n} (-1)^{s+j} \binom{i-1}{s-r-1} \binom{j-i-1}{r-1} S_{j-i,j:j}. \tag{3.10}$$

PROOF: The relations in (3.8) — (3.10) are easily proved by repeated application of the recurrence relation in Result 3, and are due to Balakrishnan, Bendre and Malik (1992). It should be mentioned here that the relations in (3.9) and (3.10) are dual in nature, and hence one can be proved from the other simply by using the duality principle established recently by Balasubramanian and Balakrishnan (1993).

Theorem 1 *If either Result 3 or 4 is used in the computation of product moments of order statistics arising from n independent and non-identically distributed variables, then the identities given in Results 1 and 2 will be automatically satisfied and hence should not be applied to check the computation of product moments of order statistics.*

PROOF: We shall prove the theorem by starting with Result 3 and the proof for Result 4 is quite similar. From Eq. (3.3), by setting $r = 2,\ s = 3, 4, \ldots, n,\ r = 3,\ s = 4, 5, \ldots, n, \ldots,$ and $r = n-1,\ s = n$, and adding the resulting $(n-2)(n-1)/2$ equations, we get

$$\sum_{r=1}^{n-1}\sum_{s=r+1}^{n} \mu_{r,s:n} = \frac{1}{n-2}\sum_{i_1=1}^{n}\sum_{r=1}^{n-2}\sum_{s=r+1}^{n-1} \mu_{r,s:n-1}^{[i_1]}. \tag{3.11}$$

Repeating the same process for the right-hand side of Eq. (3.11) by using the relation in (3.3), we get

$$\sum_{r=1}^{n-1}\sum_{s=r+1}^{n} \mu_{r,s:n} = \frac{1}{(n-2)(n-3)}\sum_{1\le i_1\ne i_2\le n}\sum_{r=1}^{n-3}\sum_{s=r+1}^{n-2} \mu_{r,s:n-2}^{[i_1,i_2]}. \tag{3.12}$$

By repeating this process, we finally obtain

$$\sum_{r=1}^{n-1}\sum_{s=r+1}^{n} \mu_{r,s:n} = \frac{1}{(n-2)!}\sum_{1\le i_1\ne\ldots\ne i_{n-2}\le n} \mu_{1,2:2}^{[i_1,\ldots,i_{n-2}]} = \sum_{1\le i_1<\ldots<i_{n-2}\le n} \mu_{1,2:2}^{[i_1,\ldots,i_{n-2}]}. \tag{3.13}$$

Eq. (3.13) when rewritten gives the identity presented in Result 2 which proves the theorem. For the i.i.d. case, Theorem 1 was proved by Balakrishnan and Malik (1986).
Result 5: For $1 \le r < s \le m \le n-1$,

$$\sum_{j=0}^{n-m}\sum_{k=j}^{n-m}\binom{r-1+j}{j}\binom{s-r-1+k-j}{k-j}\binom{n-s-k}{n-m-k}\mu_{r+j,s+k:n} = S_{r,s:m}. \tag{3.14}$$

PROOF: Let us consider the right-hand side of Eq. (3.14) and write it equivalently as

$$S_{r,s:m} = \frac{1}{(n-m)!}\sum_{1\le i_1\ne\ldots\ne i_{n-m}\le n} \mu_{r,s:m}^{[i_1,\ldots,i_{n-m}]}. \tag{3.15}$$

Upon using Result 3 in the above expression, we obtain from (3.15) that

$$\begin{aligned} S_{r,s:m} = {} & \frac{r}{(n-m)!}\sum_{1\le i_1\ne\ldots\ne i_{n-m-1}\le n} \mu_{r+1,s+1:m+1}^{[i_1,\ldots,i_{n-m-1}]} + \frac{(s-r)}{(n-m)!}\sum_{1\le i_1\ne\ldots\ne i_{n-m-1}\le n} \mu_{r,s+1:m+1}^{[i_1,\ldots,i_{n-m-1}]} \\ & + \frac{(m-s+1)}{(n-m)!}\sum_{1\le i_1\ne\ldots\ne i_{n-m-1}\le n} \mu_{r,s:m+1}^{[i_1,\ldots,i_{n-m-1}]}. \end{aligned} \tag{3.16}$$

Upon using Result 3 once again in the above expression, we obtain from (3.16) that

$$\begin{aligned} S_{r,s:m} = {} & \frac{2!}{(n-m)!}\sum_{i=0}^{2}\sum_{j=0}^{2-i}\binom{r-1+i}{i}\binom{s-r-1+j}{j}\binom{m-s+2-i-j}{2-i-j} \cdot \\ & \cdot \sum_{1\le i_1\ne\ldots\ne i_{n-m-2}\le n} \mu_{r+i,s+i+j:m+2}^{[i_1,\ldots,i_{n-m-2}]}. \end{aligned}$$

Proceeding similarly, we derive the relation in (3.14) at the $(n-m)^{th}$ step. For the i.i.d. case, Result 5 generalizes a result due to Downton (1966); also see David (1981) and Arnold and Balakrishnan (1989). Note that Result 3 may be deduced from Result 5 simply by setting $m = n-1$.

By using a method similar to the one adopted by Balakrishnan (1992a) and Bapat and Beg (1989b), Balakrishnan, Bendre and Malik (1992) established the following result.

Result 6: For $n \geq 2$,

$$\sum_{r=1}^{n-1} \sum_{s=r+1}^{n} \frac{1}{r} \mu_{r,s:n} = \sum_{r=1}^{n-1} \sum_{s=r+1}^{n} B(s-1, n-s+1) S_{1,r+1:s}, \tag{3.17}$$

$$\sum_{r=1}^{n-1} \sum_{s=r+1}^{n} \frac{1}{s-r} \mu_{r,s:n} = \sum_{r=1}^{n-1} \sum_{s=r+1}^{n} B(s-1, n-s+1) S_{r,r+1:s}, \tag{3.18}$$

and

$$\sum_{r=1}^{n-1} \sum_{s=r+1}^{n} \frac{1}{n-s+1} \mu_{r,s:n} = \sum_{r=1}^{n-1} \sum_{s=r+1}^{n} B(s-1, n-s+1) S_{r,s:s}. \tag{3.19}$$

Result 7: For $1 \leq r < s \leq n$,

$$\begin{aligned} \mu_{r,s:n} \quad &+ \sum_{j=0}^{r-1} \sum_{k=0}^{n-s} (-1)^{n-j-k} \binom{j+k}{j} S_{n-s-k+1, n-r-k+1:n-j-k} = \\ &= \sum_{j=1}^{s-r} (-1)^{s-r-j} \binom{s-1-j}{r-1} T_{s-j:s-j,j:n-s+j}. \end{aligned} \tag{3.20}$$

PROOF: For $1 \leq r < s \leq n$, let us consider

$$\begin{aligned} I \quad = \quad & \frac{1}{(r-1)!(s-r-1)!(n-s)!} \sum_{P} \int_{\mathbf{R}^2} \int xy F_{i_1}(x) \ldots F_{i_{r-1}}(x) f_{i_r}(x) \cdot \\ & \cdot \left(F_{i_{r+1}}(y) - F_{i_{r+1}}(x)\right) \ldots \left(F_{i_{s-1}}(y) - F_{i_{s-1}}(x)\right) \cdot \\ & \cdot f_{i_s}(y) \left(1 - F_{i_{s+1}}(y)\right) \ldots (1 - F_{i_n}(y)) \; dy \, dx. \end{aligned} \tag{3.21}$$

By using the fact that

$$\begin{aligned} & \left(F_{i_{r+1}}(y) - F_{i_{r+1}}(x)\right) \ldots \left(F_{i_{s-1}}(y) - F_{i_{s-1}}(x)\right) \\ & = \sum_{j=0}^{s-r-1} (-1)^{s-r-1-j} \sum_{C_j^{s-r-1}} F_{\ell_1}(y) \ldots F_{\ell_j}(y) F_{\ell_{j+1}}(x) \ldots F_{\ell_{s-r-1}}(x), \end{aligned}$$

where $\sum_{C_j^{s-r-1}}$ denotes the sum over all $\binom{s-r-1}{j}$ combinations $(\ell_1, \ell_2, ..., \ell_j)$ of $(i_{r+1}, i_{r+2}, \ldots, i_{s-1})$, we obtain from (3.21) that

$$I = \frac{1}{(r-1)!(s-r-1)!(n-s)!} \sum_{P} \sum_{j=0}^{s-r-1} (-1)^{s-r-1-j} \int_{-\infty}^{\infty} x F_{i_1}(x) \ldots F_{i_{r-1}}(x) F_{\ell_{j+1}}(x) \ldots$$

$$\ldots F_{\ell_{s-r-1}}(x) f_{i_r}(x)\,dx \int_{-\infty}^{\infty} y F_{\ell_1}(y) \ldots F_{\ell_j}(y) f_{i_s}(y)(1 - F_{i_{s+1}}(y)) \ldots (1 - F_{i_n}(y))\,dy =$$

$$= \frac{1}{(r-1)!(s-r-1)!(n-s)!} \sum_{j=0}^{s-r-1} (-1)^{s-r-1-j} \binom{s-r-1}{j} \cdot$$

$$\cdot (s-j-2)!j!(n-s)! \sum_{1 \le i_1 < \ldots < i_{n-s+j+1} \le n} \mu_{s-1-j:s-1-j}^{[i_1,\ldots,i_{n-s+j+1}]} \mu_{j+1:n-s+j+1}^{[i_{n-s+j+2},\ldots,i_n]}$$

which, when simplified, yields the right-hand side of (3.20). Alternatively, by noting that $\mathbf{R}^2 = \mathbf{R}_1 \cup \mathbf{R}_2$ we can write from Eq. (3.21) that

$$I = \mu_{r,s:n} + J, \tag{3.22}$$

where

$$\begin{aligned} J \;=\; & \frac{1}{(r-1)!(s-r-1)!(n-s)!} \sum_{P} \int_{\mathbf{R}_2}\int xy F_{i_1}(x) \ldots F_{i_{r-1}}(x) \cdot \\ & \cdot f_{i_r}(x) \left(F_{i_{r+1}}(y) - F_{i_{r+1}}(x)\right) \ldots \left(F_{i_{s-1}}(y) - F_{i_{s-1}}(x)\right) \cdot \\ & \cdot f_{i_s}(y) \left(1 - F_{i_{s+1}}(y)\right) \ldots (1 - F_{i_n}(y))\; dy\,dx. \end{aligned} \tag{3.23}$$

Upon using the facts that

$$\begin{aligned} F_{i_1}(x) \ldots F_{i_{r-1}}(x) \;&=\; \{1 - (1 - F_{i_1}(x))\} \ldots \left\{1 - \left(1 - F_{i_{r-1}}(x)\right)\right\} = \\ &=\; \sum_{j=0}^{r-1} (-1)^{r-1-j} \sum_{C_j^{r-1}} \left(1 - F_{\ell_1}(x)\right) \ldots \left(1 - F_{\ell_j}(x)\right) \end{aligned}$$

and

$$(1 - F_{i_{s+1}}(y)) \ldots (1 - F_{i_n}(y)) = \sum_{k=0}^{n-s} (-1)^{n-s-k} \sum_{C_k^{n-s}} F_{m_1}(y) \ldots F_{m_k}(y),$$

where $\sum_{C_j^{r-1}}$ denotes the sum over all $\binom{r-1}{j}$ combinations $(\ell_1, \ell_2, \ldots, \ell_j)$ of $(i_1, i_2, \ldots, i_{r-1})$ and $\sum_{C_k^{n-s}}$ denotes the sum over all $\binom{n-s}{k}$ combinations $(m_1, m_2, \ldots, m_k)$ of $(i_{s+1}, i_{s+2}, \ldots, i_n)$, we obtain from (3.23) that

$$\begin{aligned} J \;=\; & \frac{1}{(r-1)!(s-r-1)!(n-s)!} \sum_{j=0}^{r-1} \sum_{k=0}^{n-s} (-1)^{n-j-k} \frac{(r-1)!}{(r-1-j)!} \times \\ & \times \frac{(n-s)!}{(n-s-k)!} \frac{(j+k)!}{j!k!} (n-s-k)!(s-r-1)!(r-1-j)! \times \\ & \times \sum_{1 \le i_1 < \ldots < i_{j+k} \le n} \mu_{n-s-k+1,n-r-k+1:n-j-k}^{[i_1,\ldots,i_{j+k}]} = \\ =\; & \sum_{j=0}^{r-1} \sum_{k=0}^{n-s} (-1)^{n-j-k} \binom{j+k}{j} S_{n-s-k+1,n-r-k+1:n-j-k} \end{aligned}$$

which, when substituted in Eq. (3.22), yields the left-hand side of (3.20). Hence, the result is proved.

Result 8: For $r = 1, 2, \ldots, n-1$,

$$\mu_{r,r+1:n} + (-1)^n \mu_{n-r,n-r+1:n} = \sum_{j=0}^{r-1} \sum_{k=1}^{n-r-1} (-1)^{n+1-j-k} \binom{j+k}{j} S_{n-r-k,n-r-k+1:n-j-k} + $$
$$+ \sum_{j=1}^{r-1} (-1)^{n+1-j} S_{n-r,n-r+1:n-j} + T_{r:r,1:n-r}. \qquad (3.24)$$

PROOF: Eq. (3.24) follows directly from Result 7 by setting $s = r + 1$.
Result 9: For $r = 1, 2, \ldots, [n/2]$,

$$(1 + (-1)^n)\, \mu_{r,n-r+1:n} = \sum_{j=0}^{r-1} \sum_{k=1}^{r-1} (-1)^{n+1-j-k} \binom{j+k}{j} S_{r-k,n-r-k+1:n-j-k} +$$
$$+ \sum_{j=1}^{r-1} (-1)^{n+1-j} S_{r,n-r+1:n-j} +$$
$$+ \sum_{j=1}^{n-2r+1} (-1)^{n+1-j} \binom{n-r-j}{r-1} T_{n-r-j+1:n-r-j+1,j:r+j-1}. \qquad (3.25)$$

PROOF: Eq. (3.25) follows directly from Result 7 by setting $s = n - r + 1$.
Result 10: For $m = 1, 2, \ldots,$

$$2\, \mu_{1,2m:2m} = \sum_{j=1}^{2m-1} (-1)^{j-1} T_{2m-j:2m-j,j:j}. \qquad (3.26)$$

PROOF: Eq. (3.26) follows simply by setting $n = 2m$ and $r = 1$ in Result 9.
Result 11: For $m = 1, 2, \ldots,$

$$2\, \mu_{m,m+1:2m} = \sum_{j=0}^{m-1} \sum_{k=1}^{m-1} (-1)^{j+k-1} \binom{j+k}{j} S_{m-k,m-k+1:2m-j-k} +$$
$$+ \sum_{j=1}^{m-1} (-1)^{j-1} S_{m,m+1:2m-j} + T_{m:m,1:m}. \qquad (3.27)$$

PROOF: Eq. (3.27) follows from Result 9 upon setting $n = 2m$ and $r = m$.

Theorem 2 *In order to find the first two single moments and the product moments of all order statistics arising from n independent and non-identically distributed variables, given these moments of order statistics arising from $n-1$ (and less) independent and non-identically distributed variables, one needs to find atmost two single moments and $(n-2)/2$ product moments when n is even, and two single moments and $(n-1)/2$ product moments when n is odd.*

PROOF: In view of the recurrence relation (see Balakrishnan, 1988c, 1992a; Bapat and Beg, 1989a) given by

$$r\, \mu^{(k)}_{r+1:n} + (n-r)\, \mu^{(k)}_{r:n} = \sum_{r=1}^{n} \mu^{(k)[i]}_{r:n-1} = S^{(k)}_{r:n-1}, \qquad (3.28)$$

it is sufficient to find two single moments in order to compute all the first two single moments of order statistics, viz., $\mu_{r:n}$ and $\mu^{(2)}_{r:n}$ for $1 \le r \le n$. Also, as pointed out in

Remark 1, the knowledge of $n-1$ immediate upper-diagonal product moments $\mu_{r,r+1:n}$ $(1 \leq r \leq n-1)$ is sufficient for the calculation of all the product moments $\mu_{r,s:n}$ $(1 \leq r < s \leq n)$. For even values of n, say $n = 2m$, Results 8 and 11 imply that the knowledge of $(n-2)/2 = m-1$ of the immediate upper-diagonal product moments, viz., $\mu_{1,2:2m}, \ldots, \mu_{m-1,m:2m}$, is sufficient for the calculation of all the product moments. Finally, for odd values of n, say $n = 2m+1$, Result 8 implies that the knowledge of $(n-1)/2 = m$ of the immediate upper-diagonal product moments, viz., $\mu_{1,2:2m+1}, \ldots, \mu_{m,m+1:2m+1}$, is sufficient for the calculation of all the product moments. Hence, the theorem is proved.

Remark 2: It is of interest to mention here that the bounds established for the number of single and product moments while determining the means, variances and covariances of order statistics arising from n independent and non-identically distributed variables are exactly the same as the bounds established by Joshi and Balakrishnan (1982) for the i.i.d. case; also see Arnold and Balakrishnan (1989).

4. RELATIONSHIPS FOR COVARIANCES

In this section we establish several recurrence relations and identities satisfied by the covariances $\sigma_{r,s:n}$. These results generalize several well-known results on covariances of order statistics for the i.i.d. case presented in detail by Arnold and Balakrishnan (1989).

Result 12: For $n \geq 2$,

$$\sum_{r=1}^{n} \sum_{s=1}^{n} \sigma_{r,s:n} = \sum_{i=1}^{n} \sigma_{1,1:1}(i). \tag{4.1}$$

PROOF: By writing

$$\sum_{r=1}^{n} \sum_{s=1}^{n} \sigma_{r,s:n} = \sum_{r=1}^{n} \sum_{s=1}^{n} \mu_{r,s:n} - \sum_{r=1}^{n} \mu_{r:n} \sum_{s=1}^{n} \mu_{s:n}$$

and then using Result 1 on the right-hand side, we get the identity in Eq. (4.1).

Result 13: For $2 \leq r < s \leq n$,

$$\begin{aligned}(r-1)\,\sigma_{r,s:n} + (s-r)\,\sigma_{r-1,s:n} + (n-s+1)\,\sigma_{r-1,s-1:n} = \\ = \sum_{i=1}^{n} \sigma_{r-1,s-1:n-1}^{[i]} + \sum_{i=1}^{n} \left(\mu_{r-1:n-1}^{[i]} - \mu_{r-1:n}\right)\left(\mu_{s-1:n-1}^{[i]} - \mu_{s:n}\right).\end{aligned} \tag{4.2}$$

PROOF: Using Result 3, we have for $2 \leq r < s \leq n$

$$(r-1)\,\sigma_{r,s:n} + (s-r)\,\sigma_{r-1,s:n} + (n-s+1)\,\sigma_{r-1,s-1:n} =$$

$$= \sum_{i=1}^{n} \sigma_{r-1,s-1:n-1}^{[i]} + \sum_{i=1}^{n} \mu_{r-1:n-1}^{[i]}\,\mu_{s-1:n-1}^{[i]} - (r-1)\,\mu_{r:n}\,\mu_{s:n} -$$

$$-(s-r)\;\mu_{r-1:n}\,\mu_{s:n} - (n-s+1)\;\mu_{r-1:n}\,\mu_{s-1:n} =$$

$$= \sum_{i=1}^{n} \sigma_{r-1,s-1:n-1}^{[i]} + \sum_{i=1}^{n} \mu_{r-1:n-1}^{[i]}\,\mu_{s-1:n-1}^{[i]} -$$

$$-\mu_{s:n}\Big\{(r-1)\,\mu_{r:n} + (n-r+1)\,\mu_{r-1:n}\Big\} - (n-s+1)\;\mu_{r-1:n}\,(\mu_{s-1:n} - \mu_{s:n}) =$$

$$= \sum_{i=1}^{n} \sigma_{r-1,s-1:n-1}^{[i]} + \sum_{i=1}^{n} \mu_{r-1:n-1}^{[i]} \mu_{s-1:n-1}^{[i]} - \mu_{s:n} \sum_{i=1}^{n} \mu_{r-1:n-1}^{[i]} -$$

$$- \mu_{r-1:n} \sum_{i=1}^{n} \left(\mu_{s-1:n-1}^{[i]} - \mu_{s:n} \right)$$

upon using the relation in (3.28). The relation in Eq. (4.2) is derived by simplifying the above equation. Result 13 was established by Balakrishnan (1989) in the i.i.d. case as well as in the general case presented above.

Result 14: For $1 \leq r \leq n-1$ and $1 \leq \ell \leq n-r$,

$$\sum_{k=r+1}^{n-\ell+1} \binom{n-k}{\ell-1} \mu_{r,k:n} + \sum_{j=1}^{r} \sum_{k=r+1}^{r+\ell} \binom{k-j-1}{k-r-1} \binom{n-k}{n-\ell-r} \mu_{j,k:n} = T_{r:n-\ell,1:\ell}. \qquad \textbf{(4.3)}$$

PROOF: For $1 \leq r \leq n-1$ and $1 \leq \ell \leq n-r$, let us consider

$$I = \frac{1}{(r-1)!(\ell-1)!(n-r-\ell)!} \sum_{P} \int_{\mathbf{R}^2}\int xy\ F_{i_1}(x) \dots F_{i_{r-1}}(x) f_{i_r}(x) \left(1 - F_{i_{r+1}}(x)\right) \dots$$
$$\dots \left(1 - F_{i_{n-\ell}}(x)\right)\ f_{i_{n-\ell+1}}(y) \left(1 - F_{i_{n-\ell+2}}(y)\right) \dots (1 - F_{i_n}(y))\ dx\, dy. \qquad \textbf{(4.4)}$$

By writing I in Eq. (4.4) as

$$I = \frac{1}{(r-1)!(\ell-1)!(n-r-\ell)!} \sum_{P} \int_{-\infty}^{\infty} x\ F_{i_1}(x) \dots F_{i_{r-1}}(x)\ f_{i_r}(x) \left(1 - F_{i_{r+1}}(x)\right) \dots$$
$$\dots \left(1 - F_{i_{n-\ell}}(x)\right) dx \int_{-\infty}^{\infty} y\ f_{i_{n-\ell+1}}(y) \left(1 - F_{i_{n-\ell+2}}(y)\right) \dots \left(1 - F_{i_n}(y)\right)\ dy =$$
$$= \sum_{1 \leq i_1 < \dots < i_\ell \leq n} \mu_{r:n-\ell}^{[i_1,\dots,i_\ell]}\ \mu_{1:\ell}^{[i_{\ell+1},\dots,i_n]}$$

which is simply the right-hand side of the relation in (4.3). On the other hand, by noting that $\mathbf{R}^2 = \mathbf{R}_1 \cup \mathbf{R}_2$ we may write I in Eq. (4.4) as

$$I = J_1 + J_2, \qquad \textbf{(4.5)}$$

where the expressions for J_1 and J_2 are same as the expression for I in (4.4) with the integration being over the regions $\mathbf{R}_1$ and $\mathbf{R}_2$ (instead of $\mathbf{R}^2$), respectively. By considering the expression for J_1 and using the fact that

$$\left(1 - F_{i_{r+1}}(x)\right) \dots \left(1 - F_{i_{n-\ell}}(x)\right) =$$
$$= \left\{ \left(F_{i_{r+1}}(y) - F_{i_{r+1}}(x)\right) + \left(1 - F_{i_{r+1}}(y)\right) \right\} \dots$$
$$\dots \left\{ \left(F_{i_{n-\ell}}(y) - F_{i_{n-\ell}}(x)\right) + \left(1 - F_{i_{n-\ell}}(y)\right) \right\} =$$
$$= \sum_{j=0}^{n-r-\ell} \sum_{C_j^{n-r-\ell}} (F_{m_1}(y) - F_{m_1}(x)) \dots \left(F_{m_j}(y) - F_{m_j}(x)\right) \times$$
$$\times \left(1 - F_{m_{j+1}}(y)\right) \dots \left(1 - F_{m_{n-r-\ell}}(y)\right),$$

where $\sum_{C_j^{n-r-\ell}}$ denotes the sum over all $\binom{n-r-\ell}{j}$ combinations $(m_1, m_2, \ldots, m_j)$ of $(i_{r+1}, i_{r+2}, \ldots, i_{n-\ell})$, we get

$$J_1 = \frac{1}{(r-1)!(\ell-1)!(n-r-\ell)!} \sum_{j=0}^{n-r-\ell} \binom{n-r-\ell}{j} (r-1)! j! (n-j-r-1)! \mu_{r,j+r+1:n} =$$

$$= \sum_{j=r+1}^{n-\ell+1} \binom{n-j}{\ell-1} \mu_{r,j:n}. \tag{4.6}$$

Similarly, by considering the expression for J_2 and using the facts that

$$F_{i_1}(x) \ldots F_{i_{r-1}}(x) = \{F_{i_1}(y) + (F_{i_1}(x) - F_{i_1}(y))\} \ldots \left\{F_{i_{r-1}}(y) + \left(F_{i_{r-1}}(x) - F_{i_{r-1}}(y)\right)\right\} =$$

$$= \sum_{j=0}^{r-1} \sum_{C_j^{r-1}} F_{p_1}(y) \ldots F_{p_j}(y) \left(F_{p_{j+1}}(x) - F_{p_{j+1}}(y)\right) \ldots \left(F_{p_{r-1}}(x) - F_{p_{r-1}}(y)\right)$$

and

$$\left(1 - F_{i_{n-\ell+2}}(y)\right) \ldots \left(1 - F_{i_n}(y)\right) =$$

$$= \left\{\left(F_{i_{n-\ell+2}}(x) - F_{i_{n-\ell+2}}(y)\right) + \left(1 - F_{i_{n-\ell+2}}(x)\right)\right\} \ldots \left\{\left(F_{i_n}(x) - F_{i_n}(y)\right) + \left(1 - F_{i_n}(x)\right)\right\} =$$

$$= \sum_{k=0}^{\ell-1} \sum_{C_k^{\ell-1}} \left(F_{q_1}(x) - F_{q_1}(y)\right) \ldots \left(F_{q_k}(x) - F_{q_k}(y)\right)\left(1 - F_{q_{k+1}}(x)\right) \ldots \left(1 - F_{q_{\ell-1}}(x)\right),$$

where $\sum_{C_j^{r-1}}$ denotes the sum over all $\binom{r-1}{j}$ combinations $(p_1, p_2, \ldots, p_j)$ of $(i_1, i_2, \ldots, i_{r-1})$ and $\sum_{C_k^{\ell-1}}$ denotes the sum over all $\binom{\ell-1}{k}$ combinations $(q_1, q_2, \ldots, q_k)$ of $(i_{n-\ell+2}, i_{n-\ell+3}, \ldots, i_n)$, we get

$$\begin{aligned} J_2 &= \frac{1}{(r-1)!(\ell-1)!(n-r-\ell)!} \sum_{j=0}^{r-1} \sum_{k=0}^{\ell-1} \binom{r-1}{j} \binom{\ell-1}{k} \times \\ &\quad \times j!(r+k-j-1)!\,(n-r-k-1)!\, \mu_{j+1,r+k+1:n} = \\ &= \sum_{j=1}^{n} \sum_{k=r+1}^{r+\ell} \binom{k-j-1}{k-r-1} \binom{n-k}{n-\ell-r} \mu_{j,k:n}. \end{aligned} \tag{4.7}$$

Upon substituting the expressions of J_1 and J_2 in (4.6) and (4.7), respectively, into Eq. (4.5) we find that I is equal to the left-hand side of the relation in (4.3). This completes the proof of the result.

Result 15: For $1 \le \ell \le n-1$,

$$\sum_{k=2}^{n-\ell+1} \binom{n-k}{\ell-1} \mu_{1,k:n} + \sum_{k=2}^{\ell+1} \binom{n-k}{n-\ell-1} \mu_{1,k:n} = T_{1:n-\ell,1:\ell}. \tag{4.8}$$

PROOF: Eq. (4.8) is derived simply by setting $r = 1$ in Result 14.

Remark 3: If we replace ℓ by $n-\ell$ in Result 15, the relation in (4.8) remains unchanged and, therefore, there are exactly $[n/2]$ equations in $n-1$ product moments, viz., $\mu_{1,2:n}, \mu_{1,3:n}, \ldots, \mu_{1,n:n}$. For even values of n, there are $n/2$ equations in $n-1$ unknowns and hence a knowledge of $(n-2)/2$ of these product moments is sufficient. For odd values of n, there are $(n-1)/2$ equations in $n-1$ unknowns and hence a knowledge of $(n-1)/2$ of these product moments is sufficient. These are exactly the same bounds as presented in Theorem 2 which is not surprising since the product moments $\mu_{1,j:n}$ $(2 \le j \le n)$ are also sufficient for the calculation of all the product moments through Result 3.

Result 16: For $1 \le r \le n-1$,

$$\sum_{k=r+1}^{n} \mu_{r,k:n} + \sum_{j=1}^{r} \mu_{j,r+1:n} = T_{r:n-1,1:1}. \tag{4.9}$$

PROOF: Eq. (4.9) is derived simply by setting $\ell = 1$ in Result 14.

Result 17: For $1 \le r \le n-1$,

$$\mu_{r,r+1:n} + \sum_{j=1}^{r} \sum_{k=r+1}^{n} \binom{k-j-1}{k-r-1} \mu_{j,k:n} = T_{r:r,1:n-r}. \tag{4.10}$$

PROOF: Eq. (4.10) is derived simply by setting $\ell = n-r$ in Result 14.

Result 18: For $1 \le r \le n-1$,

$$\sum_{k=r+1}^{n} \sigma_{r,k:n} + \sum_{j=1}^{r} \sigma_{j,r+1:n} =$$

$$= \sum_{1 \le i_1 < \ldots < i_{n-1} \le n} \mu_{1:1}^{[i_1,\ldots,i_{n-1}]} \left(\mu_{r:n-1}^{[i_n]} - \mu_{r:n}\right) - \left(\sum_{j=1}^{r} \mu_{j:n}\right) \left(\mu_{r+1:n} - \mu_{r:n}\right). \tag{4.11}$$

PROOF: From Result 16 we have for $1 \le r \le n-1$

$$\begin{aligned} \sum_{k=r+1}^{n} \sigma_{r,k:n} + \sum_{j=1}^{r} \sigma_{j,r+1:n} &= T_{r:n-1,1:1} - \mu_{r:n} \sum_{k=r+1}^{n} \mu_{k:n} - \mu_{r+1:n} \sum_{j=1}^{r} \mu_{j:n} = \\ &= T_{r:n-1,1:1} - \mu_{r:n}\, S_{1:1} - \left(\mu_{r+1:n} - \mu_{r:n}\right) \sum_{j=1}^{r} \mu_{j:n}. \end{aligned}$$

The relation in (4.11) then follows by realizing that

$$T_{r:n-1,1:1} - \mu_{r:n}\, S_{1:1} = \sum_{1 \le i_1 < \ldots < i_{n-1} \le n} \mu_{1:1}^{[i_1,\ldots,i_{n-1}]} \left(\mu_{r:n-1}^{[i_n]} - \mu_{r:n}\right).$$

Result 19: For $n \ge 3$,

$$2\,\sigma_{1,2:n} + \sum_{k=3}^{n} \sigma_{1,k:n} = \sum_{1 \le i_1 < \ldots < i_{n-1} \le n} \mu_{1:1}^{[i_1,\ldots,i_{n-1}]} \left(\mu_{1:n-1}^{[i_n]} - \mu_{1:n}\right) - \mu_{1:n}\left(\mu_{2:n} - \mu_{1:n}\right). \tag{4.12}$$

PROOF: Eq. (4.12) is derived simply by setting $r = 1$ in Result 18.

Result 20: For $n \ge 3$,

$$\sum_{j=1}^{n-2} \sigma_{j,n:n} + 2\,\sigma_{n-1,n:n} =$$

$$= \sum\nolimits_{1 \le i_1 < \ldots < i_{n-1} \le n} \mu_{1:1}^{[i_1,\ldots,i_{n-1}]} \left(\mu_{n-1:n-1}^{[i_n]} - \mu_{n-1:n}\right) - \left(\mu_{n:n} - \mu_{n-1:n}\right)\left(S_{1:1} - \mu_{n:n}\right). \tag{4.13}$$

PROOF: Eq. (4.13) follows from Result 18 if we simply set $r = n-1$.

5. RELATIONSHIPS FOR SYMMETRIC VARIABLES

In this section we consider the special case when the variables X_i ($i = 1, 2, \ldots, n$) are all symmetric about zero and derive some recurrence relations and identities satisfied by the single and the product moments of order statistics. These results enable us to determine improved bounds (than the ones presented in Theorem 2) for the number of single and product moments to be evaluated for the calculation of means, variances and covariances of order statistics arising from n independent and non-identically symmetrically distributed variables, assuming these quantities to be known for order statistics arising from $n-1$ (and less) symmetric variables. These results generalize several well-known results for the i.i.d. case developed by Govindarajulu (1963), Joshi (1971) and Joshi and Balakrishnan (1982), and presented in detail by Arnold and Balakrishnan (1989).
From Eqs. (2.1) and (2.2), we easily observe in this case that

$$f_{r:n}(-x) = \frac{1}{(r-1)!(n-r)!} \sum_P F_{i_{r+1}}(x) \ldots F_{i_n}(x) f_{i_r}(x)\left(1 - F_{i_1}(x)\right) \ldots \left(1 - F_{i_{r-1}}(x)\right)$$

$$= f_{n-r+1:n}(x) \tag{5.1}$$

and

$$\begin{aligned} f_{r,s:n}(-x,-y) &= \frac{1}{(r-1)!(s-r-1)!(n-s)!} \sum_P F_{i_{s+1}}(y) \ldots F_{i_n}(y) f_{i_s}(y) \cdot \\ &\quad \cdot \left(F_{i_{r+1}}(x) - F_{i_{r+1}}(y)\right) \ldots \left(F_{i_{s-1}}(x) - F_{i_{s-1}}(y)\right) \times \\ &\quad \times f_{i_r}(x)\left(1 - F_{i_1}(x)\right) \ldots \left(1 - F_{i_{r-1}}(x)\right) \\ &= f_{n-s+1,n-r+1:n}(y,x). \end{aligned} \tag{5.2}$$

From Eqs. (5.1) and (5.2), it is clear that

$$\mu_{r:n}^{(k)} = (-1)^k \, \mu_{n-r+1:n}^{(k)}, \tag{5.3}$$

$$\mu_{r,s:n} = \mu_{n-s+1,n-r+1:n}, \tag{5.4}$$

$$\sigma_{r,r:n} = \mu_{r:n}^{(2)} - \mu_{r:n}^2 = \mu_{n-r+1:n}^{(2)} - \mu_{n-r+1:n}^2 = \sigma_{n-r+1,n-r+1:n} \tag{5.5}$$

and

$$\begin{aligned} \sigma_{r,s:n} &= \mu_{r,s:n} - \mu_{r:n}\,\mu_{s:n} = \mu_{n-s+1,n-r+1:n} - \mu_{n-s+1:n}\mu_{n-r+1:n} \\ &= \sigma_{n-s+1,n-r+1:n}. \end{aligned} \tag{5.6}$$

Result 21: For $m \geq 1$ and $k = 1, 2, \ldots,$

$$\mu_{m:2m-1}^{(k)[i]} = 0 \qquad \text{for odd values of } k \tag{5.7}$$

and

$$\mu_{m:2m}^{(k)} = \bar{\mu}_{m:2m-1}^{(k)} \qquad \text{for even values of } k. \tag{5.8}$$

PROOF: Eq. (5.7) follows simply from (5.3). Eq. (5.8) follows from Eq. (3.28) upon using the fact that

$$\mu_{m+1:2m}^{(k)} = \mu_{m:2m}^{(k)} \qquad \text{for even values of } k$$

observed easily from Eq. (5.3).

Result 22: For $1 \le r < s \le n$,

$$\left(1+(-1)^n\right)\mu_{r,s:n} = \sum_{j=0}^{r-1}\sum_{k=1}^{n-s}(-1)^{n-j-k-1}\binom{j+k}{j} S_{n-s-k+1,n-r-k+1:n-j-k}+$$

$$+\sum_{j=1}^{r-1}(-1)^{n-j-1}S_{n-s+1,n-r+1:n-j} + \sum_{j=1}^{s-r}(-1)^{s-r-j-1}\binom{s-1-j}{r-1}T_{1:s-j,j:n-s+j}. \quad \textbf{(5.9)}$$

PROOF: Eq. (5.9) follows directly from Result 7 upon using the symmetry relations in (5.3) and (5.4).

Result 23: For even values of n and $1 \le r < s \le n$,

$$2\,\mu_{r,s:n} = 2\,\mu_{n-s+1,n-r+1:n} = \sum_{j=0}^{r-1}\sum_{k=1}^{n-s}(-1)^{j+k-1}\binom{j+k}{j} S_{n-s-k+1,n-r-k+1:n-j-k} +$$

$$+\sum_{j=1}^{r-1}(-1)^{j-1}S_{n-s+1,n-r+1:n-j} + \sum_{j=1}^{s-r}(-1)^{s-r-j-1}\binom{s-1-j}{r-1}T_{1:s-j,j:n-s+j}. \quad \textbf{(5.10)}$$

PROOF: Eq. (5.10) follows simply from Result 22 when n is even.

Result 24: For even values of n,

$$2\,\mu_{1,2:n} = \sum_{k=1}^{n-2}(-1)^{k-1}S_{1,2:n-k}. \quad \textbf{(5.11)}$$

PROOF: Eq. (5.11) follows from Result 23 when we set $r = 1$ and $s = 2$ and use the facts that

$$S_{n-k-1,n-k:n-k} = S_{1,2:n-k} \qquad \text{and} \qquad T_{1:1,1:n-1} = 0$$

since

$$\mu_{1:1}^{[i_2,\ldots,i_n]} = \mu_{1:1}(i_1) = 0 \text{ for } i_1 = 1, 2, \ldots, n.$$

Theorem 3 *In order to find the first two single moments and the product moments of all order statistics arising from n independent and non-identically distributed symmetric (about zero) variables, given these moments of order statistics arising from $n-1$ (and less) independent and non-identically distributed symmetric variables, one needs to find atmost one single moment when n is even, and one single moment and $(n-1)/2$ product moments when n is odd.*

PROOF: In view of the relation in (3.28) and Result 21, it is sufficient to find one single moment ($\mu_{n:n}^{(2)}$ for odd values of n and $\mu_{n:n}$ for even values of n) in order to compute all the first two single moments of order statistics $\mu_{r:n}$ and $\mu_{r:n}^{(2)}$ for $1 \le r \le n$. The theorem is then proved by simply noting that there is no need to find any product moment when n is even due to Result 23.

Remark 4: It is of interest to mention here that the bounds established for the number of single and product moments while determining the means, variances and covariances of order statistics arising from n independent and non-identically distributed symmetric variables are exactly the same as the bounds established by Joshi (1971) for the i.i.d. case; also see Arnold and Balakrishnan (1989).

6. APPLICATIONS IN ROBUSTNESS STUDIES

Order statistics from outlier models and their moments have been used by many authors to examine the robustness features of linear estimators of parameters for some specific distributions. For example, Shu (1978), David and Shu (1978) and David (1979) used the means, variances and covariances of order statistics from a single-outlier normal model tabulated by David et al. (1977) to study the bias and the mean square error of various linear estimators of the mean of a normal population. They assumed in their study that the single outlier in the sample is either a normal location-outlier or a normal scale-outlier; also see Arnold and Balakrishnan (1989).

Recently, Balakrishnan (1992b) used the means, variances and covariances of order statistics from a single-outlier logistic model tabulated by Balakrishnan et al. (1992) to study the bias and the mean square error of various linear estimators of the mean and standard deviation of a logistic population. Once again, these authors have assumed in their studies that the single outlier in the sample is either a logistic location-outlier or a logistic scale-outlier.

For the exponential distribution, several authors including Kale and Sinha (1971) and Joshi (1972) studied the effects of a single outlier on the estimator for the expected lifetime; also refer to Barnett and Lewis (1984) for more references. By using the means, variances and covariances of order statistics from a single-outlier exponential model and by applying some relationships, Balakrishnan and Ambagaspitiya (1988) determined the means, variances and covariances of order statistics from a single scale-outlier double exponential model. They then used these values to study the bias and the mean square error of various linear estimators of the location and scale parameters of a double exponential distribution.

The results developed in this paper will enable one to significantly reduce the amount of computation required to be performed while determining the means, variances and covariances of order statistics arising from a sample containing more than one outlier. These results can then be used to examine the robustness properties of any linear estimator by studying its bias and mean square error (determined exactly). Work in this direction is currently in progress.

7. CONCLUSIONS

It is important to mention here that Sathe and Dixit (1990) and Balasubramanian and Balakrishnan (1992) recently proved the relation in Result 3 for the most general case, viz., non-independent and non-identically distributed variables; see also David (1993). So, either by directly using their result or by adopting their method of proof, it may be possible to prove some of the results presented in this paper for the most general case when the order statistics arise from non-independent and non-identically distributed variables; for example, see Balakrishnan, Bendre and Malik (1992). However, it is not clear, at this point of time, how their approach could be used to prove Results 7-11 (for example) involving both single and product moments of order statistics which (as mentioned earlier) are essential to claim Theorem 2. Some work in this regard is currently being carried out and we hope to report these findings in a future paper.

Acknowledgments
The author would like to thank the Natural Sciences and Engineering Research Council of Canada for funding this research and Mrs. Edna Pathmanathan and Mrs. Debbie Iscoe for the excellent typing of the manuscript.

REFERENCES

[1] ARNOLD, B. C. & BALAKRISHNAN, N. (1989). *Relations, Bounds and Approximations for Order Statistics*, Lecture Notes in Statistics No. 53, Springer–Verlag, New York.

[2] BALAKRISHNAN, N. (1987). Two Identities Involving Order Statistics in the Presence of an Outlier, *Commun. Statist.-Theor. Meth.*, **16**, 8, 2385-2389.

[3] BALAKRISHNAN, N. (1988a). Relations and Identities for the Moments of Order Statistics from a Sample Containing a Single Outlier, *Commun. Statist.- Theor. Meth.*, **17**, 7, 2173-2190.

[4] BALAKRISHNAN, N. (1988b). Recurrence Relations among Moments of Order Statistics from Two Related Outlier Models, *Biometr. J.*, **30**, 741-746.

[5] BALAKRISHNAN, N. (1988c). Recurrence Relations for Order Statistics from n Independent and Non-identically Distributed Random Variables, *Ann. Inst. Statist. Math.*, **40**, 273-277.

[6] BALAKRISHNAN, N. (1989). A Relation for the Covariances of Order Statistics from n Independent and Non-identically Distributed Random Variables, *Statist. Hefte*, **30**, 141-146.

[7] BALAKRISHNAN, N. (1992a). Relationships between Single Moments of Order Statistics from Non-identically Distributed Variables, In *Order Statistics and Nonparametrics: Theory and Applications* (Eds., P.K. Sen & I.A. Salama), North-Holland, Amsterdam. (To appear).

[8] BALAKRISHNAN, N. (1992b). Outlier and Robustness of Estimators, In *Handbook of the Logistic Distribution* (Ed., N. Balakrishnan), 263-290, Marcel Dekker, New York.

[9] BALAKRISHNAN, N. & AMBAGASPITIYA, R. S. (1988). Relationships among Moments of Order Statistics in Samples from Two Related Outlier Models and Some Applications, *Commun. Statist.-Theor. Meth.*, **17(7)**, 2327-2341.

[10] BALAKRISHNAN, N., BENDRE, S. M., & MALIK, H. J. (1992). General Relations and Identities for Order Statistics from Non-independent Non-identical Variables, *Ann. Inst. Statist. Math.*, **44**, 177-183.

[11] BALAKRISHNAN, N., CHAN, P. S., HO, K. L., AND LO, K. K. (1992). Means, Variances, and Covariances of Logistic Order Statistics in the Presence of an Outlier, *Submitted for publication.*

[12] BALAKRISHNAN, N. & MALIK, H. J. (1986). A Note on Moments of Order Statistics, *Amer. Statist.*, **40**, 147-148.

[13] BALASUBRAMANIAN, K. & BALAKRISHNAN, N. (1992). Indicator Method for a Recurrence Relation for Order Statistics, *Statist. Probab. Lett.*, **14**, 67-69.

[14] BALASUBRAMANIAN, K. & BALAKRISHNAN, N. (1993). Duality Principle in Order Statistics,*J. Roy. Statist. Soc.- Ser B.*. (To appear).

[15] BALASUBRAMANIAN, K., BAPAT, R. B., & BEG, M. I. (1991). On Families of Distributions Closed under Extrema, *Sankhyā*, Ser. A, **53**, 375-388.

[16] BALASUBRAMANIAN, K. & BEG, M. I. (1990). On Expectations of Functions of Order Statistics, *Sankhyā*, Ser. A, **52**, 103-114.

[17] BAPAT, R. B. & BEG, M. I. (1989a). Order Statistics for Non-identically Distributed Variables and Permanents, *Sankhyā*, Ser. A, **51**, 79-93.

[18] BAPAT, R. B. & BEG, M. I. (1989b). Identities and Recurrence Relations for Order Statistics Corresponding to Non-identically Distributed Variables, *Commun. Statist.-Theor. Meth.*, **18**, 1993-2004.

[19] BARNETT, V. & LEWIS, T. (1984). *Outliers in Statistical Data*, John Wiley and Sons, New York. (Second edition).

[20] BEG, M. I. (1991). Recurrence Relations and Identities for Product Moments of Order Statistics Corresponding to Non-identically Distributed Variables, *Sankhyā*, Ser. A, **53**, 365-374.

[21] DAVID, H.A (1979). Robust Estimation in the Presence of Outliers, In *Robustness in Statistics* (Eds., R.L. Launer & G.N. Wilkinson), 61-74, Academic Press, New York.

[22] DAVID, H. A. (1981). *Order Statistics*, Second edition, John Wiley and Sons, New York.

[23] DAVID, H. A. (1993). A Note on Order Statistics for Dependent Variates, *Amer. Statist.* (to appear).

[24] DAVID, H. A., KENNEDY, W. J., & KNIGHT, R. D. (1977). Means, Variances, and Covariances of Normal Order Statistics in the Presence of an Outlier, *Selected Tables in Mathematical Statistics*, **5**, 75-204.

[25] DAVID, H. A. & SHU, V. S. (1978). Robustness of Location Estimators in the Presence of an Outlier, In *Contributions to Survey Sampling and Applied Statistics: Papers in Honour of H. O. Hartley* (Ed., H. A. David), 235-250, Academic Press, New York.

[26] DOWNTON, F. (1966). Linear Estimates with Polynomial Coefficients, *Biometrika*, **53**, 129-141.

[27] GOVINDARAJULU, Z. (1963). On Moments of Order Statistics and Quasi-ranges from Normal Populations, *Ann. Math. Statist.*, **34**, 633-651.

[28] JOSHI, P. C. (1971). Recurrence Relations for the Mixed Moments of Order Statistics, *Ann. Math. Statist.*, **42**, 1096-1098.

[29] JOSHI, P. C. (1972). Efficient Estimation of the Mean of an Exponential Distribution when an Outlier Is Present, *Technometrics*, **14**, 137-144.

[30] JOSHI, P. C. & BALAKRISHNAN, N. (1982). Recurrence Relations and Identities for the Product Moments of Order Statistics, *Sankhyā*, Ser. B, **44**, 39-49.

[31] KALE, B. K. & SINHA, S. K. (1971). Estimation of Expected Life in the Presence of an Outlier Observation, *Technometrics*, **13**, 755-759.

[32] SATHE, Y. S. & DIXIT, U. J. (1990). On a Recurrence Relation for Order Statistics, *Statist. Probab. Lett.*, **9**, 1-4.

[33] SHU, V. S. (1978). Robust Estimation of a Location Parameter in the Presence of Outliers, *Ph.D. Thesis*, Iowa State University, Ames, Iowa, U.S.A.

[34] TEICHROEW, D. (1956). Tables of Expected Values of Order Statistics and Products of Order Statistics for Samples of Size Twenty and Less from the Normal Distribution, *Ann. Math. Statist.*, **27**, 410-426.

[35] VAUGHAN, R. J. & VENABLES, W. N. (1972). Permanent Expressions for Order Statistics Densities, *J. Roy. Statist. Soc., Ser. B*, **34**, 308-310.

Recent Adv. in Stat. and Prob., pp. 91-98
J. Pérez Vilaplana and M. L. Puri (Eds)

Recent Developments in Elemental Regression Methods

D. BRADU
Department of Statistics, TwW Bldg 7-126, University of South Africa, P. O. Box 392, Pretoria 0001, South Africa

Abstract – The problem of identifying multiple outliers in regression can be solved by means of a robust regression. Such a regression is obtained as a minimizer of an appropriate norm defined on the set of all possible regression pertinent to the data set in question. Difficulties with such an optimization and poor properties of the estimators obtained in the first proposals led to the use of elemental sets, that is sets of p cases where p is the number of predictors, and the corresponding elemental regressions which exactly fit the cases of each elemental set. Such subsets of cases have the maximal chance of being 'clean' that is free of outliers. In a typical method of this kind, as Rousseeuw's LMS (Least Median of Squares), one takes a sample of N elemental regressions, and one obtains for each one the value of the selection statistic specific to the method. The output will be the elemental regression for which this statistic takes the minimal value. It is hoped that examination of the residuals from the regression yielded by the procedure will pinpoint the outliers.
The LMS method is computationally efficient, has a maximal breakdown point and has become widely used, and as a consequence has received much attention. Recent results concerning the LMS will be mentioned, especially stressing the work of Stromberg who shows that the LMS can be reduced to the Chebyshev norm, and as it follows, that optimization over all possible regressions can be replaced by optimization over the elemental regressions fitting not p, but $p+1$ cases.
Another topic which will be treated is a recent method similar in schema to the LMS, but based on the ideas of outlier rejection rules seen as insurance policies, rather than in the context of hypotheses testing, ideas which were developed three decades ago by Anscombe. This method uses also the backwards elimination following a forward testing, an idea introduced by Rosner.

1. INTRODUCTION

We consider the standard situation of data given as a $n(p+1)$ matrix $(y\ X)$ where n is the number of cases. The first column y is the vector of the dependent variate values and X is the matrix of values of the p predictors, possibly with a column of ones for the intercept.
The data are assumed to satisfy a regression cum outliers model:

$$y = X\beta + I_n\delta + e \tag{1.1}$$

where:

β is the vector of unknown slope parameters;
δ is a vector of deviations from the regression, of zero elements except for the unknown cases which are outliers;
e is a vector of errors, assumed *i.i.d.* $N(0;\sigma^2)$.

Our concern are outlier location techniques, that is methods of finding the unknown outlying cases. Outlier location can be achieved via robust regression. Such a regression $\tilde{\beta}$ say, is found as a minimizer of an appropriate statistic $W(\beta)$, over the R^p space of all regressions β, or over a subset thereof. This statistic will be computed from the residuals $r = y - X\beta$ *w.r.* to that regression. More exactly, it will be obtained by separating the residuals into a cluster of 'small' residuals and the set of the remaining 'large' ones, and calculating the statistic $W(\beta)$ from the small residuals only.

In the Least Median of Squares (LMS) method of Rousseeuw (1984), the small residuals are those whose squares do not exceed $r^2_{(h)}$, the hth largest order statistic of the squared residuals, where $h = INT(n/2) + INT((p+1)/2)$ Rousseeuw's statistic, the Median of Squares (MS) is $r^2_{(h)}$ itself. In the related Least Trimmed Squares (LTS) method (Rousseeuw and Leroy 1986), the definition of small residuals is the same, but the statistic TS is obtained as the sum of squares, instead of the maximum of the small residuals.

The aim of any reasonable technique will be to find a regression $\tilde{\beta}$ which is a minimizer of $W(\beta)$, or at least a useful suboptimum point, that is a point helpful in outlier location. More precisely, a regression β will be called good if the residuals $r_i(\beta)$ display total separation (Bradu and Hawkins, 1982, 1991), that is if:

$$r_i > r_j \quad \text{whenever } i \text{ is an outlier, } j \text{ an inlier.}$$

This is a working definition, which however may be too weak in that it covers also regressions for which the outliers' absolute residuals are largest, but do not sufficiently stick out. Stricter definitions are possible, where the required condition is $r_i > K\ r_j$, for some $K > 1$.

A regression which is not good, is bad by default. We say that a run of a location procedure is successful if its output is a good regression.

The criterion function has usually multiple local minima and general optimization methods fail to yield a global minimizer or a $\tilde{\beta}$ close to such one.

A first restriction is made by searching for $\tilde{\beta}$ only among the 'elemental regressions' suggested by the data itself. Let J denote a subset of p cases and let Y_J and X_J be the corresponding submatrices of y and X. The elemental regression estimator

$$\beta_J = X_J^{-1} Y_J$$

exactly fits these p cases and gives a fitted residual vector $r_J = y - X\beta_J$. One will assume that the n data points *are in the general position* (Rousseeuw and Leroy 1986), that is, the elemental regression estimator β_J always exists. A given elemental set can be *clean*, that is free of outliers, or *contaminated*, in which case it will be automatically a bad one. Although the set of elemental *sets/regressions* is finite, its cardinality $\binom{n}{p}$ rapidly increases with n and makes exhaustive search impractical. One naturally turns to sampling, and finally, a version of the search was adopted (Hawkins, Bradu and Kass (1984), Rousseeuw (1984)) where a preset number N of elemental sets are randomly formed and the elemental regression with the smallest value of the statistic W specific to the procedure is taken as result. This sampling optimization will target on the global minimum of W and thus avoid the problems raised by multiple local minima, raising however others, because now the minimum is taken over a more restricted set.

2. RESULTS CONNECTED WITH THE LMS METHOD

The LMS is a powerful outlier location method. It retains the good properties of the one dimension median, in particular the almost 50% breakdown point property, and keeps inspiring relevant research. Here are a few quite recent results.

(i) Bradu and Hawkins (1991) deal with the problem of Probability of Success $PS(N)$, that is the probability of obtaining a good elemental regression by means of a given technique, when a sample of N elemental sets is used. Let P_1 be the probability that an elemental set drawn at random is good, and $P_2 = 1 - P_1$, the probability that it is bad. Let further F_1 and F_2 denote the cdfs of the relevant statistic W over the good- and bad elemental regressions respectively, and let $F = P_1F_1 + P_2F_2$ be the cdf of W over all the elemental regressions. Then the main result is contained in the formulas

$$PS(N) = NP_1 \int (1 - F(u))^{N-1}\, dF_1(u) = 1 - NP_2 \int (1 - F(u))^{N-1}\, dF_2(u).$$

This permits one to calculate, for a given statistic W, and for a data set of known structure, a table of $PS(N)$ for a set of values of N, and to determine what should be the sample size N in order to obtain a decent value for $PS(N)$, e.g. .99 or .999.

The applications concerned mainly the evaluation of the LMS performance on particular data sets. The formulas for $PS(N)$ however can be used for any technique based on elemental sets, and can serve for comparing different such methods.

(ii) In the case of the LMS, the minimization of the statistic $W = MS$ can be done in principle over the whole space R^p of β, or over the set of elemental regression slopes β_J, or over a finite sample thereof.

The connection between the results of these three operations of minimization was not clear, although it was believed that each of them may be a good approximation of the previous one. The study of Hawkins (1991a) confirms the truth of this conjecture, especially for large data sets.

There existed however no way to exactly calculate the LMS, even in the case of small data sets. Stromberg (1992), who relates to the work of Tichawsky (1991), is the first to have seen the LMS in the context of Chebyshev norm, and to have provided a way to obtain the exact value of the LMS.

Given a regression data set (y, X) and an arbitrary slope vector β, the *Chebyshev Residual Norm* is defined as

$$||r||_{ch} = \max_i\ |r|_i\ ,$$

where $r = y - X\beta$ is the vector of residuals. The *Chebyshev Slope* $\tilde{\beta}$ is the minimizer of the Chebyshev Residual Norm over all possible slopes β.

Let now redefine an elemental set as any subset of $p+1$ cases, and assume that the data are in the general position (in the context of Chebyshev regression, this is known as the Haar condition). Cheney (1962) gives two fundamental results.

RESULT 1. Let $\tilde{\beta}_J$ be the Chebyshev slope for the subset (y_J, X_J) of the data, formed of the cases of elemental set J. Then, the *Chebyshev slope* $\tilde{\beta}$ of the whole data set (y, X) can be obtained as the minimizer of the *Chebyshev residual norm* over the finite set of the slopes $\tilde{\beta}_J$.

RESULT 2. The result 1 reduces the problem to the calculation of the Chebyshev slope for a data subset (y_J, X_J) of $p+1$ cases and p predictors. It turns out that in this situation, the Chebyshev slope $\tilde{\beta}_J$ can be obtained by the following simple algorithm, involving only LS estimation.

(*a*) One obtains the LS slope $\hat{\beta}_J = (X_J^T X_J)^{-1} X_J^T y_J$,and the LS residuals vector $r_J = y_J - X_J \hat{\beta}_J$, which are used to form the vector S whose coordinates are the signs of the residuals, and the coefficient $\in = \|r_J\|^2{}_2 / \|r_J\|_1$, where $\|.\|_2$ and $\|.\|_1$ are the L_2 and the L_1 norms respectively.
(*b*) One updates y to $y^* = y - \in S$ and one finally obtains the Chebyshev slope as the LS slope of (y^*, X), that is by means of $\tilde{\beta}_J = (X_J^T X_J)^{-1} X_J^T y_J^*$.
One notes that obtaining the Chebyshev slope for a $p+1$ cases elemental set will take only about twice the time needed to obtain an exact fit to an elemental set of size p.
Stromberg examines further the *Least* q^{th} *Squared Residual* statistic (Rousseeuw and Leroy 1986), which for $q = INT(n/2) + INT((p+1)/2)$ gives the LMS.
For a given slope β, the largest q^{th} squared residual $r^2_{(q)}$ is by definition such that $|r_{(q)}(\beta)| = \max_{j \le q} |r_{(j)}(\beta)|$, that is $r_{(q)}(\beta)$ is the Chebyshev residual norm for the data subset formed of the q cases corresponding to the involved residuals. If one takes $\beta = \tilde{\beta}$, the Chebyshev slope for this data subset, then the qth squared residual will be minimal for this subset of q cases. It follows that the Least q^{th} Squared Residual can be obtained by screening all the subsets of q cases, calculating for each one the Chebyshev slope, and taking the minimum of the corresponding Chebyshev residuals. Results 1 and 2 entail that the operation can be done on elemental sets only, and one finally obtains the following *algorithm for the obtention of the exact Least* q^{th} *Squared Residual.*
All the elemental sets (of $p+1$ cases) are screened. For each one the Chebyshev slope is calculated, as well as the residuals of all the n cases from the Chebyshev regression, and finally the largest q^{th} squared residual is obtained. The minimization of the latter over all screened elemental sets will give the Least q^{th} Squared Residual.
One notices that the same material, the elemental sets, is used no matter which is the index q. This suggests simultaneously computing the Least q^{th} Squared Residual for several, or even for all of the values of q. This can be achieved at almost no additional cost, since most of the computational effort is required by the calculation and ordering of the residuals.
The elegant result of Stromberg is however translated into a recipe which is hardly practical unless the data set is quite small. Hawkins (1991b) proposes a probabilistic algorithm which produces only trial values satisfying the necessary condition for the optimum and which provides the exact LMS solution with probability one as the number of iterations increases. This approach uses throughout Chebyshev fits to elemental sets of size $p+1$, and proceeds from one such set to a better one by a mechanism based on the linear programming algorithm for Chebyshev fits.

3. A RESULT BASED ON ANSCOMBE'S IDEAS.

In both Stromberg (1992) and Hettmansperger and Sheather (1992) a potential lack of stability of the LMS is discussed. A related problem is that the definition of small residuals underlying this method is somewhat rigid, and some improvement in performance may be expected from possible alternative approaches, where the clustering technique used may coopt not a prespecified proportion, but as many residuals as possible into the 'small' cluster.
Bradu and Hawkins (1992) propose a method which retains the elemental set sampling schema of Rousseeuw's (1984) technique, but replaces the Median of Squares statistic, with the statistic WA to be described.

Note first that once β is given, the construction of $W(\beta)$, for any statistic W, becomes a one dimensional one sample problem, namely that of isolating the small residuals, and then using them to calculate $W(\beta)$.

Let now $x_1, ..., x_m$ be independent normal homoscedastic observations, of which at least $m-k$ come from one population, while at most k are outliers. The model describing this sample, devised in view of future application to regression residuals, can be written as follows:

$$x_i \sim N(0;\sigma^2) \qquad i \in I_o ,$$
$$x_i \sim N(\delta_i;\sigma^2) \qquad i \in I_1 , \qquad \#I_1 \leq k$$

where I_o and I_1 are the unknown subsets of $\{1, ..., m\}$ containing the indices of inliers and outliers respectively.

To locate the outliers. Once this is done, the parameter σ^2 will be estimated by means of the mean square s^2 of the inlying observations.

Anscombe (1960) regards an *outlier rejection rule*, in view of the subsequent estimation of parameters from the nonrejected observations, as an *insurance policy* for which a premium is paid in the form of a poorer estimation when no outlier is present which buys protection as improvement in estimation, in the presence of outliers. Precise quantitative definitions of premium and protection can be given.

Anscombe's rule consists mainly in examining the k 'most extreme' observations, where more extreme means of a larger square, and rejecting those for which a given statistic exceeds a chosen cutoff value C. Each possible value of C will determine corresponding values of the premium and of the protection. The optimal choice will be a value of C giving a maximal ratio *protection/premium*, *i.e.* allowing to buy the largest amount of protection per unit of paid premium.

Rosner(1983) brings the idea of backwards elimination, to follow forward testing, idea which will be later used in a wider context by Simonoff(1988). According to backwards elimination, will be excluded also the observations which escaped the initial screening, but are more extreme than at least one of the already rejected ones.

Assuming that one has already chosen the gauge C , the proposed procedure, which is Anscombe's, with incorporation of Rosner's backwards elimination, can be described as follows.

Assume that the observations are ordered in increasing square, that is $x_1^2 \leq \leq ... \leq x_m^2$, and form the decreasing sequence of samples $S_o \supset S_1 \supset \cdots \supset S_k$, where $S_j = \{x_1 ,..., x_{m-j}\}$. For each sample S_j , one calculates

$$s_j^2 = \sum_{i=1}^{m-j} x_i^2/(m-j)$$

and

$$R_j = x_{m-j}^2/s_j^2$$

One notes the monotonicity relations:

$$s_j^2 \geq s_{j+1}^2, \qquad \text{for any } j$$

and the bounds

$$R_j \leq m-j, \qquad \text{for any } j$$

The *Decision and Estimation Rule* (A) based on statistic R is as follows:

i) If $R_\ell \leq C$, $\ell = 0, 1, ..., k-1$, then one declares 0 outliers and one retains the estimate s_o^2 for σ^2 , based on the entire sample.

ii) If $R_{\ell-1} > C$, but $R_{\ell+i-1} \leq C$, $1 \leq i \leq k-\ell$, then one declares ℓ outliers x_m, $x_{m-1} \ldots$, $x_{m-\ell+1}$, and one retains the estimate s_ℓ^2 for σ^2 based on sample S_ℓ.

In order to take care of the case where $\sigma^2 = 0$ in the assumed model, one introduces case *iii*:

iii) If $R_{\ell-1} > C$ but $s_\ell = 0$, then one declares ℓ outliers x_m, $x_{m-1}, ..., x_{m-\ell+1}$, and one retains $s_\ell^2 = 0$ as estimate of σ^2.

Note that in this last case, all the observations x_1 , ..., $x_{m-\ell}$ are zero.
Assuming again that C was fixed, let us return to the problem of defining and computing a premium and a protection.
In order to define a premium measure, one assumes that there are no outliers. Applying the Rule (A) to this sample, a number ℓ of outliers will be declared, and an estimate s_ℓ^2 for σ^2 will be obtained. In view of the monotonicity of the s_j^2, s_ℓ^2 will be an underestimate, unless $\ell = 0$. The quantity s_o^2/s_ℓ^2 will be a convenient measure for the deterioration of estimation caused by the use of Rule (A); for a given sample, it will take the value one for a correct decision, and will generally increase with the number of – incorrectly – declared outliers. The premium for a given value of C will be defined as

$$\text{Premium}(C) = E(s_o^2/s_\ell^2)$$

We want now to define a measure of protection, which will equal one in a situation where outliers are present, and will decrease as one departs from it.
We take as reference situation that where we have in model (2.1) the maximal number k of outliers, and their expectations are all a multiple of σ, that is $\delta_i = M\sigma$. In practical applications, one will take $M = 10$ or 15.
For a given sample, if the Rule (A) returns exactly k outliers, then s_k^2 will be the correct estimate of σ^2. Else, one will err one way or the other. If ℓ is the declared number of outliers, and s_ℓ^2 the corresponding estimate of σ^2, then the goodness of the result will be evaluated by means of

$$2\Big/\left[(s_k^2/s_\ell^2) + (s_\ell^2/s_k^2)\right]$$

quantity which equals one if $\ell = k$, and otherwise decreases as ℓ departs from k in either direction. The protection for given C, M is accordingly defined as

$$\text{Protection}(C; M) = E\left\{2\Big/\left[(s_k^2/s_\ell^2) + (s_\ell^2/s_k^2)\right]\right\}$$

The criterion for choosing C , for given M , will be the maximization of the quantity

$$\text{Ratio}(C; M) = \text{Protection}(C; M)/\text{Premium}(C)$$

The maximizer C_o of Ratio will lead to the maximal amount of Protection being bought for a unit of Premium.
For given C, M the Premium, the Protection and consequently the Ratio can be estimated by sampling.
Then follows the optimization of Ratio$(C; M)$ as a function of C. As was shown, an upper bound for R_o, hence for C, is m; the search for the minimizer C_o can be restricted to interval $(0, m)$. Any linear search routine starting with this interval can be used, but

even calculating a restricted table of values of Ratio for a set of integer values of C will give a satisfactory half-integer value of C_o.
This choice of C_o is optimal provided there are k outliers. Experiments with various m and k, show however that the value C_o obtained assuming k outliers gives practically the same value of Ratio when the number of outliers is smaller than k, hence one can use this value C_o for any number of outliers from 1 to k .
As an example, if $m = 20$, $k = 7$, $M = 10$, the following table gives the Estimated Premium-(1), Protection-(2) and Ratio-(3) for a few values of C :

Table 1.

C	9.0	9.1	9.2	9.3	9.4	9.5	9.6	9.7	9.8	9.9
(1)	1.013	1.012	1.012	1.010	1.009	1.008	1.008	1.008	1.007	1.006
(2)	1	1	1	1	1	.999	.998	.998	.997	.994
(3)	.987	.988	.988	.990	.991	.990	.990	.990	.988	.991

which points to the minimizer $C_o = 9.45$, with Premium= 1.009, Protection= .999 and Ratio= .991 (with three exact decimals).
Returning to the regression problem, let $\{r_i\}$ be the n residuals *w.r.* to a regression β . One will assume that the number of outliers in the regression does not exceed $k = n - h$, where $h = (n/2) + ((p+1)/2)$ and this value of k will be used when analyzing the sample of $m = n$ residuals. Using Anscombe's approach, a value C_o is obtained and Rule (A) is applied to the sample of residuals $\{r_i\}$. This rule will yield a number ℓ of declared outliers and, what is of interest, a final estimate s_ℓ^2 from the declared inliers.
The Anscombe type statistic WA is defined as $WA = s_\ell^2$. It is this statistic that defines the new proposal. The term of *Regression WA Procedure* is used for the minimization of WA over a sample of elemental regressions (or over all $\beta \in R^p$); the term of *Sample WA Procedure* is used for the above technique for obtaining WA from a linear sample.
A limited experience seems to point to this method requiring a smaller number of elemental sets than the LMS to reach a good likelihood of success, and to its being able to solve some cases refractary to the LMS.
Now, one more remark. One can avoid searching for C_o by taking a collection of values of C (between 0 and n) and simultaneously performing the Regression WA Procedure for all these values, over a sample of elemental sets. Most of the computational effort being dedicated to the obtention of the residuals for each elemental set, residuals which will serve for all the values of C, the total computer time spent does not increase much. The end product will be a set of candidate regression lines, one for each C , or - which is the same - a corresponding set of residual vectors. This final set to analyze may be further reduced by using the correlations between the residual vectors for clustering them into classes of highly correlated ones. The same can be done of course for the LMS, where as an additional advantage, there is a natural set of statistics: the largest $h^{th}, ..., n^{th}$ squared residuals.

4. FINAL REMARKS

The recent results mentioned or described above show that the research in robust regression by means of elemental sets is a domain in full developement. No attempt will be made to justify the choice of the material or the accent put on the different topics. It is a subjective and to an extent an arbitrary one, which fits mainly the taste, the knowledge and the personal involvement of the author.

REFERENCES

[1] ANSCOMBE, F.J., (1960). Rejection of Outliers, *Technometrics*, **2**, 123-147.

[2] BRADU, D., AND HAWKINS, D.M., (1982). Location of Outliers in Two-way Tables Using Tetrads, *Technometrics*, **24**, 103-108.

[3] BRADU, D., AND HAWKINS, D.M., (1985). *Location of Multiple Outliers in Regression - Use of Some EM Algorithms.* Annual Conference of the South African Statistical Association, Kruger Park.

[4] BRADU, D. AND HAWKINS, D.M., (1991). *Sample Size Requirements for Multiple Outlier Location Techniques Based on Elemental Sets.* Computational Statistics and Data Analysis 16(1993) 257-270

[5] BRADU, D. AND HAWKINS, D.M., (1992). *An Anscombe Type Robust Regression Statistic.* Research Report 92/6, University of South Africa.

[6] CHENEY, E.W., (1962). *Introduction to Approximation Theory.* New York: McGraw-Hill.

[7] HAWKINS, D.M., (1991a). *The Accuracy of Elemental Sets Approximations for Regression.* University of Minnesota, St. Paul.

[8] HAWKINS, D.M., (1991b). *A Fast Exact Probabilistic Algorithm for Least Median of Squares Regression.* University of Minnesota, St. Paul.

[9] HAWKINS, D.M., BRADU, D., AND KASS, G.V., (1984). Location of Several Outliers in Multiple Regression Data Using Elemental Sets. *Technometrics*, **26**, 197-208.

[10] HETTMANSPERGER, T.P., SHEATER, S.J. (1992). A Cautionary Note on the Method of Least Median Squares, *The American Statistician*, **46**, 2, 79-83.

[11] ROSNER, B. (1983). Percentage Points for a Generalized ESD Many Outlier Procedure, *Technometrics*, **25**, 2,165-172.

[12] ROUSSEEUW, P.J. (1984). Least Median of Squares Regression, *Journal of the American Statistical Association*, **79**, 871-880.

[13] ROUSSEEUW, P.J., LEROY, A.M., (1987). *Robust Regression and Outlier Detection.* New York, Chichester: John Wiley and Sons.

[14] SIMONOFF, J.S., (1988). *General Approaches to Stepwise Identification*, New York University.

[15] STROMBERG, A.J., (1992). *Detecting Instability in the Least Median of Squares Estimate in Multiple Linear Regression.* University of Kentucky.

[16] TICHAWSKY, P., (1981). Algorithms for - and a Geometrical Characterization of Solutions in the LMS and LTS Linear Regression, *Computational Statistics Quarterly*, **6**, 139-151.

Recent Adv. in Stat. and Prob., pp. 99-103
J. Pérez Vilaplana and M. L. Puri (Eds)

Consistency of a Linear Regression Estimate with Panel Data, Obtained by Preliminary Nonparametric Estimation

C. FERNÁNDEZ-JARDÓN FERNÁNDEZ[1] and X. M. MARTÍNEZ FILGUEIRA[2]
[1] Departamento de Economía y Estadística, Facultad de Ciencias Económicas y Empresariales, Universidad de Navarra, E-31080 Pamplona, España
[2] Departamento de Economía Aplicada, Facultad de Ciencias Económicas, Universidad de La Coruña, E-15071 La Coruña, España

Abstract – In this paper we prove the almost sure consistency of a linear regression estimate with panel data. We start from a preliminary nonparametric estimation of the density and regression function and search for the parameters which minimize the distance between the proposed model and the preliminary nonparametric estimation of the regression.

1. THE MODEL

The sample data consist of observations of m entities over T time periods (the same number of periods for each entity). We have therefore mT data distributed in m subsamples, one for each entity. Consider the model

$$Y = R(X) + \epsilon$$

where Y is the dependent variable, X is a $p \times 1$ vector of exogenous variable, ϵ is a random variable of mean 0 and variance σ^2 and $R(X) = E[Y/X]$ is an unspecified regression function. The joint density is a function f whose support is compact and contained in a open set S.

The general nonparametric model is to find the function $R(X)$ which best fits the data (Ullah 1989).

In the proposed model for panel data this function will be a linear combination of the data observed, both temporary and sectionally, and a parameter Θ, so with that the model would be:

$$Y = A^t(X)\Theta + \epsilon$$

where $A(X) : \mathbf{R}^p \rightarrow \mathbf{R}^c$ is the following functional:

$$A(X) = (I_1(X) \cdot x_1, I_1(X) \cdot x_2, .., I_1(X) \cdot x_q, I_2(X) \cdot x_1, .., I_m(X)x_q, x_{q+1}, .., x_p)^t$$

$$I_i(X) = \begin{cases} 1 & \text{if } X \text{ is in the subsample } i \\ 0 & \text{if } X \text{ is not in the subsample } i \quad i = 1, 2, \ldots, n \end{cases}$$

$\Theta = (\theta_1^1, \theta_2^1, .., \theta_q^1, \theta_1^2, .., \theta_q^m, \theta_{q+1}, .., \theta_p)^t$ a $c \times 1$ vector of parameters and

$c = (m-1)q + p = mq + (p-q)$.
The particularity of this model is that the coefficients of q components of $X, (0 \leq q \leq p)$, can vary, depending of the subsample considered, while the $(p-q)$ remaining coefficients are the same for all subsamples.
We denote by θ_j^r, $r = 1,2,..,m$, $j = 1,2,..,q$, the coefficient of x_j in the rth sample, while θ_j, is the coefficient of $x_j, j = q+1,..,p$, independent of the sample.

2. THE ESTIMATE

The estimation method will be an extension of the proposed for Faraldo Roca-González Manteiga (1987) for the linear regression and later developed for Cristóbal Cristóbal-Faraldo Roca-González Manteiga (1987) for the multiple linear regression and for González Manteiga-Vilar Fernández (1987) for time series.
This estimator is obtained by minimizing of the functional

$$\Psi_1(X) = \int (\hat{\alpha}_n(x) - A^t(X)\Theta)^2 d\Omega_n(x)$$

where $\hat{\alpha}_n(x)$ is a nonparametric estimator of the regression, and $\Omega_n(x) a$ weighting function constructed starting from the sample.
The $\hat{\alpha}_n(x)$ and the $\Omega_n(x)$ chosen for this paper are:

$$\hat{\alpha}_n(x) = \frac{\sum_{i=1}^{n} Y_i \delta_s(x, X_i)}{\sum_{r=1}^{s} \delta_s(x, X_r)}; \qquad \Omega_n(x) = \int_{-\infty}^{x} \hat{f}_n(t) = \frac{\sum_{i=1}^{n} \int_{-\infty}^{x} \delta_s(t, X_i)}{n}$$

where $\hat{f}_n$ is a nonparametric estimator of f, and $\{\delta_s : \mathbf{R}^p \times \mathbf{R}^p \to \mathbf{R}\}_{s=s(n)\to\infty}$ is a sequence of measurable functions, which are generalization of the regression of the Kernel estimators and other nonparametric estimators (González Manteiga 1982).
The estimator $\hat{\Theta}$ will be given by the solution of the minimizing problem of Ψ_1, i.e.:

$$\hat{\Theta} = [\int A(x)A^t(x)d\Omega_n(x)]^{-1} \int \hat{\alpha}_n(x)A(x)d\Omega_n(x) = C^{-1}D$$

The array C has one line and one column for each component of Θ.
The lines corresponding to some θ_j^r, $r = 1,2,..,m$, $j = 1,2,..,q$ are of the form:

$$(0, \overset{q}{\ldots}, 0, 0, \overset{q}{\ldots}, 0, \ldots, \overline{V}^r_{j1}, \overset{q}{\ldots}, \overline{V}^r_{jq}, \overset{q}{\ldots}, 0, \overset{q}{\ldots}, \overline{V}^r jq+1, \overset{p-q}{\ldots}, \overline{V}^r_{jp})$$

while the lines corresponding to some $\theta_j, j = q+1,..,p$ are:

$$(\overline{V}^1_{j1}, \ldots, \overline{V}^1_{jq}, \ldots, \overline{V}^2_{j1}, \ldots, \overline{V}^2_{jq}, \ldots, \overline{V}^m_{j1}, \ldots, \overline{V}^m_{jq}, \ldots, \overline{V}_{jq+1}, \ldots, \overline{V}_{jp})$$

where

$$\overline{V}^r_{ij} = \frac{1}{n}\left(\sum_{k=1}^{n} \int x_i x_j I_r Y_k \delta_s(x, X_k)dx\right);$$

$$\overline{V}^r_{ij} = \frac{1}{n}\left(\sum_{k=1}^{n} \int x_i x_j Y_k \delta_s(x, X_k)dx\right)$$

also we have what $\overline{V}^r_{ij} = \overline{V}^r_{ji}$.
We have also $D = (\overline{T}^1_1, .., \overline{T}^1_q, \overline{T}^2_1, ..., \overline{T}^m_q, \overline{T}_{q+1}, .., \overline{T}_p)$ where

$$\overline{T}^r_i = \frac{1}{n}\left(\sum_{k=1}^{n} \int x_i I_r Y_k \delta_s(x, X_k)dx\right); \qquad \overline{T}_i = \frac{1}{n}\left(\sum_{k=1}^{n} \int x_i Y_k \delta_s(x, X_k)dx\right);$$

3. CONSISTENCY

To prove the a.s. consistency, the functions $\delta_s(x,u)$ must fulfills the following conditions:

i) $\sup_u \delta_s(x,u) = O(s^p), \forall x \in \mathbf{R}$;

ii) $\delta_s(x,u) = \delta_s(u,x), \forall(x,u) \in \mathbf{R}^{2p}$;

iii) $\delta_s(x,u)du = 1$ and $\delta_s(x,u) = 0$ if $||x-u|| > c\epsilon(n)$, where $c \in \mathbf{R}^+$ and $\epsilon(n) \to 0$ when $n \to \infty$ $[s = O(1/\epsilon(n))]$;

Also we need the following previous results:

Proposition: *Let* $g : \mathbf{R}^p \to \mathbf{R}$ *be a linear function satisfying* $||g(x)|| \le ||x||$ *, and let* $\{\delta_s : \mathbf{R}^p \to \mathbf{R}\}_{s=s(n)\to\infty}$ *be a sequence of measurable functions satisfying the conditions i), ii) and iii). Then* $\int g(x)\delta_s(u,x)dx \to g(u)$ *when* $n \to \infty$.

PROOF: Doing $t = x - u$ and substituting we will have:

$$\int g(x)\delta_s(u,x)dx = \int g(t)\delta_s(u,t+u)dt + \int g(u)\delta_s(u,t+u)dt = \Delta_1 + \Delta_2$$

Applying *iii)* we have that $\Delta_2 = g(u)$ and $\Delta_1 = \int_{||x-u||>c\epsilon(n)} g(x-u)\delta_s(u,x)dx \Rightarrow$ (applying the premises of $g(x)$ and again *iii* $\Rightarrow ||\Delta_1|| < c\epsilon(n) \to 0$ when $n \to \infty$ ■

Lemma: *If the fourth moment of X exists and the premises* i), ii), iii) *are satisfied, and if Y is bounded with probability 1 and* $\overline{S}$ *is a compact set, then:*

$$\overline{V}^r_{ij} \overset{a.s}{\to} V^r_{0,ij} = E[I^r x_i x_j]; \qquad \overline{V}_{ij} \overset{a.s}{\to} V_{0,ij} = E[x_i x_j];$$

$$\overline{T}^s_k \overset{a.s}{\to} \overline{T}^s_{0,k} = \sum_{h=1}^{q} \theta^s_h \overline{V}^s_{0.kh} + \sum_{h=q+1}^{p} \theta^s_h V_{0.kh} \; ;$$

$$\overline{T}_k \overset{a.s}{\to} \overline{T}_{0,k} = \sum_{h=1}^{q}\sum_{t=1}^{m} \theta^t_h V^t_{0.kh} + \sum_{h=q+1}^{p} \theta_h V_{0.kh}$$

PROOF: We will prove it for the elements of the form $\overline{T}^s_i$, since it will be similar for the other forms.

Consider the decomposition:

$$\overline{T}^s_i - \overline{T}^s_{0,i} = \overline{T}^s_i - E[\overline{T}^s_i] + E[\overline{T}^s_i] - \overline{T}^s_{0,i} = \Delta_1 + \Delta_2$$

$$\begin{aligned}\Delta_2 &= E\left[\frac{1}{n}\sum_{h=1}^{n} T^s_{i,h}\right] - T^s_{0,i} = E\left[\int x_i I^s Y \delta(x,X)dx\right] - T^s_{0,i} = \\ &= E\left\{E\left\{\left[\int x_i I^s Y \delta(x,X)dx\right]/X\right\}\right\} - T^s_{0,i} = \\ &= E\left\{\left[\int x_i I^s \delta(x,X)dx\right] E[Y/X]\right\} - T^s_{0,i}\end{aligned}$$

Using $E[^Y/X] = A^t(X)\Theta = \sum_{h=1}^{q}\sum_{t=1}^{m} \theta^t_h I^t x_h + \sum_{h=q+1}^{p} \theta_h x_h$ we obtain

$$\Delta_2 = \int \left[\int u_i I^s \delta(x,u)dx\right] \left[\sum_{h=1}^{q}\sum_{t=1}^{m} \theta^t_h I^t u_h + \sum_{h=q+1}^{p} \theta_h u_h\right] f(u)du - T^s_{0,i}$$

Using the previous proposition and applying the dominated convergence theorem it is easy to see that $\Delta_2 \to 0$.

To prove that $\Delta_1 \to 0$ a.s., we would apply the Bennett's inequality (Bennett, 1962) to the variables $R_h = T_{i.h} - E[T_{i.h}]$, $h = 1,2,..,T$; which are bounded with probability 1, because Y is also bounded and the support of f is contained in a compact set $\overline{S}$.
Applying the inequality we obtain that $P\{\overline{R} \geq \epsilon\} \leq 2\exp\{-n\epsilon^2 O(1)\}$. Then applying the Borel-Cantelli's lemma we complete the proof. ■

Theorem: *If the support of f is bounded and if $E[Y^4] < \infty$, then $\hat{\Theta} \xrightarrow{a.s.} \Theta$*

PROOF: The components of $\hat{\Theta}$ are construed operating with the elements of C and D; from the previous lemma; this elements converge a.s., the components of the estimator do it to the same operations done with the limits of these elements. We will have therefore:

$$\hat{\Theta} = C^{-1}D \xrightarrow{a.s.} C_0^{-1}D_0$$

being C_0 and D_0 the arrays obtained from the replacing of the elements of C and D by its limits. Now it only remains to prove that $C_0^{-1}D_0 = \Theta$.
The proof will be done only for the first component, $\hat{\theta}_1^1$:

$$\hat{\theta}_1^1 = (1^{st} \text{ line of } C^{-1}) \times D \xrightarrow{a.s.} (1^{st} \text{ line of } C_0^{-1}) \times D_0 =$$
$$= |C_0|^{-1}(W_0^{1.11} \dots W_0^{1.1q} B_0^{1.1/2.1} \dots B_0^{1.1/2.q} B_0^{1.1/m.1} \dots B_0^{1.1/m.q} W_0^{1.1q+1} \dots W0^{1.1p}) \times D_0$$

being $W_0^{1.1j}$ the adjoint of the element $V_{0,1j}^1$, and the element $B_0^{1.1/s.j}$ the adjoint of the zero of C_0, in the position corresponding to the line θ_1^1 and the column θ_j^s. The result of this product will be:

$$|C_0|^{-1}\left(\sum_{h=1}^{q}\theta_h^1\left[\sum_{k=1}^{q}W_0^{1.1k}V_{0.kh}^1 + \sum_{k=q+1}^{p}W_0^{1.1k}V_{0,kh}^1\right] + \right.$$
$$+\sum_{\substack{i=1\\i\neq r}}^{m}\sum_{h=1}^{q}\theta_h^t\left[\sum_{k=1}^{q}B_0^{1.1/t.k}V_{0.kh}^t + \sum_{k=q+1}^{p}W_0^{1.1k}V_{0.kh}^t\right] +$$
$$\left. + \sum_{h=q+1}^{p}\theta_h\left[\sum_{k=1}^{q}W_0^{1.1k}V_{0.kh}^1 + \sum_{\substack{t\neq r\\t=1}}^{m}\sum_{k=1}^{q}B_0^{1.1/t.k}V_{0,kh}^t + \sum_{k=q+1}^{p}W_0^{1.1k}V_{0,kh}\right]\right) =$$
$$= |C_0|^{-1}\left(\sum_{h=1}^{q}\theta_h^1(A) + \sum_{\substack{t\neq r\\t=1}}^{m}\sum_{h=1}^{q}\theta_h^t(B) + \sum_{h=q+1}^{p}\theta_h(C)\right)$$

(A), (B), (C) are equal to products resulting of substituting a line of C_0 in the position of her first line, therefore the result will be zero except in the case that the line replaced is the first one, in this case the result will be $|\, C_0 \,|$.
Applying this, it follows that $(1^{st} \text{ line of } C_0^{-1}) \times D_0 = \theta_1^1$
Operating in this way for all the components of $\hat{\Theta}$ we would prove that:

$$\hat{\Theta} \xrightarrow{a.s} \Theta$$ ■

REFERENCES

[1] BARNET *et al.* (Editors) (1992). *Nonparametric and Semiparametric Methods in Economics and Statistics.* Cambridge University Press.

[2] BENNET, G. (1962). Probability Inequalities for the Sum of Independent Random Variables. *J. Amer. Statist. Assoc.*, 57, 33-45.

[3] CRISTOBAL CRISTOBAL, J.A., FARALDO ROCA, P. & GONZÁLEZ MANTEIGA, W. (1987). A Class of Linear Regression Parameter Estimators Constructed by Nonparametric Estimation. *Ann. of Stat.*, **15**, 2 603-609.

[4] FARALDO ROCA, P. & GONZÁLEZ MANTEIGA, W. (1987). "On Efficiency of a New Class of Linear Regression Estimators Obtained by Preliminary Non-parametric Estimation". In *New Perspectives in Theoretical and Applied Statistics.* [M.L. Puri, J.P. Vilaplana and W. Wertz (Editors)]. John Wiley and Sons, New York.

[5] GONZÁLEZ MANTEIGA, W. (1982). *Obtención del Sesgo, Varianza y Error Medio Cuadrático de una Familia Axiomática de Estimadores No Paramétricos para Funciones de Densidad y de Regresión.* Ph. D. Dissertation. Universidad de Santiago.

[6] GONZÁLEZ MANTEIGA, W. (1988). Una Perspectiva General con Nuevos Resultados de la Aplicación de la Estimación No Paramétrica a la Estimación Paramétrica en Teoría de la Regresión. *Estadística Española*, **30**.

[7] GONZÁLEZ MANTEIGA, W. & VILAR FERNÁNDEZ, J.M. (1987). *A Class of Non-parametrically Constructed Parameter Estimators for a Stationary Autoregressive Model.* Paper presented at the 6th Pannonian Symposium on Mathematical Statistics held in Bad Tatzmannsdorf.

[8] NEWEY, W.K. (1991). *The Asymptotic Variance of Semiparametric Estimators.* Working paper M.I.T., Department of Economics.

[9] ROBINSON, P.M. 1988. Semiparametric Econometrics: A Survey. *The Journal of Applied econometrics.*, **3**, 35-51.

[10] ULLAH, A. (Editor) (1989). *Semiparametric and Nonparametric Econometrics.* Springer-Verlag. Berlin.

[11] WHIEE, H (1989). *Asymptotic Theory for Econometricians.* John Wiley and Sons, New York.

Recent Adv. in Stat. and Prob., pp. 105-126
J. Pérez Vilaplana and M. L. Puri (Eds)

Nonrecursive Procedures for Detecting Change in Simple Regression Models

M. HUŠKOVÁ
Katedra Pravdepodobnosti a Matematické Statistiky, Matematicko-Fyzikální Fakulta, University Karlovy, 18600 Praha 8, Czech Republic

Abstract – The paper concerns two classes of procedures connected with the problem known as "testing the constancy of the regression relationship over a time" or sometimes shortly "change–point–problem" in the simple regression model (1.1) below. Namely, we shall deal nonrecursive test procedures based on the least squares estimators and the M-estimators. Moreover, estimators of the change point of the model and the magnitude of the change of the parameters are introduced and their properties are derived.

1. INTRODUCTION

Let $X_1, \dots, X_n$ be independent observations obtained at time ordered points $t_1 < \dots < t_n$ and assume that they follow the model

$$X_i = \alpha + \beta c_i + I\{i > m\}\,(\delta_{n1} + \delta_{n2} c_i) + e_i, \qquad i = 1, \dots, n, \tag{1.1}$$

where α, β, δ_{n1}, δ_{n2} and m are unknown parameters, $c_1, \dots, c_n$ are known regression constants, $e_1, \dots, e_n$ are independent identically distributed (i. i. d.) random errors with common distribution F satisfying certain regularity conditions, unknown otherwise.
We shall be interested in test procedures for the following three testing problems:

$$H_0 : \delta_{n1} = \delta_{n2} = 0 \quad \text{against} \quad H_{n1} : \delta_{n1} \neq 0,\ \delta_{n2} = 0 \tag{1.2}$$
$$H_0 : \delta_{n1} = \delta_{n2} = 0 \quad \text{against} \quad H_{n2} : \delta_{n1} = 0,\ \delta_{n2} \neq 0 \tag{1.3}$$
$$H_0 : \delta_{n1} = \delta_{n2} = 0 \quad \text{against} \quad H_n : (\delta_{n1}, \delta_{n2})' \neq (0,0)'. \tag{1.4}$$

In the testing problem (1.4) one should decide whether the observations follow the same regression model (with the intercept α and the slope β) during the whole observational period or whether at some unknown time point $\tau \in (t_m, t_{m+1}\rangle$ at least one of the parameters α and β has changed. Testing problems (1.2) and (1.3) admit a change in only the intercept α or the slope β, respectively.
The closely related problems are to estimate the change point τ (or equivalently m), δ_{n1} and δ_{n2}.
These problems arise in many applications; in industry (the problem of stability of some critical product characteristics within statistical control), in meteorology (the problem to detect possible changes in climate or in weather), in economy (the problem of identification

of economic structural changes), in biology and medicine (detection of a change in behavior or state of a subject), etc.

There exists an extensive literature treating these problems from different point of view. A number of the survey papers can provide information on the existing results, e. g., Shaban (1980), Csörgő, Horváth (1988), Zacks (1983, 1991), Krishnaiah, Mio (1988), Hušková, Sen (1989). The well-known is the paper by Brown, Durbin and Evan (1975) concerning cumulative sum statistics developed for tests of change in the general linear regression model structure over time.

The present paper is related to the paper by Sen (1980), where test procedures for H_0 & H_{n2} based on ranks are developed and studied for the problem (1.3), to the paper by Jandhyala and Mac Neil (1991) dealing with the problem (1.2 – 1.3) from the Bayesian point of view, and to that of Kim and Siegmund (1989), concerning the maximum likelihood ratio test procedures for (1.2) and (1.3) under the assumption of normally distributed random errors.

In the present paper we shall concentrate on the nonrecursive procedures (which means that we use in our statistical study all n observations) related to the *least squares estimators* and to the *M-estimators*. In Section 2 and 3 the test procedures based on the least squares estimators are studied, in other words, the test procedures are developed and their limit behavior under the null hypothesis and some local alternatives is derived. Section 4 is devoted to the test procedures based on M-estimators. The estimators of the parameters m, α, β, δ_{n1}, δ_{n2} are introduced and studied in Section 5. The sketchs of the proofs are contained in Section 6.

It is known that the test procedures and estimators based on the least squares estimators are sensitive to the departure from the normal distribution. The M-estimators were introduced as a class of estimators behaving reasonably under quite general assumptions (for detail see e.g. Hampel et al. (1985)). Simulations made by Antoch (1991) pointed out the same is true for the problems considered in the present paper.

At last we remind the definition the M-estimators and the M-residuals. The *M-estimator* $(\alpha_n(\psi), \beta_n(\psi))$ *of* (α, β) generated by a score-function ψ based on $X_1, \dots, X_n$ following the model (1.1) with $\delta_{n1} = \delta_{n2} = 0$ are defined as a solution of the equations:

$$\sum_{i=1}^{n} \psi(X_i - \alpha - \beta c_i) = 0 \tag{1.5}$$

$$\sum_{i=1}^{n} c_i \psi(X_i - \alpha - \beta c_i) = 0. \tag{1.6}$$

The *M-residuals* are defined as

$$e_{ni}(\psi) = \psi(x_i - \alpha_n(\psi) - \beta_n(\psi) c_i), \qquad i = 1, \dots, n. \tag{1.7}$$

The most usual choices are:

$$\psi(x) = x \qquad x \in \mathbb{R}^1, \tag{1.8}$$

which leads to the usual *least squares estimators*, and

$$\psi(x) = \operatorname{sign}(x) \cdot \min(|x|, K), \qquad x \in \mathbb{R}^1, \tag{1.9}$$

which is so called Huber's ψ-function (typically one puts $K = 1,7$).

2. CLASSICAL PROCEDURES FOR H_0 AGAINST H_{N1} OR H_{N2}

At first, assume that the random errors $e_1, \ldots, e_n$ in the model (1.1) are i. i. d. random variables with normal distribution with parameters $(0, \sigma^2)$, $\sigma^2 > 0$ known. Then the maximum likelihood ratio method for testing problem H_0 & H_{1n}, formulated in (1.2), leads after long but straightforward computations (see Kim and Siegmund (1989)) to the test statistics

$$T_{n1}(\sigma) = \max_{1 \le k \le n-1} \left\{ \left| \sum_{i=1}^{k} \hat{e}_{ni} \right| \left(\mathrm{var}_{H_0} \left\{ \sum_{i=1}^{k} \hat{e}_{ni} \right\} \right)^{-1/2} \right\}, \tag{2.1}$$

where $\hat{e}_{ni}$ is the i-th residual:

$$\hat{e}_{ni} = X_i - \hat{\alpha}_n - \hat{\beta}_n c_i \qquad i = 1, \ldots, n \tag{2.2}$$

with $\hat{\alpha}_n$, $\hat{\beta}_n$ being the least squares estimators of α and β, respectively in the model (1.1) with $\delta_{ni} = 0$, $i = 1, 2$, i. e.

$$\hat{\alpha}_n = \overline{X}_n - \overline{c}_n \hat{\beta}_n \tag{2.3}$$

$$\hat{\beta}_n = \sum_{i=1}^{n} (c_i - \overline{c}_n) X_i \Big/ \sum_{j=1}^{n} (c_i - \overline{c}_n)^2 \tag{2.4}$$

$$\overline{X}_n = \frac{1}{n} \sum_{i=1}^{n} X_i, \qquad \overline{c}_n = \frac{1}{n} \sum_{i=1}^{n} c_i. \tag{2.5}$$

The same method for testing H_0 against H_{2n} (1.3) gives the test statistics

$$T_{n2}(\sigma) = \max_{1 \le k < n} \left\{ \left| \sum_{i=1}^{k} \hat{e}_{ni} c_i \right| \left(\mathrm{var}_{H_0} \left\{ \sum_{i=1}^{k} \hat{e}_{ni} c_i \right\} \right)^{-1/2} \right\}. \tag{2.6}$$

In this particular case (random errors normally distributed) $\left\{ \sum_{i=1}^{k} \hat{e}_{ni},\ k = 1, \ldots, n \right\}$ have under H_0 multivariate normal distribution with zero means and variance matrix with the elements

$$\mathrm{cov}_{H_0} \left(\sum_{i=1}^{k} \hat{e}_{ni}, \sum_{i=1}^{r} \hat{e}_{ni} \right) = \left(\frac{k(n-r)}{n} + \frac{\sum_{i=1}^{k} (c_i - \overline{c}_n) \sum_{j=r+1}^{n} (c_j - \overline{c}_n)}{\sum_{i=1}^{n} (c_j - \overline{c}_n)^2} \right) \sigma^2 \tag{2.7}$$

for $1 \le k \le r \le n-1$.

Similarly, $\left\{ \sum_{i=1}^{k} \hat{e}_{ni} c_i,\ k = 1, \ldots n-1 \right\}$ have under H_0 multidimensional normal distribution with zero means and the elements of the variance matrix:

$$\mathrm{cov}_{H_0} \left(\sum_{i=1}^{k} \hat{e}_{ni} c_j, \sum_{i=1}^{r} c_i \hat{e}_{ni} \right) = \left(\frac{1}{n} \sum_{i=r+1}^{n} c_i \sum_{j=1}^{k} c_j + \frac{\sum_{i=r+1}^{n} (c_i - \overline{c}_n) c_i \sum_{j=1}^{k} (c_j - \overline{c}_n) c_j}{\sum_{j=1}^{n} (c_j - \overline{c}_n)^2} \right) \sigma^2 \tag{2.8}$$

for $1 \le k \le r \le n-1$.

The exact distribution, even under H_0, of both test statistics is unknown. Hence at least reasonable approximation of them is needed in order to construct a critical region. The following theorem provides good approximation which is valid under quite mild conditions (the random errors need not be normally distributed):

Theorem 2.1. *Let $X_1,\dots,X_n$ follow the model* (1.1) *with $e_1,\dots,e_n$ being i.i.d. random variables with $\mathsf{E}\,e_1 = 0$, $\operatorname{var} e_1 = \sigma^2 > 0$, $\mathsf{E}|e_1|^{2+\eta} < +\infty$ for some $\eta > 0$. Let the regression constants $c_1,\dots,c_n$ fulfill:*

$$n^{-1}\sum_{i=1}^{n} c_i \to c_1^\star,\ (n-k)^{-1}\sum_{i=k+1}^{n} c_i \to c_1^\star \quad \text{as } n\to\infty,\ n-k\to\infty, \tag{2.9}$$

$$n^{-1}\sum_{i=1}^{n} c_i^2 \to c_2^\star \text{ as } n\to\infty, \qquad \limsup_{n\to\infty} n^{-1}\sum_{i=1}^{n}|c_i|^{2+\eta} < +\infty \tag{2.10}$$

for some $c_2^\star > c_1^{\star 2}$, $\eta > 0$. Then under H_0 ($\delta_{n1} = \delta_{n2} = 0$):

$$\lim_{n\to\infty} P_{H_0}\Big(T_{n1}(\sigma) \le b(y,\log n)\Big) = \exp\Big\{-2\exp\{-y\}\Big\} \quad y\in\mathbb{R}^1 \tag{2.11}$$

$$\lim_{n\to\infty} P_{H_0}\Big(T_{n2}(\sigma) \le b\Big(y,\log\Big(\sum_{i=1}^{n} c_i^2\Big)\Big)\Big) = \exp\Big\{-2\exp\{-y\}\Big\} \quad y\in\mathbb{R}^1 \tag{2.12}$$

where

$$b(y,\log n) = \Big(y + 2\log\log n + \frac{1}{2}\log\log\log n - \frac{1}{2}\log\pi\Big)\,(2\log\log n)^{-1/2}. \tag{2.13}$$

The assertion remains true if σ in (2.11) *and* (2.12) *is replaced by*

$$\hat\sigma_n^2 = \frac{1}{n-2}\sum_{i=1}^{n}\hat e_{ni}^2. \tag{2.14}$$

The *proof* is postponed to Section 5.
From the proof of this theorem, (see (6.1 – 6.10), (6.12 – 6.13) below), one can conclude that the limit distributions of $T_{n1}(\sigma)$ and $T_{n2}(\sigma)$ remain unchanged if $\operatorname{var}_{H_0}\left\{\sum_{i=1}^{k}\hat e_{ni}\right\}$ and $\operatorname{var}_{H_0}\left\{\sum_{i=1}^{k} c_i\hat e_{ni}\right\}$ are replaced by $\sigma^2 k(n-k)/n$ and by $\sigma^2\sum_{i=1}^{k} c_i^2\sum_{j=k+1}^{n} c_j^2/\sum_{v=1}^{n} c_v^2$, respectively.
Moreover, one observes that max of $\left|\sum_{i=1}^{k}\hat e_{ni}\right|\left(\operatorname{var}\left\{\sum_{i=1}^{k}\hat e_{ni}\right\}\right)^{-1/2}$ over $k = 1,\dots,n-1$ can be asymptotically reached only for $k\in(\log n,\ n/\log n)\cup(n - n/\log n,\ n-\log n)$ or in other words only the terms $\sum_{i=1}^{k}\hat e_{ni}\left(\operatorname{var}\left\{\sum_{i=1}^{k}\hat e_{ni}\right\}\right)^{-1/2}$ with k small or close to n have influence on the limit distribution of $T_{n1}(\sigma)$. Similar facts hold true for $T_{n2}(\sigma)$.
These arguments lead to a modification of the test statistics $T_{n1}(\sigma)$ and $T_{n2}(\sigma)$:

$$T_{n1}^\star(\sigma) = \max_{1\le k<n}\left\{\left|\sum_{i=1}^{k}\hat e_{ni}\right|\left(\sqrt{n}\,\sigma\right)^{-1}\right\} \tag{2.15}$$

$$T_{n2}^\star(\sigma) = \max_{1\le k<n}\left\{\left|\sum_{i=1}^{k} c_i\hat e_{ni}\right|\left(\sum_{j=1}^{n} c_j^2\sigma^2\right)^{-1/2}\right\}. \tag{2.16}$$

The assertion on their limit distribution is formulated in the following theorem.

Theorem 2.2. *Let $X_1,\dots,X_n$ follow model* (1.1) *with $e_1,\dots,e_n$ being i.i.d. random variables with $\mathsf{E}\,e_1 = 0$, $\operatorname{var} e_1 = \sigma^2 > 0$ and $\mathsf{E}\,|e_1|^{2+\eta} < +\infty$ for some $\eta > 0$. Let the* *regression constants $c_1, \dots, c_n$ fulfill:*

$$\lim_{n\to\infty} n^{-1}\sum_{i=1}^{[nt]} c_i^2 = t\, c_2^\star, \qquad \lim_{n\to\infty} n^{-1}\sum_{i=1}^{[nt]} c_i^2 = t\, c_1^\star, \qquad t \in \langle 0,1\rangle \tag{2.17}$$

$$\limsup_{n\to\infty} n^{-1}\sum_{i=1}^{n} c_i^4 < +\infty \tag{2.18}$$

for some $c_2^\star > c_1^{\star^2}$. Then under H_0

$$\lim_{n\to\infty} P_{H_0}\left(T_{n1}^\star(\sigma) \le x\right) = P\left(\max_{0<t<1} |B(t)| \le x\right) \qquad x \in \mathbb{R}^1 \tag{2.19}$$

$$\lim_{n\to\infty} P_{H_0}\left(T_{n2}^\star(\sigma) \le x\right) = P\left(\max_{0<t<1} |B(t)| \le x\right) \qquad x \in \mathbb{R}^1, \tag{2.20}$$

where $\{B(t),\ t \in (0,1)\}$ is the Brownian bridge. The assertion remains true if σ is replaced by $\sqrt{\sigma_n^2}$ with σ_n^2 given by (2.14).

The *proof* is based on Theorem 6.1 with $\psi(x) \equiv x,\ x \in \mathbb{R}^1$ and proceeds in the same line as that of (2.22) in Sen (1984), therefore is omitted. *Q. E. D.*

It follows, from the proof of this theorem, that the terms

$$\sum_{i=1}^{k} \hat{e}_{ni}\left(\operatorname{var}\left\{\sum_{j=1}^{k} \hat{e}_{ni}\right\}\right)^{-1/2}$$

for k small or close to n have no influence on the limit distribution.

Theorems 2.1 and 2.2 provide the critical regions for testing H_0 against H_{n1} with asymptotic level α°:

$$T_{n1}(\hat{\sigma}_n) > b\left(-\log\log(1-\alpha^\circ)^{-1/2}, \log n\right) \tag{2.21}$$

$$T_{n1}^\star(\hat{\sigma}_n) > b_{\alpha^\circ}, \tag{2.22}$$

where b_{α° is defined as

$$P\left(\sup_{t\in(0,1)} |B(t)| \ge b_{\alpha^\circ}\right) = \alpha^\circ. \tag{2.23}$$

If σ^2 is known then $\hat{\sigma}_n$ can be replaced by σ. As for the problem what critical region should be used, from the remarks below Theorems 2.1 and 2.2 we have: if we expect a change for k small or close to n then preferably (2.21) should be applied, otherwise (2.22) is recommended.

Similarly, one receives from Theorems 2.1 and 2.2 the following critical regions for H_0 & H_{n2} with asymptotic level α°:

$$T_{n2}(\hat{\sigma}_n) > b\left(-\log\log(1-\alpha^\circ)^{-1/2}, \log\left(\sum_{1}^{n} c_i^2\right)\right) \tag{2.24}$$

$$T_{n2}^\star(\hat{\sigma}_n) > b_{\alpha^\circ}, \tag{2.25}$$

where b_{α° is defined by (2.23).

Table of some critical values b_{α° and $b_{\alpha^\circ}(n) = b\left(-\log\log(1-\alpha^\circ)^{-1/2},\ \log n\right)$:

	$\alpha^\circ = 0,01$	$\alpha^\circ = 0,05$	$\alpha^\circ = 0,1$
b_{α°	1,28	1,16	1,11
$b_{\alpha^\circ}(50)$	4,61	3,62	3,19
$b_{\alpha^\circ}(100)$	4,56	3,64	3,23
$b_{\alpha^\circ}(1000)$	4,54	3,71	3,35

Now, we shall turn our attention to the behavior of $T_{ni}(\sigma)$ and $T^{\star}_{ni}(\sigma)$, $i = 1,2$, under local alternatives. Let us assume that for $m \to \infty$ as $n \to \infty$ and that

$$\frac{m}{n} \longrightarrow \theta \in (0,1) \qquad \delta_{ni} \longrightarrow 0,\ i = 1,2 \quad \text{as } n \to \infty. \tag{2.26}$$

Remark. To be precise we should consider a triangular array: $X_{n1}, \ldots, X_{nm_n}$, $X_{nm_n+1}, \ldots, X_{nn}$ instead of $X_1, \ldots, X_m$, $X_{m+1}, \ldots, X_n$. However, we shall continue in writing $X_1, \ldots, X_m$, $X_{m+1}, \ldots, X_n$ for short.

Then one has that under $H_{n1}(\delta_{n1})$

$$\begin{aligned} \mathsf{E}_{H_{n1}}\left(\sum_{i=1}^{k} \hat{e}_{ni}\right) &= \\ &= \delta_{n1}\left\{\sum_{i=1}^{k} I\{i>m\} - \frac{k(n-m)}{n} - \sum_{i=1}^{k}(c_i-\overline{c}_n) \sum_{j=m+1}^{n}(c_j-\overline{c}_n)\Big/\sum_{v=1}^{n}(c_v-\overline{c}_n)^2\right\} = \\ &= -\delta_{n1}\Big\{h_{n,m}(k)\Big\}\Big(1+o(1)\Big) \end{aligned} \tag{2.27}$$

$$\text{as } n \to \infty,\ k = 1, \ldots, n-1\,.$$

where

$$h_{n,m}(k) = \frac{k(n-m)}{n} I\{i \le m\} + \frac{m(n-k)}{n} I\{i >\}. \tag{2.28}$$

Similarly, one observes that under H_{n2} (δ_{n2})

$$\begin{aligned} \mathsf{E}_{H_{n2}}\left(\sum_{i=1}^{k} \hat{e}_{ni} c_i\right) &= \\ &= \delta_{n2}\left\{\sum_{i=1}^{k} c_i^2 I\{i>m\} - n^{-1}\sum_{i=1}^{k} c_i \sum_{j=m+1}^{n} c_j - \right. \\ &\quad \left. - \sum_{i=1}^{k} c_i(c_i-\overline{c}_n) \sum_{j=m+1}^{n} c_j(c_j-\overline{c}_n) \Big/ \sum_{v=1}^{n}(c_v-\overline{c}_n)^2\right\} = \\ &= -\delta_{n2}\left\{h_{n,m}(k)\cdot c_2^{\star}\right\}(1+o(1)) \end{aligned} \tag{2.29}$$

$$\text{as } n \to \infty,\ n-k \to \infty$$

Asymptotic maximum of the absolute value in (2.27) and (2.29) is reached for $k = m$. Next, we shall consider contiguous alternatives, i.e. we shall assume that

$$\delta_{ni}\sqrt{n} \longrightarrow \delta_i \qquad \text{as} \quad n \to \infty,\ |\delta_i| < +\infty,\ i = 1,2. \tag{2.30}$$

Theorem 2.3. *Let the assumptions of Theorem 2.2 be satisfied and let (2.26) be fulfilled.*

a) *If (2.30) is fulfilled for $i = 1$, then under H_{n1}*

$$\lim_{n\to\infty} P_{H_{n1}}\left(T^{\star}_{n1}(\hat{\sigma}_n) \le x\right) = P\left(\sup_{0<t<1}\{|B(t) - \mu_1(t)/\sigma|\} \le x\right) \quad x \in \mathbb{R}^1, \tag{2.31}$$

where $\{B(t),\ t \in (0,1)\}$ is the Brownian bridge and

$$\begin{aligned} \mu_1(t) &= -t(1-\theta)\,\delta_1 \qquad 0 < t \le \theta \\ &= -(1-t)\,\theta\,\delta_1 \qquad \theta < t < 1. \end{aligned} \tag{2.32}$$

b) *If (2.30) is fulfilled for $i = 2$, then under H_{n2}*

$$\lim_{n\to\infty} P_{H_{n2}}\left(T^{\star}_{n2}(\hat{\sigma}_n) \le x\right) = P\left(\sup_{0<t<1}\{|B(t) - \mu_2(t)/\sigma\} \le x\right) \quad x \in \mathbb{R}^1, \tag{2.33}$$

where

$$\begin{aligned} \mu_2(t) &= -t(1-\theta)\,c_2^{\star}\delta_2 \qquad 0 < t \le \theta \\ &= -(1-t)\,\theta\,c_2^{\star}\delta_2 \qquad \theta < t < 1. \end{aligned} \tag{2.34}$$

The *proof* is based on Theorem 6.1 with $\psi(x) \equiv x,\ x \in \mathbb{R}^1$ and proceeds in the same way as of (2.28) in Sen (1984) therefore is omitted. *Q. E. D.*

Under local (but noncontiguous) alternatives satisfying:

$$\delta_{ni}\sqrt{n}(2\log\log n)^{-1/2} \longrightarrow \delta_i^{\star} \qquad \text{as } n \to \infty,\ |\delta_i^{\star}| < +\infty,\ i = 1,2 \tag{2.35}$$

one can state on the limit distribution of $T_{n1}(\hat{\sigma}_n)$ and $T_{n2}(\hat{\sigma}_n)$:

Theorem 2.4. *Let assumptions of Theorem 2.1 be satisfied and let (2.26) be fulfilled.*

a) *If (2.35) is fulfilled for $i = 1$ with $\delta_1^{\star}(\theta(1-\theta))^{1/2}/\sigma < 1$, then under H_{n1}*

$$\lim_{n\to\infty} P_{H_{n1}}\left(T_{n1}(\hat{\sigma}_n) \le b(y, \log n)\right) = \exp\{-2\exp\{-y\}\} \quad y \in \mathbb{R}^1. \tag{2.36}$$

b) *If (2.35) is fulfilled for $i = 1$ with $\delta_1^{\star}(\theta(1-\theta))^{1/2}/\sigma > 1$, then under H_{n1}*

$$\lim_{n\to\infty} T_{n1}(\hat{\sigma}_n)\,(2\log\log n)^{-1/2} = \delta_1^{\star}(\theta(1-\theta))^{1/2}/\sigma \quad \text{in probability.} \tag{2.37}$$

c) *If (2.35) if fulfilled for $i = 2$ with $\delta_2^{\star}\,c_2^{\star}(\theta(1-\theta))^{1/2}/\sigma < 1$, then under H_{n2}*

$$\lim_{n\to\infty} P_{H_{n2}}\left(T_{n2}(\hat{\sigma}_n) \le b(y, \log(\sum_{i=1}^{n} c_i^2))\right) = \exp\{-2\exp\{-y\}\} \quad y \in \mathbb{R}^1. \tag{2.38}$$

d) *If (2.35) is fulfilled for $i = 2$ with $\delta_2^{\star}\,c_2^{\star}(\theta(1-\theta))^{1/2}/\sigma > 1$ then under H_{n2}*

$$\lim_{n\to\infty} T_{n2}(\hat{\sigma}_n)\,(2\log\log n)^{-1/2} = \delta_2^{\star}\,c_2^{\star}(\theta(1-\theta))^{1/2}/\sigma \quad \text{in probability.} \tag{2.39}$$

The *proof* is postponed to Section 6.

From these two theorems one can conclude that the limit distribution of $T_{n1}(\hat{\sigma}_n)$ and $T_{n2}(\hat{\sigma}_n)$ under the contiguous alternatives (2.30) and the null hypothesis H_0 coincides, however, this is not the case of $T^\star_{n1}(\hat{\sigma}_n)$ and $T^\star_{n2}(\hat{\sigma}_n)$. Under the alternatives (2.35) with $\delta_i^\star \neq 0$

$$T^\star_{ni}(\hat{\sigma}_n) \longrightarrow \infty \qquad \text{in probability as } n \to \infty,\ i = 1, 2 \tag{2.40}$$

In other words the tests based on $T^\star_{n1}(\hat{\sigma}_n)$ and $T^\star_{n2}(\hat{\sigma}_n)$ are more sensitive to the local alternatives than $T_{n1}(\hat{\sigma}_n)$ and $T_{n2}(\hat{\sigma}_n)$.

Conclusion following from the above considerations is that the test statistics $T^\star_{n1}(\hat{\sigma}_n)$ and $T^\star_{n2}(\hat{\sigma}_n)$ are preferable to $T_{n1}(\hat{\sigma}_n)$ and $T_{n2}(\hat{\sigma}_n)$ unless possible change is expected for k small or close to n.

3. CLASSICAL PROCEDURES FOR H_0 AGAINST H_N

In this section we focus on the procedures related to least squares estimators for testing H_0 & H_n introduced in (1.4). We will proceed similarly as in the previous section.

If the random errors $e_1, \ldots, e_n$ are i.i.d. random variables with normal distributon $N(0, \sigma^2)$, $\sigma^2 > 0$ (known), then the maximum likelihood ratio method for H_0 & H_n (1.4) leads to the test statistics:

$$T_n(\sigma) = \tag{3.1}$$

$$= \max_{1\le k\le n} \left\{ \left(\sum_{i=1}^{k} (c_i - \overline{c}_k)\, \hat{e}_{ni} \right)^2 \Big/ \sum_{j=1}^{k} (c_i - \overline{c}_k)^2 + \left(\sum_{i=k+1}^{n} (c_i - \overline{c}^\star_{n-k})\, \hat{e}_{ni} \right)^2 \Big/ \right.$$

$$\left. \Big/ \sum_{j=k+1}^{n} (c_j - \overline{c}^\star_{n-k})^2 + \left(\sum_{i=1}^{k} \hat{e}_{ni} \right)^2 \Big/ k + \left(\sum_{i=k+1}^{n} \hat{e}_{ni} \right)^2 \Big/ (n-k) \right\} \Big/ \sigma^2$$

where $\hat{e}_{ni}$ and $\overline{c}_k$ are defined by (2.2) and (2.5), resp. and

$$\overline{c}^\star_{n-k} = \sum_{i=k+1}^{n} c_i \,/\, (n-k) \qquad k = 1, \ldots, n-1. \tag{3.2}$$

Using analogous arguments as in the previous section one may propose as a test statistic for testing H_0 against H_n the following modification of $T_n(\sigma)$:

$$T^\star_n(\sigma) = \max_{1\le k\le n} \left\{ \left(\sum_{i=1}^{k} \hat{e}_{ni} \right)^2 \Big/ n + \left(\sum_{i=1}^{k} (c_i - \overline{c}_n)\, \hat{e}_{ni} \right)^2 \Big/ \sum_{j=1}^{n} (c_j - \overline{c}_n)^2 \right\} \Big/ \sigma^2. \tag{3.3}$$

Proceeding similarly as in the case of the alternatives H_{n1} and H_{n2} we observe that the test statistic $T^\star_n(\sigma)$ is more sensitive to the local alternatives than $T_n(\sigma)$. The approximation of the limit distribution of $T_n(\sigma)$ under H_n was derived only of $e_1, \ldots, e_n$ normally distributed (see Kim and Siegmund (1989) however, they consider different assumptions on regression constants). It is expected that the same approximation is reasonable even for random errors not necessarily normally distributed.

The assertions on the limit distribution of $T^\star_n(\sigma)$ can be stated in the following form:

Theorem 3.1. *Let assumptions of Theorem 2.2 be satisfied. Moreover, let (2.26) and (2.30) be true. Then under H_n*

$$\lim_{n\to\infty} P_{H_n}(T_n^\star(\sigma) \le x) =$$
$$= P\left(\sup_{0<t<1}\left\{(B_1(t) - \mu_1^\star(t)/\sigma)^2 + (B_2(t) - \mu_2^\star(t)/\sigma)^2\right\} \le x\right) \tag{3.4}$$

$x \in \mathbf{R}^1$, where $\{B_i(t),\ t \in (0,1)\}$, $i = 1,2$ are independent Brownian bridges and

$$\begin{aligned} \mu_1^\star(t) &= -t(1-\theta)\,(\delta_1^\star + \delta_2^\star c_1^\star) && 0 < t \le \theta \\ &= -(1-t)\,\theta\,(\delta_1^\star + \delta_2^\star c_1^\star) && \theta < t < 1 \end{aligned} \tag{3.5}$$

$$\begin{aligned} \mu_2^\star(t) &= -t(1-\theta)\,\delta_2^\star\left(c_2^\star - c_1^{\star^2}\right) && 0 < t \le \theta \\ &= -(1-t)\,\theta\,\delta_2^\star\left(c_2^\star - c_1^{\star^2}\right) && \theta < t < 1. \end{aligned} \tag{3.6}$$

The assertion (3.4) remains true if σ on the l. h. s. is replaced by $\sqrt{\hat\sigma_n^2}$, where $\hat\sigma_n^2$ is defined by (2.12).

The *proof* follows from Theorem 6.1 and proceeds in the same way as that of (2.28) Sen (1984), hence it is omitted. *Q. E. D.*

Notice that the assertions of Theorem 3.1 hold true for random errors not necessarily normally distributed. This theorem provides also the critical region for testing H_0 against H_n with asymptotic level α°:

$$T_n^\star(\hat\sigma_n) > v_{\alpha^\circ}, \tag{3.7}$$

where v_{α° is defined as:

$$P\left(\max_{0<t<1}\left\{B_1^2(t) + B_2^2(t)\right\} \ge v_{\alpha^\circ}\right) = \alpha^\circ. \tag{3.8}$$

If $\delta_{n1}\sqrt{n} \longrightarrow \infty$ or $\delta_{n2}\sqrt{n} \longrightarrow \infty$ then

$$T_n^\star(\hat\sigma_n) \longrightarrow \infty \qquad \text{in probability as } n \to \infty. \tag{3.9}$$

Now, we shall have a look at the expectations under H_n. Direct computations give for $k \to \infty$, $n - k \to \infty$

$$E_{H_n}\sum_{i=1}^{k} \hat e_{ni} = -h_{n,m}(k)\,(\delta_{n1} + \delta_{n2}c_1^\star)\,\big(1 + o(1)\big), \tag{3.10}$$

$$E_{H_n}\sum_{i=1}^{k} \hat e_{ni}c_i = -h_{n,m}(k)\,(\delta_{n1}c_1^\star + \delta_{n2}c_2^\star)\,\big(1 + o(1)\big), \tag{3.11}$$

where $h_{n,m}(k)$ is defined by (2.28). Since $k(n-m)\,I\{k \le m\} + m(n-k)\,I\{k > m\}$ is increasing for $k \le m$ and decreasing for $k \ge m$, the asymptotic maximum of the absolute values of the expectations in (3.10) and (3.11) is reach for $m = k$. In other words, the expectations will have highest influence for k close to m.

4. ROBUST PROCEDURES FOR H_0 AGAINST H_N (H_{N1}, H_{N2})

Here we shall deal with test procedures for testing problems H_0 & H_n, H_0 & H_{n1}, H & H_{n2} based on the M-residuals and the M-estimators introduced in (1.7) and (1.5 – 1.6), resp. Assume that the score-function ψ fulfills:

(4.1) ψ is nondecreasing, $\int|\psi(x)|^{2+\eta}dF(x)$ for some $\eta>0$, there exist $D_1>0$, $D_2>0$ such that

$$\int\Big(\psi(x-a)-\psi(x-b)\Big)^2 dF(x)\le D_1|a-b| \qquad |a|\le D_2,\ |b|\le D_2,$$

where F is the distribution function of e_1.

(4.2) The function $\lambda(t)=-\int\psi(x-t)dF(x)$, has the properties: $\lambda(0)=0$, $\lambda'(0)>0$, there exist $D_3>0$, $D_4>0$ and $\delta>0$ such that

$$\Big|\lambda'(t)-\lambda'(a)\Big|\le D_3|b-a|^\delta \qquad |b|\le D_4,\ |a|\le D_4.$$

Notice that if the $\mathsf{E}|e_1|^{2+\eta}<+\infty$ and $\mathsf{E}\,e_1=0$ than $\psi(x)=x$, $x\in\mathbb{R}^1$, fulfills (4.1) and (4.2). If the distribution function F is continuous at $\pm K$, then Huber's ψ-function defined in (1.8) satisfies (4.1) – (4.2). Next, we introduce M-modification of the test statistics studied in Sections 2 and 3. We define

$$V_{n1}(\psi)=\max_{1\le k<n}\left\{\sqrt{\frac{n}{k(n-k)}}\left|\sum_{i=1}^k e_{ni}(\psi)\right|\hat\sigma_n^{-1}(\psi)\right\}, \tag{4.3}$$

$$V_{n2}(\psi)=\max_{1\le k<n}\left\{\left(\sum_{v=1}^n c_v^2\Big/\Big(\sum_{s=1}^k c_s^2\sum_{j=k+1}^n c_j^2\Big)\right)^{1/2}\left|\sum_{i=1}^k c_i e_{ni}(\psi)\right|\hat\sigma_n^{-1}(\psi)\right\} \tag{4.4}$$

$$V_{n1}^\star(\psi)=\max_{1\le k<n}\left\{n^{-1/2}\left|\sum_{i=1}^k e_{ni}(\psi)\right|\hat\sigma_n^{-1}(\psi)\right\}, \tag{4.5}$$

$$V_{n2}^\star(\psi)=\max_{1\le k<n}\left\{\left(\sum_{j=1}^n c_j^2\right)^{-1/2}\left|\sum_{i=1}^k c_i e_{ni}(\psi)\right|\hat\sigma_n^{-1}(\psi)\right\}, \tag{4.6}$$

$$V_n^\star(\psi)=\max_{1\le k<n}\left\{\left(\left(\sum_{i=1}^k e_{ni}(\psi)\right)^2/n+\left(\sum_{i=1}^k(c_i-\bar c_n)\,e_{ni}(\psi)\right)^2\Big/\sum_{j=1}^n(c_j-\bar c_n)^2\right)\cdot\hat\sigma_n^{-1}(\psi)\right\}, \tag{4.7}$$

where

$$e_{ni}(\psi)=\psi\left(X_i-\alpha_n(\psi)-\beta_n(\psi)\,c_i\right) \qquad i=1,\dots,n, \tag{4.8}$$

$$\hat\sigma_n^2(\psi)=\frac1n\sum_{i=1}^n\psi^2\left(X_i-\alpha_n(\psi)-\beta_n(\psi)\,c_i\right) \tag{4.9}$$

with $(\alpha_n(\psi),\beta_n(\psi))$ defined by (1.5 – 1.6). The limit behavior of the statistics in (4.3 – 4.7) under H_0 coincides with that of their counterparts in Section 2 and 3. As for the behavior under alternatives there the drift functions are different μ's. More precisely, we can assert:

Theorem 4.1. *Let $X_1,\dots,X_n$ follow the model* (1.1) *with $e_1,\dots,e_n$ being i.i.d. random variables and let* (4.1), (4.2), (2.9), (2.10), (2.26) *and* (2.35) *be satisfied.*

a) *If $\lambda'(0),(\theta(1-\theta))^{1/2}\,|\delta_1^\star+c_1^\star\delta_2^\star|\,/\,\sigma(\psi)<1$ then under H_n*

$$\lim_{n\to\infty}P_{H_n}\Big(V_{n1}(\psi)\le b(y,\log n)\Big)=\exp\Big\{2\exp\{-y\}\Big\} \quad y\in\mathbb{R}^1, \tag{4.10}$$

where $b(y,\log n)$ is defined by (2.13) *and*

$$\sigma^2(\psi)=\int\psi^2(x)\,dF(x). \tag{4.11}$$

b) *If* $\lambda'(0)\,(\theta(1-\theta))^{1/2}\,|\delta_1^\star + c_1^\star\,\delta_2^\star|\,/\,\sigma(\psi) > 1$ *then under* H_n

$$\lim_{n\to\infty} V_{n1}(\psi)\,(2\log\log n)^{-1/2} = \lambda'(0)\big(\theta(1-\theta)\big)^{1/2}|\delta_1^\star + c_1^\star\,\delta_2^\star|\,/\,\sigma(\psi) \tag{4.12}$$

in probability.

c) *If* $\lambda'(0)\,(\theta(1-\theta))^{1/2}\,|\delta_1^\star c_1^\star + c_2^\star\,\delta_2^\star|\,/\,\sigma(\psi) < 1$ *then under* H_n

$$\lim_{n\to\infty} P_{H_n}\left(V_{n2}(\psi) \le b\left(y, \log(\sum_{i=1}^{n} c_i^2)\right)\right) = \exp\left\{-2\exp\{-y\}\right\} \quad y \in \mathbb{R}^1, \tag{4.13}$$

d) *If* $\lambda'(0)\,(\theta(1-\theta))^{1/2}\,|\delta_1^\star c_1^\star + c_2^\star\delta_2^\star|\,/\,\sigma(\psi) > 1$ *then under* H_n

$$\lim_{n\to\infty} V_{n2}(\psi)\,(2\log\log n)^{-1/2} = \lambda'(0)\big(\theta(1-\theta)\big)^{1/2}|\delta_1^\star\, c_1^\star + c_2^\star\,\delta_2^\star|/\sigma(\psi) \tag{4.14}$$

in probability.

The *proof* is postponed to Section 6.

Theorem 4.2. *Let* $X_1,\dots,X_n$ *follow the model* (1.1) *with* $e_1,\dots,e_n$ *being i. i. d. random variables and let* (4.1), (4.2), (2.26), (2.30), (2.17) *and* (2.18) *be satisfied. Then*

$$\lim_{n\to\infty} P\,(V_{n1}^\star(\psi) \le x) = P\left(\sup_{t\in(0,1)} \left\{|B_1(t) - \lambda'(0)\,\mu_1^\star(t)/\sigma(\psi)|\right\} \le x\right), \tag{4.15}$$

$$\lim_{n\to\infty} P\,(V_{n2}^\star(\psi) \le x) = P\left(\sup_{t\in(0,1)} \left\{|B_2(t) - \lambda'(0)\,\mu_2(t)/\sigma(\psi)|\right\} \le x\right), \tag{4.16}$$

$$\lim_{n\to\infty} P\,(V_n^\star(\psi) \le x) = P\left(\sup_{t\in(0,1)} \left\{(B_1(t) - \lambda'(0)\,\mu_1^\star(t)/\sigma(\psi))^2 + \right.\right. \tag{4.17}$$
$$\left.\left. + (B_2(t) - \lambda'(0)\,\mu_2^\star(t)/\sigma(\psi))^2\right\} \le x\right),$$

$x \in \mathbf{R}^1$, *where* $\{B_i(t),\ t\in(0,1)\}$, $i = 1,2$ *are independent Brownian bridges,* $\mu_1^\star(t)$, $\mu_2^\star(t)$, $\mu(t)$ *are defined by* (3.5), (3.6) *and* (2.33), *respectively and* $\sigma^2(\psi)$ *is defined by* (4.11).

PROOF. It is a modification of the proof of relation (2.28) in Sen (1984) therefore the proof is omitted. We only point out the crucial relations in the proofs are (6.21), (6.29) and (6.30). *Q. E. D.*

Theorem 4.1 and 4.2 provide the critical regions for the considered testing problem. Namely, for testing H_0 & H_{n1}:

$$V_{n1}(\psi) > b\left(-\log\log(1-\alpha^\circ)^{-1/2},\ \log n\right) \tag{4.18}$$

or

$$V_{n1}^\star(\psi) > b_{\alpha^\circ}, \tag{4.19}$$

where α° is asymptotic level, $b(\cdot,\cdot)$ is defined by (2.13) and b_{α° by (2.23). Similarly for testing H_0 & H_{n2}:

$$V_{n2}(\psi) > b\left(-\log\log(1-\alpha^\circ)^{-1/2},\ \log(\sum_{i=1}^{n} c_i^2)\right), \tag{4.20}$$

$$V_{n2}^\star(\psi) > b_{\alpha^\circ}. \tag{4.21}$$

Finally, for testing H_0 & H_n we have the critical region (with asymptotic level α°):

$$V_n^\star(\psi) > v_{\alpha^\circ}, \tag{4.22}$$

where v_{α° is defined by (3.8). Notice that the critical points do not depend on the choice of ψ.

Theorem 4.1 says also that the limit distribution of $V_{ni}(\psi)$, $i = 1,2$, is the same both under H_0 and contiguous alternatives.

Next, we should remark that under the assumptions of Theorem 4.1 and 4.2 (see (6.29) and (6.30)) below that as $n \to \infty$

$$\begin{aligned} n(\alpha_n(\psi) - \alpha) &= \\ &= \frac{1}{\lambda'(0)} \sum_{i=1}^{n} \psi(e_i) - \bar{c}_n\big(\beta_n(\psi) - \beta\big) + \sum_{i=m+1}^{n} (\delta_{n1} + c_i\,\delta_{n2}) + O_P\left(n^{1/2-v}\right) \end{aligned} \tag{4.23}$$

$$\begin{aligned} \sum_{i=1}^{n} (c_i - \bar{c}_n)^2 (\beta_n(\psi) - \beta) &= \\ &= \frac{1}{\lambda'(0)} \sum_{i=1}^{n} (c_i - \bar{c}_n)\,\psi(e_i) + \sum_{i=m+1}^{n} (\delta_{n1} + \delta_{n2}\,c_i)\,(c_i - \bar{c}_n) + O_P\left(n^{1/2-v}\right) \end{aligned} \tag{4.24}$$

for some $v > 0$, which further implies (for the proof see considerations below (6.30)) that under H_n with δ_{n1}, δ_{n2} fulfilling (2.26) and (2.35) as $k \to \infty$, $n - k \to \infty$

$$\begin{aligned} &\left(\sum_{v=1}^{n} d_v^2 \Big/ \left(\sum_{q=1}^{k} d_q^2 \sum_{j=k+1}^{n} d_j^2\right)\right)^{1/2} \sum_{i=1}^{k} d_i e_{ni}(\psi) = \\ &= \left(\sum_{v=1}^{n} d_v^2 \Big/ \left(\sum_{q=1}^{k} d_q^2 \sum_{j=k+1}^{n} d_j\right)\right)^{-1} \left\{\sum_{i=1}^{k} d_i \left(\psi(e_i) - \frac{1}{n}\sum_{s=1}^{n}\psi(e_s) - \right.\right. \\ &\left.\left. - (c_i - \bar{c}_n) \sum_{r=1}^{n} (c_r - \bar{c}_n)\,\psi(e_r) \Big/ \sum_{\ell=1}^{n} (c_\ell - \bar{c}_n)^2\right)\right\} \big(1 + o_p(1)\big) - \\ &- \lambda'(0)\sqrt{\log\log n}\left(I\{k \le m\}\,(n-m)\sqrt{\frac{k}{n-k}} + I\{k > m\}\,m\sqrt{\frac{n-k}{k}}\right) \cdot \\ &\cdot (\delta_1^\star d_1^\star + \delta_2^\star d_2^\star)\big(1 + o(1)\big), \end{aligned} \tag{4.25}$$

where either $d_i = c_i$, $i = 1, \ldots, n$ and $d_1^\star = c_1^\star$, $d_2^\star = c_2^\star$ or $d_i = 1$, $i = 1, \ldots, n$ and $d_1^\star = 1$, $d_2^\star = c_1^\star$. Clearly, in (4.25) the nonrandom terms dominate for $k \in (n \log n,\ n - n/\log n)$ while the random term d_0 for $k \in (\log n,\ n/\log n) \cup (n - n/\log n,\ n - \log n)$.

5. ESTIMATORS OF M, δ_{N1} AND δ_{N2}

Motivated by the behavior of $V_{ni}(\psi)$, $V_{ni}^\star(\psi)$, $V_n^\star(\psi)$ under the alternative hypothesis H_n (or H_{n1}, H_{n2}), formulated in Theorems 4.1 and 4.2 we introduce the estimators of m as follows:

$$m_1(\psi) = \operatorname{argmax}\left\{\left|\sum_{i=1}^{k} e_{ni}(\psi)\right|,\quad k = 1, \ldots, n\right\}, \tag{5.1}$$

$$m_2(\psi) = \operatorname{argmax}\left\{\left|\sum_{i=1}^{k} c_i\, e_{ni}(\psi)\right|, \quad k=1,\dots,n\right\}, \tag{5.2}$$

$$\hat{m}(\psi) = \operatorname{argmax}\left\{\left(\sum_{i=1}^{k} e_{ni}(\psi)\right)^2 \Big/ n + \right. \tag{5.3}$$

$$\left. + \left(\sum_{i=1}^{k}(c_i - \bar{c}_n)\, e_{ni}(\psi)\right)^2 \Big/ \sum_{j=1}^{n}(c_j - \bar{c}_n)^2\right), \quad k=1,\dots,n\right\}$$

where $e_{ni}(\psi)$ is defined by (4.8). The estimators $m_1(\psi)$ of m should be used if we expect a change only in the percept or in the slope, respectively. The estimator $m(\psi)$ is convenient for the situation when a change occurs either in both α and β or only in one of them.

Theorem 5.1. *Let $X_1,\dots,X_n$ follows the model (1.1) with $e_1,\dots,e_n$ being i. i. d. random variables and let assumptions* (4.1), (4.2) *and* (2.26) *be satisfied and let, moreover,*

$$|\delta_{ni}|\,\sqrt{n} \longrightarrow \infty \qquad \text{as } n \to \infty \tag{5.4}$$

$$r^{-1}\sum_{i=k+1}^{k+r} c_i^j \longrightarrow c_j^* \qquad \text{as } r \to \infty,\ k \to \infty,\ j=1,2. \tag{5.5}$$

Then under H_n

$$\left(m_i(\psi) - m\right)\left(|\delta_{n1}| + |\delta_{n2}|\right)^2 = O_P(1) \qquad \text{as } n \to \infty,\ i=1,2 \tag{5.6}$$

$$\left(m_(\psi) - m\right)\left(|\delta_{n1}| + |\delta_{n2}|\right)^2 = O_P(1) \qquad \text{as } n \to \infty. \tag{5.7}$$

The *proof* is in Section 6.
From (6.29 – 6.32) below we can deduce that under the assumptions of Theorem 5.1 as $k \to \infty$, $n-k \to \infty$

$$\sum_{i=1}^{k} c_i\, e_{ni}(\psi) = -h_{n,m}(k)\,(c_1^*\,\delta_{n1} + c_2^*\,\delta_{n2})\left(1+o(1)\right) \tag{5.8}$$

$$\sum_{i=1}^{k} e_{ni}(\psi) = -h_{n,m}(k)\,(c_1^*\,\delta_{n2})\left(1+o(1)\right) \tag{5.9}$$

where $h_{n,m}(k)$ is defined by (2.28). Notice that under (5.4) the terms with the asymptotic influence are nonrandom and the maximum of $\left|\sum_{i=1}^{k} e_{ni}(\psi)\, c_i\right|$ and $\left|\sum_{i=1}^{k} e_{ni}(\psi)\right|$ should be reached for k close to m.
Also, one can see that, $\hat{m}(\psi)$ can be used as an estimator of m even if we expect a change only either in the or intercept in the slope. Under additional mild assumptions $m_1(\psi)$ can be used as an estimator of a change under H_{n2} and vice versa.
Assertions (5.6) and (5.7) say that $m_1(\psi)/n$, $m_2(\psi)/n$ and $m(\psi)/n$ are consistent estimators of m/n. The order of consistency is related to the magnitude of δ_{n_1} and δ_{n2}. Particularly, under (2.35) we have

$$\left(\hat{m}(\psi) - m\right)\Big/ n = O_P\left((\log\log n)^{-1}\right) \qquad \text{as } n \to \infty \tag{5.10}$$

and if

$$|\delta_{ni}|\; n^{a} \longrightarrow \delta_i(a) \qquad \text{as } n \to \infty \tag{5.11}$$

for some $a \in (0,1/2)$, $\delta_1(a)$, $\delta_2(a)$ finite, we have

$$\left(\hat{m}(\psi) - m\right)/n = O_P\left(n^{2a-1}\right) \qquad \text{as } n \to \infty. \tag{5.12}$$

Deeper results will be published in a paper by Veraverbeke and the author (1992).

Having an estimator of m we may propose estimators of δ_{n1} and δ_{n2}. Naturally, one may estimate the parameters α, β as M-estimators based on $\hat{m}(\psi)$. Towards this we introduce M-estimators $(\alpha_m(\psi), \beta_m(\psi))$ of parameters (α, β) as a solution of the equations

$$\sum_{i=1}^{\hat{m}} \psi(X_i - \alpha - \beta c_i) = 0 \tag{5.13}$$

$$\sum_{i=1}^{\hat{m}} c_i \psi(X_i - \alpha - \beta c_i) = 0 \tag{5.14}$$

and, similarly, we introduce M-estimators $\left(\alpha^\star_{n-\hat{m}}(\psi), \beta^\star_{n-\hat{m}}(\psi)\right)$ as a solution of the equations

$$\sum_{i=\hat{m}+1}^{n} \psi(X_i - \alpha - \beta c_i) = 0 \tag{5.15}$$

$$\sum_{i=\hat{m}+1}^{n} c_i \psi(X_i - \alpha - \beta c_i) = 0. \tag{5.16}$$

Then δ_{n1} and δ_{n2} can be estimated by

$$\delta_{n1}(\psi) = \alpha^\star_{n-\hat{m}}(\psi) - \alpha_{\hat{m}}(\psi) \tag{5.17}$$

$$\delta_{n2}(\psi) = \beta^\star_{n-\hat{m}}(\psi) - \beta_{\hat{m}}(\psi). \tag{5.18}$$

These estimators are under the assumptions of Theorem 5.1 consistent. Deeper results will be studied in the paper by Veraverbeke and the author (1992).

6. PROOFS

In this section assertions from previous part of the paper will be proved.

PROOF OF THEOREM 2.1. Since the proof has some features common with the proof of Theorem 2.3 in Csörgő and Horváth (1988 b) we shall be brief. Decompose the set $\{1, \dots, n-1\}$ into six subsets $A_{n1}, A_{n2}, \dots, A_{n6}$, where

$$\begin{aligned}
A_{n1} &= \{1, \dots, [\log n]\} \\
A_{n2} &= \{[\log n] + 1, \dots, [n/\log n]\} \\
A_{n3} &= \{[n/\log n] + 1, \dots, [n/2]\} \\
A_{n4} &= \{[n/2] + 1, \dots, n - [n/\log n]\} \\
A_{n5} &= \{n - [n/\log n] + 1, \dots, n - [n/\log n]\} \\
A_{n6} &= \{n - [n/\log n] + 1, \dots, n\},
\end{aligned}$$

when $[a]$ denotes the integer part of a. It is easy to realize that for $k = 1, \dots, n-1$

$$\begin{aligned}
\sum_{i=1}^{k} c_i \hat{e}_{ni} &= \sum_{i=1}^{k} c_i e_i - \sum_{j=1}^{n} c_j e_j G_{nk} - \bar{e}_n Q_{nk} = - \sum_{i=k+1}^{n} c_i \hat{e}_{ni} = \\
&= - \sum_{i=k+1}^{n} c_i e_i + \sum_{i=1}^{n} c_i e_i (1 - G_{nk}) - \bar{e}_n Q_{nk},
\end{aligned}$$

where

$$\bar{e}_n = n^{-1}\sum_{i=1}^{n} e_i \tag{6.1}$$

$$G_{nk} = \sum_{i=1}^{k}(c_i - \bar{c}_n)\,c_i \Big/ \sum_{j=1}^{n}(c_j - \bar{c}_n)^2 \tag{6.2}$$

$$Q_{nk} = \left(\sum_{i=1}^{k} c_i \sum_{j=k+1}^{n} c_j^2 - \sum_{i=1}^{k} c_i^2 \sum_{j=k+1}^{n} c_j\right) \Big/ \sum_{v=1}^{n}(c_v - \bar{c}_n)^2, \quad k = 1,\dots,n-1. \tag{6.3}$$

Then using (6.1), the assumptions and applying the Kolmogorov- or the Chow-inequality one receives for $n \to \infty$:

$$\max_{k\in A_{n1}}\left\{\left(\sum_{i=1}^{k} c_i^2\right)^{-1/2}\left(\left|\sum_{j=1}^{n} c_j\,e_j\right| G_{nk} + |\bar{e}_n|\cdot|Q_{nk}|\right)\right\} = O_P\left(n^{-1/2}(\log n)^{1/2}\right) \tag{6.4}$$

$$\max_{k\in A_{n2}}\left\{\left(\sum_{i=1}^{k} c_i^2\right)^{-1/2}\left(\left|\sum_{j=1}^{n} c_j\,e_j\right| G_{nk} + |\bar{e}_n|\cdot|Q_{nk}|\right)\right\} = O_P\left(\log^{-1/2} n\right) \tag{6.5}$$

$$\max_{k\in A_{n3}}\left\{\left(\sum_{i=1}^{k} c_i^2\right)^{-1/2}\left|\sum_{j=1}^{k} c_j\,\hat{e}_{nj}\right|\right\} = O_P(1) \tag{6.6}$$

$$\max_{k\in A_{n4}}\left\{\left(\sum_{i=k+1}^{n} c_i^2\right)^{-1/2}\left|\sum_{j=k+1}^{n} c_j\,\hat{e}_{nj}\right|\right\} = O_P(1) \tag{6.7}$$

$$\max_{k\in A_{n5}}\left\{\left(\sum_{i=k+1}^{n} c_i^2\right)^{-1/2}\left(\left|\sum_{j=1}^{n} c_j\,e_j\right|(1 - G_{nk}) + |\bar{e}_n|\cdot|Q_{nk}|\right)\right\} = O_P\left(\log^{-1/2} n\right) \tag{6.8}$$

$$\max_{k\in A_{n6}}\left\{\left(\sum_{i=k+1}^{n} c_i^2\right)^{-1/2}\left(\left|\sum_{j=1}^{n} c_j\,e_j\right|(1 - G_{nk}) + |\bar{e}_n|\cdot|H_{nk}|\right)\right\} = O_P\left(n^{-1/2}\log^{1/2} n\right). \tag{6.9}$$

Here we also used the fact that under (2.9), (2.10) for $k \in A_{n2} \cup A_{n5}$

$$\begin{aligned}
\operatorname{var}_{H_0}&\left\{\sum_{i=1}^{k} c_i\,\hat{e}_{ni}\right\} = \qquad (6.10)\\
&= \left(\frac{1}{n}\sum_{i=k+1}^{n} c_i \sum_{j=1}^{k} c_j + \sum_{i=k+1}^{n}(c_i - \bar{c}_n)\,c_i \sum_{j=1}^{k}(c_j - \bar{c}_n)\,c_j \Big/ \sum_{v=1}^{n}(c_v - \bar{c}_n)^2\right)\sigma^2 = \\
&= \left(\sum_{i=1}^{k} c_i^2\,\sigma^2\, I\{k \in A_{n2}\} + \sum_{i=k+1}^{n} c_i^2\,\sigma^2\, I\{k \in A_{n5}\}\right)\left(1 + O\left(\frac{1}{\log n}\right)\right).
\end{aligned}$$

Next, by Theorem 2.3 in Hušková (1992)

$$\begin{aligned}
\lim_{n\to\infty} P_{H_0}&\left(\max_{k\in A_{n2}}\left\{\left(\sum_{i=1}^{k} c_i^2\right)^{-1/2}\left|\sum_{j=1}^{k} c_j\,e_j\right|\right\} \le \right. \qquad (6.11)\\
&\le b\left(y, \log\left(\sum_{i=1}^{n} c_i^2\right)\right)\Bigg) = \exp\{-\exp\{-y\}\} \qquad y \in \mathbb{R}^1
\end{aligned}$$

$$\lim_{n\to\infty} P_{H_0}\left(\max_{k\in A_{n5}}\left\{\left(\sum_{i=k+1}^{n} c_i^2\right)^{-1/2}\left|\sum_{j=k+1}^{n} c_j\,e_j\right|\right\} \le \right. \tag{6.12}$$

$$\leq b\left(y, \log\left(\sum_{i=1}^{n} c_i^2\right)\right)\right) = \exp\{-\exp\{-y\}\} \qquad y \in \mathbb{R}^1.$$

Since

$$\max_{k\in A_{n2}}\left\{\left(\sum_{i=1}^{k} c_i^2\right)^{-1/2}\left|\sum_{j=1}^{k} c_j\, e_j\right|\right\} \quad \text{and} \quad \max_{k\in A_{n5}}\left\{\left(\sum_{i=k+1}^{n} c_i^2\right)^{-1/2}\left|\sum_{j=k+1}^{n} c_j\, e_j\right|\right\}$$

are independent random variables we obtain the assertion (2.12) from (6.5 – 6.13). The assertion (2.11) can be shown in the same way, therefore its proof is omitted.
It remains to show that σ can be replaced by $\sqrt{\hat{\sigma}_n^2}$ given by (2.14). Towards this notice that

$$\hat{\sigma}_n^2 = \frac{1}{n-2}\left\{\sum_{i=1}^{n}(e_i - \bar{e}_n)^2 - \left[\left(\sum_{i=1}^{n}(c_i - \bar{c}_n)\, e_i\right)^2 \Big/ \sum_{j=1}^{n}(c_j - \bar{c}_n)^2\right]\right\}$$

and then by the law of iterated logarithm one observes that

$$\lim_{n\to\infty}\left(\hat{\sigma}_n^2 - \sigma^2\right)\sqrt{\log\log n} = 0 \qquad \text{a.s.,} \tag{6.13}$$

which immediately implies that σ can be replaced by $\sqrt{\hat{\sigma}_n^2}$. *Q. E. D.*

PROOF OF THEOREM 2.3. We shall show only (2.37) and (2.38) for the proof of (2.35) and (2.36) is quite parallel.
Notice that under the alternative H_{n2} (1.3) one can write

$$\sum_{i=1}^{k} c_i\, \hat{e}_{ni} = \sum_{i=1}^{k} c_i\, e_i - \sum_{i=1}^{k} c_i\, e_i\, G_{nk} - \bar{e}_n\, Q_{nk} + \tag{6.14}$$
$$+\ \delta_{n2}\left\{\sum_{i=1}^{k} c_i\, I\{i > m\} - \sum_{i=m+1}^{n} c_i\, G_{nk} - \frac{n-m}{n} Q_{nk}\right\}$$

where G_{nk} and Q_{nk} are defined by (6.3) and (6.4), respectively. Clearly, for δ_{n2} fulfilling (2.26) and (2.35) and $\delta_{n1} = 0$

$$\max_{k\in A_{n1}\cup A_{n2}}\left\{\left(\sum_{i=1}^{k} c_i^2\right)^{-1/2}|\delta_{n2}|\left|\sum_{i=m+1}^{n} c_i\, G_{nk} - \frac{n-m}{n} Q_{nk}\right|\right\} = \tag{6.15}$$
$$= O\left(|\delta_{n2} \max_{k\in A_{n1}\cup A_{n2}}\{\sqrt{k}\}\right) = O\left(|\delta_{n2}|\,(n/\log n)^{1/2}\right) =$$
$$= o\left((\log\log n)^{-1/2}\right) \qquad \text{as } n\to\infty$$

$$\max_{k\in A_{n3}\cup A_{n4}}\left\{\left(\sum_{j=1}^{k} c_j^2\right)^{-1/2}|\delta_{n2}|\cdot\left|\sum_{i=1}^{k} c_i\, I\{i > m\} - \sum_{i=m+1}^{n} c_i\, G_{nk} - \frac{n-m}{n} Q_{nk}\right|\right\} = \tag{6.16}$$
$$= |\delta_{n2}|\, c_2^\star\left(\frac{m(n-m)}{n}\right)^{1/2}\left(1 + o(1)\right) =$$
$$= |\delta_2^\star|\, c_2^\star\left(\theta(1-\theta)\right)^{1/2}(2\log\log n)^{1/2}\left(1 + o(1)\right) \qquad \text{as } n\to\infty$$

$$\max_{k\in A_{n5}\cup A_{n6}}\left\{\left(\sum_{i=k+1}^{n} c_i^2\right)^{-1/2}|\delta_{n2}|\left|\sum_{i=m+1}^{k} c_i G_{nk}-\frac{n-m}{n}Q_{nk}\right|\right\}= \tag{6.17}$$
$$= O\left(|\delta_{n2}|\,(n\log n)^{1/2}\right)=o\left((\log\log n)^{-1/2}\right) \qquad \text{as } n\to\infty.$$

Combining (6.5 – 6.10), (6.11 – 6.13) together with (6.16 – 6.18) we obtain the assertion (2.37). As for the behavior of $\hat{\sigma}_n^2$ defined in (2.14) we realize that under H_{n2}

$$\hat{\sigma}_n^2 = \frac{1}{n-2}\sum_{i=1}^{n}\left((e_i-\bar{e}_n)-\frac{(c_i-\bar{c}_n)}{\sum_{v=1}^{n}(c_v-\bar{c}_n)^2}\sum_{j=1}^{n}(c_j-\bar{c}_n)\,c_je_j+\right. \tag{6.18}$$
$$\left.+\delta_{n2}\left(I\{i>m\}\,c_i-\frac{1}{n}\sum_{j=m+1}^{n}c_j^2-(c_i-\bar{c}_n)\sum_{j=m+1}^{n}(c_j-\bar{c}_n)\,c_j\Big/\sum_{v=1}^{n}(c_v-\bar{c}_n)^2\right)\right)^2.$$

Since under H_0 (6.14) is true and

$$\frac{\delta_{n2}^2}{n-2}\sum_{i=1}^{n}\left(I\{i>m\}\,c_i-\frac{1}{n}\sum_{j=m+1}^{n}c_j^2-(c_i-\bar{c}_n)\sum_{j=m+1}^{n}(c_i-\bar{c}_n)\,c_j\Big/\sum_{v=1}^{n}(c_v-\bar{c}_n)^2\right)^2\leq M_1\,\delta_{n2}^2$$

holds for some M_1 (not depending on n) thus (6.14) holds true even under H_{n2} with δ_{n2} fulfilling (2.35). *Q. E. D.*

The following theorem plays the crucial role in the proofs of Theorem 4.1 and 5.2.

Theorem 6.1. *Let $e_1,\dots,e_n$ be i. i. d. random variables. Let assumptions* (4.1), (4.2) *and* (2.18) *be satisfied and let g_{ni}, $i=1,\dots,n$ be constants fulfilling:*

$$\max_{1\leq i\leq n}|g_{ni}|=o(1) \qquad \text{as } n\to\infty. \tag{6.19}$$

Then for any bounded sequence $\{a_n\}$ of real numbers:

$$\max_{1\leq k\leq n}\sup_{\substack{|t_1|\leq 1\\|t_2|\leq 1}}\left\{\left|\sum_{i=1}^{k}d_i\Big(\psi\big(e_i-g_{ni}-(t_1+t_2c_i)\,a_n\big)-\psi(e_i-g_{ni})\right.\right. \tag{6.20}$$
$$\left.\left.+\lambda\big(g_{ni}+(t_1+t_2c_i)\,a_n\big)-\lambda(g_{ni})\Big)\right|\right\}=O_p\left(n^{3/4}|a_n|^{3/4}\right) \qquad \text{as } n\to\infty,$$

$$\max_{1\leq k\leq n}\sup_{\substack{|t_1|\leq 1\\|t_2|\leq 1}}\left\{k^{-1/2}\left|\sum_{i=1}^{k}d_i\Big(\psi\big(e_i-g_{ni}-(t_1+t_2c_i)\,a_n\big)-\psi(e_i-g_{ni})\right.\right. \tag{6.21}$$
$$\left.\left.+\lambda\big(g_{ni}+(t_1+t_2c_i)\,a_n\big)-\lambda(g_{ni})\Big)\right|\right\}=O_p\left(n^{1/4}|a_n|^{3/4}\log n^{1/4}\right) \qquad \text{as } n\to\infty,$$

$$\max_{1\leq k\leq n}\sup_{\substack{|t_1|\leq 1\\|t_2|\leq 1}}\left\{k^{-1}\left|\sum_{i=1}^{k}d_i\Big(\psi\big(e_i-g_{ni}-(t_1+t_2c_i)\,a_n\big)-\psi(e_i-g_{ni})\right.\right. \tag{6.22}$$
$$\left.\left.+\lambda\big(g_{ni}+(t_1+t_2c_i)\,a_n\big)-\lambda(g_{ni})\Big)\right|\right\}=O_p\left(|a_n|^{3/4}q_n^{-1/4}\right) \qquad \text{as } n\to\infty,\ q_n\to\infty,$$

where we put either $d_i=c_i$, $i=1,\dots,n$, or $d_i=1$, $i=1,\dots,n$.

PROOF. All three assertions can be shown in the same way. They are slight modification of the proofs on the asymptotic linearity related to M-estimators. A number of these

theorems was proved, e.g. by Jurečková (1977), Jurečková and Sen (1987), Sen (1984 (Lemma 2.1) and by the author (1989). Therefore we give only the sketch of the proof of (6.22) with $d_i = c_i$, $i = 1, \dots, n$.
Due to the monotonicity of ψ one can write for $k = 1, \dots, n$

$$\sup_{|t_1|\le 1, |t_2|\le 1} \Big\{ k^{-1/2} \Big| \sum_{i=1}^{k} c_i \Big(\psi\big(e_i - g_{ni} - (t_1 + t_2 c_i) a_n\big) - \psi\big(e_i - g_{ni}\big) + \tag{6.23}$$

$$+\lambda\big(g_{ni} + (t_1 + t_2 c_i)\, a_n\big) - \lambda(g_{ni})\Big) \Big| \Big\} \le$$

$$\le \max_{0\le j, v\le w} \Big\{ k^{-1/2} \Big| \sum_{i=1}^{k} c_i\, I\{c_i > 0\} \Big(\psi\big(e_i - g_{ni} - (t_{1j} + t_{2v} c_i)\, a_n\big) - \psi\big(e_i - g_{ni}\big) +$$

$$+\lambda\big(g_{ni} + (t_{1j} + t_{2v} c_i)\, a_n\big) + \lambda(g_{ni})\Big) \Big| \Big\} +$$

$$+ \max_{0\le j, v\le w} \Big\{ k^{-1/2} \Big| \sum_{i=1}^{k} c_i\, I\{c_i < 0\} \Big(\psi\big(e_i - g_{ni} - (t_{1j} + t_{2v} c_i)\, a_n\big) - \psi(e_i - g_{ni}) +$$

$$+\lambda\big(g_{ni} + (t_{1j} + t_{2v} c_i)\, a_n\big) + \lambda(g_{ni})\Big) \Big| \Big\} +$$

$$+ \max_{0\le j, v\le w} \max_{\substack{s=0,1\\ r=0,1}} \Big\{ k^{-1/2} \sum_{i=1}^{k} |c_i| \Big| \Big(\lambda\big(g_{ni} - (t_{1j} + t_{2v} c_i)\, a_n\big) - \psi(e_i - g_{ni}) +$$

$$+\lambda\big(g_{ni} + (t_{1,j+s} + t_{2,v+r} c_i)\, a_n\big) - \lambda(g_{ni})\Big) \Big| \Big\} =$$

$$= B_{nk1} + B_{nk2} + B_{nk3},$$

where $t_{1j} = j/w$, $j = -w \dots, w$; $t_{2v} = v/w$, $v = -w \dots, w$; $t_{i,w+1} = t_i$, $w = 1$, $i = 1, 2$. By the assumptions

$$|B_{nk3}| \le M_2\, w^{-1} |a_n| \sqrt{k} \qquad k = 1, 2, \dots, n \tag{6.24}$$

for some $M_2 > 0$. By the Chow inequality for $b_n > 0$:

$$P\Big(\max_{1\le k\le n} |B_{nkj}| \ge b_n \Big) \le \tag{6.25}$$

$$\le \sum_{j=-w}^{w} \sum_{v=-w}^{w} \sum_{k=1}^{n} \Big(\frac{1}{k} - \frac{1}{k+1} \Big) \sum_{i=1}^{k} c_i^2 \int \Big(\psi\big(x - g_{ni} - (t_{1j} + t_{2v} c_i)\, a_n\big) - \psi(x - g_{ni}) +$$

$$+\lambda\big(g_{ni} + (t_{1j} + t_{2v} c_i)\, a_n\big) - \lambda(g_{ni})\Big)^2 dF(x)\, b_n^{-2} \le$$

$$\le M_3\, w^2 (\log n) \cdot |a_n|\, b_n^{-2} \qquad j = 1, 2$$

for some $M_3 > 0$. Putting $w = \big[(n\, |a_n| \,/\, \log n)^{-1/4} \big]$ and $b_n = w^{-1} |a_n|\, n^{1/2}$ in (6.26) and taking into account (6.24), (6.25) we receive assertion (6.22). *Q. E. D.*

PROOF OF THEOREM 4.1. At first we derive a useful representation of the estimators $(\alpha_n(\psi), \beta_n(\psi))$ defined by (1.5) Due to the assumptions under H_n (1.4) with δ_{n1}, δ_{n2} fulfilling (2.26) and (2.35) $\alpha_n(\psi) - \alpha = o_p(1)$ for some and $\beta_n(\psi) - \beta = o_p(1)$ as $n \to \infty$. Then applying (6.21) with $|a_n| = D > 0$ and

$$g_{ni} = \{i > m\}\, (\delta_{n1} + \delta_{n2}\, c_i) \qquad i = 1, \dots, n \tag{6.26}$$

we receive that

$$\sum_{i=1}^{n}\binom{1}{c_i}\Big[\psi\Big(X_i-\alpha_n(\psi)-\beta_n(\psi)\,c_i\Big)-\psi(e_i+g_{ni})+ \qquad (6.27)$$
$$+\lambda\Big(\alpha_n(\psi)-\alpha+\Big(\beta_n(\psi)-\beta\Big)\,c_i-g_{ni}\Big)-\lambda(g_{ni})\Big]=O_P(n^{3/4})$$

which further ensures as $n\to\infty$

$$\begin{pmatrix}\alpha_n(\psi)-\alpha\\ \beta_n(\psi)-\beta\end{pmatrix}=O_P(n^{-1/4}).$$

Repeating the same procedure with $a_n=n^{-1/4}$ and g_{ni} defined by (6.26) we get as $n\to\infty$

$$\begin{pmatrix}\alpha_n(\psi)-\alpha\\ \beta_n(\psi)-\beta\end{pmatrix}=O_P(n^{-7/16})$$

and after one more repetition (with $a_n=n^{-7/16}$) we obtain as $n\to\infty$

$$n\Big(\alpha_n(\psi)-\alpha\Big)= \qquad (6.28)$$
$$=\frac{1}{\lambda'(0)}\sum_{j=1}^{n}\psi(e_j)-\bar{c}_n\Big(\beta_n(\psi)-\beta\Big)+\sum_{i=m+1}^{n}(\delta_{n1}+c_i\,\delta_{n2})+O_P(n^{1/2-v}),$$

$$\sum_{i=1}^{n}(c_j-\bar{c}_n)^2\Big(\beta_n(\psi)-\beta\Big)= \qquad (6.29)$$
$$=\frac{1}{\lambda'(0)}\sum_{j=1}^{n}(c_j-\bar{c}_n)\psi(e_j)+\sum_{i=m+1}^{n}(\delta_{n1}+c_i\,\delta_{n2})+O_P(n^{1/2-v}),$$

for some $v>0$. Next applying (6.22) with

$$g_{ni}=-I\{i>m\}(\delta_{n1}+\delta_{n2}\,c_i)-\sum_{j=m+1}^{n}(\delta_{n1}+\delta_{n2}\,c_j)\Big(1+(c_j-\bar{c}_n)(c_j-\bar{c}_v)\Big) \qquad (6.30)$$

we receive after some standard steps

$$\max_{1\le k\le n}\Bigg\{\Big(\frac{n}{k(n-k)}\Big)^{1/2}\Bigg|\sum_{i=1}^{k}d_i\Big[\psi\Big(X_i-\alpha_n(\psi)-\beta_n(\psi)\Big)-\psi(e_i-g_{ni})+ \qquad (6.31)$$
$$+\lambda\Big(\alpha_n(\psi)-\alpha+\Big(\beta_n(\psi)-\beta\Big)\,c_j-I\{i>m\}\,(\delta_{n1}+\delta_{n2}\,c_i)\Big)-\lambda(g_{ni})\Big]\Bigg|\Bigg\}=$$
$$=O_P\left(n^{-1/8}\log^{1/4}n\right).$$

This relation implies that under the assumptions of Theorem 4.1 the asymptotic distribution of

$$\max_{1\le k\le n}\left\{\left(\sum_{v=1}^{n}d_v^2\left(\sum_{i=1}^{k}d_i^2\sum_{j=k+1}^{n}d_j^2\right)^{-1}\right)^{1/2}\left|\sum_{i=1}^{k}d_ie_{ni}(\psi)\right|\right\}$$

is the same as that of

$$\max_{1\le k\le n}\left\{\left[\sum_{v=1}^{n} d_v^2\left(\sum_{i=1}^{k} d_i^2 \sum_{j=k+1}^{n} d_j^2\right)^{-1}\right]^{1/2} \times \right. \tag{6.32}$$

$$\times \sum_{i=1}^{k} d_i\left(\psi(e_i) - \frac{1}{n}\sum_{j=1}^{n}\psi(e_j)\,(c_i-\bar{c}_n)\sum_{j=1}^{n}(c_j-\bar{c}_n)\,\psi(e_j)\Big|\sum_{v=1}^{n}(c_v-\bar{c}_n)^2\Big|\right) +$$

$$+\lambda'(0)\sum_{i=1}^{k} d_i\Big(I\{i>m\}\,(\delta_{n1}+\delta_{n2}\,c_i) - \frac{1}{n}\sum_{j=m+1}^{n}(\delta_{n1}+c_j\,\delta_{n2}) -$$

$$\left. -\,(c_i-\bar{c}_n)\sum_{j=m+1}^{n}(\delta_{n1}+c_j\,\delta_{n2})\,(c_j-\bar{c}_n)\Big|\sum_{v=1}^{n}(c_v-\bar{c}_n)^2\Big|\Big)\right\}$$

with $d_i = c_i,\ i=1,\dots,n$ or $d_i = 1,\ i=1,\dots,n$. Standard considerations enable to state

$$\left(\hat{\sigma}_n^2(\psi)-\sigma^2\right)\sqrt{\log\log n} = o_p(1) \qquad \text{as } n\to\infty \tag{6.33}$$

where $\hat{\sigma}_n^2(\psi)$ is defined by (4.10). The assertions of Theorem 4.1 can be concluded from (6.34), the facts above (6.33) and Theorem 2.3 applied to the random variable in (6.33).

Q. E. D.

PROOF OF THEOREM 5.1. We shall show only (5.5) for $i=2$. The proofs of (5.5) for $i=1$ and (5.6) are quite similar.

Notice that

$$\hat{m}_2(\psi) = \operatorname{argmax}\left\{\left(\sum_{i=1}^{k} c_i\,e_{ni}(\psi)\right)^2 - \left(\sum_{i=1}^{m} c_i\,e_{ni}(\psi)\right)^2\right\}. \tag{6.34}$$

$i=1,\dots,m$. Clearly,

$$\max\left\{\left(\sum_{i=1}^{k} c_i\,e_{ni}(\psi)\right)^2 - \left(\sum_{i=1}^{m} c_i\,e_{ni}(\psi)\right)^2\right\} \ge 0, \tag{6.35}$$

$$\left(\sum_{i=1}^{k} c_i\,e_{ni}(\psi)\right)^2 - \left(\sum_{i=1}^{m} c_i\,e_{ni}(\psi)\right)^2 = \tag{6.36}$$

$$= \sum_{i=k+1}^{m} c_i\,e_{ni}(\psi)\left(2\sum_{i=1}^{m} c_i\,e_{ni}(\psi) - \sum_{i=k+1}^{m} c_i\,e_{ni}(\psi)\right), \qquad k=1,\dots,n..$$

By the assumptions, (6.21), (6.29) and (6.30) one has

$$\sum_{i=1}^{m} c_i\,e_{ni}(\psi) = -\frac{m(n-m)}{n}\lambda'(0)\,(c_1^\star\delta_{n1}+c_2^\star\delta_{n2})\,(1+o(1)). \tag{6.37}$$

Next, by (6.29), (6.30) and (6.23) applied for $n=m$ we receive for $q_n\to\infty,\ n\to\infty$

$$\max_{1\le k\le m-q_n}\frac{1}{m-k}\left|\sum_{i=k+1}^{m} c_i\Big[e_{ni}(\psi)-\psi(e_i)+\right. \tag{6.38}$$

$$\left.+\lambda\Big(\alpha_n-\alpha+(\beta_n\beta)\,c_i - I\{i>m\}(\delta_{n1}+\delta_{n2}\,c_i)\Big)\Big]\right| =$$

$$= O_P\Big(q_n^{-1/4}(|\delta_{n1}| + |\delta_{n2}|)^{3/4}\Big)$$

$$\max_{1\le k\le m-q_n} \frac{1}{m-k}\left|\sum_{i=k+1}^{m} \psi(e_{ni})\right| = O_P\left(q_n^{-1/4}\right) \tag{6.39}$$

and, moreover, we can yield

$$\frac{1}{m-k}\sum_{i=k+1}^{m} \lambda\Big(\alpha_n(\psi) - \alpha + \big(\beta_n(\psi) - \beta\big)(c_i - \bar{c}_n)\Big) = \tag{6.40}$$

$$= \lambda'(0)\,\frac{n-m}{n}\cdot\frac{1}{m-k}\sum_{i=k+1}^{m} c_i\,(\delta_{n1} + \delta_{n2}\,c_i)\Big(1 + o(1)\Big) =$$

$$= \lambda'(0)\,\frac{n-m}{n}\,c_i\,(\delta_{n1} + \delta_{n2}\,c_i)\Big(1 + o(1)\Big) \qquad \text{as } n \to \infty,\ m - n \to \infty.$$

Analogous relations to (6.39 – 6.40) can be shown for $m \le k \le m - q_n$.
Assertion (5.5) for $i = 2$ follows this fact, (6.35 – 6.42) with $q_n = O\left(|\delta_{n1}|^2 + |\delta_{n2}|^2\right)$ and similar relations for max over $k = m, \ldots, m + q_n$. *Q. E. D.*

REFERENCES

[1] Antoch, J. (1991). *M-tests for the Change Point Problem: Monte Carlo Study.* Submitted.

[2] Broemling, L. D. and Tsurumi, H. (1986). *Econometrics and Structural Change.* Serie Statistics, vol. 74, M. Dekker, New York.

[3] Brown, R. L., Durbin, J. and Evans, J. M. (1975). Techniques for Testing the Constancy of Regression Relationships over Time (with discussion). *Journal of Royal Statistical Society B*, **37**, 149–192.

[4] Csőrgő, M., Horváth, L. (1987). Nonparametric Tests for the Change Point Problem. *Journal of Statistical Planning and Inference*, **17**, 1–9.

[5] Csőrgő, M., Horváth, L. (1988a). Nonparametric Methods for the Change Point Problem. In: *Handbook of Statistics*, vol. 7 [P. R. Krishnaiah and C. R. Rao, (eds.)]. North-Holland, Amsterdam, 403–425.

[6] Csőrgő, M., Horváth, L. (1988b). Invariance Principles for Change Point Problems. *Journal of Multivariate Analysis*, **27**, 151–168.

[7] Farley, J. U. and Hinich, M. Y. (1970). A Test for a Shifting Slope Coefficient in a Linear Model. *JASA*, **65**, 1320–1329.

[8] Hampel, F. R., Ronchetti, E. M., Rousseeuw, P. J. and Stahel W. R. (1985). *Robust Statistics: The Approach Based on Influence Functions.* J. Wiley, New York.

[9] Hušková, M. (1989). *Nonparametric Tests for Change in Regression Models at an Unknown Time Point.* International Workshop on Theory and Practice in Data Analysis, Report R-MATH-01/89, Akademie der Wissenschaften der DDR, Berlin, 101–124.

[10] Hušková, M., Sen, P. K. (1989): Nonparametric Tests for Shift and Change in Regression at an Unknown Time Point. In: *Statistical Analysis and Forecasting of Economic Structural Change* [P. Hackl, (ed.)]. Springer, New York, 71–85.

[11] Hušková, M., Veraberbeke, N. (1992). *Change-point Problem and Bootstrap.* In preparation.

[12] James, B., James, K. L. and Siegmund, D. (1987). Test for a Change-point. *Biometrika*, **74**, 71–84.

[13] Jandhyala, V. K., MacNeil, I. B. (1991). Tests for Parameters Changes at Unknown Times in Linear Regression Models. *Journal of Statistical Planning and Inference*, **27**, 291–316.

[14] Jurečková, J. (1977). Asymptotic Relations of M-estimates and R-estimates in Linear Regression Model, *Ann. Statist.*, **5**, 464–472.

[15] Jurečková, J., Sen, P. K. (1987). An Extension of Billingsley's Uniform Boundedness Theorem to Higher Dimensional M-processes. *Kybernetika*, **23**, 382–387.

[16] Kiefer, J. (1959). K-sample Analogues of the Kolmogorov–Smirnov and Cramér–von Mises Tests. *Annals of Mathematical Statistics*, **30**, 420–447.

[17] Kim, H. J., Siegmund, D. (1989). The Likelihood Ratio Test for a Change-point in Simple Linear Regression, *Biometrika*, **76**, 409–423.

[18] Krishnaiah, P. R., Miao, B. Q. (1988). Review about Estimation of Change Points. In: *Handbook of Statistics*, vol. 7 [P.R. Krishnaiah and C.R. Rao, (eds.)]. North-Holland, Amsterdam, 375–402.

[19] Sen, P. K. (1980). Asymptotic Theory of Some Tests for a Possible Change in the Regression Slope Occurring at an Unknown Time Point, *Zeitschrift für Wahrscheinlichkeitstheorie und verw. Gebiete*, **52**, 203–218.

[20] Sen, P. K. (1982). Asymptotic Theory of Some Tests for Constancy of Regression Relationships over Time, *Statistics*, **13**, 21–31.

[21] Sen, P. K. (1984). Recursive M-tests for the Constancy of Multivariate Regression Relationships over Time, *Sequential Analysis*, **3**, 191–211.

[22] Shaban, S. A. (1980). Change Point Problem and Two Phase Regression: An Annotated Bibliography, *International Statistical Review*, **48**, 83–93.

[23] Zacks, S. (1983). Survey of Classical and Bayesian Approaches to the Change Point Problem: Fixed Sample and Sequential Procedures of Testing and Estimation. In: *Recent Advances in Statistics. Papers in Honor of Herman Chernoff's Sixtieth Birthday*, [M.H. Rizvi, (ed.)]. Academic Press, New York, 245–269.

[24] Zacks , S. (1991). Detection and Change-Point Problem. In: *Handbook of Sequential Analysis*, [B.K. Ghosh and P.K. Sen, (eds.)], Serie Statistics, vol. 118. M. Dekker, New York, 531–562.

Recent Adv. in Stat. and Prob., pp. 127-138
J. Pérez Vilaplana and M. L. Puri (Eds)

Conditional Rank Tests for the Two-sample Problem under Random Censorship: Treatment of Ties

G. NEUHAUS
Institut für Mathematische Stochastik, Universität Hamburg, Bundesstrasse 55, D-2000, Hamburg 13, Germany

Abstract – For the two - sample problem with randomly censored data where ties are possible a conditional theory of linear rank tests and sup-norm tests is developed. The basis for deriving the asymptotics of the various test statistics is discrete time martingale theory. An essential technical novelty is a method for reducing the case where ties occur to the continuous case which has been handled previously by the author.

1. INTRODUCTION

We deal with the two - sample problem of testing the null hypothesis of randomness, i.e., the equality of both sample distribution functions (df's) under random right censoring with special attention to treatment of ties. Some wellknown tests for this problem are, e.g., the log rank test of Peto and Peto (1972), the generalized Wilcoxon test of Gehan (1965) and the tests of Harrington and Fleming (1982).

These tests belong to the class of generalized linear rank tests of Aalen (1978) and Gill (1980) which are functionals of certain empirical processes on $[0, \infty)$ being stochastic integrals with respect to basic martingales. By methods of continuous time martingale theory their asymptotic behavior has been derived by Gill (1980), see also Leurgans (1984) and the survey of Andersen et al. (1982). Another class of sup norm functionals leads to Renyi type statistics (RS), see Gill (1980), Fleming et al.(1987), and Schumacher (1984). All tests mentioned above are usually applied using critical values of the asymptotic null distribution of standardized versions of the various test statistics. One major advantage of this approach is that these test are asymptotically valid even if the censoring distributions are different in both samples. On the other hand the validity is an asymptotic one and it has been reported in the literature that for finite sample numbers the nominal and actual level may differ substantially, see Latta (1981), Janssen and Brenner (1991) and Schumacher (1984).

In case of equal censoring there exists another sort of rank tests being even finite sample distribution free under the null hypothesis: conditional permutation tests. In fact, the idea of permutation tests is an old one and has already been used in the framework of survival analysis under censoring by Gehan (1965) and Mantel (1967), see also Gill (1980), sec. 3.3, and Andersen et al.(1982), sec. 3.5.

Gehan standardized his linear rank statistic by the exact permutation variance and proposed a normal approximation for large samples. Yet, under unequal censoring distributions this test is usually not distribution free, neither finite sample nor asymptotically. Neuhaus (1988) and Janssen (1991) made a systematic study of conditional rank statistics under random censoring in the framework of local asymptotic decision theory. Again this theory is restricted to equal censoring.

Recently Neuhaus (1992) showed in the case of continuous distributions that the permutation method works also under unequal censoring. The idea is simple: Take the unconditional tests statistic T, say, standardized to make it asymptotically distribution free under the null hypothesis (even under unequal censoring). Then, keeping the observed *censoring pattern* fixed, perform a test based on the permutation distribution of T. It turnes out that the unconditional and the conditional test are asymptotically equivalent under very general circumstances including unequal censoring, while, by construction, the conditional tests are finite sample distribution free under the null hypothesis with equal censoring. Simulation results in Neuhaus(1992) showed the excellent behavior of the conditional tests with respect to size under equal *and* unequal censoring. At the same time power behavior of the conditional and unconditional tests was very similar in most cases. Here we will treat the general case where ties may occur. The main technical aspect is a device for reducing the general case to the continuous one so that the results of Neuhaus (1992) may be applied.

In the presence of ties the finite sample distribution freeness under equal censoring of the conditional tests remains true and for *linear rank tests* likewise the asymptotic equivalence holds true even under unequal censoring, entailing asymptotic distribution freeness of conditional linear rank tests in this case, too. If ties occur the usual unconditional *sup norm tests* are wellknown to be neither finite sample nor asymptotically distribution free under the null hypothesis, not even for equal censoring, rather they are conservative, while their conditional counterparts *are* finite sample distribution free for equal censoring.

By introducing a certain randomization device we will construct *sup norm tests* which are likewise asymptotically distribution free under the null hypothesis with unequal censoring when ties are present.

2. THE MODEL, EMPIRICAL PROCESSES AND THE TEST STATISTICS

We assume the two-sample general censorship model. Observations are made on n_k individuals from population k $(k = 1,2)$ and all $n := n_1 + n_2$ observations are independent.($a := b$ or $b =: a$ means that a equals b by definition.) The j'th subject in sample k has nonnegative, independent latent survival and censoring times X_{ki}^0 and U_{ki}, resp., whose not necessarily continuous distribution functions (df's) are F_k and G_k. Let $S_k := (1 - F_k)$ and the cumulative hazard function (chf's) $\Lambda_k(x) = \int_0^x (1 - F_{k-})^{-1}\, dF_k$, where $f_-(f_+)$ denotes the left continuous (right continuous) version of a function f and $\Delta f = f_+ - f_-$ the jump function. The observable random variables (rv's) are $X_{ki} := \min(X_{ki}^0, U_{ki})$ and $\Delta_{ki} := 1\{X_{ki}^0 \le U_{ki}\}$, where $1\{E\}$ is 1 if the event E occurs and 0 otherwise.

Always $0/0 := 0$. Let us remark that, similarly as in Gill(1980), most of the considerations below may be extended to the case where the X_{ki}^0 and U_{ki} have distributions depending also on i s.th. (with obvious notation) $(F_k^1 + \ldots + F_k^{n_k})/n_k \to F_k$ and similarly for the G's.

Moreover, put $(X_1, \ldots, X_n) := (X_{11}, \ldots, X_{1n_1}, X_{21}, \ldots, X_{2n_2})$, and let $X_{1:n} \le \ldots \le X_{n:n}$

denote the corresponding order statistics and let r be the number of different X_i's. In the continuous case $r = n$ but in general r is random. Throughout the paper 'r' has the above meaning. Define integers $T_0 := 0 < T_1 < \ldots < T_r = n$ and $\mathbf{T}_n = \{T_0/n, T_1/n, \ldots, T_r/n\}$ by

$$X_{1:n} = \ldots = X_{T_1:n} < X_{T_1+1:n} = \ldots = X_{T_2:n} < \ldots < X_{T_{r-1}+1:n} = \ldots = X_{T_r:n}.$$

The observations X_i with $i \in I_j := \{k : X_k = X_{q:n}\}, q := T_j$, form the *j'th tie group*. In the j'th tie group the uncensored observations are ranked ahead the censored ones. For $1 \le j \le r$ let D_j be the number of uncensored observations in the j'th tie group, $A_j[B_j]$ be the number of uncensored [censored] observations from sample 1 in the j'th tie group, and $p_j := (A_j + B_j + \ldots + A_r + B_r) \times (n - T_{j-1})^{-1}$, $q_j := 1 - p_j$. Put $\mathbf{D}_n := (D_1, \ldots, D_r)$ and define $\mathbf{A}_n, \mathbf{B}_n$ similarly.
Moreover, define $\Delta_n = (\Delta_{1:n}, \ldots, \Delta_{n:n})$ by

$$\Delta_{k:n} := 1\,[0] \text{ if } T_{l-1} < k \le T_l + D_l, \qquad \text{for some [for none] } l = 1, \ldots, \text{r}. \tag{2.1}$$

Apparently $(\mathbf{T}_n, \mathbf{D}_n)$ is equivalent to $(\mathbf{T}_n, \Delta_n)$.
Our hypothesis of interest is $\mathcal{H}_0 : F_1 = F_2 (=: F)$, versus either the omnibus alternative $\mathcal{A}_0 : F_1 \neq F_2$ or the one sided alternative $\mathcal{A}_1 : F_1 \ge F_2, F_1 \neq F_2$. If the censoring distributions G_1 and G_2 are equal we speak of the restricted null hypothesis $\bar{\mathcal{H}}_0$.
Now we are in the position to define empirical processes $\mathtt{W}_n$ and $\mathtt{V}_n$ in the space $D[0,1]$ from which various test statistics will be derived. The processes will be constant in $[T_{j-1}/n, T_j/n)$.

$$\mathtt{W}_n(t) = \left(\frac{n}{n_1 n_2}\right)^{1/2} \sum_{j=1}^{l} w_{nj}(A_j - D_j p_j), \tag{2.2}$$

and

$$\mathtt{V}_n(t) = \frac{n}{n_1 n_2} \sum_{j=1}^{l} w_{nj}^2 p_j q_j D_j \cdot \frac{n - T_{j-1} - D_j}{n - T_{j-1} - 1}, \tag{2.3}$$

for $T_l \le n \cdot t < T_{l+1}$, $l = 0, \ldots, r$, $T_{r+1} := \infty$, where w_{nj} are stochastic weights which may depend on $(\mathbf{T}_n, \mathbf{D}_n)$ (equivalently on $(\mathbf{T}_n, \Delta_n)$) and $A_1, B_1, \ldots, A_{j-1}, B_{j-1}$; empty sums are understood to be zero. If $\mathtt{H}_n$ denotes the empirical *df* of the $X_{i:n}$*'s* then $\mathtt{W}_n \circ \mathtt{H}_n$ is the process W of Gill (1980), formula (3.3.8), and $\mathtt{V}_n \circ \mathtt{H}_n$ is his estimator V_2, formula (3.3.12), loc.cit., with suitable K in Gill's notation. Many functionals of $(\mathtt{W}_n, \mathtt{V}_n)$ yield sensible test statistics for $\mathcal{H}_0$. E.g., for testing versus the one sided alternative $\mathcal{A}_1$ the simplest functional is (for some $0 < \tau \le 1$)

$$\mathtt{W}_n(\tau)/\mathtt{V}_n^{1/2}(\tau), \tag{2.4}$$

where $\mathtt{W}_n(1)$ is known as a *generalized linear rank statistic*. For testing versus the omnibus alternative $\mathcal{A}_0$ we will use as an example the Renyi type (RS) statistic

$$RS_n^\tau := \sup(|\,\mathtt{W}_n(t)\,| : t \in \mathbf{T}_n \cap [0,1], t \le \tau)/\mathtt{V}_n^{1/2}(\tau) \tag{2.5}$$

These statistics depend solely on $(\mathbf{A}_n, \mathbf{B}_n, \mathbf{T}_n, \mathbf{D}_n)$. An important class of weights is $(\rho, \kappa, \lambda \ge 0)$

$$w_{nj} = \mathtt{S}_n^\rho\left(\frac{i}{n}\right) \cdot \left(\frac{n-i}{n}\right)^{\kappa - 2\lambda} \cdot \left(p_j q_j n^2/(n_1 n_2)\right)^{-\lambda} \cdot 1\{p_j q_j > 0\}, \tag{2.6}$$

with $i = T_{j-1}$, $1 \le j \le r$, and, for $\nu = 1, \dots, n$,

$$S_n(\nu/n) = \prod_{1 \le k \le \nu} (1 - \Delta_{k:n}/(n-k+1)), S_n(0/n) := 1. \tag{2.7}$$

If $\mathbf{F}_n$ is the Kaplan-Meier estimator of F in the pooled sample then $1 - \mathbf{F}_n(X_{i+1:n}-) = S_n(i/n)$ for $i = T_{j-1}$. For $\kappa = \lambda = 0$ this is the class introduced by Harrington and Fleming (1982) containing the log rank ($\rho = 0$) and Prentice's (1978) Wilcoxon ($\rho = 1$) cases. The case $\rho = \lambda = 0$ and $\kappa = 1$ corresponds to the Gehan-Wilcoxon test while $\rho \ge 0.5$ and $\kappa = \lambda = 0.5$ lead to "approximately distribution free" statistics in Fleming et al.(1987). Finally, for $\rho = 0, \kappa = \lambda = 1$, the process $W_n \circ H_n$ is the standardized Nelson estimator of $\Lambda_1 - \Lambda_2$ (stopped at $\hat{x} := \min(\max\{X_{1i} : i = 1, \dots, n_1\}, \max\{X_{2i} : i = 1, \dots, n_2\})$); therefore, W_n is called *hazard process* in this case.
The common way of using the above test statistics is to reject the null hypothesis if the observed value exceeds a critical value computed from the asymptotic null distribution which may be derived from convergence properties of $W_n \circ H_n$ and $V_n \circ H_n$. By using continuous time martingale theory Gill(1980) established convergence in distribution ($\to_{\mathcal{D}}$) of $W_n \circ H_n$ and convergence in probability ($\to_p$) of $V_n \circ H_n$ in the Skorohod space $D[0, d]$ for suitable $d \in (0, \infty]$. The corresponding linear rank tests are asymptotically distribution free under $\mathcal{H}_0$ even for unequal censoring, while RS tests are in general asymptotically distribution free under $\mathcal{H}_0$ only for continuous F and are conservative for discontinuous F, see e.g. Fleming et al.(1987).
For continuous distributions Neuhaus (1992) developed a corresponding theory of conditional tests which are finite sample distribution free under $\bar{\mathcal{H}}_0$(equal censoring) and are asymptotically equivalent to their unconditional counterparts under the general null hypothesis $\mathcal{H}_0$ (possibly unequal censoring).
By reduction to the continuous case we will in the next section extend these results to possibly discontinuous distributions.

3. CONDITIONAL RANK TESTS WHEN TIES ARE PRESENT

Let $\mathbf{Z}_n = (Z_1, \dots, Z_n) \in \{0,1\}^n$ with $Z_1 + \dots + Z_n = n_1$ be a random vector on some probability space $(\Omega, \mathcal{A}, P)$ s.th. $\mathbf{Z}_n$ is distributed as a random sample without replacement taken from a population consisting of n_1 members "1" and n_2 members "0". Then $\mathbf{Z}_n$ is called to have *permutation distribution.*
By simple symmetry considerations one obtains that under $\bar{\mathcal{H}}_0$ and given $(\mathbf{T}_n, \mathbf{D}_n) = (\mathbf{t}_n, \mathbf{d}_n)$, the vector $(\mathbf{A}_n, \mathbf{B}_n)$ is distributed according to

$$A_j =_{\mathcal{D}} Z_{t_{j-1}+1} + \dots + Z_{t_{j-1}+d_j}, B_j =_{\mathcal{D}} Z_{t_{j-1}+d_j+1} + \dots + Z_{t_j}, \tag{3.1}$$

for $j = 1, \dots, r$, where empty sums are understood to be zero. In the sequel the phrase "under the permutation distribution" means that for fixed values $(\mathbf{t}_N, \mathbf{d}_N)$ of $(\mathbf{T}_N, \mathbf{D}_N)$ the random vector $(\mathbf{A}_n, \mathbf{B}_n)$ fulfilles (3.1). In fact we will assume then w.l.o.g. that in (3.1) equality in distribution ($=_{\mathcal{D}}$) is strict equality "=". Notice that the permutation distribution is the correct one in general only under the restricted null hypothesis $\bar{\mathcal{H}}_0$ but not necessarily under $\mathcal{H}_0$ with unequal censoring. Let $\mathcal{T}_n = \mathcal{T}_n(\mathbf{A}_n, \mathbf{B}_n, \mathbf{T}_n, \mathbf{D}_n)$ be one of the statistics (2.4) or (2.5). Now, having observed $(\mathbf{T}_n, \mathbf{D}_n) = (\mathbf{t}_n, \mathbf{d}_n)$, the conditional counterpart of the unconditional test based on $\mathcal{T}_n$ rejects the null hypothesis if the observed $\mathcal{T}_n$ exceeds the $(1 - \alpha)$ -quantile $k_n(\alpha, \mathbf{t}_n, \mathbf{d}_n)$, say, of the distribution of $\mathcal{T}_n(\mathbf{A}_n^*, \mathbf{B}_n^*, \mathbf{t}_n, \mathbf{d}_n)$, where $(\mathbf{A}_n^*, \mathbf{B}_n^*)$ is some *rv* having permutation distribution in the above sense. By construction, the conditional test is distribution free under the restricted null hypothesis $\bar{\mathcal{H}}_0$.

By means of conditional functional limit theorems from Neuhaus(1992) for related processes $\mathbf{W}_n^c$ and $\mathbf{V}_n^c$ in the continuous situation, see (3.7) and (3.8), we will study the relationship of the conditional and the unconditional tests. The result is that the conditional and the unconditional versions are in many situations asymptotically equivalent even under the general null hypothesis $\mathcal{H}_0$ with possibly unequal censoring, $G_1 \neq G_2$, for which $(\mathbf{A}_n, \mathbf{B}_n)$ given $(\mathbf{T}_n, \mathbf{D}_n) = (\mathbf{t}_n, \mathbf{d}_n)$ usually don't have permutation distribution. In fact, we will show in many cases that $k_n(\alpha, (\mathbf{T}_n, \mathbf{D}_n)) \rightarrow_p k(\alpha)$, under $\mathcal{H}_0$, where $k(\alpha)$ is the $(1-\alpha)$-quantile of the corresponding *unconditional* limiting null distribution. This entails the asserted equivalence under $\mathcal{H}_0$ and under contiguous alternatives.

In order to formulate our results we need some more notation and assumptions. We always will assume that $n \rightarrow \infty$ s.th.

$$n_1/n \rightarrow \eta \in (0,1). \tag{3.2}$$

Put $G := \eta G_1 + (1-\eta)G_2$, $H := 1-(1-F)(1-G)$, $H^1 := \int_{[0,\cdot]}(1-G_-)\,dF$ and

$$\bar{H}^1(t) := H^1(x-) + \min((t - H(x-)), \Delta H^1(x)), \tag{3.3}$$

with $x := H^{-1}(t) := \inf\{y : H(y) \geq t\}$. $\bar{H}^1$ may be shown to be absolutely continuous with Lebesgue density bounded by 1. Finally, for sets $A, B \subset [0,1]$ define their Hausdorff - distance $d(A,B) := \inf\{\epsilon > 0 : S_\epsilon(A) \supset B, S_\epsilon(B) \supset A\}$ with $S_\epsilon(A) := \{x \in [0,1] : | x - y | < \epsilon \text{ for some } y \in A\}$. In the next lemma we state the crucial properties of conditioning sequences.

Lemma 3.1 *Under the general null hypothesis $\mathcal{H}_0$ almost all sequences $(\mathbf{T}_n, \Delta_n) = (\mathbf{t}_n, \delta_n), n \geq 1$, have the properties*

$$d(\mathbf{t}_n, \mathbf{T}) \rightarrow 0, \mathbf{T} := H(\mathbb{R}), \tag{3.4}$$

and, for all continuity points $t \in (0,1)$ of $\mathbf{T}(\cdot) := \sup\{u : u \in \mathbf{T}, u \leq \cdot\}$,

$$\mathbf{T}_n(t) := \sup\{u : u \in \mathbf{t}_n, u \leq t\} \rightarrow \mathbf{T}(t) = \sup\{u : u \in \mathbf{T}, u \leq t\} \tag{3.5}$$

Moreover,

$$\bar{\mathbf{H}}_n^1(t) \rightarrow \bar{H}^1(t), 0 \leq t \leq 1, \tag{3.6}$$

with the function $\bar{H}_n^1$ given by $\bar{H}_n^1(t) = \int_0^t \delta_{[ns]+1:n} ds$.

All *proofs* will be given in section 4.

Now we introduce the above mentioned counterparts $\mathbf{W}_n^c$ and $\mathbf{V}_n^c$ of the processes $\mathbf{W}_n$ and $\mathbf{V}_n$, see (2.2) and (2.3), having continuous paths in the limit. For $0 \leq t \leq 1$ put

$$\mathbf{W}_n^c(t) = \left(\frac{n}{n_1 n_2}\right)^{1/2} \sum_{i=1}^{[nt]} w_n\left(\frac{i}{n+1}\right) u_n\left(\frac{i}{n+1}\right)(Z_i - \pi_i)\Delta_{i:n}, \tag{3.7}$$

and

$$\mathbf{V}_n^c(t) = \frac{n}{n_1 n_2} \sum_{i=1}^{[nt]} w_n^2\left(\frac{i}{n+1}\right) u_n^2\left(\frac{i}{n+1}\right) \pi_i(1-\pi_i)\Delta_{i:n}, \tag{3.8}$$

where $[x]$ is the integer part of $x, \pi_i := (Z_i + \ldots + Z_n)/(n-i+1)$ and

$$w_n(t) = w_{nj} \text{ if } T_{j-1} \leq nt < T_j, j = 1, \ldots, r, \tag{3.9}$$

and

$$u_n(t) = (n - T_{j-1} - D_j)/(n-i) \text{ if } T_{j-1} \leq i - 1 \leq nt < i \leq T_{j-1} + D_j, \tag{3.10}$$

$j = 1, \ldots, r$, and $u_n(t) = 1$ elsewhere. Notice that for $T_{r-1} + D_r = n$ one has $u_n(t) = 0$ for $T_{r-1}/n \leq t < 1$. The trick in the definition of u_n is that it depends only on the conditioning observations $(\mathbf{T}_n, \Delta_n)$ and that it is chosen s.th. for each $j = 1, \ldots, r$ the equation

$$\sum_{i=a}^{b} u_n\left(\frac{i}{n+1}\right)(Z_i - \pi_i) = (A_j - D_j p_j) \tag{3.11}$$

holds true, with $a = T_{j-1} + 1$ and $b = T_{j-1} + D_j, D_j > 0$. In fact, after guessing the proper weights $u_n(i/(n+1))$, see (3.10), it is a straightforward calculation to show (3.11). In the case $T_{r-1} + D_r = n$ (3.11) means 0=0. Consequently,

$$\mathtt{W}_n^c(T_j/n) = \mathtt{W}_n(T_j/n) \qquad \text{for } j = 0, \ldots, r. \tag{3.12}$$

For $\mathbf{V}_n^c$ and $\mathbf{V}_n$ one cannot expect an analogue strict equality, but it holds asymptotically on each subinterval $[0, \tau], 0 < \tau < 1$, by the following lemma.

Lemma 3.2. *Let $(\mathbf{T}_n, \Delta_n) = (\mathbf{t}_n, \delta_n), n \geq 1$, be a sequence as in Lemma 3.1. Then, under the permutation distribution, and for $0 < \tau < 1$,*

$$R_n(\tau) := \sup\{| \mathtt{V}_n^c(t) - \mathtt{V}_n(t) | : t \in \mathbf{t}_n, t \leq \tau\} \to_p 0. \tag{3.13}$$

By (3.12) and (3.13) the treatment of statistics (2.4) and (2.5) may be reduced to that of the processes $\mathtt{W}_n^c$ and $\mathtt{V}_n^c$, which have been studied in Neuhaus (1992) by discrete parameter martingale methods. Based on Theorem 3.2 of Neuhaus, loc.cit.,we will derive our following main convergence criterion.

Theorem 3.3. (Conditional Convergence Criterion) *Assume $n_1/n \to \eta \in (0,1)$ and let $\mathbf{Z}_n, n \geq 1$, have permutation distribution. Convergence in probability and in distribution below refer to this assumption. Moreover, let $(\mathbf{T}_n, \Delta_n) = (\mathbf{t}_n, \delta_n), n \geq 1$, be a sequence as in Lemma 3.1. Finally, let w_{nj} be real weights depending on $(\mathbf{t}_n, \delta_n)$ and $Z_i, 1 \leq i \leq T_{j-1}$, s.th. the jump functions w_n in (3.9), converge in quadratic mean in probability, i.e., with λ Lebesgue measure,*

$$\int_0^t (w_n - w)^2 d\lambda \to_p 0 \qquad \text{for } 0 < t < 1, \tag{3.14}$$

where w is some nonrandom function on [0,1] being square integrable on each subinterval $[0, t], 0 < t < 1$.

a) *Then, conditionally under $(\mathbf{T}_n, \Delta_n) = (\mathbf{t}_n, \delta_n), n \geq 1$,*

$$\mathtt{W}_n^c \to_{\mathcal{D}} \mathbf{B} \circ V^c \qquad \text{in } D[0,1) \tag{3.15}$$

where $\mathbf{B}$ is Brownian motion on [0,1) and

$$\mathtt{V}_n^c(t) \to_p V^c(t) \qquad \text{for } t \in [0,1), \tag{3.16}$$

with V^c given by

$$V^c(t) = \int_{[0,t]} w^2 u^2 \, d\bar{H}^1, \tag{3.17}$$

where u is defined by

$$u(t) := (1 - H(x-) - \Delta H^1(x))/(1-t), \tag{3.18}$$

if $H(x-) < t < H(x-) + \Delta H^1(x), x := H^{-1}(t) := \inf\{y : H(y) \geq t\}$, and $u(t) = 1$ elsewhere.

b) *If the weights w_{nj} are uniformly bounded by some nonrandom constant, (3.15) is true even in the Skorohod space $D[0,1]$ and (3.16) holds true also for $t = 1$.*

From the above theorem we will easily derive the following corollary which applies to the general class of weights in (2.6).

Corollary 3.4. *Retain the assumptions of Theorem 3.3 where we fix again a sequence $(\mathbf{T}_n, \Delta_n) = (\mathbf{t}_n, \delta_n), n \geq 1$, as in Lemma 3.1.*

a) *Then, for every $\tau \in (0,1)$ with $V^c(\tau^-) > 0, \tau^- := H(H^{-1}(\tau)-) = \mathrm{T}(\tau-)$, for linear rank statistics, conditionally under $(\mathbf{T}_n, \Delta_n) = (\mathbf{t}_n, \delta_n), n \geq 1$,*

$$\mathtt{W}_n(\tau)/\mathtt{V}_n^{1/2}(\tau) \to_{\mathcal{D}} \mathcal{N}(0,1), \tag{3.19}$$

where $\mathcal{N}(0,1)$ is standard normal distribution. If, additionally, $\Delta\mathrm{T}(\tau) = 0$, then for Renyi type statistics one has

$$\mathcal{L}(RS_n^\tau) \to_{\mathcal{D}} \mathcal{L}(\sup\{| \,\mathrm{B}(t)\, | : t \in A^\tau\}), \tag{3.20}$$

with $A^\tau := \{V^c(t)/V^c(\tau) : t \in H(\mathrm{R}), t \leq \tau\}$.

b) *If the weights w_{nj} fulfill $w_{nj} = v_n(T_{j-1}/n), j = 1, \ldots, r$, for suitable random functions $v_n, n \geq 1$, on [0,1), and if $\sup\{| \, v_n(t) - v(t) \, | : 0 \leq t \leq \tau\} \to_p 0$ for all $\tau \in (0,1)$, where v is a nonrandom, continuous function on [0,1), then condition (3.14) is fulfilled. If, additionally, the w_{nj}'s are uniformly bounded by some nonrandom constant and if $V^c(1) > 0$ then (3.19) and (3.20) hold true for $\tau = 1$.*

Remark. In the situation of Corollary 3.4 b) under equal censoring $G_1 = G_2$ it is easy to see that for $H(x) < 1$ Gill's (1980) variance estimator $\mathrm{V}_n \circ \mathrm{H}_n(x)$ converges to $V_{Gill}(x) = \int_{[0,x]} v^2 \circ H_-(1-F)/(1-F_-) \, dH^1$ which coincides by (3.16) with $V^c \circ H(x)$. By using $H^1 \circ H^{-1}(t) = \bar{H}^1(t)$ if H is continuous at $H^{-1}(t)$ and a computation similar to (4.9), below, the equality $V_{Gill}(x) = V^c \circ H(x)$ may also be checked directly.

Example 3.5. *Under $n_1/n \to \eta \in (0,1)$ and the permutation distribution, condition (3.14), is fulfilled for weight functions w_n with weights w_{nj} from (2.6) with arbitrary $\rho, \kappa, \lambda \geq 0$ and limiting function $w = v \circ \mathrm{T}$ with*

$$v(s) = S^\rho(s) \cdot (1-s)^{\kappa - 2\lambda}, 0 \leq s < 1, \tag{3.21}$$

$S(s) := \exp\{-\int_0^s (1 - id)^{-1} \, d\bar{H}^1\}, id :=$ identity, and $\mathrm{T}(t) = \sup\{u : u \in H(\mathrm{R}), u \leq t\}$. The limiting variance function $V^c(t)$ becomes

$$V^c(t) = \int_0^t v^2 \circ \mathrm{T} \cdot u^2 \, d\bar{H}^1 \leq \int_0^t \max\{1, (1-s)^{2\kappa - 4\lambda}\} \, ds. \tag{3.22}$$

Therefore, part a) of Corollary 3.4. applies to all weights (2.6). If $\lambda = 0$, the weights (2.6) are uniformly bounded by 1 for $\rho, \kappa \geq 0$, and part b) of Corollary 3.4 applies if $V^c(1) > 0$. Since $S \geq 1 - id$, one has for $\rho + \kappa - 2\lambda \geq 0$ that $V^c(1) \geq \int_0^1 (1-id)^{\rho+\kappa-2\lambda} d\bar{H}^1$, the latter expression being strictly positive if and only if the probability of no censoring, $\bar{H}^1(1) = (1 - \int G_- dF)$, is strictly positive.

DISCUSSION. It is an immediate consequence, that under $\mathcal{H}_0$ for almost all sequences $(\mathbf{T}_n, \Delta_n) = (\mathbf{t}_n, \delta_n), n \geq 1$, the $(1-\alpha)$-quantiles $k_n(\alpha, \mathbf{t}_n, \mathbf{d}_n)$, of the distributions in (3.19) (linear rank statistics) converge to the quantile $k(\alpha)$ of the standard normal distribution being the true asymptotical critical value of the corresponding unconditional test. Hence, in this way the conditional and the unconditional tests are asymptotically equivalent under $\mathcal{H}_0$ and under contiguous alternatives also for the discontinuous case.
An analogue statement for RS statistics is in general not possible. While for continuous H one has $A^\tau = [0,1]$, the set A^τ and, consequently, the distribution of $\sup\{| \mathrm{B}(t) | : t \in A^\tau\}$ depends on H for discontinuous H. Hence, if H is discontinuous, the corresponding $(1-\alpha)$-quantiles $k_n(\alpha, \mathbf{t}_n, \mathbf{d}_n)$ of the conditional Renyi type tests depend asymptotically under $\mathcal{H}_0$ on F, G_1 and G_2, just as for the unconditional Renyi type tests, though having different limits if $G_1 \neq G_2$. But recall that the conditional tests are finite sample distribution free under the restricted null hypothesis.
By introducing randomized Renyi type statistics we get tests which are asymptotically distribution free under $\mathcal{H}_0$ with unequal censoring.

Corollary 3.6. (Randomized Renyi Type Tests) *For the randomized version $RS_n^{\tau c}$ of RS_n^τ, given by*

$$RS_n^{\tau c} = \sup\left(\left|\mathrm{W}_n^c\left(\frac{i}{n}\right)\right| : \frac{i}{n} \leq \tau\right) \Big/ \left(\mathrm{V}_n^c(\tau)\right)^{1/2}, \tag{3.23}$$

one gets in the situation of Corollary 3.4 *b),*

$$RS_n^{\tau c} \to_{\mathcal{D}} \sup\{| \mathrm{B}(t) | : 0 \leq t \leq 1\}. \tag{3.24}$$

Computation of $RS_n^{\tau c}$ requires an external randomization. Given $(\mathbf{A}_n, \mathbf{B}_n, \mathbf{T}_n, \mathbf{D}_n)$, the Z_i's are obtained by placing at random $A_j[B_j]$ "ones" at $i \in \{T_{j-1}+1, \ldots, T_{j-1}+D_j\}[i \in \{T_{j-1}+D_j+1, \ldots, T_j\}]$ and "zeroes" at the remaining places $i \in \{T_{j-1}, \ldots, T_j\}$. Of course, if there are no ties among the uncensored observations, randomization is not needed since then $RS_n^{\tau c} = RS_n^\tau$. Hence, at the cost of randomization the conditional limiting distribution of the Renyi type statistics are the same as in the continuous case. Gill [(1980), p. 75,] introduced a certain randomization method in order to reduce in his unconditional setting the discontinuous case to the continuous one. By replacing his randomization method for removing too big jumps by ours, the result is that the unconditional and the conditional limiting distribution of $RS_n^{\tau c}$ under $\mathcal{H}_0$ coincide. Hence, in this case the randomized conditional and the unconditional $RS_n^{\tau c}$ tests are asymptotically equivalent under $\mathcal{H}_0$ and under contiguous alternatives. Of course, the conditional tests remain to be finite sample distribution free under $\bar{H}_0$.

4. PROOFS

PROOF OF LEMMA 3.1 (3.4) and (3.5) follow immediately from the Glivenko-Cantelli theorem for H_n. Let H_n^1 be the empirical sub-df of the uncensored $X_{i:n}$'s , i.e., $\mathrm{H}_n^1(x) =$

$\frac{1}{n}\sum_{i=1}^{n}\Delta_{i:n}1\{X_{i:n}\le x\}$. For proving (3.6) let H_n^1 and $\mathrm{H}_n, n\ge 1$, be a fixed sequences of realizations, s.th. $\mathrm{H}_n^1 - H^1 \to 0$ and $\mathrm{H}_n - H \to 0$ uniformly on $[0,\infty)$, which is possible by Glivenko-Cantelli theorems for H_n^1 and H_n. For $t\in(0,1)$ define $x := H^{-1}(t)$ and let $\epsilon>0$ be fixed. Put $t_{1n} := \mathrm{H}_n(x-), t_{2n} := \mathrm{H}_n(x), t_1 := H(x-), t_2 := H(x)$ and choose n_0 s.th. $|\ t_{1n}-t_1\ |<\epsilon, |\ t_{2n}-t_2\ |<\epsilon \forall n\ge n_0$. Then $\bar{\mathrm{H}}_n^1(t_{1n}) = \mathrm{H}_n^1(x-), \bar{\mathrm{H}}_n^1(t_{2n}) = \mathrm{H}_n^1(x)$. Hence, $\bar{\mathrm{H}}_n^1(t_{1n}) \to H^1(x-) = \bar{H}^1(t_1)$ and $\bar{H}_n^1(t_{2n}) \to H^1(x) = \bar{H}^1(t_2)$. Since $|\ \bar{\mathrm{H}}_n^1(t_{1n}) - \bar{\mathrm{H}}_n^1(t_1)\ |\le|\ t_{1n}-t_1\ |<\epsilon$ and $|\ \bar{\mathrm{H}}_n^1(t_{2n}) - \bar{\mathrm{H}}_n^1(t_2)\ |\le|\ t_{2n}-t_2\ |\ \forall n\ge n_0$, and since $\epsilon>0$ is arbitrary, one gets $\bar{\mathrm{H}}_n^1(t_1)\to\bar{\mathrm{H}}^1(t_1)$ and $\bar{\mathrm{H}}_n^1(t_2)\to\bar{H}^1(t_2)$. If $t=t_1$ or $t=t_2$, then (3.5) follows. If $t_1<t_2$ and $t_1<t<t_2$, then, for $n\ge n_0, t_{1n} = \mathrm{H}_n(x-) < t < \mathrm{H}_n(x) = t_{2n}$ and according to the definition of $\bar{\mathrm{H}}_n^1 : \bar{\mathrm{H}}_n^1(t) = \mathrm{H}_n^1(x-) + \min((t-\mathrm{H}_n(x-)), \Delta\mathrm{H}_n^1(x)) \to H^1(x-) + \min((t-H(x-)), \Delta H^1(x)) = \bar{H}^1(t)$. ∎

Proof of Lemma 3.2 For given $\tau\in(0,1)$ denote by $c_n^0[c_n^1]$ the minimum [maximum] over the values $e_i := \pi_i(1-\pi_i)n^2/(n_1n_2)$, $1\le i\le n\tau$. Fix $j\in\{1,\dots,r\}$ s.th. $T_j\le n\tau$ and put $a := T_{j-1}, b := T_j$ and $d := D_j$. Then, assuming $w_{nj}^2>0$, one has for $E_n := (\mathrm{V}_n(b/n) - \mathrm{V}_n(a/n))/w_{nj}^2$ and E_n^c similarly defined for V_n^c that

$$E_n = \frac{n-a-d}{n-a-1}\cdot\frac{d}{n}\cdot e_a \tag{4.1}$$

and

$$E_n^c = \sum_{i=a+1}^{a+d}\left(\frac{n-a-d}{n-i}\right)^2\cdot\frac{1}{n}\cdot e_i. \tag{4.2}$$

We have $E_n = E_n^c$ for $d=1$ and for $n-a-d=0$. Therefore, we assume in the sequel that $d\ge 2$ and $n-a-d>0$. Using the general inequality

$$\frac{(\beta-\alpha+1)}{(n-\beta)(n-\alpha+1)} \le \sum_{i=\alpha}^{\beta}(n-i)^{-2} \le \frac{\beta-\alpha+1}{(n-\beta-1)(n-\alpha)}, \tag{4.3}$$

and $(n-a-d)/(n-a-d-1)\le 1+z/n$, $(n-a-1)/(n-a)\ge 1-z/n$ with $z := 2/(1-\tau)$ if $n\ge z$ one obtains in the latter case

$$E_n \ge E_n^c\left(c_n^0/c_n^1\right)\left(1+\frac{z}{n}\right)^{-1} \tag{4.4}$$

and

$$E_n \le E_n^c\left(c_n^1/c_n^0\right)\left(1-\frac{z}{n}\right)^{-1}, \tag{4.5}$$

hence $|\ E_n^c - E_n\ |\le E_c^n\cdot\epsilon_n^\tau, \epsilon_n^\tau := \max\left\{\left(c_n^1/c_n^0\right)\left(1-\frac{z}{n}\right)^{-1} - 1;\ 1-\left(c_n^0/c_n^1\right)\left(1+\frac{z}{n}\right)^{-1}\right\}$ and, consequently, $\sup\{|\ \mathrm{V}_n^c(t) - \mathrm{V}_n(t)\ | : t\in\mathrm{t}_n, t\le\tau\} \le \mathrm{V}_n^c(\tau)\cdot\epsilon_n^\tau$. Since $c_n^i \to_p 1, i = 0,1$, according to Theorem 24.1 of Billingsley (1968), one gets $\epsilon_n^\tau\to_p 0$. Combined with convergence in distribution of $\mathrm{V}_n^c(\tau)$, see Theorem 3.3, (3.13) follows. ∎

Proof of Theorem 3.3.

a) According to Theorem 3.2 in Neuhaus (1992) we have merely to show

$$\int_0^1 (w_nu_n - wu)^2\, d\lambda \to_p 0, \tag{4.6}$$

in order to get (3.15) and (3.16). Because of assumption (3.14) and of $| u_n | \leq 1, | u | \leq 1$, it is apparently enough to show

$$u_n \to u\,[\lambda]. \tag{4.7}$$

Let $x_1, x_2, \ldots$ be an enumeration of the jump points of H and put $t_i^{\pm} := H(x_i\pm)$, $d_i := \Delta H^1(x_i)$. If $d_i > 0$ and $t \in (t_i^-, t_i^- + d_i)$ we may assume w.l.o.g. that H_n and H_n^1 are realizations s.th. $\mathrm{H}_n(x_i\pm) \to t_i^{\pm}, \Delta\mathrm{H}_n^1(x_i) \to d_i$, hence $u_n(t) \to u(t)$. Likewise, for $t \in (t_i^- + d_i, t_i^+) \neq \emptyset$ (possibly $d_i = 0$) one gets $u_n(t) \to u(t) = 1$.
Finally, assume that $t = H(x)$ for some x with $\Delta H(x) = 0$ and $H(y) \neq t\,\forall y \neq$ x. Such we exclude at most countably many of the points not treated so far. Given $\epsilon > 0$, choose $y < x < z$ s.th. $t - \epsilon < H(y-) \leq H(y) < t < H(z) < t + \epsilon$. For n large enough, $\mathrm{H}_n(y\pm)$ and $\mathrm{H}_n(z)$ fulfill the same inequality. Consequently,

$$\begin{aligned} u_n(t) &\geq \min\left\{\frac{n - T_j - D_j}{n - T_j - 1} : \frac{T_j}{n} \in \left[\mathrm{H}_n(y-), \mathrm{H}_n(z)\right]r = \right\} \geq \\ &\geq \left(1 - \mathrm{H}_n(z)\right) / \left(1 - \mathrm{H}_n(y-) - \frac{1}{n}\right) \geq (1 - t - \epsilon) / (1 - t + \epsilon + 1/n), \end{aligned}$$

implying $u_n(t) \to 1 = u(t)$.
b) If the weights w_n are uniformly bounded by some constant c, say, then $V_n^c(1) < \infty$ and $V^c(1) < \infty$, and $\mathrm{V}_n^c(1) - V_n(t) \leq n^2/(n_1 \cdot n_2) \cdot c^2 \cdot (1 - \frac{[nt]}{n}) = 0(1 - t)$. According to part *b*) Theorem 3.2 in Neuhaus (1992), this yields (3.15) in the Skorohod space $D[0, 1]$ and (3.16) also for $t = 1$. ■

PROOF OF COROLLARY 3.4.
a) Note that $\mathrm{W}_n(t) = \mathrm{W}_n^c(\mathrm{T}_n(t))\,\forall t \in [0, 1]$ and $R_n(\tau) := \sup\{| \mathrm{V}_n(t) - V_n^c(\mathrm{T}_n(t)) | : 0 \leq t \leq \tau\} \to_p 0\,\forall \tau \in (0, 1)$, the latter property following from (3.13). If $\Delta\mathrm{T}(\tau) = 0$, then $\mathrm{T}_n(\tau) \to \mathrm{T}(\tau) = \tau^-$and (3.19) resp (3.20) follow immediately from (3.15) resp.(3.16). If $\Delta\mathrm{T}(\tau) > 0$, then $\{\mathrm{T}_n(\tau)\}$ has two possible accumulation points, namely τ^- and τ. Along all converging subsequences of $\{\mathrm{T}_n(\tau)\}$ assertion (3.19) holds true and, therefore, it is true in general. Note that for Renyi type statistics this argument does not work since then the limiting distributions may differ along different subsequences.
b) Note that $w_n = v_n \circ \mathrm{T}_n$.Hence, for $t \in [0, \tau]$ with $\Delta\mathrm{T}(t) = 0$ we get $| w_n(t) - v(\mathrm{T}(t)) | \leq \sup\{| v_n(u) - v(u) | : 0 \leq u \leq \tau\} + | v \circ \mathrm{T}_n(t) - v \circ \mathrm{T}(t) | = o_p(1) + o(1)$. Since v is bounded on $[0, \tau]$, Lebesgue's theorem applies and yields (3.14). By part *b*) of Theorem 3.3 one gets that (3.15) and (3.16) hold true in the Skorohod space $D[0, 1]$ resp. on [0,1]. It remains to show that (3.13) is fulfilled for $\tau = 1$. If $\mathrm{T}(t)$ is left continuous at $t = 1$, we use the fact that $\mathrm{V}_n(1) - V_n(\mathrm{T}_n(\tau)) \leq c(1 - \mathrm{T}_n(\tau))$, for $0 < \tau < 1$, and an analogue inequality for V_n^c, where c is some generic constant. Hence $R_n(1) \leq 2c(1 - \mathrm{T}_n(\tau)) + R_n(\tau)$. By $R_n(\tau) \to_p 0, \mathrm{T}_n(\tau) \to \mathrm{T}(\tau)$ for $\Delta\mathrm{T}(\tau) = 0$, and $\mathrm{T}(\tau) \to 1$ if $\tau \to 1$, we get $R_n(1) \to_p 0$. Secondly, if $\Delta\mathrm{T}(1) > 0$, then there exists some x_0 s.th.$H(x_0) = 1$ and $H(x_0-) =: t_0^- < 1$. Then $T_{r-1}/n = \mathrm{H}_n(x_0-) \to t_0^-$ and $D_r/n = \mathrm{H}_n(1) - \mathrm{H}_n(T_{r-1}/n)$. We will use the inequality

$$R_n(1) \leq R_n(\tau) + | a_n - a_n^c | \tag{4.8}$$

for $0 < \tau < 1$, with

$$a_n := \frac{n}{n_1 n_2} v_n^2(T_{r-1}/n) p_r q_r D_r \cdot \frac{n - T_{r-1} - D_r}{n - T_{r-1} - 1} \text{and} a_n^c := \mathrm{V}_n^c(1) - V_n^c(T_{r-1}/n).$$

Now one has $a_n \to_p a$ with $a := v^2(t_0^-)d_0(1-t_0^- - d_0)/(1-t_0^-)$, $d_0 := \bar{H}^1(1) - \bar{H}^1(t_0^-)$ and $a_n^c \to V^c(1) - V^c(t_0)$ which equals also 'a' by

$$V^c(1) - V^c(t_0) = \int_{t_0^-}^{1} w^2u^2\, d\bar{H}^1 = v^2(t_0^-)(1-t_0^- - d_0)^2 \int_{t_0^-}^{t_0^- + d_0} (1-t)^{-2} dt \qquad (4.9)$$

Since $R_n(\tau) \to_p 0$, (4.8) yields $R_n(1) \to_p 0$. ∎

PROOF OF THE CONDITIONS IN EXAMPLE 3.5. Define functions v_n by

$$v_n(t) := \mathrm{S}_n^\rho\left(\frac{i}{n}\right)\left(\frac{n-i}{n}\right)^{\kappa - 2\lambda}\left(p_jq_jn^2/(n_1n_2)\right)^{-\lambda} \cdot 1\{p_jq_j > 0\},$$

with $i = [nt]$ and $j := \max\{\nu : T_{\nu-1} \le i\}$. Then $w_{nj} = v_n(T_{j-1}/n)$ and it follows from Neuhaus(1992) that $\mathrm{S}_n \to S$ uniformly on [0,1). Argueing as in the proof of Lemma 3.2 the terms $p_jq_jn^2/(n_1n_2)$ tend to 1, hence $\sup\{|\, v_n(t) - v(t)\,| : 0 \le t \le \tau\} \to_p 0$ for all $\tau \in (0,1)$. The remaining assertions are clear. ∎

REFERENCES

[1] AALEN, O.O. (1978). Nonparametric Inference for a Family of Counting Processes, *Ann. Statist.*, **46**, 77–85.

[2] ANDERSEN, P.K., BORGAN, O., GILL, R. AND KEIDING, N. (1982). Linear Nonparametric Tests for Comparison of Counting Processes, with Applications to Censored Survival Data, *Int. Stat. Rev.*, **50**, 219–258.

[3] BILLINGSLEY, P. (1968). *Convergence of Probability Measures.* John Wiley and Sons, New York.

[4] FLEMING, TH.R., HARRINGTON, D.P. AND O'SULLIVAN, M. (1987). Supremum Versions of the Log-rank and Generalized Wilcoxon Statistics, *J. Amer. Statist. Assoc.*, **82**, 312–320.

[5] GEHAN, E. (1965). A Generalized Wilcoxon Test for Comparing Arbitrarily Single Censored Samples, *Biometrika*, **52**, 203–223.

[6] GILL, R.D. (1980). *Censoring and Stochastic Integrals. Math. Centre Tracts 124.* Mathematisch Centrum, Amsterdam.

[7] HARRINGTON, D.P. AND FLEMING, T.R. (1982). A Class of Rank Test Procedures for Censored Survival Data, *Biometrika*, **69**, 553–566.

[8] JACOD, J. AND SHIRYAEV, A.N. (1987). *Limit Theorems for Stochastic Processes.* Springer Verlag, Berlin.

[9] JANSSEN, A. (1991). Conditional Rank Tests for Randomly Censored Data, *Ann. Statist.*, **19**, 1434–1456.

[10] JANSSEN, A. AND BRENNER, S. (1991). Monte Carlo Results for Conditional Survival Tests under Randomly Censored Data, *J. Statist. Comput. Simul.*, **39**, 47–62.

[11] LATTA, R.B. (1981). A Monte Carlo Study of Some Two-sample Rank Tests with Censored Data, *Journal of the American Statistical Association*, **76**, 713–719.

[12] LEURGANS, S. (1984). Asymptotic Behavior of Two-sample Rank Tests in the Presence of Random Censoring, *Ann. Statist.*, **12**, 572–589.

[13] MANTEL, N. (1967). Ranking Procedures for Arbitrarily Restricted Observation, *Biometrics*, **23**, 165–178.

[14] NEUHAUS, G. (1988). Asymptotically Optimal Rank Tests for the Two-sample Problem with Randomly Censored Data, *Commun. Statist. Theory Methods*, **17**, 2037–2058.

[15] NEUHAUS, G. (1992/93). Conditional Rank Tests for the TwO-sample Problem under Random Censorship, *Ann. Statist.* (Tentatively accepted)

[16] PETO, R. AND PETO, J. (1972). Asymptotically Efficient Rank Invariant Test Procedures, *J. R. Statist. Soc. A*, **135**, 1185–206.

[17] PRENTICE, R.L. (1978). Linear Rank Tests with Right Censored Data, *Biometrika*, **65**, 165–179.

[18] SCHUMACHER, M. (1984). Two Sample Tests of Cramér–von Mises and Kolmogorov–Smirnov Type for Randomly Censored Data, *Int. Statist. Review*, **52**, 263–281.

Recent Adv. in Stat. and Prob., pp. 139-154
J. Pérez Vilaplana and M. L. Puri (Eds)

Nonparametric Estimation of Regression Functions based on Dependent Data

A. QUINTELA DEL RÍO
Departamento de Matemáticas, Facultad de Informática, Universidad de La Coruña, Campus da Zapateira, 15071 La Coruña, Espana

Abstract – The Gasser-Müller Kernel estimate is considered for constructing a nonparametric estimate of a regression function, in a fixed design context, in presence of strongly mixing errors. Some asymptotic properties, as strong consistency and asymptotic normality are derived. The bandwidth selection problem is also investigated, and the classical leave-one-out technique of cross-validation is extended to incorporate the dependence assumptions. The method leads to a selected parameter which is asymptotically optimal with respect to some quadratic measures of error.

1. INTRODUCTION

Let us assume a situation where a real random variable Y is recorded which depends on a design parameter t, *e.g.* time or age. This is the *fixed design regression* $Y_i = m(t_i) + \varepsilon_i$ where $m(\cdot)$ is an unspecified function not restricted in its functional form.
A kernel estimate for fixed design was proposed by Priestley and Chao (1972) and further studied by Benedetti (1977). Schuster and Yakowitz (1979) proved uniform convergence of the Priestley-Chao estimate and also of derivatives of order v of the smoothed curve. A further kernel estimate was introduced and studied by Gasser and Müller (1979) and later generalized to estimating derivatives (1984).
In this paper, the model considered is the following:

$$Y_i = m(t_i) + \varepsilon_i \qquad i = 1, \ldots, n$$

Here m is a smooth unknown regression function defined on the interval $[0,1]$ (without loss of generality), t_i are design points depend on n $(0 \leq t_1 \leq \ldots \leq t_n \leq 1)$ and ε_i are the regression errors. The estimate proposed in Gasser and Müller (1979) for estimating m is defined as follows:

$$\hat{m}_h(t) = \sum_{j=1}^{n} Y_j \int_{S_{j-1}}^{S_j} K_h(t-u)du \tag{1}$$

where $0 = S_0 \leq S_1 \leq \ldots \leq S_n = 1, \quad t_j \leq S_j \leq t_{j+1}, \quad \max_j |S_j - S_{j-1}| = O(n^{-1})$ and where $K_h(z) = h^{-1}K(z/h)$ with K the *kernel function* and h the *smoothing parameter*, or *bandwidth*.
Asymptotic properties, as strong consistency, and asymptotic normality, are derived in Gasser and Müller (1979), when the regression errors ε_i are independent. We prove

similar results when the errors are not independent. We assume that the observations are generated by a stationary strong mixing process. This is a less restrictive condition than the ones usually assumed for time series.
In section 2 we discuss the dependence structure for the regression errors, and give conditions that ensure the asymptotic properties of the estimator. In section 3, we define a data driven bandwidth that is asymptotically optimal, because minimizes some quadratic measures of error. The bandwidth selector is defined by cross-validation, modifying the leave-out technique involved in the method. Recent works in nonparametric regression involving time series are, for example, Diggle and Hutchinson (1989), Györfi et al (1990), Truong (1991), Truong and Stone (1992), and Roussas et al (1992), among others.

2. DEPENDENCE STRUCTURES AND ASYMPTOTIC OPTIMALITY.

2.1. Dependence Structures.

Let N denote the set of positive integers, and for any i and j in $N \cup \{\infty\} (i \leq j)$ define $\mathcal{F}_i^j$ to be the σ-algebra spanned by the variables $\varepsilon_i, \ldots, \varepsilon_j$. The dependence among the observations is usually modelled by some mixing condition. The most often studied of these is the so-called α-mixing or strong mixing condition, due to Rosenblatt (1956). The sequence $\{\varepsilon_i\}$ is said to be α-mixing if there exist mixing coefficients $\alpha(m)$ such that $lim_{m\to\infty}\alpha(m) = 0$ and for positive integers k and m and for any sets A and B that are, respectively, $\mathcal{F}_1^k$-measurable and $\mathcal{F}_{k+m}^{\infty}$-measurable,

$$|P(A \cap B) - P(A)P(B)| \leq \alpha(m)$$

When the dependent observations are considered in nonparametric regression, this is a convenient dependence structure, because contains the class of ARMA processes with continuous white noise (see Bradley (1985) for more complete discussions of various dependence structures).

2.2. Asymptotic Results.

We make the following assumptions:

About the S_j :
(S1) $\max_{1\leq j\leq n} |S_j - S_{j-1}| = O(n^{-1})$

About the errors and the mixing coefficients:

- **(E1)** $\{\varepsilon_i\}_{i=1}^n$ are stationary, identically distributed, strong mixing, with $E(\varepsilon_i) = 0$ and $Var(\varepsilon_i) = \sigma^2 < \infty$.
- **(E2)** There exists $\delta > 0$ such that $E|\varepsilon_i|^{2+\delta} < \infty$.
- **(E3)** $\sum_{n=1}^{\infty} \alpha(n)^{\delta/2+\delta} < \infty$, where $\alpha(n)$ are the mixing coefficients.
- **(E4)** Let $z = z(n) \in N$ and let $\mu = \mu(n)$ the largest positive integer for which $2z\mu \leq n$. Then

$$\limsup \left[1 + 6\sqrt{e}(\alpha(z))^{1/1+\mu}\right]^{\mu} < \infty$$

About the kernel function K :

(K1) K is symmetric and has compact support

(K2) K is Hölder continuous:

$$|K(x) - K(y)| \leq C_k |x-y|^{\psi}, \quad \psi > 0, \quad C_k \in R$$

(K3) $\int x^q K(x)dx = 0, \quad q = 1, \ldots, k-1$ and $0 < \int x^k K(x)dx < \infty$

About the function m :

(M1) m is Hölder continuous.

(M2) m has k continuous derivatives.

2.3. Results

Theorem 1 *Assume that conditions* **(S1)** - **(E1)** - **(K1)** - **(M1)** *hold and that* $h \to 0$ *and* $nh \to \infty$. *Then we have*

$$E(\hat{m}_h(t)) \longrightarrow m(t)$$

if m *is continuous at* t.

PROOF: The proof is the same to that of theorem 1 of Gasser and Müller (1984), for the case of independent errors.

2.4. Pointwise Strong Consistency.

Theorem 2 *Under assumptions* **(S1)** - **(E1)** - **(E2)** - **(E4)** - **(K1)** - **(M1)**, *and if* h, z *are chosen, so that*

(E5) $\sum_{n=1}^{\infty} \exp(-Cn^{1-2\lambda}h^2) < \infty \quad \left(C > 0, \lambda > \frac{1}{2+\delta}\right)$ *and* $\{n^{\lambda}/zh\}$ *is bounded away from 0, then we have*

$$\hat{m}_h(t) \longrightarrow m(t) \qquad a.s.$$

for any continuity point t *of* m.,

Remark. In order to gain some insight as to what kind of h, z and λ satisfy the assumptions, let us suppose that the choice of z equal to the integral part of $n^r/2, z = [n^r/2]$ for some $0 < r < 1$ satisfies assumption **(E4)**, in which case $\mu = [n^{1-r}]$. Then, for $h = n^{-\theta}, (\theta > 0)$, $\sum_{n=1}^{\infty} \exp(-Cn^{1-2\lambda}h^2) < \infty$, and $\{n^{\lambda}/zh\}$ is bounded away from 0, provided $r \leq \theta$ and $\lambda + \theta < 0.5$. (See Roussas (1989)).

PROOF: We can write

$$\hat{m}_h(t) - m(t) = \hat{m}_h(t) - E\hat{m}_h(t) + [E\hat{m}_h(t) - m(t)]$$

By Theorem 1, the second term tends to zero. Let us study the first term:

$$\hat{m}_h(t) - E\hat{m}_h(t) = h^{-1} \sum_{j=1}^{n} \varepsilon_j \int_{S_{j-1}}^{S_j} K\left(\frac{t-u}{h}\right) du$$

The regression errors ε_i are not assumed to be bounded. The problems arising from this unboundedness are dealt with by means of suitable truncation. Denote by

$$\bar{\varepsilon}_j = \varepsilon_j 1_{\{|\varepsilon_j| \leq B_n\}}, \quad \text{with} \quad B_n = n^{\lambda}, \quad \lambda > \frac{1}{2+\delta} \tag{2}$$

Clearly, $\varepsilon_j = \overline{\varepsilon}_j + (\varepsilon_j - \overline{\varepsilon}_j)$. Then, we have

$$\begin{aligned} R_n(t) &= h^{-1}\sum_{j=1}^{n}\varepsilon_j \int_{S_{j-1}}^{S_j} K\left(\frac{t-u}{h}\right) du = \\ &= (R_n(t) - R_n^B(t)) + (R_n^B(t) - ER_n^B(t)) + (ER_n^B(t) - ER_n(t)) \end{aligned} \quad \textbf{(3)}$$

with

$$R_n^B(t) = h^{-1}\sum_{j=1}^{n}\overline{\varepsilon_j}\int_{S_{j-1}}^{S_j} K(\frac{t-u}{h})du \quad \textbf{(4)}$$

Previous result: Let us note that

$$\begin{aligned} P(|\varepsilon_n| > B_n) \leq B_n^{-(2+\delta)} E|\varepsilon_n|^{2+\delta} &\Longrightarrow \sum_{n=1}^{\infty} P(|\varepsilon_n| > B_n) \leq \sum_{n=1}^{\infty} B_n^{-(2+\delta)} E|\varepsilon_n|^{2+\delta} = \\ &= E|\varepsilon_1|^{2+\delta}\sum_{n=1}^{\infty} n^{-\lambda(2+\delta)} < \infty \end{aligned}$$

by (2). We obtain that, with probability one, $|\varepsilon_n| \leq B_n$ for n large enough, and we can also obtain that

$$|\varepsilon_j| \leq B_n \quad \forall j \leq n \quad \textbf{(5)}$$

Now, we study the almost sure convergence to zero for the three terms of (3).
First term:

$$R_n(t) - R_n^B(t) = h^{-1}\sum_{j=1}^{n}(\varepsilon_j - \overline{\varepsilon}_j)\int_{S_{j-1}}^{S_j} K\left(\frac{t-u}{h}\right) du = 0$$

for n large enough, using (5).
Third term:

$$\begin{aligned} \left|ER_n(t) - ER_n^B(t)\right| &= \left|E\left\{h^{-1}\sum_{j=1}^{n}(\varepsilon_j - \overline{\varepsilon}_j)\int_{S_{j-1}}^{S_j} K\left(\frac{t-u}{h}\right) du\right\}\right| \leq \\ &\leq h^{-1}\sum_{j=1}^{n}\left|\int_{S_{j-1}}^{S_j} K\left(\frac{t-u}{h}\right) du\right| E\left\{\left|\varepsilon_j 1_{\{|\varepsilon_j| > B_n\}}\right|\right\} \end{aligned}$$

Let us compute this last expectation

$$E\left\{\left|\varepsilon_j 1_{\{|\varepsilon_j| > B_n\}}\right|\right\} = \int_{\{|x| > B_n\}} |x|^{1-s}|x|^s dF_\varepsilon(x)dx \leq (\star)$$

where F_ε are the distribution of the errors, and $s = 2+\delta$. Then, since $|x| > B_n$

$$(\star) \leq \int_{\{|x|>B_n\}} B_n^{1-s}|x|^s dF_\varepsilon(x)dx \leq B_n^{1-s}\int_R |x|^s dF_\varepsilon(x)dx = B_n^{1-s}E|\varepsilon_j|^s.$$

Also, we have

$$h^{-1}\sum_{j=1}^{n}\left|\int_{S_{j-1}}^{S_j} K\left(\frac{t-u}{h}\right) du\right| = O(1)$$

Therefore,

$$\left|ER_n(t) - ER_n^B(t)\right| = O\left(B_n^{1-s}\right) = O\left(n^{\lambda(1-(2+\delta))}\right)$$

Second term:

$$\begin{aligned} R_n^B(t) - ER_n^B(t) &= h^{-1}\sum_{j=1}^{n}\int_{S_{j-1}}^{S_j} K\left(\frac{t-u}{h}\right) du\,\{\bar{\varepsilon}_j - E\bar{\varepsilon}_j\} = \\ &= n^{-1}\sum_{j=1}^{n} nh^{-1}\int_{S_{j-1}}^{S_j} K\left(\frac{t-u}{h}\right) du\,\{\bar{\varepsilon}_j - E\bar{\varepsilon}_j\} = n^{-1}\sum_{j=1}^{n} Z_j \end{aligned}$$

We now investigate the probability $P(|n^{-1}\sum_{j=1}^{n} Z_j| > \varepsilon)$. For this, we use Theorem A of Roussas (1989) and we have

$$\begin{aligned} EZ_j &= 0 \\ |Z_j| &\le nh^{-1}|S_j - S_{j-1}| \cdot ||K||_\infty 2B_n = O(h^{-1}B_n) \end{aligned}$$

because, using the mean value theorem of integration, we have

$$\int_{S_{j-1}}^{S_j} K\Big((t-u)/h\Big)m(u)\,du = (S_j - S_{j-1})K\Big((t-c_j)\;/\;h\Big)m(c_j)$$

with c_j a point in (S_{j-1}, S_j). It follows that

$$P\Big(|n^{-1}\sum_{j=1}^{n} Z_j| > \varepsilon\Big) \le C\exp\Big\{-c\varepsilon^2 n(h^{-1}B_n)^{-2}\Big\} = a_n$$

and $\sum a_n < \infty$ by means of **(E5)**.

2.5. Uniform Strong Consistency.

Conditions in Theorem 2 are not suffice for uniform strong consistency. They are to be supplemented as follows:

Theorem 3 *Under assumptions* **(S1)** - **(E1)** - **(E2)** - **(E4)** - **(K1)** - **(K2)** - **(M1)** *and*
(E5'), *h, z and N (to use in the proof) are chosen so that*
$n^\lambda h^{-2} N^{-1} \to 0$,
$\sum_{n=1}^{\infty} N^{-1}\exp\Big\{-Cn^{1-2\lambda}h^{-2}\Big\} < \infty$
$\left\{\frac{n^\lambda}{zh}\right\}$ *is bounded away from zero.* $\Big(C > 0, \lambda > \frac{1}{2+\delta}\Big)$
then we have

$$\sup_t |\hat{m}_h(t) - m(t)| \longrightarrow 0 \qquad a.s.$$

Proof: Is similar to the proof of theorem above. We can write

$$\hat{m}_h(t) - m(t) = \hat{m}_h(t) - E\hat{m}_h(t) + [E\hat{m}_h(t) - m(t)]$$

and the first term is

$$\hat{m}_h(t) - E\hat{m}_h(t) = h^{-1}\sum_{j=1}^{n}\varepsilon_j\int_{S_{j-1}}^{S_j} K\Big((t-u)/h\Big)\,du = R_n(t)$$

All that remains is to prove

$$\sup_{t\in[0,1]} |R_n(t)| \longrightarrow 0 \qquad a.s.$$

For this, let N be as in **(E5')**, and let us divide $[0,1]$ into N subintervals J_{nl}, each of length $1/N$. Let t_{nl} be arbitrary points in J_{nl}. Now, we decompose $R_n(t)$ in the following form

$$\begin{aligned}\left|R_n(t)\right| \;\le\; & \left|R_n(t)-R_n^B(t)\right| + \left|ER_n(t)-ER_n^B(t)\right| + \left|R_n^B(t)-R_n^B(t_{nl})\right| + \\ & + \left|ER_n^B(t)-ER_n^B(t_{nl})\right| + \left|ER_n(t)\right| + \left|R_n^B(t_{nl})-ER_n^B(t_{nl})\right| \qquad (6)\end{aligned}$$

with $R_n^B(t)$ as in (4).
We have that

$$\sup_{t\in[0,1]} |R_n(t)| = \max_{l=1\ldots N} \sup_{t\in J_{nl}} |R_n(t)|$$

Now, we study, one by one, the terms of (6):

(1) $|R_n(t)-R_n^B(t)| = 0$ a.s. for all sufficiently large n (as theorem above).

(2) $\forall t |ER_n(t)-ER_n^B(t)| \le CB_n^{1-s}$ (See theorem above.) $(B_n = n^\lambda)$

(3) We have

$$\left|R_n^B(t)-R_n^B(t_{nl})\right| \le CB_n h^{-2}|t-t_{nl}|$$

PROOF: $\forall t, t' \in [0,1]$,

$$\begin{aligned}&\left|R_n^B(t)-R_n^B(t')\right| \le \\ &\le h^{-1}\sum_{j=1}^{n}\left|\bar{\varepsilon}_j\left[\int_{S_{j-1}}^{S_j} K\big((t-u)/h\big)\,du - \int_{S_{j-1}}^{S_j} K\big((t'-u)/h\big)\,du\right]\right| \le \\ &\le h^{-1}\sum_{j=1}^{n}\left|\bar{\varepsilon}_j\right|\int_{S_{j-1}}^{S_j} Ch^{-1}|t-t'|\,du \le CB_n h^{-2}|t-t'|\end{aligned}$$

because K is Hölder continuous. Then

$$\left|R_n^B(t)-R_n^B(t_{nl})\right| \le CB_n h^{-2}|t-t_{nl}| \le CB_n h^{-2}N^{-1} \longrightarrow 0 \qquad \text{by } \mathbf{(E5')}$$

(4) For this term, **(3)** is also true.

(5) This term is trivially zero.

(6) We have

$$\sum_{n=1}^{\infty} P\left[\max_l \left|R_n^B(t_{nl})-ER_n^B(t_{nl})\right| > \epsilon\right] \le \sum_{n=1}^{\infty}\sum_{l=1}^{N} P\left[\left|R_n^B(t_{nl})-ER_n^B(t_{nl})\right| > \epsilon\right] \qquad (7)$$

The term

$$\left|R_n^B(t_{nl})-ER_n^B(t_{nl})\right| = \left|h^{-1}\sum_{j=1}^{n}\int_{S_{j-1}}^{S_j} K\big((t_{nl}-u)/h\big)\,du\,\{\bar{\varepsilon}_j - E\bar{\varepsilon}_j\}\right| = \left|n^{-1}\sum_{j=1}^{n} Z_j\right|$$

with $EZ_j = 0\ \forall j$ and $|Z_j| \le Ch^{-1}B_n$ Now, using again Theorem A of Roussas (1989), we have that (7) is least or equal than $\sum a_n < \infty$ by **(E5')**. This completes the proof of the theorem.

2.6. Asymptotic Normality.

Asymptotic normality of the Gasser-Müller estimator (1) when the errors are dependent is achieved by employing the classics device of proving a central limit theorem for dependent random variables, that is, by writing the sum as a sum of big blocks separated by small blocks; the contribution due to the small blocks is shown to be negligible and the big blocks are shown to be approximately independent. The standard Lindeberg-Feller central limit theorem for independent random variables is then used to get asymptotic normality.
We wish to show that

$$\frac{\hat{m}_h(t) - E\hat{m}_h(t)}{\sigma(\hat{m}_h(t))} \longrightarrow N(0,1)$$

where $\sigma(\hat{m}_h(t))$ is the standard deviation of the estimator at a point t. The variance take the form

$$Var(\hat{m}_h(t)) = \frac{\sigma^2}{nh}\int K(u)^2 du + O((nh)^{-1}) \tag{8}$$

Gasser-Müller (1984) proved result (8) when the errors are independent. In Section 3 we establish a similar result in strong mixing case. Now, we use it to prove asymptotic normality.

Theorem 4 *We assume that* $S_j - S_{j-1} = 1/n$ *and let* K *be a support compact kernel. Assume that* **(E1)**, **(E2)**, **(E3)** *holds and*
(N1) *A sequence of integer* $\{q_n\} \to \infty$, *with* $q_n = o((nh^3)^{1/2})$ *exist such that*

$$\left(\frac{n}{h}\right)^{1/2} \sum_{i=q_n}^{\infty} [\alpha(i)]^{1-\gamma} \longrightarrow 0, \qquad if\ n \to \infty$$

where $\gamma = 2/(2+\delta)$. *Then, we have*

$$\frac{\hat{m}_h(t) - E\hat{m}_h(t)}{\sigma(\hat{m}_h(t))} \longrightarrow N(0,1)$$

PROOF: Define

$$Z_{i,n} = (\sigma_n)^{-1}h^{-1}\varepsilon_i \int_{S_{i-1}}^{S_i} K\Big((t-u)/h\Big)\,du, \qquad \sigma_n = \sigma(\hat{m}_h(t))$$

We want to study the asymptotic normality of the sum $S_n = \sum_{i=1}^{n} Z_{i,n}$. The sum S_n is written in terms of small and large blocks as follows. Partition the set $\Delta_n = \{1, 2, \ldots, n\}$ into $2k+1$ subsets

$$\Delta_n = \bigcup_{i=0}^{k-1} \Delta'_{i,n} + \bigcup_{i=0}^{k-1} \Delta''_{i,n} + \Delta''_{k,n}$$

where the subsets $\left\{\Delta'_{i,n}\right\}_{i=0}^{k-1}$ have size p (big block), and the subsets $\left\{\Delta''_{i,n}\right\}_{i=0}^{k-1}$ have size q (small block).

$$\Delta'_{i,n} = \{i(p+q)+1, \ldots, i(p+q)+p\} \qquad i = 0, \ldots, k-1$$
$$\Delta''_{i,n} = \{i(p+q)+p+1, \ldots, (i+1)(p+q)\} \qquad i = 0, \ldots, k-1$$

and $\Delta''_{k,n} = \{k(p+q)+1,\ldots,n\}$ is the remaining term at the end not included in the big or small blocks. Then, S_n can be written as

$$S_n = \sum_{j=0}^{k-1} \eta_j + \sum_{j=0}^{k-1} \xi_j + \xi_k = S'_n + S''_n + S'''_n$$

where

$$\begin{aligned} \eta_j &= \sum \left\{ Z_i : i \in \Delta'_{j,n} \right\} \\ \xi_j &= \sum \left\{ Z_i : i \in \Delta''_{j,n} \right\} \qquad j = 0,\ldots,k-1 \\ \xi_k &= \sum \left\{ Z_i : i \in \Delta''_{k,n} \right\} \end{aligned}$$

We shall prove that :

Lemma 1

$$E\,[S''_n]^2 \longrightarrow 0$$

Lemma 2

$$E\,[S'''_n]^2 \longrightarrow 0$$

Lemma 3

$$\left| E\left[e^{iuS'_n}\right] - \prod_{j=0}^{k-1} E\left[e^{iu\eta_j}\right] \right| \longrightarrow 0$$

Lemma 4

$$\sum_{j=0}^{k-1} E\left[\eta_j^2\right] \longrightarrow 1$$

Lemma 5

$$\sum_{j=0}^{k-1} E\left[\eta_j^2 1_{\{|\eta_j|>\epsilon\}}\right] \longrightarrow 0$$

Asymptotic normality of S_n is a direct consequence from Lemma 1 - Lemma 5.
First, let's take into account that **(N1)** imply that integers $r_n \to \infty$ exist such that

$$r_n q_n = o\left((nh^3)^{1/2}\right) \tag{9}$$

and

$$r_n (nh^{-1})^{1/2} \sum_{i=q_n}^{\infty} [\alpha(i)]^{1-\gamma} \longrightarrow 0 \tag{10}$$

Define $p_n = \left[(nh)^{1/2}/r_n\right]$. Then the following properties trivially hold as $n \to \infty$.

$$\frac{q_n}{p_n h} \longrightarrow 0 \ \left(\Rightarrow \frac{q_n}{p_n} \to 0\right) \tag{11}$$

$$\frac{p_n}{nh} \longrightarrow 0 \ \left(\Rightarrow \frac{p_n}{n} \to 0\right) \tag{12}$$

$$\frac{p_n}{(nh)^{1/2}} \longrightarrow 0 \tag{13}$$

$$\frac{1}{h} \sum_{i=q_n}^{\infty} [\alpha(i)]^{1-\gamma} \longrightarrow 0 \tag{14}$$

$$\frac{n}{p_n} \alpha(q_n) \longrightarrow 0 \tag{15}$$

2.6.1. Proof of Lemmas

Lemma 1

Because $E[Z_j] = 0, \forall j$, we have that $E[\xi_j] = E[\eta_j] = E[\xi_k] = 0$. Then

$$E[S_n'']^2 = Var\left\{\sum_{j=0}^{k-1} \xi_j\right\} = \sum_{j=0}^{k-1} Var(\xi_j) + 2\sum_{\substack{i=0\\ i>j}}^{k-1}\sum_{j=0}^{k-1} Cov(\xi_i, \xi_j) \equiv A_1 + A_2 \tag{16}$$

Now with $m_j = j(p+q)+p$, we have

$$Var(\xi_j) = \sum_{i=1}^{q} Var(Z_{m_j+i}) + \sum_{\substack{i=1\\ i\neq l}}^{q}\sum_{l=1}^{q} Cov(Z_{m_j+i}, Z_{m_j+l}) \tag{17}$$

and then, $A_1 = O(kq/nh)$, since that

$$\left|Cov(Z_{m_j+i}, Z_{m_j+l})\right| \leq 8\alpha(|i-l|)^{1-\gamma}\left(E|Z_{m_j+i}|^{2+\delta}E|Z_{m_j+l}|^{2+\delta}\right)^{1/(2+\delta)} \tag{18}$$

by Davydov's Inequality (See Hall-Heyde, (1980)), and since $\sum_{i=1}^{\infty}\alpha(i)^{1-\gamma} < \infty$ by **(E3)**, and $\sigma_n = \sigma(\hat{m}_h(t)) \sim (nh)^{-1}$.
Next, for A_2 in (16) we have

$$\begin{aligned} A_2 &\leq 2\sum_{\substack{i=0\\ i\neq j}}^{k-1}\sum_{j=0}^{k-1}\sum_{l_1\in\Delta''_{i,n}}\sum_{l_2\in\Delta''_{j,n}} |Cov(Z_{l_1}, Z_{l_2})| \leq \\ &\leq 4\sum_{l_1=1}^{n-p}\sum_{l_2=l_1+p}^{n} |Cov(Z_{l_1}, Z_{l_2})| \leq C\frac{1}{nh}\sum_{l_1=1}^{n-p}\sum_{l_2=l_1+p}^{n}\alpha(|l_2-l_1|)^{1-\gamma} \end{aligned} \tag{19}$$

for n large enough, using again Davydov's Inequality. Hence, by (18) and (19)

$$E[S_n'']^2 = O\left(\frac{kq}{nh} + \frac{1}{h}\sum_{l=p}^{\infty}\alpha(l)^{1-\gamma}\right)$$

using (11) and (14) and the fact that $q \leq p$.

Lemma 2

We have

$$E[S_n''']^2 = Var(\xi_k) = Var\left(\sum_{i\in\Delta''_{k,n}} Z_i\right) \leq C\cdot(\#\Delta''_{k,n})\frac{1}{nh}$$

for n large enough, where $\#\Delta''_{k,n}$ is the size of $\Delta''_{k,n}$. It's easy to check that

$$\#\Delta''_{k,n} \leq p+q$$

and then $E[S_n''']^2 = O\left((p+q)/nh\right) = O\left(p/nh\right)$, that tends to zero, by (12).

Lemma 3

Is an immediate consequence of the following Lemma of Volkonskii and Rozanov (1959): *Let $V_1, V_2, \ldots, V_m$ be r.v.-s. measurable with respect to $\mathcal{F}_{i_1}^{j_1}, \ldots, \mathcal{F}_{i_m}^{j_m}$, respectively, where $1 \leq i_1 < j_1 < i_2 < \ldots < j_m \leq n$, $i_{r+1} - j_r \geq s \geq 1$ and $|V_r| \leq 1$, $r = 1, \ldots, m$. Then,*

$$\left|E\left[\prod_{r=1}^{m} V_r\right] - \prod_{r=1}^{m} E[V_r]\right| \leq 16(m-1)\alpha(s)$$

Using this Lemma with $\mathcal{F}_i^j = \sigma\{X_l, i \leq l \leq j\}$, $V_r = e^{iu\eta_r}$, and $s = q$, we have

$$\left| E\left[e^{iuS_n'}\right] - \prod_{j=0}^{k-1} E\left[e^{iu\eta_j}\right] \right| \leq 16(k-1)\,\alpha(q) \sim \frac{n}{p+q}\alpha(q) \sim \frac{n}{p_n}\alpha(q)$$

and this term tends to zero by (15).

Lemma 4

Note that $Var(S_n') \to 1$ by Lemmas 1-2 and since $Var(S_n) \to 1$. Then, since

$$E[S_n']^2 = Var(S_n') = \sum_{j=0}^{k-1} Var(\eta_j) + \sum_{\substack{i=0 \\ i \neq j}}^{k-1} \sum_{j=0}^{k-1} Cov(\eta_i, \eta_j) \tag{20}$$

Using the same arguments as above for Lemma 1, the second term in (20) is bounded by

$$cte \cdot \frac{1}{h} \sum_{r=q_n}^{\infty} \alpha(r)^{1-\gamma} \longrightarrow 0 \ qquadby \ (14)$$

Lemma 5

We have

$$\sum_{j=0}^{k-1} E\left[\eta_j^2 1_{\{|\eta_j|>\epsilon\}}\right] \leq Cte \cdot kp_n^2 E\left[Z_l^2 1_{\{|\eta_j|>\epsilon\}}\right]$$

where $\eta_j = \sum_{i=1}^{p_n} Z_{j(p+q)+i}$. Then, since that $E[Z_l^2] \leq Cte(nh)^{-1}$ for n large enough, and

$$\max_{0 \leq j \leq k-1} P\left[|\eta_j| > \epsilon\right] \leq O\left(p_n (nh)^{-1/2}\right)$$

by (13), we have

$$\sum_{j=0}^{k-1} E\left[\eta_j^2 1_{\{|\eta_j|>\epsilon\}}\right] \leq Cte \cdot \frac{p_n^2 k}{nh} \max_{0 \leq j \leq k-1} P\left[|\eta_j| > \epsilon\right]$$

that tends to zero, since $p_n^2 (nh)^{-1}$ tends to zero, by (13).

3. CHOOSING THE SMOOTHING PARAMETER

A basic problem of applying the Gasser-Müller estimator to a given sample data is the choice of the bandwidth h. The selection of this smoothing parameter decides about the form of the function $\hat{m}_h$. It's necessary to choose the bandwidth so that some criterion of error is minimized. We consider the following quadratic measures of accuracy for the estimate

- Average Squared Error (ASE):

$$ASE(h) = n^{-1} \sum_{j=1}^{n} \left[\hat{m}_h(t_j) - m(t_j)\right]^2 w(t_j)$$

- Integrated Squared Error (ISE):

$$ISE(h) = \int \left(\hat{m}_h(t) - m(t)\right)^2 w(t)\,dt$$

– and his mean (MISE):

$$MISE(h) = E\int (\hat{m}_h(t) - m(t))^2\, w(t)dt$$

w being some nonnegative weight function, that is introduced to allow elimination of boundary effects [See Gasser and Müller, (1979)].
Below, we establish asymptotic equivalence of ASE, ISE and $MISE$, and asymptotic decomposition of the form

$$d(h) = C_1 h^{2k} + C_2(nh)^{-1} + o\left(h^{2k} + (nh)^{-1}\right)$$

where $d = ASE, ISE$ or $MISE$,
The bandwidth of $\hat{m}_h$ will be selected to minimize the cross-validation criterion

$$CV(h) = n^{-1}\sum_{j=1}^{n} [Y_j - \hat{m}_{h,j}(t_j)]^2\, w(t_j) \tag{21}$$

where the leave-out estimator $\hat{m}_{h,j}$ is defined by

$$\hat{m}_{h,j}(x) = \sum_{i=1}^{n} h^{-1} Y_i \frac{n}{\overline{\gamma}}\left(\int_{S_{i-1}}^{S_i} K((t-u)/h)du\right)\gamma(|i-j|)$$

and γ is a function that verifies

$$\gamma(u) = \begin{cases} 0 & \text{if } u = 0 \\ 1 & \text{if } u > l_n \end{cases}$$

$$0 \le \gamma(u) \le 1 \text{ if } u \le l_n \quad \text{and} \quad \overline{\gamma} = \textstyle\sum_{i=1}^{n} \gamma(|i-j|)$$

The choice of $\gamma(u)$ is very important since it classifies the data in relation to their dependence (proximity in time). A typical function is $\gamma(u) = 1 - I_{[-l_n, l_n]}(u)$, with l_n a positive integer. l_n say us when two data can be treated as if they were independent. By this way, the classical leave-one out technique is extended. See the papers of Altman (1991), Chiu (1989), Chu and Marron (1991), Hart and Wehrly (1986), Hart (1991), and Härdle and Vieu (1990), for a detailed discussion of the efect of dependence on bandwidth selection. In this section, we have to give the following additional conditions to those of Section 2:

About the weight function w :
(W1) w is bounded and has compact support in the inner of $[0,1]$.

The bandwidth h is in the interval:
(H1) $H_n = [An^{-a}, Bn^{-b}]$ with $0 < b \le 1/(2k+1) \le a < 2/(1+4k)$, and $0 < A < B < \infty$.

About the mixing coefficients:
We assume that for

$$r_1 \in (0, s), \qquad \text{with} \qquad s = \frac{1 - \left(a + \frac{2}{2+\delta}\right)}{2}$$

and

$$q' = q - 1 + 2\lambda(2 + \frac{U}{V}) \qquad q = U + V + (2a + 4ka)(2 + \frac{U}{V})$$
$$U = 1 + 2a + 2ka - b \qquad V = (2 - a(1 + 4k) - 2r_1) + 2\lambda$$

then
(L1) $\overline{\alpha}(n^{r_1}) = \sup_{j\geq n^{r_1}} \alpha(j) = o(n^{-q'})$ and $\sum_{k=0}^{n} k\alpha(k)^{\delta/(2+\delta)} = o((nh)^{-1})$
with δ as in **(E2)**, and $\lambda > 1/(2+\delta)$.

Theorem 5 *Let us assume that $S_j - S_{j-1} = 1/n$. Under conditions* **(E1) - (K1) - (K2) - (K3) - (M2) - (W1) - (H1) - (L1)** *we have*

$$\begin{aligned} MISE(h) &= \frac{1}{k!^2}h^{2k}\left(\int u^k K(u)du\right)^2 \left(\int \left(m^k(t)Big\right)^2 w(t)dt\right) + \\ &+ \frac{\sigma^2}{nh}\left(\int K^2(u)\,du\right)\left(\int w(t)\,dt\right) + o\big(MISE(h)\big) \end{aligned}$$

Theorem 6 *Under conditions* **(S1) - (E1) - (K1) - (K2) - (K3) - (M2) - (W1) - (H1) - (L1)** *we have*

$$\lim_{n\to\infty} \sup_{h\in H_n} \frac{|ASE(h) - MISE(h)|}{MISE(h)} = 0 \qquad a.s.$$

$$\lim_{n\to\infty} \sup_{h\in H_n} \frac{|ISE(h) - MISE(h)|}{MISE(h)} = 0 \qquad a.s.$$

Theorem 7 *Under conditions* **(S1) - (E1) - (K1) - (K2) - (K3) - (M2) - (W1) - (H1) - (L1)** *and taking l_n a positive integer least or equal than n^{r_1}* **(H1)**, *and $h_{CV} = \arg\min CV(h)$, we have*

$$\frac{d(h_{CV})}{\inf_{h\in H_n} d(h)} \longrightarrow 1 \quad a.s. \tag{22}$$

with $d(h) = MISE(h)$, $ASE(h)$ or $ISE(h)$.

3.1. Sketches of Proofs

For Theorems 5 and 6, similar results are given by Vieu (1991), for the Nadaraya-Watson Kernel estimates

$$\hat{m}_h(x) = \frac{n^{-1}\sum_{i=1}^{n} K_h(x - X_i)Y_i}{n^{-1}\sum_{i=1}^{n} K_h(x - X_i)}$$

Our proofs use similar arguments than those of Lemmas 8.1 and 9.2 of Vieu (1991), and for the sake of brevity, these are omitted. See Quintela (1992) for details.
Proof of Theorem 7 involve long and tedious computations, and these are based on the techniques used in Hart and Vieu (1990). First, we point out that the regression errors ε_i are not assumed to be bounded, as in Section 2, and we shall have to consider again suitable truncation.
First, we define a new average squared error based on the leave-out estimator, as follows:

$$\widehat{ASE}(h) = n^{-1}\sum_{j=1}^{n} \left[\hat{m}_{h,j}(t_j) - m(t_j)\right]^2 w(t_j)$$

We have:

Lemma 6 *Under the conditions of theorem:*

$$\lim_{n\to\infty}\sup_{h\in H_n}\frac{|\widehat{ASE}(h)-MISE(h)|}{MISE(h)}=0 \qquad a.s.$$

The statement (22) follows from Lemma 6 and

$$\sup_{h,h'\in H_n}\frac{|\widehat{ASE}(h)-\widehat{ASE}(h')-(CV(h)-CV(h'))|}{d(h)+d(h')}\longrightarrow 0 \qquad a.s.$$

and this is true if

$$\sup_{h\in H_n}\frac{CS(h)}{MISE(h)}\longrightarrow 0 \qquad a.s. \tag{23}$$

with

$$CS(h)=n^{-1}\sum_{j=1}^{n}[\hat{m}_{h,j}(t_j)-m(t_j)]\,[m(t_j)-Y_j]\,w(t_j) \tag{24}$$

Because of the following lemma, it is enough to prove (24) for $l_n = l_n^*$

Lemma 7 *Under the conditions of theorem we have*

$$\lim_{n\to\infty}\sup_{h\in H_n'}\left|\frac{CS_{l_n}(h)-CS_{l_n^*}(h)}{MISE(h)}\right|=0 \quad a.s.$$

Now denote by H_n' a finite subset of H_n composed of equally spaced elements, and such that $(\#H_n')\le n^{\xi}$, $\xi>0$

$$\xi = 2a+2ka-b+\rho, \qquad 2\lambda<\rho<V \tag{25}$$

The following lemmas complete the proof:

Lemma 8

$$\lim_{n\to\infty}\sup_{h\in H_n'}\left|\frac{CS_{l_n^*}(h)}{MISE(h)}\right|=0 \qquad a.s.$$

Lemma 9

$$\lim_{n\to\infty}\sup_{h\in H_n}\left|\frac{CS_{l_n}(h)-CS_{l_n}(h^*)}{MISE(h)}\right|=0 \qquad a.s.$$

where, for each $h\in H_n$, h^* denotes the element of H_n' that is closest to h.

PROOF OF LEMMA 6:
To prove Lemma 6, first observe that

$$\hat{m}_{h,j}(t)-\hat{m}_h(t)=\frac{n}{\overline{\gamma}}\sum_{|i-j|\le l_n}Y_i\left(\int_{S_{i-1}}^{S_i}K_h(t-u)\,du\right)\gamma(|i-j|)+$$
$$+\frac{n-\overline{\gamma}}{\overline{\gamma}}\sum_{|i-j|>l_n}Y_i\left(\int_{S_{i-1}}^{S_i}K_h(t-u)\,du\right)-\sum_{|i-j|\le l_n}Y_i\left(\int_{S_{i-1}}^{S_i}K_h(t-u)\,du\right)$$

Now write

$$\widehat{ASE}(h)=ASE(h)+A(h)+B(h)$$

with

$$\begin{aligned} A(h) &= n^{-1}\sum_{j=1}^{n}\left[\hat{m}_{h,j}(t_j)-\hat{m}_h(t_j)\right]^2 w(t_j) \\ B(h) &= 2n^{-1}\sum_{j=1}^{n}\left[\hat{m}_{h,j}(t_j)-\hat{m}_h(t_j)\right]\left[\hat{m}_h(t_j)-m(t_j)\right] w(t_j) \end{aligned}$$

Asymptotic equivalence in the case of $\widehat{ASE}$ follows from the case of ASE, from Theorem 5, and conditions **(H1)**, **(L1)**, **(E2)**, **(S1)**, **(K1)**, **(M2)** and **(W1)**.

PROOF OF LEMMA 7:

Let us write

$$|CS_{l_n}(h)-CS_{l_n^*}(h)| = \left|n^{-1}\sum_{j=1}^{n}\left[\hat{m}_{j,l_n}(t_j)-\hat{m}_{j,l_n^*}(t_j)\right]\varepsilon_j w(t_j)\right| \le$$

$$\le \left(n^{-1}\sum_{j=1}^{n}\left[\hat{m}_{j,l_n}(t_j)-\hat{m}_{j,l_n^*}(t_j)\right]^2 w(t_j)\right)^{1/2}\left(n^{-1}\sum_{j=1}^{n}\varepsilon_j^2 w(t_j)\right)^{1/2}$$

by Schwartz inequality. Now, we can write

$$\left|\hat{m}_{j,l_n}(t_j)-\hat{m}_{j,l_n^*}(t_j)\right| \le |R_1(h)|+|R_2(h)|$$

with

$$\begin{aligned} R_1(h) &= \sum\sum_{l_n<|i-j|\le l_n^*} Y_i\left[\frac{n}{\overline{\gamma}}-\frac{n}{\overline{\gamma}^*}\right]\left[\gamma(|i-j|)-1\right]\int_{S_{i-1}}^{S_i} K_h(t_j-u)du \\ R_2(h) &= \sum\sum_{|i-j|>l_n^*} Y_i\left[\frac{n}{\overline{\gamma}}-\frac{n}{\overline{\gamma}^*}\right]\int_{S_{i-1}}^{S_i} K_h(t_j-u)du \end{aligned}$$

Using the fact that K is a support compact kernel, that $\overline{\gamma}=\overline{\gamma}^*=O(n)$, and using **(S1)**, **(L1)** and **(H1)**, we have

$$\sup_{h\in H_n'}|R_2(h)| = \sup_{h\in H_n'}|R_1(h)| = O(n^{-1}l_n^* B_n) \qquad a.s. \tag{26}$$

Also, we have

$$n^{-1}\sum_{j=1}^{n}\varepsilon_j^2 w(t_j) \longrightarrow E\left(\varepsilon_j^2 w(t_j)\right) \qquad a.s. \tag{27}$$

using again Theorem A of Roussas (1989). Lemma 7 is a consequence of (26) and (27).

PROOF OF LEMMA 8:

Let us write

$$\begin{aligned} CS_{l_n^*}(h) &= n^{-1}\sum_{j=1}^{n}\sum_{i=1}^{n}\Phi(i,j) = n^{-1}\sum\sum_{|i-j|>l_n^*}\Phi(i,j)+n^{-1}\sum\sum_{|i-j|\le l_n^*}\Phi(i,j) = \\ &= T_1(h)+T_2(h) \end{aligned}$$

All we have to prove is that

$$\sup_{h\in H_n'}\frac{|T_1(h)|}{MISE(h)} \longrightarrow 0 \qquad a.s. \tag{28}$$

$$\sup_{h\in H'_n} \frac{|T_2(h)|}{MISE(h)} \longrightarrow 0 \qquad a.s. \tag{29}$$

Proof of (29) is similar to that of Lemma 6. Proof of (28) is long and fairly complex, and follows a similar reasoning carried out by Hart and Vieu [(1990), Lemma 2], and makes use of the truncation (2).

PROOF OF LEMMA 9:

Let us write

$$\begin{aligned}\left|CS_{l_n^*}(h) - CS_{l_n^*}(h^*)\right| \;\le\; & \left|n^{-1}\sum_{j=1}^{n}\left[\hat{m}_{h,j}(t_j) - \hat{m}_{h^*,j}(t_j)\right]\overline{\varepsilon}_j w(t_j)\right| + \\ & + \left|n^{-1}\sum_{j=1}^{n}\left[\hat{m}_{h,j}(t_j) - \hat{m}_{h^*,j}(t_j)\right](\varepsilon_j - \overline{\varepsilon}_j)w(t_j)\right|\end{aligned} \tag{30}$$

The second term in (30) is zero, for n large enough, since we have

Lemma 10 *Under conditions of theorem*

$$\sup_{1\le j\le n}\sup_{h\in H_n} |\hat{m}_{h,j}(t_j) - \hat{m}_{h^*,j}(t_j)| \longrightarrow 0 \qquad a.s.$$

Lemma 10 is a consequence of Lemma 6.

For the first term of (30), we use Hölder continuity of K, **(W1)**, **(H1)** and the facts that $\overline{\gamma} = O(n)$ and $\gamma(u) \le 1, \forall u$, to obtain

$$\sup_{h\in H_n} \frac{\left|CS_{l_n^*}(h) - CS_{l_n^*}(h^*)\right|}{MISE(h)} = O(n^{-\rho}B_n^2)$$

This completes the proof.

REFERENCES

[1] ALTMAN, N.S. (1990). Kernel Smoothing of Data with Correlated Errors. *Journal of the American Statistical Association,* **411**, 749-759.

[2] BENEDETTI, J.K. (1977). On the Nonparametric Estimation of Regression Functions. *Journal of the Royal Statistical Society, Ser. B,* **39**, 248-253.

[3] BRADLEY, R. (1985). Basic Properties of Strong Mixing Condition. In: *Dependence in probability and statistics,* [Ernst Eberlein, Murad S. Taqqu, (eds.)]. Birkhäuser.

[4] CHENG, K.F. & LIN, P.E. (1981). Nonparametric Estimation of a Regression Function. *Zeitschrift für Wahrscheinlichkeitstheorie und verwandte Gebiete,* **59**, 223-233.

[5] CHIU, S.T. (1989). Bandwidth Selection for Kernel Estimation with Correlated Noise. *Statistics and Probability Letters,* **8**, 347-354.

[6] CHU, C.K. & MARRON, J.S. (1991). Comparison of Two Bandwidth Selectors with Dependent Errors. *Annals of Statistics,* **4**, 1906-1918.

[7] DIGGLE, P. & HUTCHINSON, M. (1989). On Spline Smoothing with Autocorrelated Errors. *Australian Journal of Statistics,* **31**, 166-182.

[8] GASSER, T. & MÜLLER, H.G. (1979). Kernel Estimation of Regression Functions. In: *Smoothing Techniques for Curve Estimation,* [T. Gasser and M. R. Rosenblatt, (eds.)]. Springer-Verlag.

[9] GASSER, T. & MÜLLER, H.G. (1984). Estimating Regression Functions and their Derivatives by the Kernel Method. *Scandinavian Journal of Statistics,* **11**, 171-185.

[10] Györfi, L., Härdle, W., Sarda, P. & Vieu, P. (1990). *Nonparametric Curve Estimation from Time Series.* Lecture Notes in Statistics, Vol. 60. Springer-Verlag.

[11] Hall, P. & Heyde, C. (1980). *Martingale Limit Theory and its Applications.* Academic Press.

[12] Härdle, W. & Vieu, P. (1990). *Kernel Regression Smoothing of Time Series.* Core discussion Paper, Number 9031.

[13] Hart, J. (1991). Kernel Regression Estimation with Time Series Errors. *Journal of the Royal Statistical Society, Ser. B,* **51**, 173-187.

[14] Hart, J. & Vieu, P. (1990). Data-driven Bandwidth Choice for Density Estimation Based on Dependent Data. *The Annals of Statistics,* **18**, 873-890.

[15] Hart, J. & Wehrly, T.E. (1986) Kernel Regression Using Repeated Measurements Data. *Journal of the American Statistical Association,* **81**, 1080-1088.

[16] Priestley, M.B. & Chao, M.T. (1972). Nonparametric Function Fitting. *Journal of the Royal Statistical Society, Series B,* **34**, 385-92.

[17] Quintela, A. (1992). *Cálculo del Parámetro de Suavización en la Estimación No Paramétrica de Curvas con Datos Dependientes.* Ph. D. Dissertation. Universidad de Santiago de Compostela. Spain.

[18] Rosenblatt, M. (1956). A Central Limit Theorem and a Strong Mixing Condition. *Proc. Nat. Acad. Sci.,* **42**, 43-47.

[19] Roussas, G. (1989). *Nonparametric Regression Estimation under Mixing Conditions.* Preprint.

[20] Roussas, G., Tran, L.T. & Ioannides, D.A. (1992). Fixed Design Regression for Time Series: Asymptotic Normality. *Journal of Multivariate Analysis,* **40**, 262-291.

[21] Schuster, E. & Yakowitz, S. (1979). Contribution to the Theory of Nonparametric Regression, with Application to System Identification. *Annals of Statistics,* **7**, 139-149.

[22] Truong, Y.K. (1991). Nonparametric Curve Estimation with Time Series Errors. *Journal of Statistical Planning and Inference,* **28**, 167-183.

[23] Truong, Y.K. & Stone, C.J. (1992). Nonparametric Function Estimation Involving Time Series. *Annals of Statistics,* **20**, 77-97.

[24] Vieu, P. (1991). Quadratic Errors for Nonparametric Estimates under Dependence. *Journal of Multivariate Analysis,* **2**, 324-347.

[25] Volkonskii, V.A. & Rozanov, Y.A. (1959). Some Limit Theorems for Random Functions I. *Theory Prob. Appl.,* **4**, 178-197.

Recent Adv. in Stat. and Prob., pp. 155-171
J. Pérez Vilaplana and M. L. Puri (Eds)

Process Tracking of Time Series with Change Points

S. ZACKS[1] and R.S. KENETT[2]
[1] Department of Mathematical Sciences, SUNY–Binghamton, P.O. Box 6000, Binghamton, NY 13902-6000, U.S.A.
[2] Department of Mathematical Sciences, SUNY–Binghamton, P.O. Box 6000, Binghamton, NY 13902-6000, U.S.A.

Abstract – The problem of industrial process control has been tackled in several ways by both theoreticians and practitioners. The most common approach was developed in the twenties by Shewhart in the form of "control charts". Major developments since the control chart include Cumulative Sum Procedures, Exponentially Weighted Moving Average charts and Fast Initial Response Startups.
We apply here a Bayesian framework for tracking a process over time. We use a Bayesian model that explicitly models the effect of assignable causes as a shift in the mean. We derive, after each observation, the posterior distribution of the process mean, given the observations collected in the past.
The fractiles of the posterior distribution of the process mean and the posterior probabilities of a shift at any given time constitute powerful control and diagnostic tools. The sensitivity of the procedure to the prior parameters and the properties of detection criteria are studied by simulation experiments.

1. INTRODUCTION

How do we keep a process under control? Obviously the decision tool to use depends on how we interpret the process control problem.
The most common interpretation is attributed to WṠhewhart [1931]. In the Shewhart framework one first identifies process variability due to common causes. Then, control limits based on estimates of the process mean and standard deviation permit us to detect significant assignable causes, thereby triggering corrective actions. The control limits are set to give fixed and preset probabilities for undercontrol and overcontrol activities. Shewhart control charts have been adapted to wear out or other known changes in distribution parameters. For a modern exposition of the subject see the book by Stephens et al (1986). In all cases, control charts carry the flavor of hypothesis testing procedures, with control limits and action rules designed to provide a known significance level, α, such as .27%, 1% or 5%. Using three sigma limits ($\alpha = 0.0027$), for instance, provides a test of stability for statistical variations over time. This approach has proven efficient and practical and gained wide range use and acceptability throughout industry.
A drawback of using control charts for process control is the nature of repeated testing of significance over time. The significance level α is determined from an individual, one time, test perspective. A long term perspective on the performance of process monitoring tools

demeans the role of α. In the long run the statistically based control scheme will always trigger an alarm. Overall significance levels are therefore one, and so is the power to detect assignable causes. Any change or non-change in the process triggers action, in the long run. This has led to investigations of Average Run Length (ARL) characteristics of control charts. In order to increase the sensitivity for detecting changes and get improved ARLs, Cumulative Sum procedures (CUSUM) were developed (Page (1954)). Further enhancements to CUSUMs including a combined Shewhart-CUSUM control scheme and Fast Initial Response (FIR) start ups, were proposed by Lucas (1982) and Lucas and Crosier (1982).

Numerical algorithms and analytical determination of the Run Length probability distribution of CUSUM procedures is presented in Zacks (1980, 1981). An extension of CUSUMs to cases were processes cannot be reset is given by Kenett and Pollak (1983). Pollak investigated optimality of an alternative sequential procedure (Pollak (1985)) labeled after Shiryayev and Roberts. Such a Shiryayev-Roberts control scheme was applied to track reliability growth in a software development effort (Kenett and Pollak (1986)).

Another approach was taken by BHoadley and his colleagues at AT&T (Hoadley 1981) resulting in the Quality Measurement Plan (QMP). The QMP mechanism dynamically estimates a quality index distribution using hierarchical Bayes methods. The index is a ratio of observed to expected defects. Posterior fractiles of the quality index are presented in a box and whiskers plot format. An alert is triggered by having the 95th percentile of the index posterior distribution above the standard level of one. When the 99th percentile is above one a "Below Normal" condition is declared. A scheme which is more sensitive to drifts in the mean was developed by Phadke (1981).

Like Hoadley, we use a Bayesian framework for tracking the process over time. The control problem we tackle is that of a continuous variable with a normal distribution. The basic reference for our study is the work of Chernoff and Zacks (1964) originally motivated by a radar tracking problem.

The uncertainty in the process is reflected by a probabilistic structure for the observations, which includes random errors around a mean drawn from a normal distribution.

We assume no initial bias in process startups so that, on the average, the mean is on target. At some unknown point in time a perturbation in the mean occurs. That time instance can be in the past, present or future. Our goal is to control fractiles of the posterior distribution of the current mean and of predictive distributions of future observations, given the past behavior of the process.

In Section 2 we present a Bayesian tracking model for the case of at most one change. The posterior distribution of the current mean, μ_n, is derived in Section 3. As shown, the posterior distribution of μ_n is a mixture of normal distribution, with mixing posterior probabilities which are nonlinear functions of the data. The central moments of the posterior distribution of μ_n are derived, and a method is given for the computation of the posterior fractiles.

Numerical illustrations of the tracking procedure, and an associated detection procedure are given in Section 4. This detection procedure is based on the posterior probability, after n observations, that the change has not yet occurred. An alert is issued when that posterior probability drops below a prescribed threshold, Π^*. At the time of an alert, an estimate of the point of change is computed. Simulation results generated from a fractional factorial computer experiment demonstrate the effects of the prior parameters on the procedure's performance. Section 5 concludes the paper with a discussion of our results.

2. THE BAYESIAN TRACKING MODEL

Observations from a continuous distribution characteristic of a process are recorded at discrete times $i = 1, 2, \ldots$ The observations, X_i, are determined by the value of a base average level, μ_0, and an unbiased error, ϵ_i. Possible system deviations at start-up are modeled by having μ_0 randomly determined according to a normal distribution with known mean μ_T (a specified target value) and a known variance σ^2. At some unknown point in time a disturbance might change the process average level to a new value, $\mu_0 + Z$. The probability of having a disturbance between any two observations, taken at fixed operational time intervals, is considered a constant, p. Accordingly we apply a model of at most one change point at an unknown location. The corresponding statistical model, for at most one change point, within n consecutive observations is as follows:

$$X_i = \mu_i + \epsilon_i, \quad i = 1, 2, \ldots, n. \tag{2.1}$$

Where μ_i is the observation's mean and ϵ_i the measurement error, the at-most-one-change (AMOC) point model is:

$$\mu_i = \begin{cases} \mu_0, & i = 1, \ldots, j-1 \\ \mu_0 + Z, & i = j, \ldots, n \end{cases} \tag{2.2}$$

Where j indicates that the disturbance occurred in the time interval between the $(j-1)$st and j-th observations. $j = 1$ indicates a disturbance before X_1, $j = n+1$ indicates no disturbance among the first n observations.

The Bayesian framework is based on the following distributional assumptions:

$$\mu_0 \sim N(\mu_T, \sigma^2) \tag{2.3}$$

$$Z \sim N(\delta, \tau^2), \tag{2.4}$$

$$\epsilon_i \sim N(0, 1) \quad i = 1, 2, \ldots \tag{2.5}$$

The parameters μ_T, σ^2, τ^2 are assumed to be known and $N(\cdot, \cdot)$ designates a random variable having a normal distribution law. Assumption (2.5) of the model requires that the raw data be scaled, to have conditional variance equal to 1.

For n observations, let J_n denote the random variable corresponding to the index of the time interval at which a change occurred. It is assumed that the prior distribution of J_n is the truncated geometric, i.e.,

$$\begin{aligned} \Pr(J_n = j \mid p) &= p(1-p)^{j-1}, \quad j = 1, 2, \ldots, n; \qquad 0 < p < 1 \\ &= (1-p)^n, \qquad j = n+1; \qquad 0 < p < 1 \end{aligned} \tag{2.6}$$

The assumption underlining (2.6) is that disturbances occur at random according to a homogeneous Poisson process. Finally, we assume that μ_0, Z, J_n and ϵ are independent, for each n.

3. THE POSTERIOR DISTRIBUTION OF THE CURRENT MEAN

In the present section we develop the formula of the posterior distribution of μ_n, given $\mathbf{X}^{(n)} = (X_1, \ldots, X_n)$, and compute its moments and fractiles.

3.1. General Derivation

For $j = 1, 2, \ldots, n+1$, let

$$\bar{X}_{j-1} = \frac{1}{j-1}\sum_{i=1}^{j-1} X_i \quad \text{and} \quad \bar{X}^*_{n-j+1} = \frac{1}{n-j+1}\sum_{i=j}^{n} X_i$$

where $\bar{X}_0 \equiv 0$ and $\bar{X}^*_0 \equiv 0$. The current mean μ_n can be written, in the AMOC model, as

$$\mu_n = \mu_0 + I\{J_n \le n\}Z, \tag{3.1}$$

where $I\{\cdot\}$ is an indicator variable, assuming the value 1 if the relation in brackets is true, and the value 0 otherwise. Thus, given $\{J_n = j, \mathbf{X}^{(n)}\}$, the Bayesian model of Section 2 implies that the posterior distribution of μ_n is normal $\mathcal{N}\left(E_{j,n}(\mathbf{X}^{(n)}),\ D^2_{j,n}\right)$, where $E_{j,n}(\mathbf{X}^{(n)})$ is the posterior mean of μ_n, given $\{J_n = j,\ \mathbf{X}^{(n)}\}$, and $D^2_{j,n}$ is the corresponding posterior variance. Formulae for $E_{j,n}(\mathbf{X}^{(n)})$ and $D^2_{j,n}$, $j = 1, 2, \ldots, n+1$, will be derived below.

Let $\Pi(J_n = j \mid \mathbf{X}^{(n)})$ denote the posterior probability function of $\{J_n = j\}$, given $\mathbf{X}^{(n)}$. Then, the posterior distribution law of μ_n, given $\mathbf{X}^{(n)}$, $\mathcal{L}(\mu_n \mid \mathbf{X}^{(n)})$, is the mixture of the normals $\mathcal{N}\left(E_{j,n}(\mathbf{X}^{(n)}),\ D^2_{j,n}\right)$, with mixing probabilities $\Pi_{j,n}(\mathbf{X}^{(n)}) = \Pi(J_n = j \mid \mathbf{X}^{(n)})$, i.e.,

$$\mathcal{L}(\mu_n \mid \mathbf{X}^{(n)}) = \sum_{j=1}^{n+1} \Pi_{j,n}(\mathbf{X}^{(n)})\ \mathcal{N}\left(E_{j,n}(\mathbf{X}^{(n)}), D^2_{j,n}\right). \tag{3.2}$$

To determine $E_{j,n}(\mathbf{X}^{(n)})$ and $D^2_{j,n}$, for $j = 1, \ldots, n+1$, notice that the joint distribution of $(\mu_n, \bar{X}_{j-1}, \bar{X}^*_{n-j+1})$, given $\{J_n = j\}$, $j = 2, \ldots, n$, is multi-normal with mean $(\mu_T + \delta, \mu_T, \mu_T + \delta)$ and covariance matrix

$$\Sigma_{\mid j} = \begin{bmatrix} \sigma^2+\tau^2 & \cdots & \sigma^2 & \sigma^2+\tau^2 \\ \vdots & \ddots & \vdots & \vdots \\ \sigma^2 & \cdots & \sigma^2 + \dfrac{1}{j-1} & \sigma^2 \\ \sigma^2+\tau^2 & \cdots & \sigma^2 & \sigma^2+\tau^2+\dfrac{1}{n-j+1} \end{bmatrix}. \tag{3.3}$$

From the common formulae of regression theory we obtain, for $j = 2, \ldots, n$.

$$E_{j,n}(\mathbf{X}^{(n)}) = \mu_0 + \delta + \tag{3.4}$$

$$+ (\sigma^2, \sigma^2+\tau^2)\begin{bmatrix} \sigma^2 + \dfrac{1}{j-1} & \sigma^2 \\ \sigma^2 & \sigma^2+\tau^2+\dfrac{1}{n-j+1} \end{bmatrix}^{-1} \begin{pmatrix} \bar{X}_{j-1} - \mu_T \\ \bar{X}^*_{n-j+1} - \mu_T - \delta \end{pmatrix},$$

and

$$D^2_{j,n} = \sigma^2+\tau^2-(\sigma^2,\sigma^2+\tau^2)\begin{bmatrix} \sigma^2 + \dfrac{1}{j-1} & \sigma^2 \\ \sigma^2 & \sigma^2+\tau^2+\dfrac{1}{n-j+1} \end{bmatrix}^{-1} \begin{pmatrix} \sigma^2 \\ \sigma^2+\tau^2 \end{pmatrix}. \tag{3.5}$$

The cases of $j = 1$ or $j = n+1$ are similar. Thus, after some algebraic manipulations we obtain the following formulae:

$$E_{j,n}(\mathbf{X}^{(n)}) = \begin{cases} \dfrac{\mu_T + \delta}{1 + n(\sigma^2 + \tau^2)} + \dfrac{n(\sigma^2 + \tau^2)}{1 + n(\sigma^2 + \tau^2)} \bar{X}_n, & j = 1 \\[2ex] W_{j,n}^{(1)}(\mu_T + \delta + (j-1)\sigma^2) + W_{j,n}^{(2)} \bar{X}_{j-1} + W_{j,n}^{(3)} \bar{X}^*_{n-j+1}, & j = 2, \dots, n \\[2ex] \dfrac{\mu_T}{1 + n\sigma^2} + \dfrac{n\sigma^2}{1 + n\sigma^2} \bar{X}_n, & j = n+1, \end{cases} \tag{3.6}$$

where

$$\begin{aligned} W_{j,n}^{(1)} &= [1 + n\sigma^2 + \tau^2(n - j + 1)(1 + \sigma^2(j-1))]^{-1} \\ W_{j,n}^{(2)} &= \sigma^2(j-1)W_{j,n}^{(1)}, \qquad\qquad j = 2, \dots, n \\ W_{j,n}^{(3)} &= 1 - W_{j,n}^{(1)} - W_{j,n}^{(2)} \end{aligned} \tag{3.7}$$

Furthermore,

$$D_{j,n}^2 = \begin{cases} \left(\sigma^2 + \tau^2 + \sigma^2\tau^2(j-1)\right) W_{n,j}^{(1)}, & j = 1, \dots, n \\[2ex] \dfrac{n\sigma^2}{1 + n\sigma^2}, & j = n+1. \end{cases} \tag{3.8}$$

We derive now the formulae of the posterior probabilities $\Pi_{j,n}(\mathbf{X}^{(n)})$.

3.2. The Likelihood Functions and Posterior Probabilities

Let $L(j, \mu_0, Z; \mathbf{X}^{(n)})$ denote the likelihood function of (j, μ_0, Z) given the first n observations $\mathbf{X}^{(n)}$. Let $L^*(j; \mathbf{X}^{(n)})$ denote the marginal (predictive) likelihood of j, given $\mathbf{X}^{(n)}$; which is the expected value of $L(j, \mu_0, Z; \mathbf{X}^{(\mathbf{n})})$ with respect to the prior distribution of (μ_0, Z). By Bayes theorem, for all $j = 1, 2, \dots$

$$\Pi_{j,n}(\mathbf{X}^{(n)}) = \frac{I\{j \le n\} p(1-p)^{j-1} L^*(j; \mathbf{X}^{(n)}) + I\{j \ge n+1\}(1-p)^n L^*(n+1; \mathbf{X}^{(n)})}{p \sum_{j=1}^{n} (1-p)^{j-1} L^*(j; \mathbf{X}^{(n)}) + (1-p)^n L^*(n+1; \mathbf{X}^{(n)})} \tag{3.9}$$

We further develop this formula. From the normal distribution of ϵ_i $(i = 1, \dots, n)$,

$$\begin{aligned} L(j, \mu_0, Z; \mathbf{X}^{(n)}) &= \frac{1}{(2\pi)^{n/2}} \exp\left\{ -\frac{1}{2}\sum_{l=1}^{j-1}\left(X_l - \mu_0\right)^2 - \frac{1}{2}\sum_{l=j}^{n}\left(X_l - \mu_0 - Z\right)^2 \right\} = \qquad (3.10) \\ &= \frac{1}{(2\pi)^{n/2}} \exp\left\{ -\frac{1}{2}Q_{j-1} - \frac{1}{2}Q^*_{n-j+1} - \frac{j-1}{2}\left(\bar{X}_{j-1} - \mu_0\right)^2 - \right. \\ &\qquad \left. -\frac{n-j+1}{2}\left(\bar{X}^*_{n-j+1} - \mu_0 - Z\right)^2 \right\}, \qquad j = 2, \dots, n \end{aligned}$$

where $Q_{j-1} = \sum_{l=1}^{j-1}(X_l - \bar{X}_{j-1})^2$ and $Q^*_{n-j+1} = \sum_{l=j}^{n}(X_l - \bar{X}^*_{n-j+1})^2$. When $j = 1$ or $j = n+1$ we substitute in (3.10) $Q_0 \equiv 0$ or $Q^*_0 \equiv 0$, respectively. Let $Q_n = \sum_{i=1}^{n}(X_i - \bar{X}_n)^2$. Simple algebraic manipulations yield

$$-\frac{1}{2}Q_{j-1}-\frac{1}{2}Q^*_{n-j+1}=-\frac{1}{2}\left[Q_n-2(j-1)\left(1-\frac{j-1}{n}\right)(\bar{X}_{j-1}-\bar{X}^*_{n-j+1})^2\right], \quad \textbf{(3.11)}$$

$j=1,\ldots,n+1$. Substituting (3.11) in (3.10) we obtain

$$L(j,\mu_0,Z;\mathbf{X}^{(n)})= \quad \textbf{(3.12)}$$
$$=\frac{1}{(2\pi)^{n/2}}\exp\left\{-\frac{1}{2}Q_n\right\}\exp\left\{(j-1)\left(1-\frac{j-1}{n}\right)\left(\bar{X}_{j-1}-\bar{X}^*_{n-j+1}\right)^2-\right.$$
$$\left.-\frac{j-1}{2}\left(\bar{X}_{j-1}-\mu_0\right)^2-\frac{n-j+1}{2}\left(\bar{X}^*_{n-j+1}-\mu_0-Z\right)^2\right\}.$$

Moreover, for $j=1,\ldots,n+1$,

$$L^*(j;\mathbf{X}^{(n)}) = \quad \textbf{(3.13)}$$
$$=\frac{1}{(2\pi)^{n/2}}\exp\left\{-\frac{1}{2}Q_n\right\}\exp\left\{(j-1)\left(1-\frac{j-1}{n}\right)\left(\bar{X}_{j-1}-\bar{X}^*_{n-j+1}\right)^2\right\}\cdot$$
$$\cdot E\left\{\exp\left\{-\frac{j-1}{2}\left(\bar{X}_{j-1}-\mu_0\right)^2-\frac{n-j+1}{2}\left(\bar{X}^*_{n-j+1}-\mu_0-Z\right)^2\right\}\right\},$$

where the expectation in (3.13) is with respect to the prior bivariate normal distribution of (μ_0,Z). Thus,

$$L^*(j;\mathbf{X}^{(n)}) = \frac{1}{(2\pi)^{\frac{n}{2}-1}}\exp\left\{-\frac{1}{2}Q_n\right\}\cdot \quad \textbf{(3.14)}$$
$$\cdot\left\{(j-1)\left(1-\frac{j-1}{n}\right)\left(\bar{X}_{j-1}-\bar{X}^*_{n-j+1}\right)^2\right\}L^{**}(j;\mathbf{X}^{(n)}),$$

$j=1,\ldots,n+1$, where

$$L^{**}(1;\mathbf{X}^{(n)})=\frac{1}{\left(1+n(\sigma^2+\tau^2)\right)^{1/2}}\cdot\exp\left\{-\frac{n}{2\left(1+n(\sigma^2+\tau^2)\right)}(\bar{X}^2_n-\mu_T-\delta)^2\right\}, \quad \textbf{(3.15)}$$

and for $j=2,\ldots,n$,

$$L^{**}(j;\mathbf{X}^{(n)})=\frac{1}{\left(1+n\sigma^2+(n-j+1)\tau^2(1+(j-1)\sigma^2)\right)^{1/2}}\cdot \quad \textbf{(3.16)}$$
$$\cdot\exp\left\{-\frac{1}{2\left(1+n\sigma^2+(n-j+1)\tau^2(1+(j-1)\sigma^2)\right)}\cdot\right.$$
$$\cdot\left[\frac{1+(n-j+1)(\sigma^2+\tau^2)}{n-j+1}\cdot\right.$$
$$\cdot\left((\bar{X}_{j-1}-\mu_t)-2\sigma^2(\bar{X}_{j-1}-\mu_T)(\bar{X}^*_{n-j+1}-\mu_T-\delta)\right)+$$
$$\left.\left.+\frac{1+(j-1)\sigma^2}{j-1}\left(\bar{X}^*_{n-j+1}-\mu_T-\delta\right)^2\right]\right\},$$

while

$$L^{**}(n+1;\mathbf{X}^{(n)})=\frac{1}{\sqrt{1+n\sigma^2}}\exp\left\{-\frac{n}{2(1+n\sigma^2)}(\bar{X}_n-\mu_T)^2\right\}. \quad \textbf{(3.17)}$$

3.3. Moments of the Posterior Distribution of μ_n

Let $m_r^*(\mathbf{X}^{(n)})$ denote the r-th central moment of the posterior distribution of μ_n, given $\mathbf{X}^{(n)}$; i.e.,

$$m_r^*(\mathbf{X}^{(n)}) = E\left\{(\mu_n - \hat{\mu}_n)^r \mid \mathbf{X}^{(n)}\right\}, \tag{3.18}$$

where $\hat{\mu}_n = \sum_{j=1}^{n+1} \Pi_{j,n}(\mathbf{X}^{(n)})\, E_{j,n}(\mathbf{X}^{(n)})$ is the mean of the posterior distribution (the Bayes estimator of μ_n for a squared-error loss). According to the mixture (3.2), the central posterior moments of μ_n can be expressed in terms of the corresponding moments of $\mathcal{N}\left(E_{j,n}(\mathbf{X}^{(n)}),\ D_{j,n}^2\right)$ and the moments of $E_{j,n}(\mathbf{X}^{(n)})$ around $\hat{\mu}_n$. Let $M^*(t)$ designate the posterior moment generating function of $(\mu_n - \hat{\mu}_n)$, i.e.,

$$\begin{aligned} M^*(t) &= E\left\{e^{t(\mu_n - \hat{\mu}_n)} \mid \mathbf{X}^{(n)}\right\} = \\ &= \sum_{j=1}^{n+1} \Pi_{j,n}(\mathbf{X}^{(n)}) e^{t\left(E_{j,n}(\mathbf{X}^{(n)}) - \hat{\mu}_n\right)}\, E\left\{e^{t\left(\mu_n - E_{j,n}(\mathbf{X}^{(n)})\right)} \mid J = j, \mathbf{X}^{(n)}\right\}. \end{aligned} \tag{3.19}$$

Since the conditional posterior distribution of μ_n, given $\{J = j, \mathbf{X}^{(n)}\}$ is $\mathcal{N}\left(E_{j,n}(\mathbf{X}^{(n)}),\right.$ $\left.D_{j,n}^2\right)$, the conditional posterior moment generating function is

$$M_j^*(t) = E\left\{e^{t\left(\mu_n - E_{j,n}(\mathbf{X}^{(n)})\right)} \mid J = j, \mathbf{X}^{(n)}\right\} = \exp\left\{\frac{t^2}{2} D_{j,n}^2\right\}, \quad j = 1, \ldots, n+1. \tag{3.20}$$

Thus,

$$M^*(t) = \sum_{j=1}^{n+1} \Pi_{j,n}(\mathbf{X}^{(n)}) \exp\left\{t\left(E_{j,n}(\mathbf{X}^{(n)}) - \hat{\mu}_n\right) + \frac{t^2}{2} D_{j,n}^2\right\}. \tag{3.21}$$

From (3.21) one can obtain the central posterior moments of μ_n, given $\mathbf{X}^{(n)}$. In particular,

$$\operatorname{Var}\left\{\mu_n \mid \mathbf{X}^{(n)}\right\} = \sum_{j=1}^{n+1} \Pi_{j,n}\left(\mathbf{X}^{(n)}\right) D_{j,n}^2 + \sum_{j=1}^{n+1} \Pi_{j,n}\left(\mathbf{X}^{(n)}\right)\left(E_{j,n}\left(\mathbf{X}^{(n)}\right) - \hat{\mu}_n\right)^2, \tag{3.22}$$

the third central moment is

$$m_3^*\left(\mathbf{X}^{(n)}\right) = \sum_{j=1}^{n+1} \Pi_{j,n}\left(\mathbf{X}^{(n)}\right)\left[E_{j,n}\left(\mathbf{X}^{(n)}\right) - \hat{\mu}_n\right]^3 + 3\sum_{j=1}^{n+1} \Pi_{j,n}\left(\mathbf{X}^{(n)}\right)\left[E_{j,n}\left(\mathbf{X}^{(n)}\right) - \hat{\mu}_n\right] D_{j,n}^2 \tag{3.23}$$

and the fourth central moment is

$$\begin{aligned} m_4^*\left(\mathbf{X}^{(n)}\right) &= \sum_{j=1}^{n+1} \Pi_{j,n}\left(\mathbf{X}^{(n)}\right)\left[E_{j,n}\left(\mathbf{X}^{(n)}\right) - \hat{\mu}_n\right]^4 + \\ &+ 6\sum_{j=1}^{n+1} \Pi_{j,n}\left(\mathbf{X}^{(n)}\right)\left[E_{j,n}\left(\mathbf{X}^{(n)}\right) - \hat{\mu}_n\right]^2 D_{j,n}^2 + 3\sum_{j=1}^{n+1} \Pi_{j,n}\left(\mathbf{X}^{(n)}\right) D_{j,n}^4. \end{aligned} \tag{3.24}$$

3.4. Posterior Fractiles of μ_n

According to (3.2), the posterior c.d.fôf μ_n, at a point x is

$$F(x \mid \mathbf{X}^{(n)}) = \sum_{j=1}^{n+1} \Pi_{j,n}(\mathbf{X}^{(n)}) \Phi\left(\frac{x - E_{j,n}(\mathbf{X}^{(n)})}{D_{j,n}}\right), \tag{3.25}$$

where $\Phi(Z)$ is the standard normal integral. Accordingly, the p-th fractile of μ_n is the root $\mu_n(p)$ of the equation

$$\sum_{j=1}^{n+1} \Pi_{j,n}(\mathbf{X}^{(n)})\Phi\left(\frac{\mu_n(p) - E_{j,n}(\mathbf{X}^{(n)})}{D_{j,n}}\right) \equiv p, \tag{3.26}$$

for $0 < p < 1$. The root $\mu_n(p)$ of the non-linear equation (3.26) can be obtained by the Newton-Raphson iterative formula:

$$\mu_n^{(i+1)}(p) = \mu_n^{(i)}(p) - \frac{\sum_{j=1}^{n+1} \Pi_{j,n}\left(\mathbf{X}^{(n)}\right) \Phi\left(\frac{\mu_n^{(i)}(p) - E_{j,n}\left(\mathbf{X}^{(n)}\right)}{D_{j,n}}\right) - p}{\sum_{j=1}^{n+1} \frac{\Pi_{j,n}\left(\mathbf{X}^{(n)}\right)}{\mathbf{D_{j,n}}} \psi\left(\frac{\mu_n^{(i)}(p) - E_{j,n}\left(\mathbf{X}^{(n)}\right)}{D_{j,n}}\right)} \tag{3.27}$$

$i = 1, 2, \ldots$, where $\psi(Z)$ is the standard normal p.d.f. One can start the iterations with the initial approximation

$$\mu_n^{(1)}(p) = \sum_{j=1}^{n} \Pi_{j,n}\left(\mathbf{X}^{(n)}\right) E_{j,n}\left(\mathbf{X}^{(n)}\right) + Z_p \sum_{j=1}^{n+1} \Pi_{j,n}\left(\mathbf{X}^{(n)}\right) D_{j,n}, \tag{3.28}$$

in which $Z_p = \Phi^{-1}(p)$.
Numerical examples show that generally (3.27) requires only a few iterations to obtain convergence within four or five significant figures.

4. NUMERICAL EXAMPLES AND SENSITIVITY ANALYSIS

4.1. Numerical Examples

In Tables 4.1 and 4.2 we present portions of two simulated sequences (SIM1 and SIM2) of 15 observations, with a change point at $J = 10$. In both sequences the prior parameters are $\mu_T = 0$, $\delta = 1.5$, $\sigma^2 = \tau^2 = 1$. The random values of μ_i $(i = 1, \ldots, 15)$ for the two sequences are:

$$\mu_i^{(1)} = \begin{cases} -1.91523, & i \leq 10 \\ 0.53421, & i \geq 11 \end{cases}$$

and

$$\mu_i^{(2)} = \begin{cases} 1.32238, & i \leq 10 \\ 0.72655, & i \geq 11 \end{cases}$$

In Tables 4.1 and 4.2 we also present the posterior probabilities of $\{J = j\}$, computed sequentially on the basis of $\mathbf{X}^{(i)}$, $i = 7, \ldots, 13$. We see in Table 4.1 (SIM1) that for $i \leq 10$, $\Pi_{i+1,i}(\mathbf{X}^{(i)})$ assumes the maximal value, which is always greater than 0.8; while immediately after the change, for $i \geq 11$, $\Pi_{11,i}(\mathbf{X}^{(i)})$ assumes the maximum (close to 0.9), while $\Pi_{i+1,i}(\mathbf{X}^{(i)})$ is close to zero. This is a case in which a detection of a change point based on $\Pi_{ji}(\mathbf{X}^{(i)})$ can be easily attained. This is not the case in Table 4.2. The simulated sequence SIM2 does not lead to a close detection of the change point. The value of the prior parameter δ is small relative to $(\tau^2 + \sigma^2)$.

Table 4.1.

Posterior Probabilities $\Pi_{j,i}(\mathbf{X}^{(i)})$, and Posterior Fractiles, Means and Standard Deviations of μ_i ($i = 7, 8, \ldots, 13$), Based on the Simulated Sequence SIM1. Change Point at $i = 11$.

i	7	8	9	10	11	12	13
X_i \ j	-2.081	-2.837	-0.731	-1.683	1.289	0.565	0.256
1	0.0027	0.0028	0.0023	0.0025	0.	0.	0.
2	0.0057	0.0066	0.0037	0.0037	0.	0.	0.
3	0.0022	0.0023	0.0016	0.0016	0.	0.	0.
4	0.0072	0.0085	0.0029	0.0026	0.	0.	0.
5	0.0050	0.0054	0.0019	0.0017	0.	0.	0.
6	0.0047	0.0043	0.0017	0.0016	0.	0.	0.
7	0.0194	0.0039	0.0051	0.0055	0.0004	0.0001	0.
8	0.9606	0.0079	0.0078	0.0074	0.0011	0.0003	0.0002
9		0.9583	0.1583	0.0643	0.0964	0.0794	0.0927
10			0.8149	0.0139	0.0155	0.0149	0.0184
11				0.8952	0.8833	0.9052	0.8886
12					0.0032	0.0001	0.0001
13						0.	0.
14							0.
$\mu_{.01}$	-2.9375	-2.9796	-2.7464	-2.6971	-1.7890	-1.0448	-0.9110
$\mu_{.25}$	-2.3287	-2.4012	-2.1939	-2.1819	-0.5875	-0.1011	-0.0948
$\mu_{.50}$	-2.0855	-2.1720	-1.9473	-1.9664	-0.1136	0.2923	0.2431
$\mu_{.75}$	-1.8399	-1.9419	-1.6270	-1.7358	0.3684	0.6907	0.5844
$\mu_{.99}$	-0.8312	-1.2333	0.8723	-0.2637	1.5813	1.6683	1.4259
$\hat{\mu}_i$	-2.0694	-2.1617	-1.7315	-1.9184	-0.1085	0.2968	0.2463
$v_i^{1/2}$	0.3830	0.3652	0.7782	0.4364	0.8368	0.5848	0.5281

It is of practical and theoretical interest to study the properties of the sampling distributions of various statistics of interest. From a Bayesian point of view, the Bayes estimators of the current mean, μ_n, are optimal as long as the Bayesian model is valid. The question is, how good and effective are the estimators when the Bayesian model is not necessarily valid. Answers to various such questions can be obtained by appropriate simulation experiments.

For example, if we consider a detection procedure which flags a warning on the basis of the stopping time

$$N_{\Pi^*} = \text{first } n \geq 2 \text{ for which } \Pi_{n+1,n}(\mathbf{X}^{(n)}) \leq \Pi^*, \tag{4.1}$$

it would be of interest to estimate the probability distribution of $(N_{\Pi^*} - J)$, where J is the actual (random) epoch of change. This probability distribution depends on the parameters of the process and on the constants we use in the detection procedure.

Table 4.2.
Posterior Probabilities $\Pi_{ji}(\mathbf{X}^{(i)})$ and Posterior Fractiles, Means and Standard Deviations of μ_i ($i = 7, \ldots, 13$), Based on the Simulated Sequence SIM2. Change Point at $i = 11$.

i	7	8	9	10	11	12	13
X_i \ j	1.483	0.959	1.057	1.419	2.614	0.046	1.626
1	0.0197	0.0187	0.0181	0.0183	0.0203	0.0185	0.0191
2	0.0286	0.0229	0.0200	0.0200	0.0268	0.0192	0.0203
3	0.0206	0.0144	0.0177	0.0115	0.0168	0.0106	0.0112
4	0.0069	0.0049	0.0041	0.0040	0.0054	0.0036	0.0038
5	0.0030	0.0025	0.0023	0.0022	0.0022	0.0020	0.0020
6	0.0083	0.0041	0.0030	0.0027	0.0039	0.0023	0.0024
7	0.0059	0.0031	0.0024	0.0021	0.0022	0.0018	0.0017
8	<u>0.9070</u>	0.0048	0.0030	0.0023	0.0023	0.0018	0.0017
9		<u>0.9246</u>	0.0047	0.0029	0.0039	0.0019	0.0018
10			<u>0.9310</u>	0.0055	0.0098	0.0024	0.0022
11				<u>0.9286</u>	0.0341	0.0030	0.0025
12					<u>0.8724</u>	0.0081	0.0030
13						<u>0.9249</u>	0.0065
14							<u>0.9219</u>
$\mu_{.01}$	0.5112	0.5142	0.5290	0.5760	0.6537	0.6240	0.6822
$\mu_{.25}$	1.1147	1.0802	1.0654	1.0886	1.2105	1.1094	1.1416
$\mu_{.50}$	1.3620	1.3107	1.2831	1.2962	1.4162	1.3007	1.3258
$\mu_{.75}$	1.6154	1.5442	1.5029	1.5059	1.6270	1.4933	1.5118
$\mu_{.99}$	2.3380	2.1641	2.0771	2.0607	2.4834	2.0004	2.0148
$\hat{\mu}_i$	1.3710	1.3149	1.2861	1.2995	1.4293	1.3024	1.3291
$v_i^{1/2}$	0.3829	0.3502	0.3296	0.3161	0.3544	0.2931	0.2839

In Section 4.2 we present results of simulation analysis, in which the parameters of the detection procedure coincide with those of the process.
In Section 4.3 we present the results of a sensitivity analysis based on simulation, in which the parameters of the procedure are different from those of the process.

4.2. The Joint Empirical Distribution of (N_{Π^*}, N_A) and Related Statistics.

We present in the Tables 4.3 to 4.5 the results of a simulation experiment, designed to study the sampling distribution of N_{Π^*} and of the mode, N_A, of the posterior distribution of J_n at the stopping time N_{Π^*}. The mode N_A provides an estimate of the location of the shift. To avoid a high proportion of false alarms, in which $N_{\Pi^*} \leq J - 1$, we used the constant $\Pi^* = .5$ for the stopping time (4.1). In the present experiment we consider the

Table 4.3.
Frequency Distribution of (N, NA) Obtained by 100 Simulation Runs, with Parameters $\mu_T = 0$, $\sigma^2 = \tau^2 = 1$, $\delta = 1.5$, $p = .01$, $\Pi^* = .5$.

N_{Π^*}	NA																					SUM
	1	2	3	4	5	6	7	8	9	10	11	12	13	14	15	16	17	18	19	20	21	
1																						
2																						
3																						
4			1																			1
5					1	1																2
6		1																				1
7							2															2
8								1														1
9							1		1													2
10										1	3											4
11									1	4	28	1										34
12									1	1	8	7										17
13											2	3	4	2								11
14										1				1								2
15												1	1			3						5
16																						
17																		1				1
18																			1			1
19																				1		1
20																					15	15
SUM		1	1		1	1	3	1	3	7	41	12	5	3		3		1	1	1	15	100

Table 4.4.
Frequency Distribution of (N_{Π^*}, NA) Obtained by 100 Simulation Runs, with Parameters $\mu_T = 0$, $\sigma^2 = \tau^2 = 1$, $\delta = 2.0$, $p = .01$, $\Pi^* = .5$.

N_{Π^*}	NA																					SUM
	1	2	3	4	5	6	7	8	9	10	11	12	13	14	15	16	17	18	19	20	21	
1																						
2		2																				2
3																						
4			1	2																		3
5		2			2																	4
6					1																	1
7						1	1															2
8							1	2	1													4
9									1													1
10										1												1
11								1	1	2	22	4										30
12					1					2	8	5	2									18
13										1	1	2	8	1								13
14											1		1	1	2							5
15											2		1									3
16															1		1					2
17																1						1
18																1			1			2
19																				1		1
20																					7	7
SUM		4	1	2	4	1	2	3	3	6	34	11	12	2	3	2	1		1	1	7	100

Table 4.5.
Frequency Distribution of (N_{Π^*}, NA) Obtained by 100 Simulation Runs,
With Parameters $\mu_T = 0$, $\sigma^2 = \tau^2 = 1$, $\delta = 2.5$, $p = .01$, $\Pi^* = .5$.

N_{Π^*}	NA																					SUM
	1	2	3	4	5	6	7	8	9	10	11	12	13	14	15	16	17	18	19	20	21	
1																						
2		1																				1
3																						
4	1		1	1																		3
5					2	1																3
6					1																	1
7					1																	1
8								2	1													3
9									1													1
10							1			3												4
11							2			1	49	2										54
12											9	6										15
13											2		1	2								5
14											1			1								2
15															1							1
16										1							1					1
17														1								1
18																		1				1
19																						
20																					3	3
SUM	1	1	1	1	4	1	3	2	2	5	61	8	1	4	1			1			3	100

case in which one knows the parameters of the process μ_T, σ^2, τ^2 and δ. We applied the values $\mu_T = 0$, $\sigma^2 = \tau^2 = 1$ and $\delta = 1.5, 2.0$ and 2.5. The joint frequency distributions of (N_{Π^*}, N_A), obtained from 100 independent simulation runs (for each case) are presented in Tables 4.3-4.5. In these simulations the epoch of change is fixed at $J = 10$. Each run is terminated after $n = 20$ observations, if it has not stopped earlier. The proportion of runs in which $N_{\Pi^*} \leq 10$ is an estimator of the "false alarms".
The statistic

$$CED = \sum_{n=11}^{20} (n - J) f_n / \sum_{n=11}^{20} f_n, \tag{4.2}$$

in which f_n $(n = 1, \ldots, 20)$ are the marginal frequencies of N_{Π^*}, is an estimator of the conditional expected delay, i.e., $E\{N_{\Pi^*} - J \mid N_{\Pi^*} \geq J\}$. Notice that this empirical estimator is biased downwards because of the truncation at $n = 20$. The conditional median delay (CMD) is, however, unaffected by the truncation. Notice that if $N_{\Pi^*} = J$ there is no delay in the warning signal and no false alarm.
Let PFA denote an estimate of the probability of false alarms; PES denote an estimate of the probability of exact stopping and CED the estimator (4.2). From Tables 4.3–4.5 we obtain the following estimates.

Table 4.6.
Estimates PFA, PES and CED for
$\mu_T = 0$, $\sigma^2 = \tau^2 = 1$, $p = 0.01$, $\Pi^* = 0.5$.

δ	1.5	2.0	2.5
PFA	0.13	0.18	0.17
PES	0.28	0.22	0.49
CED	2.540	2.098	0.964
CMD	1.00	1.00	0.00

As anticipated, there is no significant difference between the PFA values of the three δ's. The critical decision value Π^* plays the important role. The values of PES do depend on δ. There is no significant difference between the estimates of PES when $\delta = 1.5$ or $\delta = 2.0$. But, when $\delta = 2.5$ the value of PES increases significantly (about doubles). A similar picture is obtained on the CED. It is interesting to know how small could Π^* be without substantially effecting the CED. In additional simulation runs we kept $\delta = 2.0$ and decreased the value of Π^*. The results are presented in Table 4.7. 100 independent simulation runs were performed for each value of Π^*.

Table 4.7.
Estimates PFA, PES and CED for
$\mu_T = 0$, $\sigma^2 = \tau^2 = 1$, $p = 0.01$, $\delta = 2.0$.

Π^*	0.4	0.3	0.2
PFA	0.17	0.11	0.11
PES	0.34	0.35	0.25
CED	1.639	1.461	2.966

We see in Table 4.7 that, decreasing Π^* from 0.5 to 0.4 or 0.3 we obtain significantly higher PES values, the PFA values tend to decrease and so are the CED values. However, further decrease to $\Pi^* = 0.22$ increases CED. Thus, we recommend the use of $\Pi^* = 0.3$ in practical applications. Observe in Tables 4.3–4.5 that, there are cases in which $N_A = N_{\Pi^*} + 1$ when $N_{\Pi^*} < 20$. These are cases of stopping without a clear indication that a change took place. In addition, there are runs in which there was no stopping before $N = 20$. If we denote by PNS the proportion of runs without stopping ($N_{\Pi^*} \geq 20$) and by PFS the proportion of false stopping (stopping without an indication of a change) then we obtain from Tables 4.3–4.5 the estimates which are presented in Table 4.8.

Table 4.8.
Estimates of PNS and PFS for $\mu_T = 0$,
$\sigma^2 = \tau^2 = 1$, $p = 0.01$, $\Pi^* = 0.50$.

δ	PNS	PFS
1.5	0.15	0.13
2.0	0.07	0.13
2.5	0.03	0.07

4.3. Robustness Study of the AMOC Model with Respect to Choice of Parameters

Obviously, if the Bayesian model is valid, and one knows the prior parameters of the process, then the Bayesian estimators of the current mean, μ_n, and other detection indices are optimal. However, in reality, even if the assumed prior distributions are the correct ones, the prior parameters may not be exactly known. It is therefore interesting to study the effects of using wrong values of the parameters.

In the present section we study the sensitivity of the tracking procedure to deviations of the assumed values of p, σ^2, τ^2 and δ from the actual parameters of the process. For this purpose we performed a simulation experiment in which the assumed values of the prior parameters were varied around the true values of the process. These simulations were planned according to a 2^4 factorial experiment, in which ten independent replicas (runs) were performed of each treatment combination. The observations were simulated from sequences of $NS = 15$ random variables, with a fixed change point at $J = 10$. The process parameters used for simulating the random variables were: $\mu_T = 0$, $\sigma^2 = \tau^2 = 1$, $\delta = 2.5$. The levels of the four factors substituted in the detection formulae were: $p = 0.067, 0.100$; $\sigma^2 = 0.8, 1.2$; $\tau^2 = 0.8, 1.2$ and $\delta = 1.5, 3.5$.

If the probability of a shift in the future, $\Pi_{n,n+1}$ drops below $\Pi^* = 0.3$ we consider it as sufficient evidence to declare that a shift already occurred and "sound an alarm". We define the experiment's response variable in each one of the 16 treatment combinations as an angular transformation of the proportion of false alarms among the 10 replicas, i.e.,

$$Y_\nu = 2 \sin^{-1} \sqrt{\hat{p}_\nu},$$

where $\hat{p}_\nu$ is the observed proportion of false alarms among the 10 replicas of the ν-th treatment combination. The other experiment's response variable measures consistency in process level estimates and is the average squared distance of $\hat{\mu}_{nl}$ from μ_n over the 10 replicas, i.e.,

$$ASD_\nu = \frac{1}{10NS} \sum_{l=1}^{10} \sum_{n=1}^{NS} (\hat{\mu}_{nl} - \mu_n)^2,$$

where $\hat{\mu}_{nl}$ is the Bayes estimate of the current mean, μ_n, at the l-th replication. The values of Y_ν and ASD_ν are given in Table 4.9. Y_ν is approximately normal with mean $\eta_\nu = 2 \sin^{-1} \sqrt{p_\nu}$ and constant variance 0.1, where p_ν is the true proportion of false alarms. In all the 160 runs, an alarm was flagged when $\Pi_{n,n+1} < \Pi^* = 0.3$.

Table 4.9.
The observed values of Y_ν and ASD_ν for the 16 treatment combinations of the four factors.

p	σ^2	τ^2	δ	Y_ν	ASD_ν
0.067	0.8	0.8	1.5	1.1593	0.0099
0.100	0.8	0.8	1.5	1.1593	0.0091
0.067	1.2	0.8	1.5	1.3694	0.0159
0.100	1.2	0.8	1.5	1.5708	0.0183
0.067	0.8	1.2	1.5	0.9273	0.0162
0.100	0.8	1.2	1.5	1.1593	0.0124
0.067	1.2	1.2	1.5	1.3694	0.0111
0.100	1.2	1.2	1.5	1.1593	0.0163
0.067	0.8	0.8	3.5	0.9273	0.0136
0.100	0.8	0.8	3.5	0.6435	0.0200
0.067	1.2	0.8	3.5	1.1593	0.0115
0.100	1.2	0.8	3.5	0.9273	0.0066
0.067	0.8	1.2	3.5	0.9273	0.0085
0.100	0.8	1.2	3.5	0.9273	0.0124
0.067	1.2	1.2	3.5	1.1593	0.0190
0.100	1.2	1.2	3.5	0.9273	0.0129

Routine analysis of the experimental results yields that the main effects of σ^2 and δ are the largest effects on y_ν, and that the third order interaction, $\sigma^2 \times \tau^2 \times \delta$, has the largest effect on ASD_ν. This implies that, deflating σ^2 and inflating δ will reduce the proportion of false alarms. The significant third order interaction of σ^2, τ^2 and δ can be exploited to produce Bayes estimates of the current mean with small ASD_ν values. For example, choosing an inflated (high) value of τ^2 yields $\text{ASD}_\nu \approx .012$, whatever the values of δ and σ^2 are. In all cases the procedure was found to be robust with respect to the choice of p.

5. DISCUSSION

There is a fundamental difference between sequential detection procedures and filtering procedures. The sequential detection procedures, like the CUSUM (non-Bayesian) or the Shiryayev-Roberts (Bayesian), are designed to monitor a process and flag as soon as sufficient evidence have accumulated that a change has occurred. A filtering procedure, on the other hand, is designed to estimate a signal (process mean level) in the presence of noise. The tracking algorithm developed and illustrated in the present paper is a filtering procedure. Unlike the Kalman filter it is a non-linear algorithm designed to detect sudden changes in the mean level of a stochastic process. The tracking algorithm provides, in addition to the fractiles (or other parameters) of the posterior distribution of the current mean, also the posterior probabilities of the change epoch, for the time interval under consideration. These posterior probabilities can be used, as illustrated in Section 4, for monitoring objectives. They provide estimates of the location of the change points.

Although our tracking algorithm is based on a model of at most one change, it is a filtering algorithm, and as such it can be applied with appropriate modification, also in the more general case of frequent or even continuous changes in the mean level. The examples presented in Section 4 fall in the class of processes in which the mean level is a step function. The estimates provided by the tracking algorithm fluctuate around these constant levels, not completely at random due to the interdependence (autocorrelations) between the estimates $\mu_{.5}$, for different n values (Tables 4.1 and 4.2).

Numerical examples have shown that the estimates of μ_n obtained by the tracking algorithm are sometimes sensitive to the choice of the parameters μ_T, δ and p. In Section 4 we study with simulation experiments how to choose these parameters. We propose that, before implementing the tracking algorithm, one should try it on historical data and "fine tune" it by means of choosing the parameters which yield robust and reasonable estimates. Guidelines for such fine tuning are provided in the robustness study presented in Section 4.3. The chosen values of the parameters can then be used for tracking the future movements of the process.

REFERENCES

[1] CHERNOFF, H. AND ZACKS, S. (1964). Estimating the Current Mean of a Normal Distribution which Is Subjected to Changes in Time. *Ann. Math. Statist*, **35**, 999–1018.

[2] HOADLEY, B. (1981). The Quality Measurement Plan. *Bell System Technical Journal*, **60**, 215–271.

[3] KENETT, R. AND POLLAK, M. (1983). On Sequential Detection of a Shift in the Probability of a Rare Event. *J. Amer. Statist. Assoc.*, **78**, 389–395.

[4] KENETT, R. AND POLLAK, M. (1986). A Semi–Parametric Approach to Testing for Reliability Growth, With Application to Software Systems. *IEEE Transactions on Reliability*, **R-35**, 304–311.

[5] LUCAS, J.M. (1982). Combined Shewhart CUSUM Quality Control Schemes. *J. of Quality Technology*, **14**, 51–59.

[6] LUCAS, J.M. AND CROSIER, R.B. (1982). Fast Initial Response of CUSUM Quality–Control Schemes: Give Your CUSUM a Head Start. *Technometrics*, *24*, 199–205.

[7] PAGE, E.S. (1954). Continuous Inspection Schemes. *Biometrika*, **41**, 100–114.

[8] PHADKE, M.S. (1981). Quality Audit Using Adaptive Kalman Filtering. *ASCQ Quality Congress Trans.*, 1045–1052.

[9] POLLAK, M. (1985). Optimal Detection of a Change in Distribution. *Ann. Statistics*, **13**, 206–227.

[10] SHEWHART, W.A. (1931). *Economic Control of Quality of Manufactured Product*. D. Van Nostrand, Princeton, NJ.

[11] STEPHENS, K.S., WADSWORTH, H.M. AND GODFREY, A.B. (1986). *Modern Methods for Quality Control and Improvement*. John Wiley, New York.

[12] ZACKS, S. (1980). Distribution of Stopping Variables In Sequential Procedures For The Detection of Shifts In Distributions of Discrete Random Variables. *Communications In Statistics, B*, **9**, 1–18.

[13] ZACKS, S. (1981). The Probability Distribution and the Expected Value of a Stopping Variable Associated with One-Sided CUSUM Procedures for Non-Negative Integer Valued Random Variables, *Communications in Statistics, A*, **10**, 2245–2258.

Statistical Decision Theory

Recent Adv. in Stat. and Prob., pp. 175-190
J. Pérez Vilaplana and M. L. Puri (Eds)

Robust Bayesian Bounds for Outlier Detection *

M.J. BAYARRI[1] and J.O. BERGER[2]
[1] Departamento de Estadística e Investigación Operativa, Universidad de Valencia, E–46089 Valencia, España
[2] Department of Statistics, Mathematical Sciences Building, Purdue University, West Lafayette, IN 47907, U.S.A.

Abstract – Outlier detection is a statistical problem that has received considerable attention and has originated a large body of research. Under suitable formulation, testing whether an observation is an outlier can be reduced to a testing problem concerning a parameter of a contaminating distribution. As in Bayesian testing of a point null, Bayes factors and posterior probabilities of the hypotheses cannot be computed here using non-informative priors. Thus we consider wide classes of priors for the relevant parameter and explore bounds on the Bayes factor as the prior varies in these classes.

1. INTRODUCTION

A universal, precise definition of the term "outlier" seems rather difficult. A generally accepted definition is the one used, for example, by Box and Tiao (1968) (also used in earlier papers, as mentioned there):

> "*an outlier being an observation which is suspected to be partially or wholly irrelevant because it is* not *generated by the stochastic model assumed.*"

Freeman (1980) adopts a similar definition without the requirement for the outlier to be irrelevant:

> "*The word 'outlier' here will mean any observation that has not been generated by the mechanism that generated the majority of observations in the data set.*"

The most common approach to outlier detection is to assess a model to explain the behavior of the (possible) outliers. This contaminating model is usually taken to be a generalization of the original model, involving an extra parameter (or parameter vector). The original model is then typically a particular case of the contaminating model, corresponding to some specific value of the extra parameter. Testing for outliers can then be reduced to testing for this specific value of the extra parameters.
Previous Bayesian work along these lines can be found in Box and Tiao (1968), Guttman (1973), Abraham and Box (1978), Guttman, Dutter and Freeman (1978), Freeman (1980), Pettit and Smith (1983, 1985), Pettit (1988, 1990, 1992), Verdinelli and Wasserman (1991), and Peña and Tiao (1992). Alternative approaches have also been explored,

*Research supported by the National Science Foundation, grant DMS–8923071, and by the Universitat de Valencia, ajuda de la convocatoria de 1992 d'Estades en Centres d'Investigació Nacionals y Estrangers.

usually requiring either the outlier to actually "outlie" or to be poorly predicted by the rest of the data. These approaches do not require a model for the outlying observations and they have been taken, for instance, by Zellner (1975), Geisser (1980, 1985, 1987, 1989, 1990), Chaloner and Brant (1988), Guttman and Peña (1988), and Giron, Martínez and Morcillo (1992). A third approach is to develop an overall model, usually with heavy tails, that automatically handles outliers in suitable ways without requiring their identification. This approach is taken by O'Hagan (1979, 1988) and West (1984, 1985); related work can be found in DeFinetti (1961), Dawid (1973), and Gómez-Villegas and Maín (1992). Reviews of previous work and discussions of the meaning of the term "outlier" can be found in Freeman (1980) and Pettit and Smith (1985).

We shall follow the usual formulation and assume that the data consists of $n+1$ independent random variables $X_0, X_1, \ldots, X_n$. In the outlier scenario, although we feel quite confident that $X_i \sim f(x_i|\theta)$, $i = 0, 1, \ldots, n$, we recognize (or fear) that we might get some (hopefully few) outlying observations. In other words, we recognize that, for any X_i, there is a (usually small) prior probability ε that it is generated by some contaminating distribution with density g. (A more general formulation would specify a possibly different g_i for each X_i.) Usually g is taken to be a density $f^*(x|\theta,\tau)$ such that the original model $f(X|\theta)$ is the special case corresponding to a specific value τ_0, so that $f(x|\theta) = f^*(x|\theta,\tau_0)$. If the goal of the statistical analysis is inference about θ, then it can be carried out using the implied mixture model, $X_i \sim (1-\varepsilon)f(x_i|\theta) + \varepsilon f^*(x_i|\theta,\tau)$. Our goal, however, will be to detect the outliers, perhaps because this is actually one of the main goals of the analysis (as in Raccine-Poon, 1992), or more usually, because we feel that our information about f^* is too vague to be of much help in inference about θ.

In this paper, we limit ourselves to study of the simplest case, that is, to detect whether a specific observation, x_0, is an outlier provided that all the rest are not. Hence, in the rest of the paper we assume the following model:

$$\begin{aligned} &X_1, X_2, \ldots, X_n \text{ i.i.d.}, \ X_i \sim f(x_i|\theta), \\ &X_0 \sim (1-\varepsilon)f(x_0|\theta) + \varepsilon f^*(x_0|\theta,\tau). \end{aligned} \tag{1.1}$$

Also, we shall let $\underline{x} = (x_1, \ldots, x_n)$ denote the non-outlying observations and $\ell(\theta)$ denote the likelihood function for θ based solely on $\underline{x}$, that is

$$\ell(\theta) = \prod_{i=1}^{n} f(x_i|\theta). \tag{1.2}$$

Prior densities for (θ,τ) are denoted by $\pi(\theta|\tau)g(\tau)$. Usually θ and τ are taken to be a priori independent, so that their joint prior would then be $\pi(\theta)g(\tau)$.

The posterior probability that x_0 is an outlier, given the data and given that $\underline{x}$ is non-outlier, can be expressed as

$$\Pr(x_0 \text{ outlier}|x_0, \underline{x}, \underline{x} \text{ non-outlier}) = \left(1 + \frac{(1-\varepsilon)}{\varepsilon}B(g)\right)^{-1}, \tag{1.3}$$

where

$$B(g) = \frac{\int \ell(\theta) f(x_0|\theta)\pi(\theta|\tau_0)d\theta}{\iint \ell(\theta) f^*(x_0|\theta,\tau)\pi(\theta|\tau)g(\tau)d\theta d\tau} \tag{1.4}$$

is the Bayes factor in favor of x_0 being non-outlier when the prior g for τ is used. An alternative expression for $B(g)$ is

$$B(g) = \frac{m^f(x_0|\underline{x})}{m^{f^*}(x_0|\underline{x})}, \tag{1.5}$$

where $m^h(x_0|x)$ is the (posterior) predictive density at x_0 assuming that the model generating x_0 is h. That is, $m^f(x_0|x) = \int f(x_0|\theta)\pi(\theta|x,\tau_0)d\theta$ and $m^{f^*}(x_0|x) = \int\int f^*(x_0|\theta,\tau)\pi(\theta|x,\tau)g(\tau)d\theta d\tau$. (For simplicity, the dependence of $m^{f^*}(x_0|x)$ on g is left implicit in the notation.) Note that Bayesian measures of "surprise" that do not require specification of f^* are typically based solely on the numerator of (1.5).

Bayes factors have become an increasingly popular way to report the experimental evidence in favor of (or against) the null hypothesis, especially in contexts of testing a sharp null hypothesis (see Berger and Delampady, 1987, and Berger and Sellke, 1987). Note that all experimental evidence for or against x_0 being an outlier enters (1.3) only through $B(g)$ and that the prior probability, ε, that x_0 is an outlier does not enter $B(g)$. This makes use of $B(g)$, rather than posterior probability, attractive in ordinary hypothesis testing.

In the outlier problem, however, the effect of ε cannot reasonably be ignored, since ε is usually quite small. As an indication of how small B needs to be in order to have reasonable belief that x_0 is an outlier, suppose $\varepsilon = 0.05$ (a typically considered value) and that we would decide that x_0 is an outlier if the posterior probability of such were larger than 0.85; this would require $B \leq .0093$. Thus, as a rough rule of thumb, $B \leq .01$ would be reasonable evidence that x_0 is an outlier. (This is compatible with the reasoning in Pettit, 1992, which suggests that reasonable evidence for an outlier is $B \leq .015$.)

A well known limitation to the use of Bayes factors for testing a point null hypothesis is that improper non-informative priors over the parameters identifying the two hypotheses cannot be used; indeed, by multiplying these *improper* priors by an arbitrary constant, the Bayes factors can be made arbitrary large or small. In our problem, this means that $g(\tau)$ in (1.4) cannot be a standard, non-informative prior. (Notice that $\pi(\theta)$ does not share this difficulty, in that multiplication of an improper $\pi(\theta)$ by a constant does not alter the Bayes factor.) This limitation is particularly disturbing in the outlier scenario since, as is almost implicit in the formulation of the problem, very little is typically known about the contaminating distribution, so that quantification of a proper, informative $g(\tau)$ is very difficult.

Pettit and Smith (1983, 1985), and Pettit (1988, 1990, 1992) deal with the problem by using the Spiegelhalter and Smith (1982) device of an imaginary training sample, in order to choose a suitable non-informative prior. In this paper, a different approach is taken. We select wide classes, Γ, of priors, g, and investigate the lower bound, $\underline{B}$, of $B(g)$ as g varies in Γ. Thus, the evidence in favor of x_0 *not* being an outlier would be *at least* $\underline{B}$, no matter what our prior g might be (provided, of course, that it belongs to the class Γ). If $\underline{B}$ is moderately large, x_0 would not be considered an outlier and further refinements of our prior knowledge about g would not be necessary.

In this paper, we confine ourselves to the most widely studied scenario for outliers, namely that in which the original distribution is a normal $N(x|\theta,\sigma^2)$ distribution and the contaminating distribution is normal $N(x|\theta,\sigma^2,\tau)$, with an extra parameter τ. This contaminating distribution will be taken to be either $N(x|\theta,\tau\sigma^2)$ with $\tau > 1$, or $N(x|\theta+\tau,\sigma^2)$. The former is usually referred to as a *scale* contamination model, while the later is usually called a *location* or *location-shift* contamination model. In Section 2, we study the scale contamination model, both for known and unknown σ^2; the same is done in Section 3 for the location contamination model. Section 4 is devoted to comparisons and generalizations.

2. SCALE CONTAMINATION

In this section we assume that $x_1, \ldots, x_n$ are non outlying observations from a $N(\theta, \sigma^2)$ distribution, and that the possible outlier, x_0, is generated by the mixture density

$$(1-\varepsilon)N(x_0|\theta,\sigma^2) + \varepsilon\, N(x_0|\theta,\tau\sigma^2) \tag{2.1}$$

with $\tau > 1$. We shall continue to let $\underline{x}$ denote the non outlying observations $\underline{x} = (x_1, x_2, \ldots, x_n)^t$ and let $\pi(\theta|\underline{x})$ denote the posterior density of θ given $\underline{x}$ (excluding x_0). Also, the posterior predictive densities (given $\underline{x}$) at x_0 under model $N(\theta, \sigma^2)$ and under model $N(\theta, \tau\sigma^2)$ for X_0 are denoted by $m(x_0|\underline{x})$ and $m^*(x_0|\underline{x}, \tau)$, respectively. We explore both the situations σ^2 known and σ^2 unknown.

2.1. σ^2 Known

Here, and in the rest of the paper, we assume that θ and τ are independent a priori, so that their joint density is given by $\pi(\theta)g(\tau)$. We take $\pi(\theta)$ to be the natural conjugate density $\pi(\theta) = N(\theta|m_0, \sigma_0^2)$, so that the posterior density is $\pi(\theta|\underline{x}) = N(\theta|m_1, \sigma_1^2)$, where

$$\begin{aligned} m_1 &= \lambda\bar{x} + (1-\lambda)m_0, \quad \sigma_1^2 = \lambda\sigma^2/n \\ \lambda &= n\sigma_0^2/(\sigma^2 + n\sigma_0^2). \end{aligned} \tag{2.2}$$

The non-informative prior $\pi(\theta) \propto 1$ results in the same posterior, but with $\lambda = 1$. Testing H_0: x_0 *is not an outlier*, versus H_1: x_0 *is an outlier*, is, with the above formulation, equivalent to testing $H_0: \tau = 1$, versus $H_1: \tau \neq 1$. The Bayes factor (1.5) in favor of H_0 is, for any given prior $g(\tau)$ for τ:

$$\begin{aligned} B(g) &= \frac{m(x_0|\underline{x})}{\int m^*(x_0|\underline{x},\tau)g(\tau)\,d\tau} = \frac{\int N(x_0|\theta,\sigma^2)N(\theta|m_1,\sigma_1^2)\,d\theta}{\iint N(x_0|\theta,\tau\sigma^2)N(\theta|m_1,\sigma_1^2)g(\tau)\,d\theta\,d\tau} = \\ &= \frac{N(x_0|m_1,\sigma_1^2+\sigma^2)}{\int N(x_0|m_1,\sigma_1^2+\tau\sigma^2)\,g(\tau)\,d\tau}. \end{aligned} \tag{2.3}$$

Following the robust Bayesian approach, we shall derive

$$\underline{B} = \inf_{g\in\Gamma} B(g) = \frac{N(x_0|m_1,\sigma_1^2+\sigma^2)}{\sup_{g\in\Gamma} \int N(x_0|m_1,\sigma_1^2+\tau\sigma^2)g(\tau)\,d\tau} \tag{2.4}$$

for two classes, Γ, of priors $g(\tau)$, namely the class of all possible prior densities and the class of all densities that are non-increasing functions of τ.

2.1.1. Case 1: $\Gamma_1 = \{\text{all priors } g(\tau)\}$

For this case, the supremum in the denominator of (2.4) is attained at a point-mass density, so that $\underline{B}$ is

$$\underline{B}_1 = \frac{N(x_0|m_1,\sigma_1^2+\sigma^2)}{\sup_{\tau} N(x_0|m_1,\sigma_1^2+\tau\sigma^2)}, \tag{2.5}$$

which is easily seen to be

$$\begin{aligned} \underline{B}_1 &= \sqrt{e}\; z\; e^{-z^2/2}, \qquad &&\text{for } z > 1, \\ &= 1, &&\text{for } z \leq 1 \end{aligned} \tag{2.6}$$

where $z = |x_0 - m_1|/s_1$, and m_1 and $s_1 = \sqrt{\sigma_1^2 + \sigma^2}$ are the mean and standard deviation, respectively, of the predictive distribution, $m(x_0|\boldsymbol{x})$, of x_0 under the non-outlier model. For the non-informative $\pi(\theta) \propto 1$, $m_1 = \bar{x}$ and $s_1^2 = (n+1)\sigma^2/n$. Thus, z in (2.6) is the distance, measured in units of predictive standard deviation, of x_0 from its predicted value m_1 under the non-outlier assumption.

2.1.2. Case 2: $\Gamma_2 = \{$all non-increasing densities g on $[1,\infty)\}$

If the class Γ_1 is deemed to be too large, a natural reduction is to consider all densities on $[1,\infty)$ that are non-increasing in τ. (Some authors maintain that τ should be much larger than 1; the analysis presented here can be trivially generalized to densities g on $[c,\infty)$, non increasing in τ, with c any positive constant.) All such densities can be represented as mixtures of uniform densities, so that now (2.4) becomes

$$\underline{B}_2 = \frac{N(x_0|m_1, \sigma_1^2 + \sigma^2)}{\sup_{\Gamma} \frac{1}{r-1}\int_1^r N(x_0|m_1, \sigma_1^2 + \tau\sigma^2)\, d\tau}, \tag{2.7}$$

which, after some algebra, can be expressed as

$$\begin{aligned} \underline{B}_2 &= \gamma^{-1/2} e^{-(1-\gamma)z^2/2}, && \text{for } z > 1, \\ &= 1, && \text{for } z \leq 1, \end{aligned} \tag{2.8}$$

where $z = |x_0 - m_1|/s_1$ and γ is the unique solution to

$$\frac{1+\gamma}{2\sqrt{\gamma}} e^{-\gamma z^2/2} - e^{-z^2/2} - z\sqrt{2\pi}\,\{\Phi(z) - \Phi(z\sqrt{\gamma})\} = 0, \tag{2.9}$$

where Φ denotes, as usual, the standard normal distribution function.
Although, obviously, $\underline{B}_2$ is greater than $\underline{B}_1$ (and in fact γ can be shown to be smaller than $1/z^2$), both lower bounds are surprisingly close to each other, even for moderate values of z. Table 1 gives the values of $\underline{B}_1$ and $\underline{B}_2$ for selected values of z. Figure 1 displays $\underline{B}_1$ and $\underline{B}_2$ as functions of z. Both Table 1 and Figure 1 assume the non-informative prior, $\pi(\theta) = 1$, for θ.

2.2. σ^2 Unknown

If we define $\delta = 1/\sigma^2$, then the usual conjugate prior for (θ,δ) has $\theta|\delta \sim N(m_0, (h_0\delta)^{-1})$, and $\delta \sim Ga(a_0, b_0)$, that is, $\pi(\delta) \propto \delta^{a_0-1}\exp\{-b_0\delta\}$. The ensuing joint distribution for (θ,δ) is the normal-gamma distribution, denoted $(\theta,\delta) \sim NGa(m_0, h_0, a_0, b_0)$. The posterior distribution of (θ,δ), given the non-outlying $\boldsymbol{x}$, is then the revised normal-gamma $(\theta,\delta|\boldsymbol{x}) \sim NGa(m_1, h_1, a_1, b_1)$, where

$$\begin{aligned} m_1 &= \lambda\bar{x} + (1-\lambda)m_0, \qquad \lambda = n/(h_0+n) \\ h_1 &= h_0 + n,\ a_1 = a_0 + b/2, \\ b_1 &= b_0 + \frac{1}{2}\sum_{i=1}^n (x_i - \bar{x})^2 + \frac{\lambda h_0}{2}(\bar{x} - m_0)^2. \end{aligned} \tag{2.10}$$

If the non-informative $\pi(\theta,\delta) \propto 1/\delta$ is used, then $m_1 = \bar{x}$, $h_1 = n$, $a_1 = (n-1)/2$, $b_1 = \Sigma(x_i - \bar{x})^2/2$.

Table 1.
Lower Bounds on the Bayes Factors for the Scale Contamination Model (σ^2 known).

z	1.5	2	2.5	3	3.5	4	4.5
$\underline{B}_1$	.8029	.4463	.1811	.0549	.0126	.0022	.0003
$\underline{B}_2$	.8305	.4832	.2026	.0628	.0146	.0026	.0003

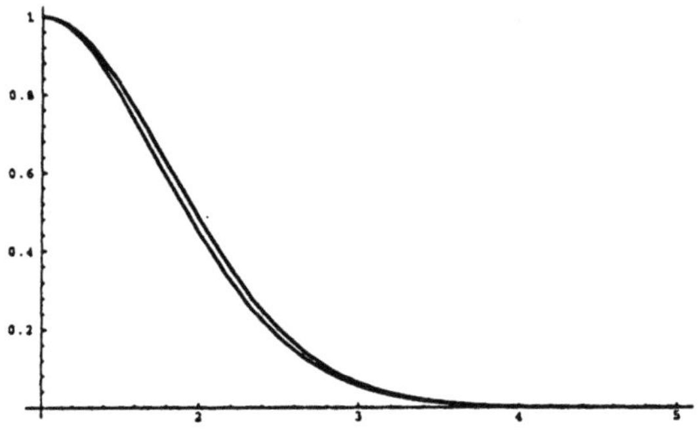

Figure 1. $\underline{B}_1$ and $\underline{B}_2$ for the Scale-Contamination Model (σ^2 known) as Functions of z.

The predictive densities that enter the computation of the Bayes factor are the posterior predictive densities of X_0 (at x_0) assuming that either X_0 is not an outlier:

$$m(x_0|x) = \iint N(x_0|\theta,\sigma^2) NGa(\theta,\sigma^2|m_1,h_1,a_1,b_1)\,d\theta\,d\sigma^2 = \\ = St\left(x_0|m_1,\left(1+\frac{1}{h_1}\right)\frac{b_1}{a_1-1},2(a_1-1)\right), \tag{2.11}$$

or that x_0 comes from the contaminating density whose extra-parameter has the value τ:

$$m^*(x_0|x,\tau) = \iint N(x_0|\theta,\tau\sigma^2) NGa(\theta,\sigma^2|m_1,h_1,a_1,b_1)\,d\theta\,d\sigma^2 \\ = St\left(x_0|m_1,\left(\tau+\frac{1}{h_1}\right)\frac{b_1}{a_1-1},2(a_1-1)\right), \tag{2.12}$$

where $St(x|m,s^2,\alpha)$ denotes a Student t density as given by

$$St(x|m,s^2,\alpha) = C\left[1+\frac{1}{\alpha s^2}(x-m)^2\right]^{-(\alpha+1)/2}, \tag{2.13}$$

where $C = \Gamma\left(\frac{\alpha+1}{2}\right) \Big/ \left[\Gamma\left(\frac{\alpha}{2}\right)\Gamma\left(\frac{1}{2}\right)\left(\alpha s^2\right)^{1/2}\right]$.

Again we shall study the infimum of the Bayes factor (in favor of H_0: *non-outlier*)

$$\underline{B} = \inf_{g\in\Gamma} \frac{m(x_0|x)}{\int m^*(x_0|x,\tau)g(\tau)\,d\tau}, \tag{2.14}$$

for the same classes Γ_1, Γ_2 considered in Subsection 2.1.

2.2.1. Case 1: $\Gamma_1 = \{\text{all priors } g(\tau)\}$

In this case, from (2.14), (2.11) and (2.12):

$$\underline{B}_1 = \frac{St\Big(x_0|m_1, (1+h_1)\, b_1 \,/\, [h_1(a_1-1)], 2(a_1-1)\Big)}{\sup\limits_{\tau} St\Big(x_0|m_1, (1+\tau h_1)\, b_1 \,/\, [h_1(a_1-1)], 2(a_1-1)\Big)}. \tag{2.15}$$

To parallel the developments in the previous subsection, let s_1 denote the scale parameter of the posterior predictive of X_0 under the non-outlier model, that is:

$$s_1^2 = \frac{(1+h_1)}{h_1} \cdot \frac{b_1}{(a_1-1)}, \tag{2.16}$$

and define

$$z = \left|\frac{x_0 - m_1}{s_1}\right|. \tag{2.17}$$

For the non-informative prior, $m_1 = \bar{x}$ and $s_1^2 = (n+1)\hat{\sigma}^2/(n-3)$, where $\hat{\sigma}^2 = \sum\limits_{i=1}^{n}(x_i - \bar{x})^2/n$ is the usual MLE of σ^2 based on the non-outlying observations. (Recall that, when σ^2 was known, $s_1^2 = (n+1)\sigma^2/n$). With this notation, the lower bound $\underline{B}_1$ on the Bayes factor can be computed from (2.15) to be

$$\begin{aligned} \underline{B}_1 &= z\left(\frac{\alpha+1}{\alpha+z^2}\right)^{(\alpha+1)/2} && \text{for } z > 1, \\ &= 1, && \text{for } z \le 1, \end{aligned} \tag{2.18}$$

where $\alpha = z(a_1 - 1)$ is the degrees of freedom of both predictive distributions. For the non-informative prior, $\alpha = n - 3$. It can be checked that this lower bound, $\underline{B}_1$, converges, as α goes to ∞, to the known σ^2 lower bound, given in (2.6).

2.2.2. Case 2: $\Gamma_2 = \{\text{all non-increasing densities, } g, \text{ on } [1, \infty)\}$

In this case, the same argument as in Subsection 2.1 gives

$$\underline{B}_2 = \frac{m(x_0|x)}{\sup\limits_{r} \frac{1}{r-1}\int_1^r m^*(x_0|x,\tau)\, d\tau}, \tag{2.19}$$

where $m(x_0|x)$ and $m^*(x_0|x,\tau)$ are the t densities given in (2.11) and (2.12), respectively. Again defining $z = |x_0 - m_1|/s_1$, with m_1 and s_1 as in *case* 1, $\underline{B}_2$ can be computed to be

$$\begin{aligned} \underline{B}_2 &= \gamma^{-1/2}\left(\frac{\alpha+z^2}{\alpha+\gamma z^2}\right)^{-(\alpha+1)/2} && \text{for } z > 1, \\ &= 1, && \text{for } z \le 1, \end{aligned} \tag{2.20}$$

where $\alpha = 2(a_1 - 1)$ and γ is the unique solution of

$$\begin{aligned} &\frac{2}{z}\left\{\frac{1}{\sqrt{\gamma}}\left(1+\frac{\gamma z^2}{\alpha}\right)^{-(\alpha+1)/2} - \left(1+\frac{z^2}{\alpha}\right)^{-(\alpha+1)/2}\right\} - \\ &\quad -\frac{\sqrt{\alpha\pi}\,\Gamma\left(\frac{\alpha}{2}\right)}{\Gamma\left(\frac{\alpha+1}{2}\right)}\left\{I_{\frac{\alpha}{\alpha+\gamma z^2}}\left(\frac{\alpha+2}{2},\frac{1}{2}\right) - I_{\frac{\alpha}{\alpha+z^2}}\left(\frac{\alpha+2}{2},\frac{1}{2}\right)\right\} = \\ &= \frac{1-\gamma}{z\sqrt{\gamma}}\left[1+\frac{\gamma z^2}{\alpha}\right]^{-(\alpha+1)/2}, \end{aligned} \tag{2.21}$$

where

$$I_x(\alpha,\beta) = \frac{\Gamma(\alpha+\beta)}{\Gamma(\alpha)\Gamma(\beta)} \int_0^x u^{\alpha-1}(1-u)^{\beta-1}du \qquad (2.22)$$

is the incomplete beta function.

As in Subsection 2.1, where σ^2 was assumed known, $\underline{B}_2$ is again surprisingly close to $\underline{B}_1$. (Obviously $\underline{B}_2 > \underline{B}_1$.) In Table 2, we display the values of $\underline{B}_1$ and $\underline{B}_2$ for certain values of z and for $\alpha = 20$. (Smaller values of α produced larger values of both lower bounds, $\underline{B}_1$ and $\underline{B}_2$.) In Figure 2, the lower bound $\underline{B}_1$ is graphed, as a function of z, for $\alpha = 10$, 20, 40, 100. (The corresponding graphs of $\underline{B}_2$ were extremely similar.) Both Table 2 and Figure 2 were derived with the non-informative prior.

Comparison of Tables 1 and 2 or Figures 1 and 2 show that, for moderate or large α, there is little difference between the known σ^2 and unknown σ^2 cases, as long as $z < 2.5$. However, for $z > 2.5$ the difference can be significant, with the lower bounds being substantially larger in the case of unknown σ^2. (It is, of course, not surprising that unknown σ^2 makes it less certain that large z implies that x_0 is an outlier). The stability of $\underline{B}_1$ to varying α, that is exhibited in Figure 2, actually held to remarkably small values (e.g., $\alpha = 3$).

Table 2.
Lower Bounds on the Bayes Factors for the
Scale Contamination Model (σ^2 unknown).

z	1.5	2	2.5	3	3.5	4	4.5	5
$\underline{B}_1$	.8174	.4922	.2401	.1012	.0387	.0139	.0049	.0017
$\underline{B}_2$	.8437	.5300	.2665	.1146	.0444	.0161	.0056	.0020

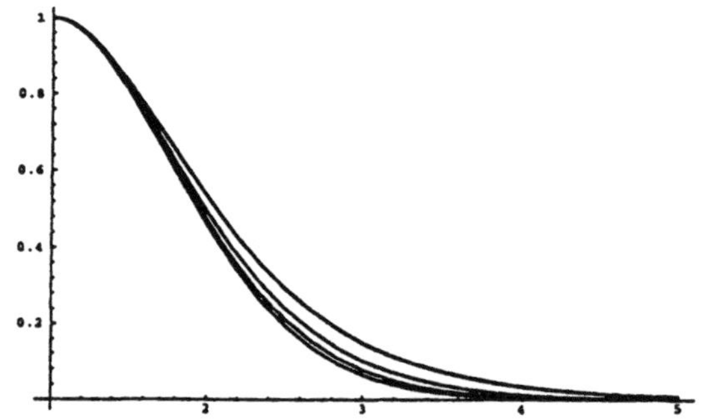

Figure 2.$\underline{B}_1$ for the Scale-Contamination Model (σ^2 unknown) as a Function of z for $\alpha = 10, 20, 40, 100$.

3. LOCATION CONTAMINATION

In this section, we assume that the non-outliers, $x = (x_1, \ldots, x_n)$, come from a $N(\theta, \sigma^2)$ distribution and that the possible outlier, x_0, is generated by the location contamination mixture

$$X_0 \sim (1-\varepsilon)N(\theta,\sigma^2) + \varepsilon N(\theta+\tau,\sigma^2). \qquad (3.1)$$

It should be noted that this outlier model has the scale contamination model as a particular case, corresponding to certain prior distributions for τ. Indeed, if $X_0 \sim N(\theta+\tau,\sigma^2)$

and τ has the $N(0,(\nu-1)\sigma^2)$ prior distribution (with $\nu > 1$), then, conditional on θ, ν and σ^2, the distribution of X_0 is given by the scale contamination model $X_0 \sim N(\theta, \nu\sigma^2)$ with $\nu > 1$; putting prior distributions on ν would thus yield the Section 2 analyses. It follows that the lower bounds to be computed in this section will all be smaller than those obtained in Section 2.

Here $\pi(\theta|x)$ and $m(x_0|x)$ are as in Section 2, while $m^*(x_0|x)$ will denote the (posterior) predictive density of X_0, assuming now that $X_0 \sim N(\theta+\tau, \sigma^2)$. Note that the symbol τ is still used to denote the contaminating parameter, though, of course, the restriction $\tau \geq 1$ no longer applies.

3.1. σ Known

Under a conjugate analysis, $\pi(\theta)$ and $\pi(\theta|x)$ are the same as in Subsection 2.1, and the Bayes factor in favor of H_0: *non-outlier* ($\tau = 0$) is

$$B(g) = \frac{\int N(x_0|\theta,\sigma^2)N(\theta|m_1,\sigma_1^2)d\theta}{\iint N(x_0|\theta+\tau,\sigma^2)N(\theta|m_1,\sigma_1^2)g(\tau)\,d\theta\,d\tau} = \frac{N(x_0|m_1,s_1^2)}{\int N(x_0|m_1+\tau,s_1^2)g(\tau)\,d\tau}, \tag{3.2}$$

where $s_1^2 = \sigma^2 + \sigma_1^2$, and m_1 and σ_1^2 are given in (2.2). We shall again study

$$\underline{B} = \inf_{g\in\Gamma} B(g) = \frac{N(x_0|m_1,s_1^2)}{\sup\limits_{g\in\Gamma} \int N(x_0|m_1+\tau,s_1^2)g(\tau)\,d\tau},$$

for two classes, Γ, of priors for τ.

3.1.1. Case 1: $\Gamma_1 = \{\text{all priors } g(\tau)\}$

In this case,

$$\underline{B}_1 = \frac{N(x_0|m_1,s_1^2)}{\sup\limits_{\tau} N(x_0|m_1+\tau,s_1^2)}, \tag{3.3}$$

which is trivially seen to be

$$\underline{B}_1 = e^{-z^2/2}, \tag{3.4}$$

where, again, $z = |x_0 - m_1|/s_1$, the standardized distance of x_0 to its predicted value under the non-outlier model.

3.1.2. Case 2: $\Gamma_2 = \{\text{all densities } g(\tau) = h(|\tau|),\ h \text{ non-increasing}\}$

Since τ represents a shift of the mean θ, a natural way to restrict the class is to consider all densities that are symmetric about $\tau = 0$, and that are non-increasing in $|\tau|$. These densities can, again, be expressed as mixtures of uniform densities, so that the lower bound on the Bayes factor is

$$\underline{B}_2 = \frac{N(x_0|m_1,s_1^2)}{\sup\limits_{r} \frac{1}{2r}\int_{-r}^{r} N(x_0|m_1+\tau,s_1^2)d\tau}, \tag{3.5}$$

which can be calculated to be

$$\begin{aligned} \underline{B}_2 &= \frac{2\exp\left\{-z^2/2\right\}}{\exp\left\{-(z+\gamma)^2/2\right\} - \exp\left\{-(z-\gamma)^2/2\right\}} && \text{for } z > 1, \\ &= 1, && \text{for } z \leq 1, \end{aligned} \tag{3.6}$$

where $z = |x_0 - m_1|/s_1$, and γ is the unique solution of

$$\gamma\left[\phi(z+\gamma) + \phi(z-\gamma)\right] = \Phi(z+\gamma) - \Phi(z-\gamma), \tag{3.7}$$

where ϕ, Φ represent the standard normal p.d.f. and c.d.f., respectively.

Table 3.
Lower Bounds on the Bayes Factor for the
Location Contamination Model (σ^2 known).

z	1.5	2	2.5	3	3.5	4	4.5
$\underline{B}_1$	.3247	.1353	.0439	.0111	.0022	.0003	.0000
$\underline{B}_2$	.7493	.3835	.1458	.0420	.0093	.0016	.0002

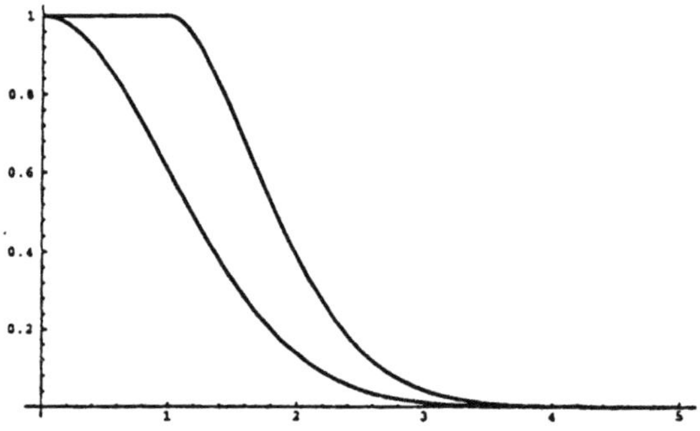

Figure 3. $\underline{B}_1$ and $\underline{B}_2$ for the Location–Contamination Model (σ^2 known) as Functions of z.

In Table 3 we give $\underline{B}_1$ and $\underline{B}_2$ for certain values of z. In Figure 3, they are graphed as functions of z. Both Table 3 and Figure 3 are, as before, derived for the non-informative prior $\pi(\theta) = 1$. In contrast to the situation in Section 2 with the scale contamination model, the differences between $\underline{B}_1$ and $\underline{B}_2$ can here be quite substantial.

3.2. σ^2 Unknown

Letting $\delta = 1/\sigma^2$, the joint prior and the joint posterior (given only $\boldsymbol{x}$) of (θ, δ) are the same as those in Subsection 2.2. Furthermore, the posterior predictive under the non-outlier $m(x_0|\boldsymbol{x})$ is as in (2.11), while the predictive under the outlier model for x_0, $m^*(x_0|\boldsymbol{x}, \tau)$, is

$$m^*(x_0|\boldsymbol{x}, \tau) = St\left(m_1 + \tau, \frac{(h_1+1)}{h_1}\frac{b_1}{(a_1-1)}, 2(a_1-1)\right). \tag{3.8}$$

Again define $z = |x_0 - m_1|/s_1$, where m_1 and s_1 are the location and scale parameter of the non-outlier predictive, $m(x_0|\boldsymbol{x})$, and are given in (2.10), together with h_1, a_1, b_1 that appear in (3.7) above. Our goal is to study

$$\underline{B} = \inf_{g \in \Gamma} B(g) = \frac{St\left(x_0|m_1, s_1^2, \alpha\right)}{\sup_{g\in\Gamma} \int St\left(x_0|m_1+\tau, s_1^2, \alpha\right) g(\tau)\, d\tau}, \tag{3.9}$$

where $\alpha = 2(a_1 - 1)$.

3.2.1. Case 1: $\Gamma_1 = \{\text{all priors } g(\tau)\}$

In this case, (3.8) becomes

$$\underline{B}_1 = \frac{St\left(x_0|m_1, s_1^2, \alpha\right)}{\sup_{\tau} St\left(x_0|m_1+\tau, s_1^2, \alpha\right)}, \tag{3.10}$$

which is trivially seen to be

$$\underline{B}_1 = \left(\frac{\alpha}{\alpha+z^2}\right)^{(\alpha+1)/2}. \tag{3.11}$$

Note that, as $\alpha \to \infty$ (recall that, with the non-informative prior, $\alpha = n-3$), (3.10) goes to the $\underline{B}_1$ given in (3.4) for known σ^2.

3.2.2. Case 2: $\Gamma_2 = \{\text{all densities } g(\tau) = h(|\tau|),\ h \text{ non increasing}\}$

In this case, (3.8) becomes

$$\underline{B}_2 = \frac{St\left(x_0|m_1, s_1^2, \alpha\right)}{\sup_{r} \frac{1}{2r}\int_{-r}^{r} St\left(x_0|m_1+\tau, s_1^2, \alpha\right)d\tau}, \tag{3.12}$$

which can be calculated to be

$$\underline{B}_2 = \frac{2\left[1+z^2/\alpha\right]^{-(\alpha+1)/2}}{\left[1+(z+\gamma)^2/\alpha\right]^{-(\alpha+1)/2} + \left[1+(z-\gamma)^2/\alpha\right]^{-(\alpha+1)/2}}, \tag{3.13}$$

where γ is the solution of

$$\gamma\left[t_\alpha(z+\gamma) + t_\alpha(z-\gamma)\right] = T_\alpha(z+\gamma) - T_\alpha(z-\gamma), \tag{3.14}$$

with t_α and T_α being the p.d.f. and c.d.f., respectively, of a $St(0,1,\alpha)$ distribution.

Table 4.
Lower Bounds on the Bayes Factors for
the Location Contamination Model with
σ^2 Unknown and α = 1, 10, 50.

	$\alpha = 1$		$\alpha = 10$		$\alpha = 50$	
z	$\underline{B}_1$	$\underline{B}_2$	$\underline{B}_1$	$\underline{B}_2$	$\underline{B}_1$	$\underline{B}_2$
1.5	.3077	.7008	.3275	.7461	.3255	.7490
2	.2000	.5476	.1571	.4398	1405	.3972
2.5	.1379	.4387	.0692	.2273	.0496	.1643
3	.1000	.3611	.0293	.1098	.0147	.0555
3.5	.0755	.3043	.0123	.0516	.0037	.0158
4	.0588	.2615	.0052	.0242	.0008	.0039
4.5	.0471	.2284	.0023	.0115	.0002	.0009
5	.0385	.2021	.0010	.0056	.0000	.0002

Table 4 gives the values of $\underline{B}_1$ and $\underline{B}_2$ for α = 1, 10 and 50 (corresponding to n = 4, 13 and 53 non-outliers in the sample). Figure 4 graphs $\underline{B}_1$, as a function of z, for $\alpha = 1$ and $\alpha = \infty$ (which corresponds to the known σ^2 case). Figure 5 graphs $\underline{B}_2$ for α = 1, 3, 5, 10. All figures and tables use the non-informative prior $\pi(\theta, \delta) \propto 1/\delta$.

As in the scale contamination scenario, there is remarkably little difference between the lower bounds for the known σ^2 and unknown σ^2 cases when $z < 2.5$ (unless α is very small, e.g. $\alpha = 1$ or $\alpha = 2$). For $z > 2.5$, however, the difference can be appreciable, with the lower bounds for the unknown σ^2 case being substantially larger.

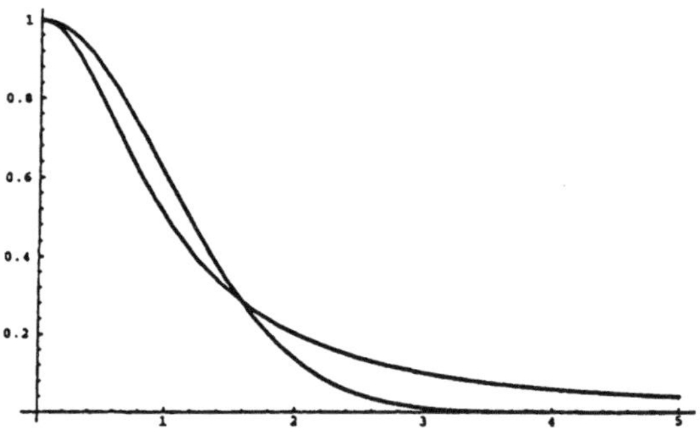

Figure 4. $\underline{B}_1$ for the Location–Contamination Model as a Function of z for σ^2 Unknown ($\alpha = 1$) and σ^2 Known.

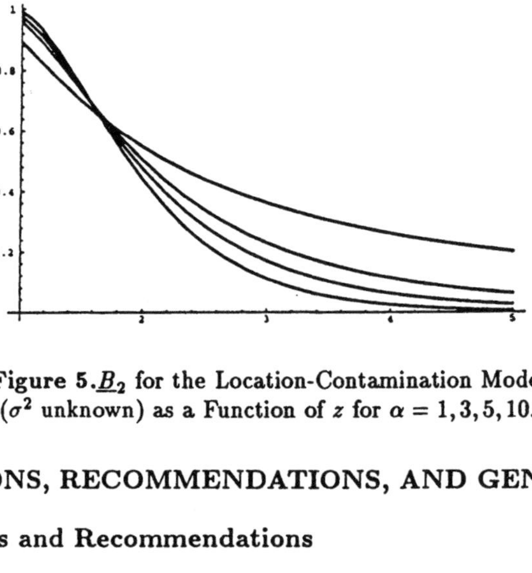

Figure 5. $\underline{B}_2$ for the Location-Contamination Model (σ^2 unknown) as a Function of z for $\alpha = 1, 3, 5, 10$.

4. COMPARISONS, RECOMMENDATIONS, AND GENERALIZATIONS

4.1. Comparisons and Recommendations

Comparison of Tables 1 and 3 or Tables 2 and 4 reveals a rather remarkable similarity between $\underline{B}_1$ (scale contamination), $\underline{B}_2$ (scale contamination) and $\underline{B}_2$ (location contamination). These lower bounds do, however, differ substantially from $\underline{B}_1$ (location contamination). Since this last lower bound was based on the most unreasonably large class of priors (*all* priors for the contaminating location parameter), it will typically be an unreasonably small lower bound, and hence is not recommended for use.

Since the other three lower bounds are very similar, it is reasonable to choose among them according to computational simplicity. In this regard, $\underline{B}_1$ (scale contamination) is the clear winner, being given by the simple closed-form formulas (2.6) and (2.18) (for the known σ^2 and unknown σ^2 cases, respectively). It is, indeed, fortuitous (and unusual in Bayesian robustness) that a simple lower bound resulting from an "all prior" model is reasonable.

In the Introduction, it was suggested that a rough rule-of-thumb for outlier detection is to judge x_0 to be an outlier if $B \leq .01$. A *necessary* condition for this to be true, in the presence of prior uncertainty as reflected by the class Γ, is

$$\underline{B} = \inf_{g\in\Gamma} B(g) \leq .01. \tag{4.1}$$

For the recommended lower bounds (2.6) and (2.18), this necessary condition becomes (using a conservative bound for (2.18))

$$\begin{aligned} \text{for known } \sigma^2 &: \ z \geq 3.57 \\ \text{for unknown } \sigma^2 &: \ z \geq (3.57)\left(1-(.982)/\sqrt{\alpha}\right)^{-1}, \end{aligned} \tag{4.2}$$

where $z = |x_0 - m_1|/s_1$ is the distance between x_0 and the (non-outlier) predictive mean m_1, in units of (non-outlier) predictive standard deviations. For the noninformative prior case, recall that $m_1 = \bar{x}$ and $s_1^2 = (n+1)\sigma^2/n$ (known σ^2 case) or $s_1^2 = (n+1)\sum_{i=1}^{n}(x_i - \bar{x})^2/[n(n-3)]$ (unknown σ^2 case). The second inequality in (4.2) is an approximation to (4.1) and (2.18).

The implications of this rule-of-thumb to practice are perhaps surprising. It would seem that x_0 needs to be at least 4 standard errors out (somewhat less for the known σ^2 case; somewhat more for unknown σ^2) before one can consider it to be a likely outlier. And, since $\underline{B}_1$ was a lower bound, even this would not conclusively prove that x_0 is an outlier. Our recommendation for practice is thus to use (4.2) as a preliminary screen; only observations for which z exceeds the indicated values need be investigated as possible outliers.

4.2. Generalizations

Recall that we based our analyses on

$$\Pr(x_0 \text{ is an outlier} \mid \text{data}, \boldsymbol{x} \text{ being non-outliers}). \tag{4.3}$$

Of perhaps more interest is

$$\Pr(x_0 \text{ is an outlier}, \boldsymbol{x} \text{ are non-outliers} \mid \text{data}), \tag{4.4}$$

which does not assume that the remaining observations are known to be non-outliers. It is possible to show, however, that (4.3) is a good approximation to (4.4), if x_0 seems to be the only likely outlier. Here is a heuristic argument to this effect.

If $\boldsymbol{x} = (x_1, \ldots, x_n)$ are not outlying (i.e., are reasonably compatible with $f(x_i|\theta)$), then their joint density under the full mixture model for all observations is

$$\prod_{i=1}^{n}\left[(1-\varepsilon)f(x_i|\theta) + \varepsilon f^*(x_i|\theta,\tau)\right] \cong (1-\varepsilon)^n \prod_{i=1}^{n} f(x_i|\theta),$$

the approximation based on observing that ε is typically small and that $f(x_i|\theta)$ will be comparable or larger than $f^*(x_i|\theta,\tau)$. But then

$$\Pr(x_0 \text{ is an outlier}, \boldsymbol{x} \text{ are non-outliers} \mid \text{data}) =$$

$$= \frac{\varepsilon(1-\varepsilon)^n \iint f^*(x_0|\theta,\tau)\left(\prod_{i=1}^n f(x_i|\theta)\right)\pi(\theta,\tau)\,d\theta d\tau}{\iint \prod_{i=0}^n \left[(1-\varepsilon)f(x_i|\theta)+\varepsilon f^*(x_i|\theta,\tau)\right]\pi(\theta,\tau)\,d\theta\,d\tau} \cong$$

$$\cong \frac{\varepsilon \iint f^*(x_0|\theta,\tau)\left(\prod_{i=1}^n f(x_i|\theta)\right)\pi(\theta,\tau)\,d\theta\,d\tau}{\iint \left[(1-\varepsilon)f(x_0|\theta)+\varepsilon f^*(x_0|\theta,\tau)\right]\left(\prod_{i=1}^n f(x_i|\theta)\right)\pi(\theta,\tau)\,d\theta\,d\tau} =$$

$$= \Pr\,(x_0 \text{ is an outlier} \mid \text{data}, \underset{\sim}{x} \text{ being non-outliers}),$$

completing the argument.

If there seems to be more than one likely outlier, the situation becomes considerably more complicated. The main complication is that the possible outliers, in combination, can perhaps provide significant information about the contaminating distribution. To properly analyze this situation, it is probably necessary to deal with the formal mixture model.

It is, however, trivial to consider the generalization to the Bayes factor for testing

$$H_0 \;:\; \underset{\sim}{x}_0 = (x_{01},\ldots,x_{0k}) \textit{ all non-outliers}, \text{ versus}$$
$$H_1 \;:\; \underset{\sim}{x}_0 \textit{ all outliers}.$$

Indeed, all expressions obtained for $k=1$ (i.e., a single outlier, x_0) are valid for the above hypotheses if z is instead defined as $z = |\bar{x}_0 - m_1|/s_1^*$, where $\bar{x}_0 = \sum_{i=1}^k x_{0i}/k$ and

$$s_1^{*2} = \begin{cases} \sigma_1^2 + \sigma^2/k & \text{if } \sigma^2 \text{ is known,} \\ \left(\dfrac{1}{k}+\dfrac{1}{n_1}\right)\dfrac{b_1}{(a_1-1)} & \text{if } \sigma^2 \text{ is unknown.} \end{cases}$$

A final generalization of interest is to the multiple outlier scenario when the outliers are thought to arise from *different* contaminating distributions. Then the outliers could be checked one-at-a-time, using (4.2), with the modification that $\underset{\sim}{x}$ now consist only of the clearly non-outlying observations.

REFERENCES

[1] ABRAHAM, B., AND BOX, G. E. P. (1978). Linear Models and Spurious Observations. *Applied Statistics* **27**, 120–130.

[2] BERGER, J. O. AND DELAMPADY, M. (1987). Testing Precise Hypotheses. *Statistical Science* **2**, 317-352.

[3] BERGER, J. O. AND SELLKE, T. (1987). Testing a Point Null Hypothesis: The Irreconcilability of Significance Levels and Evidence. *Journal of the American Statistical Association* **82**, 112–122.

[4] BOX, G. E. P., AND TIAO, G. C. (1968). A Bayesian Approach to Some Outlier Problems. *Biometrika* **55**, 119–129.

[5] CHALONER, K., AND BRANT, R. (1988). A Bayesian Approach to Outlier Detection and Residual Analysis. *Biometrika* **75**, 651–659.

[6] DAWID, A. P. (1973). Posterior Expectations for Large Observations. *Biometrika* **60**, 664–667.

[7] DEFINETTI, B. (1961). The Bayesian Approach to the Rejection of Outliers. *Proceedings of the Fourth Berkeley Symposium on Probability and Statistics* **1**, 199–210. Berkeley: University Press.

[8] FREEMAN, P. R. (1980). On the Number of Outliers in Data from a Linear Model. In *Bayesian Statistics* (J. M. Bernardo, M. H. DeGroot, A. F. M. Smith, D. V. Lindley, eds.), 349–365. Valencia: Valencia University Press.

[9] GEISSER, S. (1980). Discussion of Sampling and Bayes' Inference in Scientific Modelling and Robustness, by G. E. P. Box. *Journal of the Royal Statistical Society A* **143**, 416–417.

[10] GEISSER, S. (1985). On the Prediction of Observables: A Selective Update. In *Bayesian Statistics 2* (J. M. Bernardo, M. H. DeGroot, A. F. M. Smith, D. V. Lindley, eds.), 203–229. Amsterdam: North-Holland.

[11] GEISSER, S. (1987). Influential Observations, Diagnostics and Discordancy Tests. *Journal of Applied Statistics* **14**, 133–142.

[12] GEISSER, S. (1989). Predictive Discordancy Testing for Exponential Observations. *Canadian Journal of Statistics* **17**, 19–26.

[13] GEISSER, S. (1990). Predictive Approaches to Discordancy Testing. In *Bayesian and Likelihood Methods in Statistics and Econometrics* (S. Geisser, J. S. Hodges, S. J. Press, A. Zellner, eds.), 321–335. Amsterdam: North Holland.

[14] GIRÓN, F. J., MARTÍNEZ, M. L., AND MORCILLO, C. (1992). A Bayesian Justification for the Analysis of Residuals and Influence Measures. In *Bayesian Statistics 4* (J. M. Bernardo, J. O. Berger, A. P. Dawid, A. F. M. Smith, eds.), 651–660. Oxford: Clarendon Press.

[15] GÓMEZ-VILLEGAS, M. A., AND MAÍN, M. (1992). The Influence of Prior and Likelihood Tail Behaviour on the Posterior Distribution. In *Bayesian Statistics 4* (J. M. Bernardo, J. O. Berger, A. P. Dawid, A. F. M. Smith, eds.), 661–667. Oxford: Clarendon Press.

[16] GUTTMAN, I. (1973). Care and Handling of Univariate or Multivariate Outliers in Detecting Spuriosity: A Bayesian Approach. *Technometrics* **15**, 723–738.

[17] GUTTMAN, I., DUTTER, R., AND FREEMAN, P. R. (1978). Care and Handling of Univariate Outliers in the General Linear Model to Detect Spuriosity: A Bayesian Approach. *Technometrics* **20**, 187–193.

[18] GUTTMAN, I., AND PEÑA, D. (1988). Outliers and Influence: Evaluation by Posteriors of Parameters in the Linear Model. In *Bayesian Statistics 3* (J. M. Bernardo, M. H. DeGroot, D. V. Lindley, A. F. M. Smith, eds.), 631–640. Oxford: Clarendon Press.

[19] JEFFREYS, H. (1983). *Theory of Probability (Third edition)*. Oxford: Clarendon Press.

[20] O'HAGAN, A. (1979). On Outlier Rejection Phenomena in Bayes Inference. *Journal of the Royal Statistical Society B* **41**, 358–367.

[21] O'HAGAN, A. (1988). Modelling with Heavy Tails. In *Bayesian Statistics 3* (J. M. Bernardo, M. H. DeGroot, D. V. Lindley, A. F. M. Smith, eds.), 345–359.

[22] PEÑA, D. AND TIAO, G. C. (1992). Bayesian Robustness Functions for Linear Models. In *Bayesian Statistics 4* (J. M. Bernardo, J. O. Berger, A. P. Dawid, A. F. M. Smith, eds.), 365–388. Oxford: Clarendon Press.

[23] PETTIT, L. J. (1988). Bayes Methods for Outliers in Exponential Samples. *Journal of the Royal Statistical Society B* **50**, 371–380.

[24] PETTIT, L. I. (1990). The Conditional Predictive Ordinate for the Normal Distribution. *Journal of the Royal Statistical Society B* **52**, 175–184.

[25] PETTIT, L. I. (1992). Bayes Factors for Outlier Models Using the Device of Imaginary Observations. *Journal of the American Statistical Association* **87**, 541–545.

[26] PETTIT, L. I., AND SMITH, A. F. M. (1983). Bayesian Model Comparison in the Presence of Outliers. *Bulletin of the International Statistical Institute* **50**, 292–309.

[27] PETTIT, L. I., AND SMITH, A. F. M. (1985). Outliers and Influential Observations in Linear Models. In *Bayesian Statistics 2* (J. M. Bernardo, M. H. DeGroot, D. V. Lindley, A. F. M. Smith, eds.), 473–494. Amsterdam: North-Holland.

[28] RACINE-POON, A. (1992). SAGA: Sample Assisted Graphical Analysis. In *Bayesian Statistics 4* (J. M. Bernardo, J. O. Berger, A. P. Dawid, A. F. M. Smith, eds.), 389–404. Oxford: Clarendon Press.

[29] SPIEGELHALTER, D. J., AND SMITH, A. F. M. (1982). Bayes Factors for Linear and Log-linear Models with Vague Prior Information. *Journal of the Royal Statistical Society B* **44**, 377–387.

[30] VERDINELLI, I., AND WASSERMAN, L. (1991). Bayesian Analysis of Outlier Problems Using the Gibbs Sampler. *Statistics and Computing* **1**, 105–117.

[31] WEST, M. (1984). Outlier Models and Prior Distributions in Bayesian Linear Regression. *Journal of the Royal Statistical Society B* **46**, 431–439.

[32] WEST, M. (1985). Generalised Linear Models; Scale Parameters, Outlier Accommodation and Prior Distributions. In *Bayesian Statistics 2* (J. M. Bernardo, M. H. DeGroot, D. V. Lindley, A. F. M. Smith, eds.), 531–558. Amsterdam: North Holland.

[33] ZELLNER, A. (1975). Bayesian Analysis of Regression Error Terms. *Journal of the American Statistical Association* **70**, 138–144.

Recent Adv. in Stat. and Prob., pp. 191-206
J. Pérez Vilaplana and M. L. Puri (Eds)

Network Designs for Monitoring Multivariate Random Spatial Fields

N.D. LE[1]; and J.V. ZIDEK[2]
[1] Biometry Section, British Columbia Cancer Agency, Vancouver V6T 1W5, Canada
[2] Department of Statistics, University of British Columbia, 2021 West Mall, Vancouver V6T 1W5, Canada

Abstract – This paper presents an approach to network design when the responses at each site are vector rather than scalar valued. The decision problem fits into the general framework of Bayesian ranking and selection theory. We suppose the designer must add sites to or remove sites from an existing environmental monitoring network. Optimality is based on entropy. It yields a natural decision criterion in the usual situation, where there is no single well defined design objective. We assume the responses have (possibly after transformation) a joint multivariate Gaussian distribution. The natural conjugate is adopted as the first stage prior for the Gaussian model parameters. An empirical Bayes strategy is used at the second stage. The result is an approximation to the Bayesian predictive and parameter posterior distribution whose entropy may be found in a reasonably explicit form. The theory presented here enables convenient and flexible temporal (and spatial) modeling.

1. INTRODUCTION

Designing experiments invariably involves the use of prior information. For example, in the first phase of a major study of surface water in the United States, Linthurst and his co-investigators chose to sample lakes from just those which would be "expected" to contain relatively low alkalinity (Linthurst et al., 1986, p. 4). The use of prior information may in fact be mandatory since the data which the experiment is designed to produce is not yet in hand.

Usually prior information is introduced in an informal way. But a Bayesian approach seems desirable. It assures coherence of design and subsequent analysis. Moreover, the role of prior beliefs is then not obscured.

Rational design requires that a design objective be formulated. In some situations this objective may be clear. An intervention (impact detection) study examplifies such a situation. The second author was involved in designing a network to determine the impacts if any from exploratory drilling in Harrison Bay on the North Slopes of Alaska (see Zidek, 1984). Some subset of a spatial array of sites had to be designated as before-and-after sampling sites. There was little background data and a Bayesian approach was inevitable. The expectation of the noncentrality parameter of the F-test of space-time interaction was chosen to be the objective function. [This noncentrality parameter was random under the Bayesian model of an uncertain, hence random impact field.] The general theory

resulting from this practical application is reported in Schumacher and Zidek (1988)
In long term environmental monitoring, fixing a design objective may well be far more difficult than it was in the above example. A number of related but different potential experimental aims can usually be discerned. The objective might be the detection of extreme values. Alternatively the aim of the investigator could be that of detecting an unfavorable trend. Even then a number of distinct aims emerge depending how "trend" is defined; this can be done in a variety of parametric or nonparametric ways. Leading to the multivariate problem of this paper would be a simultaneous interest in a number of different environmental attributes. The optimal design would then depend on the weights attached to the different attributes depending on ultimate purposes. Alternatively interest might lie in any of a variety of indices produced from the measurements of these attributes. The level of temporal aggregration might be a design factor as might the corresponding level of spatial aggregration. Permutations of the factors mentioned above would lead to countless different possible design objectives and optimal designs. Making the situation even more complex are the many important but unforeseen future uses to which the network data might be put.
The cost of establishing and maintaining a long term monitoring program is considerable. The agencies responsible for doing so are therefore concerned that the data be of high quality. But quality means fitness for intended use. And, paradoxically, as argued above, there is no well-defined purpose for the network. Is there a way out of this morass?
One solution is proposed by Caselton and Husain (1984), Caselton and Zidek (1984, hereafter CZ) and again by Shewry and Wynn (1987). In fact, the idea of using entropy in experimental design goes back at least to Lindley (1956). There is a substantial body of work on optimal design in the Bayesian context although none as far as we are aware covers the application which is made here. For a recent review of Bayesian design see Verdinelli (1991).
The entropy approach to design was implemented by Caselton, Kan and Zidek (1990, hereafter, CKZ) to obtain a method of ranking stations for possible elimination from an existing network; refinements are added by Wu and Zidek (1992). Guttorp, Le, Sampson and Zidek (1992) tackle the complementary problem of extending an existing network. But these last cited papers deal ony with scalar valued responses. In this paper the corresponding design problem for multivariate responses is addressed.
The basic idea underlying the just cited work is that all data have a fundamental purpose, that of reducing uncertainty about some aspect of the world. Uncertainty according to the postulates of Bayesian theory is quantifiable in terms of probability distributions. And the postulates of entropy theory, in turn, imply that the uncertainty in any distribution is indexed by its entropy. Therefore, an optimal design must minimize residual entropy after data has been collected.
To describe our results more precisely, we associate a K dimensional random vector with every site in a spatial random field. The vectors corresponding to the sites in the discrete random field may be "stacked" to obtain a random vector which represents the discrete random field. For simplicity of exposition, we may then assume $K = 1$. And this allows us to borrow from another context, the distribution theoretical results of Le and Zidek (1992). But we will be forced by substantive issues to return to the case $K \geq 1$ in Section 4.
The random vector field is observed at g discrete, "gauged" sites at sampling times, $j = 1, \cdots, n$, to yield $g \times 1$ data vectors, $X_j^{(2)} = (X_j^{(21)}, \cdots, X_j^{(2g)})'$. Of interest is a

$u \times 1$ vector, $X_{n+1}^{(1)} = (X_{n+1}^{(11)}, \cdots, X_{n+1}^{(1u)})'$, of unmeasured future values at u "ungauged" sites at time $n+1$. The spatial field is over the $p = u + g$ discrete sites. The unknown covariance of $X_j = (X_j^{(1)'}, X_j^{(2)'})'$ is Σ for all j. It is assumed to have an inverted Wishart prior distribution. Finally, the mean of X_j is allowed to depend (linearly) on a vector of observable covariates z_j. This makes the assumed conditional independence of the X_j, $j = 1, \cdots, n+1$, more realistic. The coefficients of the linear mean model are taken to have a multivariate normal distribution conditional on the unspecified covariance matrix. In short, a conjugate prior distribution is assumed for the parameters of the multivariate Gaussian distribution.

The principal product of our analysis is a formula for the entropy of the posterior distribution of the unobserved responses and model parameters. This is given in Section 5. It turns out that the optimal design does not depend on the future covariates, z. This is an important finding. It means the optimal design will be the same whatever future time is selected.

In Section 2, we develop the entropy approach in detail. The needed distribution theory is found in Le and Zidek (1992), but summarized for completeness in Section 3 which also includes the needed entropy criterion. Although we specifically deal with scalar valued responses in these sections, the same results hold for multidimensional vector valued responses. Then in Section 4 we discuss the problem of specifying the prior hyperparameters. In Section 5, we will suppose at the second level of the model hierarchy that the prior mean of the covariance matrix for the $Kp \times 1$ response vector has the form of a Kronecker product. The substantive motivations for this choice are given in there along with a brief description of the method used for obtaining an estimate of this hypercovariance. Primarily, the Kronecker product is chosen because it greatly reduces the number of parameters which have to be specified while at the same time preserving a good deal of model flexibility. With this choice it proves possible to find an iterative method of estimating the Kronecker product, thereby achieving an empirical Bayes approximant to posterior distributions of interest. This too is described in Section 5. Concluding remarks are in Section 6.

2. ENTROPY-BASED DESIGN

To put our work in perspective, we would mention that there are, broadly speaking, two distinct paradigms in the modern theory of optimal design (Federov and Mueller, 1988, 1989). They might be called the "regression" and "random fields" approaches. The former comes from the work of Smith (1918) with further development by Elfving (1952), Keifer (1959), and others (c.f. Silvey, 1980, for a review). Optimal designs for fixed effect regression experiments are those that maximize the accuracy of estimates of functions of the regression coefficients. This has led to work on algorithms for finding the optimum or approximately optimum designs (see Pazman, 1986, for a recent review).

However, the regression approach does not solve the problem addressed in this paper. The myriad of possible design objectives cannot be reduced to the estimation of a combination of the coefficients of a fixed effects regression model. So the second of the two paradigms seems more appropriate for our purpose.

Federov and Mueller (1988, 1989, hereafter FM), attempt to bridge the gap between the regression and random field approaches by making the regression effects random. Their random effects model, as a description of a spatial random field, seems somewhat tenuous. The FM method does yield a design criterion which is formally similar to, but more structured than, that of Caselton and Zidek (1984). However, the theory remains oriented towards estimation.

We now describe our approach. Let X_j be the random vector defined in the last section at time $j = 1, \cdots, n+1$. Suppose X_j has the joint probability density function, f_j, for all j. The total uncertainty about X_j may be expressed by the entropy of its distribution, *i.e.* $H_j(X_j) = E[-log f_j(X_j)/h(X_j)]$, where $h(\cdot)$ is a not necessarily integrable reference density (see Jaynes 1963). The inclusion of $h(\cdot)$ in $H_j(\cdot)$ ensures the latter is invariant under one–to–one transformations of the scale of X_j.

Given the network's mission to monitor the environment, we regard the next value, X_{n+1}, as being of primary interest. However, X_{n+1}'s probability density function, $f_{(n+1)}(\cdot) = f_{(n+1)}(\cdot \mid \theta)$, depends on a vector of unspecified model parameters, say θ, and so cannot be used directly in computing its uncertainty, $H_{n+1}(X_{n+1})$. Uncertainty about θ could be absorbed by averaging $f_{(n+1)}(\cdot \mid \theta)$ with respect to θ's distribution to obtain X_{n+1}'s marginal distribution and hence H_{n+1}. However, θ is of interest in its own right; in our theory θ includes the (spatial) covariance matrix of X_{n+1}, Σ, which has potential use in Kriging. Therefore we, like Caselton, Kan and Zidek (1992), include among the network's objectives, that of reducing the uncertainty about θ. The total entropy is then $H_{n+1}(X, \theta)$ conditional on the data, $D \overset{defn}{=} \{X_j^{(2)}, j = 1, \cdots, n\}$.

CKZ consider the problem of reducing the number of sites in a network which has been providing data for some time. They address the problem of optimally partitioning X so that after appropriately relabeling the coordinates of X, it can be partitioned as $X' = (X^{(1)\prime}, X^{(2)\prime})$ where $X^{(1)}$ and $X^{(2)}$ are u and g dimensional vectors respectively, $u + g = p$, of the coordinates of X corresponding to the sites which will be *ungauged* and *gauged* in the future. Wu and Zidek (1992) implement this approach in an analysis of 81 selected sites from the NADP/NTN network, an existing network of wet deposition monitoring stations in the USA, using 48 months of available data. For brevity and since these results are published elsewhere, we will not discuss the problem of network reduction further here, but this type of redesign should be considered part of the general theory which is completed in this paper. Here, only the complementary problem of extending an existing network will be addressed.

The objective addressed here is that of augmenting the network by gauging a specified number, u_2, of sites corresponding to coordinates of $X^{(1)}$. That is, we seek an optimal partitioning of $X^{(1)}$ which, after reordering its coordinates, yields $X^{(1)} = (X^{(rem)\prime}, X^{(add)\prime})$ where $X^{(rem)\prime}$ is a u_1-dimensional vector representing the future ungauged sites and $X^{(add)\prime}$ is a u_2-dimensional vector representing the future gauged sites. The resulting network will consist of the sites corresponding to the coordinates of $(X^{(add)\prime}, X^{(2)\prime}) \equiv G$, which is of dimension $(g + u_2)$.

The total a priori uncertainty conditional on D may, for certain purposes, be usefully decomposed as

$$H(X, \theta) = H(X \mid \theta) + H(\theta)$$

where, assuming the reference density to be $h(X, \theta) = h_1(X)h_2(\theta)$,

$$H(X \mid \theta) = E[-\log(f(X \mid \theta, D)/h_1(X)) \mid D]$$

and

$$H(\theta) = E[-\log(f(\theta \mid D)/h_2(\theta)) \mid D].$$

But for our immediate purposes of optimizing design, we need a different decomposition that reflects the partitioning of future observations into ungauged and gauged sites, $X' = (U, G)$ where $U \equiv X^{(rem)'}$. We express the total uncertainty $H(X \mid \theta)$ as

$$TOT = PRED + MODEL + MEAS$$

where, assuming $h_1(X) = h_{11}(U)h_{12}(G)$,

$$\begin{aligned} PRED &= E[-\log(f(U \mid G, \theta, D)/h_{11}(U)) \mid D], \\ MODEL &= E[-\log(f(\theta \mid G, D)/h_2(\theta)) \mid D], \end{aligned}$$

and

$$MEAS = E[-\log(f(G \mid D)/h_{12}(G)) \mid D].$$

If we assume that measurement error is negligible, eliminating all uncertainty about G by observing it will lead to an expected reduction in uncertainty given by $MEAS$. Thus, it will be optimal to augment the present network with sites represented in $X^{(add)'}$ so as to maximize $MEAS$. Since TOT is fixed, it follows that the same selection of additional sites will meet another design objective, that of minimizing $PRED + MODEL$. This latter entropy represents the residual uncertainty about the model parameters and the values of the random field at the ungauged sites, after observing G.
Incidentally, it is easily seen that had we started with $H(X)$ instead of $H(X, \theta)$, and made a decomposition analogous to that given above, we would have arrived at the same optimization criterion: maximization of $MEAS$.

3. POSTERIOR DISTRIBUTION AND ENTROPY CRITERION

3.1. Posterior Distribution

To proceed further it is necessary to specify the joint distribution of (X, θ) and this will be done in this section. Assume that conditional on θ and the covariates $\{z_j,\ j = 1, \cdots, n\}$, the successive X's are independent random vectors with a Gaussian distribution. Here $\theta = (B, \Sigma)$ where B is a matrix of regression coefficients linking the site variables with the covariates and Σ is the spatial covariance matrix. We propose to represent prior information about Σ by a p-dimensional inverted Wishart distribution, $\Sigma \sim W_p^{-1}(\Psi, m)$ with parameters Ψ and m (c.f. Anderson, 1984).
To be more precise, we recall that $X_j = \left(X_j^{(1)'}, X_j^{(2)'}\right)'$ is a p-dimensional random vector where $X_j^{(1)}$ and $X_j^{(2)}$ are u- and g-dimensional vectors, respectively ($p = u + g$). Assume

$$X_j \mid z_j, B, \Sigma \overset{\text{independent}}{\sim} N_p(Bz_j, \Sigma) \tag{1}$$

where B denotes a $(p \times k)$ matrix of regression coefficients,

$$B = \begin{pmatrix} \beta_{11} & \cdots & \beta_{ik} \\ & \vdots & \\ \beta_{p1} & \cdots & \beta_{pk} \end{pmatrix} \equiv \begin{pmatrix} B_1 \\ B_2 \end{pmatrix}.$$

The covariance matrix, Σ, is partitioned accordingly as

$$\Sigma = \begin{pmatrix} \Sigma_{11} & \Sigma_{12} \\ \Sigma_{21} & \Sigma_{22} \end{pmatrix}$$

where in general for any two random vectors R and S,

$$\Sigma_{RS} \equiv Cov(R, S') \equiv E\,[R - E(R)][S - E(S)]'.$$

Reparametrize Σ as $(\Sigma_{22}, \Sigma_{1|2}, \tau)$ where $\Sigma_{1|2}$ is the $(u \times u)$ residual covariance matrix given by

$$\Sigma_{1|2} \equiv \Sigma_{11} - \Sigma_{12}\Sigma_{22}^{-1}\Sigma_{21}.$$

The $(g \times g)$ matrix Σ_{22} is the covariance matrix of $X_j^{(2)}$ and τ is a $(u \times g)$ matrix representing the slope of the optimal linear predictor of $X_j^{(1)}$ based on $X_j^{(2)}$ given by

$$\tau \equiv \Sigma_{12}\Sigma_{22}^{-1}.$$

This transformation is achieved through the Bartlett (1933) decomposition, $\Sigma = T\Delta T'$, where

$$\Delta = \begin{pmatrix} \Sigma_{1|2} & 0 \\ 0 & \Sigma_{22} \end{pmatrix} \quad \text{and} \quad T = \begin{pmatrix} I & \tau \\ 0 & I \end{pmatrix}.$$

We suppose the uncertainty about B and Σ can be represented by the conjugate prior distribution (c.f. Press 1972, page 234),

$$B \mid B^o, \Sigma, F \sim N_{kp}(B^o, \Sigma \otimes F^{-1}) \quad (2)$$
$$\Sigma \mid \Psi, m \sim W_p^{-1}(\Psi, m), \quad (3)$$

where $W_p^{-1}(\Psi, m)$ denotes the p-dimensional inverted Wishart distribution with scale matrix Ψ and m degrees of freedom; it is proper if $p < m$. B^o and Ψ are the prior parameter matrices.
The conjugate prior distribution (3) for Σ can be equivalently presented in terms of the new parameters $(\Sigma_{22}, \Sigma_{1|2}, \tau)$ as

$$\begin{aligned} \Sigma_{22} \mid \Psi, m &\sim W_g^{-1}(\Psi_{22}, m - u) \quad (4)\\ \Sigma_{1|2} \mid \Psi, m &\sim W_u^{-1}(\Psi_{1|2}, m) \\ \tau \mid \Sigma_{1|2}, \Psi &\sim N_{ug}(\eta, \Sigma_{1|2} \otimes \Psi_{22}^{-1}). \end{aligned}$$

Here $(\Psi_{22}, \Psi_{1|2}, \eta)$ denotes the decomposition of the prior parameter matrix Ψ analogous to that of Σ; that is, $\eta = \Psi_{12}\Psi_{22}^{-1}$. Moreover, Σ_{22} is independent of $(\Sigma_{1|2}, \tau)$ when the prior distribution is proper. See Caselton et al. (1992) for more details on this decomposition.
The observed $\{x_j^{(2)}\}$'s are, given the covariate z_j's, independent realizations of

$$X_j^{(2)} \mid B, \Sigma, z_j \sim N_g(B_2 z_j, \Sigma_{22}). \quad (5)$$

Note that this represents only the second component of the X_j in model (1). Hence, D, defined above, represents only the partially observed data; i.e. there are no observations for the first u-coordinates.
Define $\hat{B}_2$ and S as follows:

$$\begin{aligned} \hat{B}_2 &= CA^{-1} \quad (6)\\ S &= \sum_{j=1}^{n}(x_j^{(2)} - \hat{B}_2 z_j)(x_j^{(2)} - \hat{B}_2 z_j)' \end{aligned}$$

where

$$C = \sum_{j=1}^{n} x_j^{(2)} z_j' \tag{7}$$

and

$$A = \sum_{j=1}^{n} z_j z_j'.$$

Anderson (1984, p291) shows that, conditional on B_2 as well as Σ_{22}, A, $\hat{B}_2$ and S are independent and

$$\hat{B}_2 \mid B_2, \Sigma_{22}, A \sim N_{gk}(B_2, \Sigma_{22} \otimes A^{-1}) \tag{8}$$

and

$$S \mid B_2, \Sigma_{22}, A \sim W_g(\Sigma_{22}, n-k).$$

Here $W_g(\Psi, m)$ denotes the g-dimensional Wishart distribution with scale matrix Ψ and m degrees of freedom.
Let (B_1^o, B_2^o) be a partition of B^o analogous to that of B. Let the prior distributions of B and $(\Sigma_{22}, \Sigma_{1|2}, \tau)$ be defined as in (2) and (4). In a different context, Le and Zidek (1992) obtain a number of implications of the above assumptions which we will present in the remainder of this section for completeness.

$$B \mid D, B^o, \Sigma_{22}, \Sigma_{1|2}, \tau \sim N_{kp}(B^*, \Sigma^*) \tag{9}$$

where

$$\begin{aligned} B^* &= B^o + \begin{pmatrix} \tau \\ I_g \end{pmatrix} \left(\hat{B}_2 - B_2^o\right) \hat{E}' & (10) \\ \Sigma^* &= \Sigma \otimes F^{-1} - \left[\begin{pmatrix} \tau \\ I_g \end{pmatrix} (\Sigma_{22}\tau', \Sigma_{22})\right] \otimes (\hat{E} F^{-1}) & (11) \\ \hat{E} &= F^{-1}(A^{-1} + F^{-1})^{-1} \end{aligned}$$

and A and $\hat{B}_2$ are defined as in (6)–(7). The posterior distribution of $(\Sigma_{22}, \Sigma_{1|2}, \tau)$ is

$$\begin{aligned} \Sigma_{22} \mid D, \Psi_{22}, m &\sim W_g^{-1}(\hat{\Psi}_{22}, m+n-u) & (12) \\ \Sigma_{1|2} \mid D, \Psi_{1|2}, m &\sim W_u^{-1}(\Psi_{1|2}, m) \\ \tau \mid D, \eta, \Sigma_{1|2} &\sim N_{ug}(\eta, \Sigma_{1|2} \otimes \Psi_{22}^{-1}), \end{aligned}$$

where

$$\hat{\Psi}_{22} = \Psi_{22} + S + (\hat{B}_2 - B_2^o)'(A^{-1} + F^{-1})^{-1}(\hat{B}_2 - B_2^o) \tag{13}$$

and S is given in (6).
The result (12) establishes the intuitive result that the posterior distributions of τ and $\Sigma_{1|2}$ are just their prior distributions; our knowledge of the covariance matrix Σ is updated by the data D only through Σ_{22}. This is reasonable since Σ_{22} is independent of $(\Sigma_{1|2}, \tau)$, *a priori*. However, although the data cannot be used to improve the prior models of $\Sigma_{1|2}$, τ, the prior covariance matrix, Σ_{11}, for the ungauged sites and the cross-covariance matrix, Σ_{12}, between the gauged and ungauged sites can be improved with the data as noted by Le and Zidek (1992).
To be precise, let Σ_o be the mean of the prior distribution of Σ given by (3); that is,

$$\Sigma_o = \frac{1}{m-p-1}\Psi.$$

The posterior mean, $\tilde{\Sigma}$, of Σ is then given by

$$\tilde{\Sigma} = \alpha\Sigma_o + (1-\alpha)\hat{\Sigma} \tag{14}$$

where $\alpha = (m-p-1)/(m+n-p-1)$ while

$$\hat{\Sigma} = \begin{pmatrix} \Upsilon\Psi_{1|2} + \eta\tilde{S}\eta' & \eta\tilde{S} \\ \tilde{S}\eta' & \tilde{S} \end{pmatrix}$$

$$\Upsilon = \frac{1 + tr(\tilde{S}\Psi_{22}^{-1})}{m-u-1}$$

and

$$\tilde{S} = \frac{1}{n}\left(\hat{\Psi}_{22} - \Psi_{22}\right).$$

Using the available data, D, and the posterior distributions of B and Σ, we obtain the predictive distribution of the random vector X with sampling distribution described by (1) given the new covariate vector z. To present the result we let, in general, $t_r(\mu, H, v)$ denote the r-variate t-distribution with density function given by

$$f(x) \propto |H|^{-\frac{1}{2}}\left[v + (x-\mu)'H^{-1}(x-\mu)\right]^{-\frac{1}{2}(v+r)}.$$

Let the posterior distributions of B and Σ be as determined above. Then the predictive distribution of $X = \left(X^{(1)'}, X^{(2)'}\right)$ given a covariate vector z and the prior parameters B^o and $(\Psi_{22}, \Psi_{1|2}, \eta)$, is

$$X^{(2)} \mid D \sim t_g\left(a_2 + b, \frac{c-d}{l}\hat{\Psi}_{22}, l\right) \tag{15}$$

$$X^{(1)} \mid X^{(2)} = x^{(2)}, D \sim t_u\left(a_1 + \eta(x^{(2)} - a_2), \frac{c + (a_2 - x^{(2)})'\Psi_{22}^{-1}(a_2 - x^{(2)})}{q}\Psi_{1|2}, q\right) \tag{16}$$

where $\hat{\Psi}_{22}$ is given by (13), $l = m+n-u-g+1$, $q = m-u+1$ and

$$\begin{aligned} \hat{E} &= F^{-1}(A^{-1} + F^{-1})^{-1} \\ a &\equiv \begin{pmatrix} a_1 \\ a_2 \end{pmatrix} = B^o z \\ b &= (\hat{B}_2 - B_2^o)\hat{E}z \\ c &= 1 + z'F^{-1}z \\ d &= z'\hat{E}F^{-1}z. \end{aligned}$$

The value of the hierarchical approach we have adopted lies in the very general covariance structure introduced at the first level. We are then free to model the covariance structure at the second level through the various hyperparametric objects in equations (2) and (3). Where possible the past and future data will update these prior models to achieve increasingly more realistic versions of their level one counterparts, B and Σ. But whether or not such updating is possible, our uncertainty about the level one parameters is reflected in the marginal likelihood of the hyperparameters. Consequently (credible) intervals derived from the heavy tailed (Student's t–) predictive distribution, will realistically account for that uncertainty.

It should be emphasized that the regression model in equation (1) allows time varying covariates. This is to enable trend and seasonality to be incorporated directly into the Bayesian analysis; there is no need to remove them in an ad hoc preliminary analysis. In this case, z_j would contain a coordinate, t_j, to represent the time while the *ith* row of B would have slope and intercept coefficients for station i. As well, z_j would contain $\cos \alpha t_j$ and $\sin \alpha t_j$ to account for seasonality. Here α determines the period and might well depend on the station, i, in which case there would have to be different cos and sin coordinates in z_j for each station. A technically awkward feature of the resulting model is that it is nonlinear in α.

We could also include, in z_j, spatial modeling coordinates, $f_l(x_i), i = 1, \cdots p,\ l = 1, \cdots L$ where the f_l's are specified functions and x_i represents the location vector of station i. However, they are of little relevance in the context of environmental analysis where location has little explanatory value (for a discussion, see Wu and Zidek, 1992).

3.2. Entropy Criterion

First we present the evaluation of entropy for the above posterior distribution which can be then used to derive the optimal criterion for extending the network. The reference measure needed for the calculation of entropy expresses baseline uncertainty. Since U are G are elements of (affine invariant) Euclidean spaces, it seems natural that they should be translation invariant; uncertainty about a rectangle of given volume should not depend on where that rectangle is situated, for otherwise this would imply that we had knowledge above the baseline to get varying levels of uncertainty about such rectangles. Thus we take $h(U) \equiv 1$, $h(G) \equiv 1$, and $h(\mu) \equiv 1$. In particular, none of these measures depend on the data or other unobserved objects such as Σ. However, if U and G or μ were transformed in some way, the resulting h's would have to include the Jacobian of the transformation to make the entropy invariant under the transformation.

Now suppose $h(\theta)d\mu d\Sigma$ with $h(\theta) = |\Sigma|^{\frac{-1}{2(p+1)}}$, were used as the differential element of the (unintegrable, but formally acceptable) prior distribution of θ. Then under the resulting posterior distribution of θ, μ and Σ (respectively) have expectations equal to the sample mean and variance. This can be seen from the results of Anderson (1984, p272) by taking the appropriate limits. We adopt $h(\theta)$ defined above as our reference measure since it yields the classical estimators of these normal parameters and so can be thought of as corresponding to no prior knowledge. This agrees with the concept underlying the proposal of Jaynes (1963).

With these specifications, the entropy of the posterior distribution can be presented as (see Guttorp et al., 1992, for more details)

$$H\left[X^{(1)}, X^{(2)} \mid D\right] = H\left[X^{(1)} \mid X^{(2)}, D\right] + H\left[X^{(2)} \mid D\right].$$

Here

$$H\left[X^{(1)} \mid X^{(2)}, D\right] = \frac{1}{2}\log|\Psi_{1|2}| + \\ + \frac{1}{2}E\left[\log\left|1 + (X^{(2)} - a_2)'(c\Psi_{22})^{-1}(X^{(2)} - a_2)\right|\right] + c_1(c, q, u)$$

and

$$H\left[X^{(2)} \mid D\right] = \frac{1}{2}\log|\hat{\Psi}_{22}| + c_2(c, d, l, g)$$

where c_1 and c_2 are constants depending on c, q, u, d, l, g defined in equations (15) and (16) which are of no importance in this application.
Hence,

$$H\left[X^{(1)}, X^{(2)} \mid D\right] = \frac{1}{2}\log|\Psi_{1|2}| + c_3(\hat{\Psi}_{22}, \Psi_{22}, c, d, l, g) \qquad \textbf{(17)}$$

where c_3 is a constant.
Using this result, the entropy criterion, which is to maximize $MEAS \equiv H(X^{(add)} \mid X^{(2)})$ can be presented as to find the new u_2 sites to augment the old netmork to maximize $\Psi_{add|2}$. Here $\Psi_{add|2}$ is a $(u_2 \times u_2)$ matrix represent the residual covariance of $X^{(add)}$ conditional on $X^{(2)}$.
Remark: Although the posterior distribution and the entropy evaluation were obtained for the univariate case (i.e $K = 1$), the same results are correct for the case with multivariate responses at each station. In this multivariate case, the dimensions of the vectors and the matrices are changed appropriately; for example, Ψ is now a $(pK \times pK)$ matrix. In the next section, we discuss the issue of specifying the hyperparameters. This is particularly important for Ψ in the multivariate case since we would like to reduce the number of parameters to a manageable size.

4. SPECIFYING THE HYPERPARAMETERS

At level two of the hierarchy, we may specify (hyper) parametric models for objects like Ψ, for example, to accommodate our remaining prior knowledge about the random field under consideration. The paper of Loader and Switzer (1989) exemplifies that approach. This will leave a small number of additional (third stage) unspecified parameters. In a strictly Bayesian approach, we would add another distribution to our hierarchy to accommodate our uncertainty about these parameters. However, as noted in the Introduction, the posterior distribution of quantities of inferential interest will not be especially sensitive to the choices made of these parameters and this suggests alternatives to a fully Bayesian approach. One possibility would be a robust Bayesian approach like that of Pilz (undated). Another would be to introduce a diffuse prior on these remaining unspecified quantities. But the simplest option is that of estimating them from the marginal distribution of the data conditional on these quantities.
The simplest special case of our theory perhaps, is that in which the station records are "exchangeable". To be precise, referring to displays (1) and (2), $k = 1$, $EB = B^0 = \mathbf{1}\otimes\mu^0$, $z_j = 1$ for all j and $F = k^0$ where $\mathbf{1}$ denotes the $p \times 1$ vector all of whose elements are 1, and μ^0 as well as k^0 are specified constants. Here μ^0, a $K \times 1$ matrix, is the overall level of the β_j's in $B = (\beta_1, \cdots, \beta_p)'$. The value chosen for k^0 would depend on how well μ^0 represents the population of the β_j's in equation (1). This would depend partly on the size of the geographical area under consideration. For this model, it would seem reasonable to choose $\Psi = \xi \times \sigma^2$ where $\xi_{ij} = 1$ or ρ according as $i = j$ or $i \neq j$, $i, j = 1, \cdots, p$, andσ^2 is a $K \times K$ covariance matrix for the site response vectors. We will not pursue this model here since it is not needed for our application.
A simple extension of the model of the last paragraph for the hypothetical application under consideration would cluster, on statistical or some other grounds, the stations into homogeneous subgroups, $l = 1, \cdots, L$. We would then have $B = (B_1', \ldots, B_L')$ where $B_l^0 = 1_{s_l} \otimes \mu^{0l}$ with s_l denoting the size of the lth cluster and μ^{0l} playing for cluster l, the role played in the last paragraph by the scalar μ^0, itself. And μ^{0l} might be estimated

from the data for that cluster, if the posterior were not deemed to be unduly sensitive to this choice.

In the situation described in the last paragraph, X_i would be partitioned to reflect the clustering. The between cluster covariances for stations might be approximated by 0 while that within clusters might be given the intraclass correlation structure with just two parameters that might also be estimated from the corresponding within cluster data sets. Alternatively, one might adopt a components-of-variance model, with one component for overall variation and independent cluster specific components to represent the individual clusters. This too would lead to a substantial reduction in the number of parameters to be fitted.

An extension of the exchangeable means models discussed above, would be the exchangeable regressions model. These exchangeable regression models like their counterpart for means, have the well-known advantage that they allow "borrowing from strength". When this is justified and the amount of data is small, this advantage can be substantial. Variations on this theme emerge when clustering is permitted. For brevity, we will not provide any further details here.

Analogous to the components-of-variance model proposed above, is a fixed effects linear model in equation (1). Now the "covariates" in (1) would be the appropriate indicator variables, to pick out the fixed effects appropriate to the individual coordinates of the X_j vectors. The results of the next section could then be used to obtain estimates of these fixed effects if they were of inferential interest.

When the responses are vector valued the need for simple hyperparameter parametric models becomes even more acute, since now we confront, potentially, a much larger set of parameters. One particularly simple level two covariance structure is provided by the Kronecker product, $\Psi = \Lambda \otimes \Omega$, say where Λ is a $p \times p$ matrix corresponding to between site covariation and Ω, that between responses at any given site. This model applies when we have at each time point, a multiplicative random effects model: $X_{sr} - \mu_{sr} = \alpha_s \times \beta_r$ where X_{sr} represents the rth response at site s, μ_{sr} denotes the mean of that response, while α_s and β_r denote the zero mean multiplicative random effects components. This could be easily be extended to include an independent additive noise term. All of the above discussion now applies to the selection of appropriate parametric models for Λ. However, obtaining an estimate for the covariance matrix with Kronecker structure presents a challenging problem which is addressed in the next section.

5. HYPERPARAMETER ESTIMATION

We now present an approach for estimating the hyperparameters in equation (3); namely, the degrees of freedom m and the scale matrix Ψ with Kronecker structure, $\Lambda \otimes \Omega$, given the available data D at the gauged sites. The approach consists of two levels. The first level is to estimate m, Ω, and Λ_g using the available data, where Λ_g denotes the between-gauged-site covariance matrix, i.e. the spatial covariance of the gauged sites. The second level is then to use the estimate of Λ_g to obtain an estimate for Λ.

In principle, the second level can be easily implemented by using either a parametric approach like that of Loader and Switzer (1989) where isotropy and stationarity of the covariance field are assumed or preferably the nonparametric approach studied by Sampson and Guttorp (1992) where neither isotropy nor stationarity is assumed. More details on these approaches can be found in these references.

On the other hand, the first level can only be achieved with a careful adaptation of the Bayesian methodology proposed by Chen (1979) for estimating a dispersion matrix. This methodology utilizes the EM algorithm developed by Dempster et al. (1977) to obtain an iterative scheme for obtaining estimates of m, Λ_g and Ω. We now describe our adaptation of the Chen methodology to the current problem.
First, we recall that the distributions of Σ_{22} conditional on (Ψ_{22}, m) and S conditional on Σ_{22} are given in equations (4) and (8) respectively. Hence, the marginal density of S given Ψ_{22}, m and D can be obtained. In principle, maximum likelihood estimates (MLE) of Ψ_{22} and m can be obtained by maximizing this marginal density of S. Direct maximization of this marginal density is very difficult, however. Fortunately, as noted by Dempster et al. (1977), this problem is equivalent of that of maximizing the complete density of (S, Σ_{22}) given (Ψ_{22}, m). This complete density can be decomposed as

$$\begin{aligned} f(S, \Sigma_{22} \mid \Psi_{22}, m) &= f_1(S \mid \Sigma_{22}, \Psi_{22}, m) f_2(\Sigma_{22} \mid \Psi_{22}, m) \\ &= f_1(S \mid \Sigma_{22}) f_2(\Sigma_{22} \mid \Psi_{22}, m) \end{aligned}$$

with the last equality being a consequence of (8). Here (see Anderson 1984),

$$f_1(S \mid \Sigma_{22}) = c(S, n, k, g) \exp\left\{-\frac{1}{2} tr \Sigma_{22}^{-1} S - \frac{n-k}{2} \log|\Sigma_{22}|\right\}$$

where $c(S, n, k, g)$ is a constant depending on S, n, k and g whose precise value is of no importance; also

$$f_2(\Sigma_{22} \mid \Psi_{22}, m) = c_o |\Psi_{22}|^{(m-u)/2} |\Sigma_{22}|^{-(m+u+g+1)/2} \exp\left\{-\frac{1}{2} tr \Psi_{22} \Sigma_{22}^{-1}\right\} \quad \textbf{(18)}$$

where

$$c_o^{-1} = 2^{g(m-u)/2} \pi^{g(g-1)/4} \prod_{i=1}^{g} \Gamma\left(\frac{m-u-i+1}{2}\right).$$

Moreover, as noted by Chen (1979), since $f_1(S \mid \Sigma_{22})$ does not involve hyperparameters, the complete density $f(S, \Sigma_{22} \mid \Psi_{22}, m)$ is of exponential type with $(\Sigma_{22}^{-1}, \log|\Sigma_{22}|)$ as the complete-data sufficient "statistic" for (Ψ_{22}, m). Hence, the EM algorithm can be used in this problem. In the E-step, the sufficient statistic $(\Sigma_{22}^{-1}, \log|\Sigma_{22}|)$ is estimated by its posterior mean given data D and the current value of $(\Psi_{22} = \Lambda_g \otimes \Omega, m)$. In the M-step, a new value of (Ψ_{22}, m) is found by maximizing $f_2(\Sigma_{22} \mid \Psi_{22}, m)$. Specifically, these steps can be described as follows.
E-step: Given the current value of (Ψ_{22}, m), the posterior means of Σ_{22} and $\log|\Sigma_{22}|$ are

$$E(\Sigma_{22}^{-1} \mid D, \Psi_{22}, m) = (m+n-k-u)\hat{\Psi}_{22}^{-1} \quad \textbf{(19)}$$

and

$$E(\log|\Sigma_{22}| \mid D, \Psi_{22}, m) = -g \log 2 - \sum_{i=1}^{g} \psi\left(\frac{m+n-k-u-i+1}{2}\right) + \log|\hat{\Psi}_{22}| \quad \textbf{(20)}$$

where $\hat{\Psi}_{22}$ is defined in equation (11) and ψ is the digamma function (see Chen, 1979). These results are obtained by utilizing a result in Theorem 2.1 of Le and Zidek (1992) where

$$\Sigma_{22} \mid D, \Psi_{22}, m \sim W_g^{-1}(\hat{\Psi}_{22}, m+n-k-u).$$

M-step: Given the current value of Σ_{22}, the MLE of $\Psi_{22} = \Lambda_g \otimes \Omega$ can be obtained by maximizing the logarithmic transformation of the density $f_2(\Sigma_{22} \mid \Psi_{22}, m)$ represented as a function of Λ and Ω. That is, Λ and Ω are obtained by maximizing the function

$$\begin{aligned} H(\Lambda_g, \Omega) &= (m-u)\log|\Psi_{22}| - tr\left(\Psi_{22}\Sigma_{22}^{-1}\right) \\ &= -(m-u)K\log|\Lambda_g^{-1}| - (m-u)gK^{-1}\log|\Omega^{-1}| - tr[(\Lambda_g \otimes \Omega)\Sigma_{22}^{-1}] \quad \textbf{(21)} \end{aligned}$$

where Σ_{22}^{-1} is replaced by its posterior mean given in (19). Here, another iterative scheme is needed to do the maximization. This can be done by repeating the following two steps until convergence.

- **Step 1**: Given the current value of Λ_g, the term $tr[(\Lambda_g \otimes \Omega)\Sigma_{22}^{-1}]$ in (21) is arranged as $tr(\Omega G)$ where G is a function of Λ_g and Σ_{22}^{-1}. With this representation, $H(\Lambda_g, \Omega)$ is maximized when $\Omega = gK^{-1}(m-u)G^{-1}$ (see Anderson, 1984, Lemma 3.2.2).
- **Step 2**: Similarly given the current value of Ω, the term $tr[(\Lambda_g \otimes \Omega)\Sigma_{22}^{-1}]$ in (21) is arranged as $tr(\Lambda_g Q)$ where Q is a function of Ω and Σ_{22}^{-1}. Then similarly, $H(\Lambda_g, \Omega)$ is maximized when $\Lambda_g = K(m-u)Q^{-1}$.

Moreover, with the current values of Ψ_{22} and Σ_{22}, an estimate of m can be obtained by solving the "likelihood" equation

$$-g\log(2) - \log|\Sigma_{22}| + \log|\Psi_{22}| - \sum_{i=1}^{g}\psi\left(\frac{m-u-i+1}{2}\right) = 0$$

where $\log|\Sigma_{22}|$ is replaced by its expectation given by (20). That is, m has to satisfy

$$\sum_{i=1}^{g}\psi\left(\frac{m+n-k-u-i+1}{2}\right) - \sum_{i=1}^{g}\psi\left(\frac{m-u-i+1}{2}\right) = \log|\hat{\Psi}_{22}| - \log|\Psi_{22}|. \quad \textbf{(22)}$$

Here given a current estimate of Ψ_{22}, the right hand side of equation (22) is a positive constant as a consequence of equation (11). The left hand side is a monotonically decreasing function with respect to m and it approaches 0 as $m \to \infty$. This shows that there is always a unique solution of m for equation (22).

Thus, repeating the EM-algorithm until convergence in conjunction with iterating the above two steps will eventually yield maximum likelihood estimates for m and Ψ_{22} which assumes a Kronecker structure.

6. DISCUSSION

This paper addresses the problem of enlarging an existing monitoring network. The context is that in which there is no well defined design objective. This is done by adopting the generic objective of minimizing the entropy of the posterior probability distribution of the quantities of interest. Roughly speaking the new network stations are those whose response vector would be the most highly unpredictable either because it is not well correlated with the remaining stations, or because of its high intrinsic variability.

The proposed design methodology uses a posterior distribution with its entropy for a random field. The posterior distribution is obtained from a Gaussian model for the field, with a conjugate prior distribution for its parameters, including the spatial covariance matrix Σ. Our uncertainty about Σ is incorporated through the inverted Wishart distribution at the first level of a Bayesian hierarchical model. Thus we can specify the spatial covariance structure at higher levels without thereby creating unrealistically high levels of certainty in the posterior distribution. [This is similar to the approach taken by Chen (1979) for similar reasons.] At level two, a Kronecker product is adopted as a covariance model for simplicity. The model also incorporates the effect of time varying covariates on the prior mean levels for the spatial field through a linear model.

However, at level two of the Bayesian hierarchical model, the residual covariance matrix is independent of the data under the inverted Wishart prior. Therefore the data from existing stations does not update that part of the hyperparameter matrix for the prior distribution of Σ associated with new stations. Another level would be needed to build in that dependence and it is not clear how that can be done. Instead an empirical approach like that of Loader and Switzer is adopted to estimate that part of the Kronecker covariance product which refers to between station covariation. It is used in conjunction with the EM algorithm as suggested by Chen (1979) albeit for other covariance structures.

There are a number of practical issues which need to be addressed in implementing our method. It must be possible to transform the data to achieve an approximately Gaussian data distribution. Adjustment with this model must yield an approximately uncorrelated data series. And computation becomes a major practical consideration for networks of realisitic size; combinatorial optimization rapidly becomes infeasible.

We are currently working on a number of improvements to the method described in this paper. In particular, we are seeking a more realistic prior distribution for the Σ than the inverted Wishart distribution.

Finally, we recognize that in practice, "good" rather than "optimal" designs are needed and optimal designs like those in this paper must be considered as tentative proposals susceptible to modification depending on the circumstances prevailing in the context of their implementation. These "optimal" designs may well be valuable starting points, however, since they can be explicated in terms of their axiomatic underpinnings and proposed changes to these optimal designs can be interpreted in terms of the axioms. This can provide a degree of confidence and clarity in the typically complex situation confronting a designer. The redesigns are founded on a coherent theory and we believe that the resulting designs will therefore be defensible in an operational context.

REFERENCES

[1] ANDERSON, T.W. (1984). *An Introduction to Multivariate Statistical Analysis.* New York: Wiley.

[2] BARTLETT, M.S. (1933). On the Theory of Statistical Regression. *Proc. Roy. Soc. Edenburgh*, **53**, 260–283.

[3] CASELTON, W.F. AND HUSAIN, T. (1980). *Hydrologic Networks: Information Transmission.* Water Resources Planning and Management Division, A.S.C.E, 106(WR2), 503–520.

[4] CASELTON, W.F. AND ZIDEK, J.V. (1984). Optimal Monitoring Network Designs. *Statist. Prob. Letters* , **2**, 223–227.

[5] CASELTON, W.F., KAN, L. AND ZIDEK, J.V. (1992). " Quality Data Networks that Minimize Entropy". In: *Statistics in the Environmental and Earth Sciences,* [P. Guttorp and A. Walden (eds)]. London: Griffin.

[6] CHEN, C.F. (1979). Bayesian Inference for a Normal Dispersion Matrix and its Application to Stochastic Multiple Regression Analysis. *J. R. Statist. Soc., Ser. B*, **41**, 235-248.

[7] DE WAAL, D.J., AND GROENEVALD, P.C.N. (undated). *Spatial Bayesian estimation.* Unpublished manuscript.

[8] DEMPSTER, A.P., LAIRD, N.M. AND RUBIN, D.B. (1977). Maximum Likelihood from Incomplete Data via the EM Algorithm (with discussion). *J. R. Statist. Soc., Ser. B* **39**, 1–38.

[9] ELFVING, G. (1952). Optimum Allocation in Linear Regression Theory. *Ann. Math. Statist.*, **23**, 255–262.

[10] FEDEROV, V. AND MULLER, W. (1988). "Two Approaches in Optimization of Observing Networks." In: *Optimal Design and Analysis of Experiments* , [Y.Dodge, V.V. Fedorov, and H.P.Wynn (Eds.)]. Elsevier Science Publishers B.V..

[11] FEDEROV, V. AND MULLER, W. (1989). Comparison of Two Approaches in the Optimal Design of an Observation Network. *Statistics*, **3**, 339–351.

[12] GUTTORP, P., SAMPSON, P.D. AND NEWMAN, K. (1991). "Nonparametric Estimation of Spatial Covariance and Application Monitoring Network Evaluation". In: *Statistics in the Environmental and Earth Sciences,* [P. Guttorp and A. Walden (eds.)]. London: Griffin.

[13] GUTTORP, P., LE, N.D., SAMPSON, P.D. AND ZIDEK, J.V. (1992). "Using Entropy in the Redesign of an Environmental Monitoring Network". Technical Report No. 116, Department of Statistics, UBC. To appear in *Multivariate Environmental Statistics,* [G.P Patil, C.R. Rao and N.P. Ross (eds.)]. North Holland/Elsevier Science, New York.

[14] JAYNES, E.T. (1963). "Information Theory and Statistical Mechanics". In: *Statistical Physics,* 3, [Ford, K.W. (ed.)]. Benjamin, New York, 102–218.

[15] KEIFER, J. (1959). Optimum Experimental Design." *J.R. Statist. Soc. Ser. B.*, **21**, 272–319.

[16] LE, N.D. AND ZIDEK, J. V. (1992). Interpolation With Uncertain Spatial Covariances: A Bayesian Alternative To Kriging". *Journal of Multivariate Analysis.* (To appear).

[17] LINDLEY, D.V. (1956). On the Measure of the Information Provided by an Experiment. *Ann. Math. Statist.*, **27**, 968–1005.

[18] LINTHURST, R.A., LANDERS, D.H., EILERS, J.M., BRAKKE, D.F., OVERTON, W.S., MEIER, E.P. AND CROWE, R.E. (1986). *Characteristics of Lakes in the Eastern United States. Volume 1. Population Descriptions and pHysico-Chemical Relationships.* EPA/600/4-86/007a, U.S. Environmental Protection Agency, Washington, DC, 1986, 136 pp.

[19] LOADER, C. AND SWITZER, P. (1989). *Spatial Covariance Estimation For Monitoring Data.* SIMS Technical Report 133. Stanford University.

[20] MATHERON, G. (1971). *The Theory of Regionalized Variables and its Applications.* Centre de Geostatistique, Ecole des Mines and Paris, 212 pp.

[21] OMRE, H., S. D. (1987). Bayesian Kriging–merging Observations and Qualified Guesses in Kriging. *Math. Geology,* **19**, 25–39.

[22] OMRE, H. AND HALVORSEN, K.B. (1989a). The Bayesian Bridge between Simple and Universal Kriging. *Math. Geology,* **21**, 767–786.

[23] OMRE, H., HALVORSEN, K.B. AND BERTIG, V. (1989b). "A Bayesian Approach to Kriging". In: *Geostatistics,* 1, [M. Armstrong (ed.)]. Klumer Academic Publishers

[24] PAZMAN, A. (1986). *Foundations of Optimum Experimental Design.* Dordrecht: D. Reidel Publishing Co.

[25] PILZ, J. (undated). *Robust Bayes Linear Prediction of Regionalized Variables.* Unpublished manuscript.

[26] PRESS, J.S. (1972). *Applied Multivariate Analysis.* Holt, Rinehart and Winston, Inc.

[27] SAMPSON, P. AND GUTTORP, P. (1992). Nonparametric Estimation of Nonstationary Spatial Covariance structure. *JASA,* **87**, 108–119.

[28] SCHUMACHER, P. AND ZIDEK, J.V. (1988). *Using Prior Information in Designing Point Impact Detection Networks.* SIMS Tech. Rep. No. 120. Department of Statistics. University of British Columbia.

[29] SHEWRY, M., AND WYNN, H. (1987). Maximum Entropy Sampling." *Jour. Applied Statist.*, **14**, 165–207.

[30] SILVEY, S.D. (1980). *Optimal Design* . London: Chapman and Hall.

[31] SMITH, K. (1918). On the Standard Deviations of Adjusted and Interpolated Values of an Observed Polynomial Function and its Constants and Guidance they Give towards a Proper Choice of the Distribution of Observations." *Biometrika* , **12**, 1–85.

[32] SWITZER, P. (1989). "Non–stationary Spatial Covariances Estimated from Monitoring Data". In: *Geostatistics,* 1. [M. Armstrong (ed.)]. Klumer Academic Publishers

[33] VERDINELLI, I. (1991). *Advances in Bayesian Experimental Design.* Unpublished manuscript.

[34] WU, S. AND ZIDEK, J.V. (1992). An Entropy Based Review of Selected NADP/NTN Network Sites for 1983–86. *Atmospheric Environment.* (To appear).

[35] ZIDEK, J.V. (1984). "Detailed Statistical Approach to Sediment Chemistry Monitoring." Appendix B of *Beaufort Sea Monitoring Program Workshop Synthesis and Sampling Design Recommendations.* Report to National Oceanic and Atmospheric Administration, National Ocean Systems, Juneau, Alaska, prepared by Dames and Moore, SEAMOcean, Inc., and University of Washington, Department of Statistics.

Recent Adv. in Stat. and Prob., pp. 207-218
J. Pérez Vilaplana and M. L. Puri (Eds)

Joint Sensitivity Analysis for Covariance Matrices in Bayesian Linear Regression *

W. POLASEK
Institut für Statistik und Ökonometrie, Universität Basel, CH-4051 Basel, Switzerland

Abstract – The paper explores the boundedness of the posterior mean in a simple Bayesian linear model with diagonal data structure and a conjugate prior distribution: $\mathbf{y} \sim \mathcal{N}(\mathbf{Xb}, \mathbf{\Sigma} = \mathbf{W} + \mathbf{V}), \mathbf{b} \sim \mathcal{N}(\mathbf{b}^*, \mathbf{H}^*)$. Choosing the class of prior distributions where $\mathbf{W}$ and the prior mean $\mathbf{b}^*$ is fixed but the matrices $\mathbf{V}$ and $\mathbf{H}^*$ are any positive definite covariance matrices, we can show that set of posterior means is bounded.

This joint robustness analysis combines Bayesian robustness with respect to the prior covariance matrix and classical sensitivity analysis for heteroscedasticity diagnostics. The robustness analysis in a two period ANOVA model leads to the so-called "correlation contract curve" while the joint analysis of the covariance matrices $\mathbf{\Sigma}$ and $\mathbf{H}^*$ leads to so-called "Pascal's limaçons".

1. INTRODUCTION

Bounding the posterior mean for classes of prior distributions was investigated by Chamberlain and Leamer (1976), Leamer (1978 and 1983), and DasGupta and Studden (1988) mainly for regression purposes. Little has been done to apply Bayesian robustness to ANOVA models, except for Grieve (1985, 1989) and Polasek and Poetzelberger (1989).

In this paper we extend the class of priors by looking at special cases of the linear model having a diagonal data structure as e.g. in simple ANOVA models. Allowing the data covariance matrix to vary over all positive definite matrices which are bounded from below, we can show that the resulting set is a limaçon. A bi-weight class of prior distributions is analyzed to deal with the classical heteroscedasticity problem and an unspecified covariance matrix.

A first Bayesian type of robustness analysis for the cross-over model was given in Grieve (1985). This paper approaches this problem from a heteroscedasticity and a diagnostic perspective.

The plan of the paper is as follows: In Section 2 we derive a general result on the ellipsoidal bound of the generalized least squares (GLS) estimator having a general covariance structure, extending the result of Leamer (1986). Section 3 reviews the results on matrix weighted averages with bounded covariance matrices. In section 4 we extend the limaçon results of DasGupta and Studden (1988) to derive the bounded set for two varying covariance matrices. A final section summarizes this approach and the appendix lists the results for the extreme bound analysis (EBA).

*Support of the NSF Grant SES 8910950 is gratefully acknowledged.

2. HETEROSCEDASTIC ELLIPSOIDAL VARIANCE BOUNDS

Consider the problem where we have partial non-informativeness for the parameters in a linear model, either in the prior distribution or in the likelihood function. Can we then derive useful robustness pro perties, e.g. sets for the posterior mean which are bounded? If we let the prior covariance matrix unspecified and we fix only the prior mean, then the resulting set is the so-called feasible ellipsoid (see Chamberlain and Leamer 1976). The feasible set of posterior means can be extended to a so-called limaçon if we let the prior mean vary in a circle around a fixed hyper-prior mean. This was first discussed in DasGupta and Studden (1988).

Now we try to reformulate the question of a partially specified data covariance matrix into a one which can be applied to these previous sensitivity results. Therefore we transform in Theorem 2.1 the problem of a unknown data covariance matrix in a first step to a problem of a matrix weighted average.

THEOREM 2.1: *GLS is a matrix weighted average*

PROOF: The generalized least squares estimator with covariance matrix $\mathbf{\Sigma} = \mathbf{W} + \mathbf{V}$ can be written as

$$\begin{aligned} \mathbf{b}(\Sigma) &= [\mathbf{X}'(\mathbf{W}+\mathbf{V})^{-1}\mathbf{X}]^{-1}\mathbf{X}'(\mathbf{W}+\mathbf{V})^{-1}\mathbf{y} = \\ &= (\mathbf{X}'\mathbf{W}^{-1}\mathbf{X})^{-1}\mathbf{X}'\mathbf{W}^{-1}(\mathbf{y}-\gamma) = \\ &= \mathbf{b} - (\mathbf{X}'\mathbf{W}^{-1}\mathbf{X})^{-1}\mathbf{X}'\mathbf{W}^{-1}\gamma , \end{aligned} \tag{1}$$

with

$$\begin{aligned} \gamma &= (\mathbf{M}+\mathbf{V}^{-1})^{-1}\mathbf{M}\mathbf{y} , \\ \mathbf{b} &= (\mathbf{X}'\mathbf{W}^{-1}\mathbf{X})^{-1}\mathbf{X}'\mathbf{W}^{-1}\mathbf{y} = \mathbf{b}_{\mathbf{W}} , \\ \mathbf{M} &= \mathbf{W}^{-1} - \mathbf{W}^{-1}\mathbf{X}(\mathbf{X}'\mathbf{W}^{-1}\mathbf{X})^{-1}\mathbf{X}'\mathbf{W}^{-1} . \end{aligned} \tag{2}$$

This is a slightly modified version of Leamer (1984). First note that $\mathbf{b}(\Sigma)$ in (1) can be interpreted as a conditional mean of $\mathbf{b}$ given the realizations $\mathbf{y}$ and γ. To show γ, we embed the GLS estimation problem in a regression model with two sets of variables $\mathbf{b}$ and γ, where γ has a prior distribution with mean zero and covariance matrix $\sigma^2\mathbf{W}$:

$$\mathbf{y} = \mathbf{X}\mathbf{b} + \mathbf{I}_n\gamma + \mathbf{u} , \qquad \mathrm{Var}(\mathbf{u}) = \sigma^2\mathbf{W} , \tag{3}$$

$$\begin{pmatrix} \mathbf{b} \\ \gamma \end{pmatrix} \sim \left\langle \begin{pmatrix} 0 \\ 0 \end{pmatrix}, \begin{pmatrix} 0 & 0 \\ 0 & \sigma^2\mathbf{V} \end{pmatrix} \right\rangle$$

where the $\langle . \rangle$ brackets denote the mean and the variance of a random variable. Now the posterior mean of this augmented regression model can be written as

$$\begin{aligned} \mathrm{E}\left[\begin{pmatrix} \mathbf{b} \\ \gamma \end{pmatrix} \mathbf{y}, \sigma^2\right] &= \left(\begin{bmatrix} \mathbf{X}'\mathbf{W}^{-1}\mathbf{X} & \mathbf{X}'\mathbf{W}^{-1} \\ \mathbf{W}^{-1}\mathbf{X} & \mathbf{W}^{-1} \end{bmatrix} + \begin{bmatrix} 0 & 0 \\ 0 & \mathbf{V}^{-1} \end{bmatrix} \right)^{-1} \begin{pmatrix} (\mathbf{X}'\mathbf{W}^{-1} & \mathbf{y} \\ \mathbf{W}^{-1} & \mathbf{y} \end{pmatrix} = \\ &= \begin{pmatrix} \mathbf{X}'\mathbf{W}^{-1}\mathbf{X} & \mathbf{X}'\mathbf{W}^{-1} \\ \mathbf{W}^{-1}\mathbf{X} & \mathbf{W}^{-1}+\mathbf{V}^{-1} \end{pmatrix}^{-1} \begin{pmatrix} (\mathbf{X}'\mathbf{W}^{-1} & \mathbf{y} \\ \mathbf{W}^{-1} & \mathbf{y} \end{pmatrix}^{-1} \end{aligned} \tag{4}$$

Using the partitioned inverse rule we find

$$\begin{pmatrix} \mathbf{X}'\mathbf{W}^{-1}\mathbf{X} & \mathbf{X}'\mathbf{W}^{-1} \\ \mathbf{W}^{-1}\mathbf{X} & \mathbf{W}^{-1}+\mathbf{V}^{-1} \end{pmatrix}^{-1} = \begin{bmatrix} \bullet & \bullet \\ -\mathbf{C}^{-1}\mathbf{W}^{-1}\mathbf{X}(\mathbf{X}'\mathbf{W}^{-1}\mathbf{X})^{-1} & \mathbf{C}^{-1} \end{bmatrix}$$

where

$$\mathbf{C} = \left(\mathbf{W}^{-1} + \mathbf{V}^{-1} - \mathbf{W}^{-1}\mathbf{X}(\mathbf{X}'\mathbf{W}^{-1}\mathbf{X})^{-1}\mathbf{X}'\mathbf{W}^{-1}\right) = \mathbf{M} + \mathbf{V}^{-1}. \tag{5}$$

Therefore we find for the conditional mean γ in the second row of (4):

$$\begin{aligned} E(\gamma \mid \mathbf{y}, \sigma^2) &= \mathbf{C}^{-1}[-\mathbf{W}^{-1}\mathbf{X}(\mathbf{X}'\mathbf{W}^{-1}\mathbf{X})^{-1} : \mathbf{I}_n] \begin{bmatrix} \mathbf{X}'\mathbf{W}^{-1} \\ \mathbf{W}^{-1} \end{bmatrix} \mathbf{y} = \\ &= (\mathbf{M} + \mathbf{V}^{-1})^{-1}[-\mathbf{W}^{-1}\mathbf{X}(\mathbf{X}'\mathbf{W}^{-1}\mathbf{X})^{-1}\mathbf{X}'\mathbf{W}^{-1} + \mathbf{W}^{-1}]\mathbf{y} = \\ &= (\mathbf{M} + \mathbf{V}^{-1})^{-1}\mathbf{M}\mathbf{y}. \end{aligned}$$

■

Note that $\mathbf{My}$ is a weighted GLS residual $\mathbf{My} = \mathbf{W}^{-1}\mathbf{e}_W = \mathbf{y} - \mathbf{Xb}$ with $\mathbf{b}$ as in (2).

THEOREM 2.2: [Heteroscedasticity (upper variance) bound]. *Given the upper bounded matrix weighted average of the form*

$$\gamma = (\mathbf{M} + \mathbf{V}^{-1})^{-1}\mathbf{My}, \quad \text{with} \quad \mathbf{V} \le \mathbf{V}_u \quad \text{p.d.}, \tag{6}$$

where $\mathbf{V}_u$ is the upper bound covariance matrix, then γ has to lie in the ellipsoid $\gamma \in$ ELL$(\mathbf{0}, \mathbf{f}, \mathbf{V}_u^{-1} + \mathbf{M})$ or

$$(\gamma - \mathbf{f}/2)'(\mathbf{V}_u^{-1} + \mathbf{M})(\gamma - \mathbf{f}/2) \le c, \tag{7}$$

where the ellipsoidal parameters are the midpoint of the origin $\mathbf{0}$ and $\mathbf{f} = (\mathbf{V}_u^{-1} + \mathbf{M})^{-1}\mathbf{My} = \mathbf{H}^{-1}\mathbf{My}$, the metric $\mathbf{H} = \mathbf{V}_u^{-1} + \mathbf{M}$, and the constant $c = \mathbf{y}'\mathbf{MH}^{-1}\mathbf{My}/4 = \mathbf{f}'\mathbf{Hf}$.

PROOF: See Leamer (1984). ■

3. JOINT VARIANCE BOUNDS - BAYESIAN HETEROSCEDASTICITY DIAGNOSTICS

This section looks into a larger class of prior distributions where we let the prior mean vary as well in a spherical or ellipsoidal set. We first start with the general ellipsoid results of Polasek (1984).

THEOREM 3.1: [Lower bound theorem] *Let $\mathbf{V}_L$ be a lower bound for the prior covariance matrix $\mathbf{H}^*$ in a normal Bayesian regression model, then for any positive definite covariance matrix $\mathbf{B}$, the posterior mean*

$$\mathbf{b}^{**} = \left(\mathbf{H} + (\mathbf{V}_L + \mathbf{B})^{-1}\right)^{-1}(\mathbf{Hb} + (\mathbf{V}_L + \mathbf{B})^{-1}\mathbf{b}^*), \tag{8}$$

has to lie in the ellipsoid ELL$(\mathbf{b}_\Pi{}^*, \mathbf{b}, \mathbf{HV}_L\mathbf{H} + \mathbf{H})$ where $\mathbf{b}_\Pi^*$ is the so-called lower bound shifted prior mean (a limiting location parameter of the ellipsoid). Explicitly, the ellipsoid is given by

$$(\mathbf{b}^{**} - \mathbf{f})'(\mathbf{HV}_L\mathbf{H} + \mathbf{H})(\mathbf{b}^{**} - \mathbf{f}) \le c, \tag{9}$$

$$\begin{aligned} 2\mathbf{f} &= (\mathbf{b} - \mathbf{b}^*) + \mathbf{V}_L(\mathbf{V}_L + \mathbf{H}^{-1})^{-1}(\mathbf{b} - \mathbf{b}^*) = \mathbf{b} + \mathbf{b}_\Pi^*, \\ c &= (\mathbf{b} - \mathbf{b}^*)'(\mathbf{V}_L + \mathbf{H}^{-1})^{-1}(\mathbf{b} - \mathbf{b}^*)/4, \\ \mathbf{b}_\Pi^* &= (\mathbf{V}_L^{-1} + \mathbf{H})^{-1}(\mathbf{V}_L^{-1}\mathbf{b}^* + \mathbf{Hb}). \end{aligned} \tag{10}$$

The ellipsoid can be considered as a robust feasible posterior mean set for the class of prior distributions $\Gamma_\Sigma = \{(\mathbf{b}^*, \Sigma) \mid \mathbf{V}_L < \Sigma^* \in \mathcal{M}_+\}$.

PROOF: See Polasek (1984) p. 242. ■

THEOREM 3.2: [Pascal's limaçon obtained by spherical restrictions] *Given the normal linear model*

$$\mathbf{y} \sim \mathcal{N}(\mathbf{Xb},\sigma^2\mathbf{I}_n), \quad \mathbf{b} \sim \mathcal{N}(\mathbf{b}^*,\sigma^2\mathbf{\Sigma}), \tag{11}$$

and consider the larger prior class

$$\mathbf{\Gamma}^*_{\mathbf{\Sigma}} = \{(\mathbf{b}^*,\mathbf{\Sigma}) \,|\, \mathbf{V}_L < \mathbf{\Sigma}^* \in \mathcal{M}_+ \, , \, (\mathbf{b}-\mathbf{b}^*)'\mathbf{H}^*(\mathbf{b}-\mathbf{b}^*) \le \epsilon^2 \, , \tag{12}$$

where $\mathbf{b}^*$ *and* σ^2 *are fixed and constant, and* $\mathbf{V}_L$ *and* $\mathbf{X'X}$ *are given by the special configurations*

$$\mathbf{V}_L^{-1} = r\mathbf{I}_K \, , \qquad \text{and} \qquad \mathbf{H} = \mathbf{X'X} = mr\mathbf{I}_K \, . \tag{13}$$

Then the boundary of the set of posterior means $\mathbf{b}^{**}$ *consists of* K*-dimensional vectors* $\mathbf{z}+\mathbf{b}$ *where* $\mathbf{z}$ *satisfies*

$$\mathbf{z'z}\gamma - 2\mathbf{z}'(\mathbf{b}_0 - \mathbf{b}) + 2\epsilon\sqrt{\mathbf{z'z}} = 0 \, , \qquad \text{with} \qquad \gamma^{-1} = 2(m+1) \, .$$

The posterior mean is constrained to lie in the limaçon "LIMA($\mathbf{b}^{**}$)", given by

$$\text{LIMA}(\mathbf{b}^{**}) = \{(\mathbf{z}+\mathbf{b}) \in \mathcal{R}^K \,|\, (1+m)\mathbf{z'z} - \epsilon\sqrt{\mathbf{z'z}} + \mathbf{z}'(\mathbf{b}-\mathbf{b}^*) = 0\} \, , \tag{14}$$

with $\mathbf{b} = (\mathbf{z'z})^{-1}\mathbf{X'y}$, *which is a set of vectors* $\mathbf{z}$ *shifted by* $\mathbf{b}$. *In general this set is a translated (Pascal's) limaçon. In two dimensions* ($K = 2$) *and for* $\|\mathbf{b}-\mathbf{b}^*\|| = \epsilon$ *the set is a cardioid, and for* $\|\mathbf{b}-\mathbf{b}^*\|| = \epsilon\mathbf{z}$ *it is a translated trisectrix* (see e.g. Lawrence 1972).

PROOF: The original proof was given in DasGupta and Studden (1988), which we restate in terms of our notation using the EBA result of Lemma A1. Substitute $\mathbf{H} = rm\mathbf{I}_K$ and $\mathbf{V}_L = \mathbf{I}_K/r$ into the lower bound theorem in (10)

$$\begin{aligned}
\mathbf{HV}_L\mathbf{H} + \mathbf{H} &= r^2m^2\,\mathbf{I}/r + rm\,\mathbf{I} = rm(m+1)\,\mathbf{I}, \\
(\mathbf{V}_L + \mathbf{H}^-1)^{-1} &= (\mathbf{I}/r + \mathbf{I}/rm)^{-1} = \mathbf{I}\left(\frac{1}{r} + \frac{1}{rm}\right)^{-1} = \mathbf{I}\,r\frac{m}{m+1}, \\
c &= (\mathbf{b}-\mathbf{b}^*)'(\mathbf{b}-\mathbf{b}^*)\frac{rm}{4(m+1)}, \\
\mathbf{b}^*{}_\Pi &= (rm\mathbf{I} + r\mathbf{I})^{-1}(rm\mathbf{b} + r\mathbf{b}^*) = \frac{1}{m+1}\,m\mathbf{b} + \mathbf{b}^*, \\
2\mathbf{f} &= \mathbf{b} + \mathbf{b}^*{}_\Pi = \frac{1}{m+1}\,[(m+1)\mathbf{b} + m\mathbf{b} + \mathbf{b}^*] = \\
&= \frac{1}{m+1}\,(\mathbf{b}+2m\mathbf{b} + \mathbf{b}^*), \\
\mathbf{f} &= \frac{\mathbf{b}(1+2m) + \mathbf{b}^*}{(2m+1)+1} = \gamma\mathbf{b}^* + (1-\gamma)\mathbf{b}
\end{aligned}$$

with $\gamma = 1/[2(m+1)]$. Putting these results together yields

$$\begin{aligned}
&(\mathbf{b}^{**} - \gamma\mathbf{b}^* - (1-\gamma)\,\mathbf{b})'\,rm\,(m+1)\,\mathbf{I}\,(\mathbf{b}^{**} - \gamma\mathbf{b}^* - (1-\gamma)\mathbf{b}) \le \\
&\qquad \le (\mathbf{b}-\mathbf{b}^*)'\frac{rm}{4(m+1)}(\mathbf{b}-\mathbf{b}^*), \\
&(\mathbf{b}^{**} - \gamma\mathbf{b}^* - (1-\gamma)\,\mathbf{b})'\,\gamma^{-2}\,(\mathbf{b}^{**} - \gamma\mathbf{b}^* - (1-\gamma)\mathbf{b}) \le c,
\end{aligned}$$

with $c = (\mathbf{b}-\mathbf{b}^*)'(\mathbf{b}-\mathbf{b}^*)$. Now

$$\begin{aligned}
&(\mathbf{b}^{**}/\gamma - \mathbf{b}^* - (1/\gamma - 1)\mathbf{b})'\,(\mathbf{b}^{**}/\gamma - \mathbf{b}^* - (1/\gamma - 1)\mathbf{b}) \le c, \\
&((\mathbf{b}^{**} - \mathbf{b})/\gamma - (\mathbf{b}^* - \mathbf{b}))'\,((\mathbf{b}^{**} - \mathbf{b})/\gamma - (\mathbf{b}^* - \mathbf{b})) - c \le 0 \, .
\end{aligned} \tag{15}$$

So we can now transform the inequality to the following minimization problem:

$$\min_{\mathbf{b}^* \in \mathcal{B}} [(\mathbf{b}^{**} - \mathbf{b})'(\mathbf{b}^{**} - \mathbf{b})/\gamma - 2(\mathbf{b}^{**} - \mathbf{b})'(\mathbf{b}^{**} - \mathbf{b})] = 0 \tag{16}$$

where $\mathcal{B}$ is the set $\mathcal{B}= \{\mathbf{b}^* \,|\, (\mathbf{b}^* - \mathbf{b}_0)'(\mathbf{b}^* - \mathbf{b}_0) = \epsilon^2\}$.
Now the problem has been reduced to minimize the linear combination $\mathbf{z}'(\mathbf{b}^* - \mathbf{b})$ as a variable of $\mathbf{b}^*$ with coefficients $\mathbf{z} = \mathbf{b}^{**} - \mathbf{b}$ over the sphere $(\mathbf{b}^* - \mathbf{b}_0)'(\mathbf{b}^* - \mathbf{b}_0) \leq \epsilon^2$. But this is just the same problem of finding the extreme bounds on the sphere, i.e.

$$\min \ \mathbf{z}'\tilde{\mathbf{b}}^* \qquad \text{s.t.} \qquad (\mathbf{b}^* - \mathbf{b}_0)'(\mathbf{b}^* - \mathbf{b}_0) \leq \epsilon^2 \tag{17}$$

with $\tilde{\mathbf{b}}^* = (\mathbf{b}^* - \mathbf{b})$ and $\tilde{\mathbf{b}}_0 = (\mathbf{b}_0 - \mathbf{b})$ and since $\tilde{\mathbf{b}}^* - \tilde{\mathbf{b}}_0 = \mathbf{b}^* - \mathbf{b}_0$.
Using the EBA results of Lemma A1 in the appendix we can express the minimum as

$$\mathbf{z}'(\mathbf{b}^* - \mathbf{b}_0) = -\sqrt{\epsilon^2 \mathbf{z}'\mathbf{z}}. \tag{18}$$

Now inserting this minimum into the reformulated argument of (16)

$$\mathbf{z}'\mathbf{z}/\gamma - 2\mathbf{z}'(\mathbf{b}^* - \mathbf{b}_0 + \mathbf{b}_0 - \mathbf{b}) = 0$$

gives

$$\gamma^{-1}\mathbf{z}'\mathbf{z} - 2\mathbf{z}'(\mathbf{b}_0 - \mathbf{b}) + 2\epsilon\sqrt{\mathbf{z}'\mathbf{z}} = 0 \tag{19}$$

This proves the theorem. ■

LEMMA 3.3: [General ellipsoid transformations] *Given the ellipsoid*

$$(\mathbf{b} - \mathbf{f})'\mathbf{H}(\mathbf{b} - \mathbf{f}) \leq c, \tag{20}$$

and consider the transform

$$\mathbf{a} = \mathbf{Q}\mathbf{b} + \mathbf{q}, \tag{21}$$

where $\mathbf{Q}$ *and* $\mathbf{q}$ *are known constants, and the* $\mathrm{m} \times \mathrm{p}$ *matrix* $\mathbf{Q}$ *has full column rank, i.e.* $\text{rank}(\mathbf{Q}) = \mathrm{p} < \mathrm{m}$. *Then the transformed vector* $\mathbf{a}$ *lies in the ellipsoid*

$$(\mathbf{a} - \tilde{\mathbf{f}})'\mathbf{Q}^{+\prime}\mathbf{H}\mathbf{Q}^{+}(\mathbf{a} - \tilde{\mathbf{f}}) \leq c, \tag{22}$$

with

$$\tilde{\mathbf{f}} = \mathbf{q} + \mathbf{Q}\mathbf{f} \qquad \textit{and} \qquad \tilde{\mathbf{H}} = \mathbf{Q}^{+\prime}\mathbf{H}\mathbf{Q}^{+}. \tag{23}$$

PROOF: Let $\mathbf{Q}^+$ be a generalized inverse with the power to solve general equations like $\mathbf{Q}\mathbf{b} = \mathbf{y}$ by $\mathbf{b} = \mathbf{Q}^+\mathbf{Q}\mathbf{b} = \mathbf{Q}^+\mathbf{y}$, then $\mathbf{Q}^+$ is given by $\mathbf{Q}^+ = (\mathbf{Q}'\mathbf{Q})^{-1}\mathbf{Q}'$ for a matrix $\mathbf{Q}$ with full column rank, and has the properties

$$\mathbf{Q}^+\mathbf{Q} = \mathbf{I}_p \quad ; \quad (\mathbf{Q}^+)' = (\mathbf{Q}')^+ \quad ; \quad (\mathbf{Q}^+)^+ = \mathbf{Q}.$$

The general linear transform (21) $\mathbf{a} = \mathbf{Q}\mathbf{b} + \mathbf{q}$ can now be reversely expressed by

$$\mathbf{b} = \mathbf{Q}^+\mathbf{a} - \mathbf{Q}^+\mathbf{q} = \mathbf{Q}^+(\mathbf{a} - \mathbf{q}) = \mathbf{Q}^+\tilde{\mathbf{a}} \qquad \text{with} \qquad \tilde{\mathbf{a}} = \mathbf{a} - \mathbf{q}$$

where $\mathbf{Q}^+$ is the generalized inverse which gives

$$\mathbf{Q}^+\mathbf{a} = \mathbf{Q}^+(\mathbf{Q}\mathbf{b} + \mathbf{q}) = \mathbf{b} + \mathbf{Q}^+\mathbf{q}.$$

Now substitute **b** into (20) to obtain:

$$
\begin{aligned}
(\mathbf{Q^+a-f})'\mathbf{H}(\mathbf{Q^+a-f}) &\le c,\\
(\mathbf{Q^+a-Q^+Qf})'\mathbf{H}(\mathbf{Q^+a-Q^+Qf}) &\le c,\\
(\mathbf{a-Qf})'\mathbf{Q^{+\prime}HQ^+}(\mathbf{a-Qf}) &\le c,\\
(\mathbf{a-q-Qf})'\mathbf{Q^{+\prime}HQ^+}(\mathbf{a-q-Qf}) &\le c,\\
(\mathbf{a-f})'\mathbf{H}(\mathbf{a-f}) &\le c,
\end{aligned}
$$

with the ellipsoidal parameters $\tilde{\mathbf{f}} = \mathbf{q} + \mathbf{Qf}$ and $\tilde{\mathbf{H}} = \mathbf{Q^{+\prime}HQ^+}$. ■

4. THE TWO PERIOD ANOVA MODEL

Consider a regression ANOVA model, where the same patient receive two treatments in two periods. We are interested to estimate the re-sponse in each period and make possibly inferences on the difference. So we can formulate the model as

$$\mathbf{y}_1 = \mathbf{X}_1 + \mathbf{e}_1 \qquad \mathbf{y}_2 = \mathbf{X}_2\mathbf{b}_2 + \mathbf{e}_2$$

$$\mathrm{Var}\begin{pmatrix}\mathbf{e}_1\\ \mathbf{e}_2\end{pmatrix} = \mathbf{R}\otimes\mathbf{I}_n \tag{24}$$

or more compactly, let $\mathbf{y}' = (\mathbf{y}_1', \mathbf{y}_2')$ then

$$\mathbf{y} \sim \mathcal{N}\left[\begin{pmatrix}\mathbf{X}_1 & 0\\ 0 & \mathbf{X}_2\end{pmatrix}\begin{pmatrix}\mathbf{b}_1\\ \mathbf{b}_2\end{pmatrix}, \Sigma\right]$$

where the general form of the covariance matrix Σ is $\Sigma = \mathbf{R}\otimes\mathbf{I}_n = \mathbf{W} + \rho\mathbf{V}$ with $\mathbf{W} = \sigma^2\mathbf{I}_{nK}$ with

$$\mathbf{R} = \sigma^2\begin{pmatrix}1 & \rho\\ \rho & 1\end{pmatrix} \quad \text{and} \quad \mathbf{V} = \sigma^2\begin{pmatrix}\mathbf{O} & \mathbf{I}_n\\ \mathbf{I}_n & \mathbf{O}\end{pmatrix}. \tag{25}$$

THEOREM 4.1: [GLS heteroscedasticity bounds for $\mathbf{b}(\Sigma)$] *The ellipsoidal bound for the GLS estimator* $\mathbf{b}(\Sigma)$ *with* $\Sigma = \mathbf{W} + \mathbf{V}$ *as in Theorem* 2.1 *and* 2.2 *is given by*

$$\mathbf{b}(\Sigma) = \mathbf{X}^+(\mathbf{y} - \mathbf{e_W}) = \mathbf{b} - \mathbf{X}^+\mathbf{e_W}, \tag{26}$$

where $\mathbf{X}^+ = (\mathbf{X}'\mathbf{X})^{-1}\mathbf{X}'$ *and* $\mathbf{e_W} = (\mathbf{M} + \mathbf{V}^{-1})^{-1}\mathbf{My}$, *has the form*

$$(\mathbf{b}(\Sigma) - \tilde{\mathbf{f}})'\tilde{\mathbf{H}}(\mathbf{b}(\Sigma) - \tilde{\mathbf{f}}) \le c, \tag{27}$$

with

$$\tilde{\mathbf{f}} = \tilde{\mathbf{b}} + \mathbf{b}, \qquad \tilde{\mathbf{b}} = \mathbf{b} - \mathbf{X}^+\mathbf{H}^{-1}\mathbf{e_W} \qquad \mathbf{e_W} = \mathbf{y} - \mathbf{Xb}, \tag{28}$$

and the metric $\tilde{\mathbf{H}} = \mathbf{X}'\mathbf{HX}$ *with* $\mathbf{H} = \mathbf{V}_u^{-1} + \mathbf{M}$ *from* (7). *Written compactly the ellipsoid is*

$$\mathbf{b}(\Sigma) \in \mathrm{ELL}(\tilde{\mathbf{b}}, \mathbf{b}, \mathbf{X}'\mathbf{HX})\,^1. \tag{29}$$

[1]The extreme bounds of this ellipsoid can be used to check if a simple covariance specification is sufficient or a richer data covariance matrix should be elicited since then the 'GLS' posterior mean can be very much different from an 'OLS' posterior mean.

PROOF: By Theorem 2.2 $\mathbf{e_W} = \gamma$ is bounded by

$$(\mathbf{e_W} - \mathbf{f}/2)'\mathbf{H}(\mathbf{e_W} - \mathbf{f}/2) \leq c,$$

and we apply the transformation Lemma 3.3 for $\mathbf{b}(\Sigma)$ in (26), we find the transformed parameters

$$\begin{aligned} \tilde{\mathbf{f}} &= \mathbf{q} + \mathbf{Q}\mathbf{f}/2 = \mathbf{b} - \mathbf{X}^+\mathbf{f}/2), \\ \tilde{\mathbf{H}} &= \mathbf{Q}^{+'}\mathbf{H}\mathbf{Q}^+ = (-\mathbf{X}^+)^{+'}\mathbf{H}(-\mathbf{X}^+)^+ = \mathbf{X}'\mathbf{H}\mathbf{X}, \end{aligned} \tag{30}$$

which constitute the ellipsoid $(\mathbf{b}(\Sigma) - \tilde{\mathbf{f}})'\tilde{\mathbf{H}}(\mathbf{b}(\Sigma) - \tilde{\mathbf{f}}) \leq c$.
Since $\mathbf{X}^+\mathbf{f} = (\mathbf{X}'\mathbf{X})^{-1}\mathbf{X}'\mathbf{H}^{-1}\mathbf{M}\mathbf{y} = \mathbf{X}^+\mathbf{H}^{-1}(\mathbf{y} - \mathbf{X}\mathbf{b_W}) = \mathbf{X}^+\mathbf{H}^{-1}\mathbf{e_W}$ we find from (30)

$$2\tilde{\mathbf{f}} = 2\mathbf{b} - \mathbf{X}^+\mathbf{f} = \mathbf{b} + \tilde{\mathbf{b}}, \qquad \text{with} \qquad \tilde{\mathbf{b}} = \mathbf{b} - \mathbf{X}^+\mathbf{H}^{-1}\mathbf{e_W}$$

and therefore $\tilde{\mathbf{f}} = (\mathbf{b} + \tilde{\mathbf{b}})/2$. ■

LEMMA 4.2. [GLS correlation bounds for $\mathbf{b}(\rho)$ in the two period model] *Consider a simple ANOVA model in* (24) *where the data covariance matrix is now* $\Sigma = \mathbf{I}_{nK} + \rho\mathbf{V}$, *i.e. a special matrix weight depending on the scalar* ρ:

$$\begin{aligned} \mathbf{b}(\rho) &= (\mathbf{X}'(\mathbf{I}_{nK} + \rho\mathbf{V})^{-1}\mathbf{X}'(\mathbf{I}_{nK} + \rho\mathbf{V})^{-1}\mathbf{y} = \\ &= \mathbf{b} - (\mathbf{X}'\mathbf{X})^{-1}\mathbf{X}'\mathbf{e}(\rho), \end{aligned} \tag{31}$$

with simplified location and orthogonal projection

$$\mathbf{b} = (\mathbf{X}'\mathbf{X})^{-1}\mathbf{X}'\mathbf{y}, \qquad \text{and} \qquad \mathbf{M} = \mathbf{I}_{nK} - \mathbf{X}(\mathbf{X}'\mathbf{X})^{-1}\mathbf{X}', \tag{32}$$

and $\mathbf{e}(\rho)$ *is the residual depending on* ρ *and the OLS residual* $\mathbf{e}$:

$$\mathbf{e}_\rho = \mathbf{e}\rho = (\mathbf{M} + \mathbf{V}^{-1}/\rho)^{-1}\mathbf{e}, \qquad \mathbf{e} = \mathbf{M}\mathbf{y} = \mathbf{y} - \mathbf{X}'\mathbf{b}. \tag{33}$$

Then the posterior mean (31) *lies on the information contract curve* (see Leamer 1978) *induced by* ρ *between* $\mathbf{b}(\rho = -1)$ *and* $\mathbf{b}(\rho = +1)$ *where* $\mathbf{b}(\rho)$ *itself is constrained to lie in the ellipsoid*

$$\mathbf{b}(\rho) \in \mathrm{ELL}(\tilde{\mathbf{b}}, \mathbf{b}, \mathbf{X}'\mathbf{H}\mathbf{X}), \tag{34}$$

with now simpler parameters $\tilde{\mathbf{b}} = \mathbf{b} - \mathbf{X}^+\mathbf{H}^{-1}\mathbf{e}$ *and* $\mathbf{H} = \mathbf{M} + \mathbf{I}$. *Therefore this curve could be also called "correlation contract" (CC) curve. Graphically this curve is shown in Figure 1.*

PROOF: A direct application of Theorem 4.1. ■

NOTE: For the two period correlation matrix

$$\mathbf{R} \otimes \mathbf{I}_n = \begin{pmatrix} \mathbf{I}_n & \rho\mathbf{I}_n \\ \rho\mathbf{I}_n & \mathbf{I}_n \end{pmatrix} = \mathbf{I}_{nK} + \rho\mathbf{V}, \qquad \text{with} \qquad \mathbf{V} = \begin{pmatrix} \mathbf{0} & \mathbf{I}_n \\ \mathbf{I}_n & \mathbf{0} \end{pmatrix},$$

or $\mathbf{V} = \mathbf{1}_K'\mathbf{1}_K \otimes \mathbf{I}_n - \mathbf{I}_{nK}$, where $\mathbf{1} = \mathbf{1}_K$ is a vector of ones of length $K = 2$; $\mathbf{V}$ can be inverted by the matrix inversion lemma:

$$\begin{aligned} \mathbf{V}^{-1} &= \mathbf{I}_{nK} - (\mathbf{1} \otimes \mathbf{I}_n)[\mathbf{1}'\mathbf{1} \otimes \mathbf{I}_n - \mathbf{I}_{nK}]^{-1}(\mathbf{1}' \otimes \mathbf{I}_n) = \\ &= \mathbf{I}_{nK} - (\mathbf{1} \otimes \mathbf{I}_n)[(\mathrm{K} - 1)\mathbf{I}_n)^{-1}\mathbf{1}' \otimes \mathbf{I}_n = \\ &= \mathbf{I}_{nK} - (\mathbf{1}\,\mathbf{1}' \otimes \mathbf{I}_n)/(\mathrm{K} - 1) = \\ &= \mathbf{I}_{nK} + \mathbf{P} \end{aligned}$$

with $\mathbf{P} = (\mathbf{1}\,\mathbf{1}' \otimes \mathbf{I}_n)/(K-1)$.

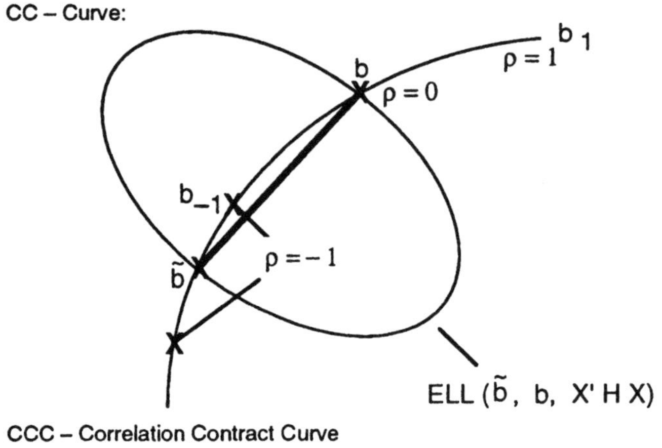

Figure 1.- Lower Bound Ellipsoid

LEMMA 4.3: [Bi-weight bounds for upper bounded posterior means] *Given the Bayesian normal linear model* (11) *as in Theorem* 3.2, *then the boundary of the set posterior mean is given by the roots of an "implicit quadratic form" of the form*

$$\mathbf{z}\,[(\mathbf{b} - \mathbf{b}^{**})]\,\mathbf{D}^{-1}\,\mathbf{z}\,[(\mathbf{b} - \mathbf{b}^{**})] = 0\,, \tag{35}$$

with $\mathbf{z}[(.)]$ *denoting a nonlinear function with argument* $\mathbf{b} - \mathbf{b}^{**}$, *and the metric of the quadratic form is the difference of metrics* $\mathbf{D}^{-1} = \mathbf{H}^{-1} - (\mathbf{H} - \mathbf{V}_u^{-1})^{-1}$.

PROOF: Taking the difference of the quadratic forms on the left hand side of the ellipsoidal inequality in (9), one can show that the problem is equivalent to

$$\min_{\mathbf{b}^* \in \mathcal{B}} \; [(\mathbf{b}^{**} + \mathbf{b}^* - 2\mathbf{b})'\,\mathbf{D}\,(\mathbf{b}^{**} + \mathbf{b}^* - 2\mathbf{b})] = 0\,, \tag{36}$$

where $\mathcal{B}$ is the set $\mathcal{B} = \{\mathbf{b}^* \mid (\mathbf{b}^* - \mathbf{b}_0)'\,(\mathbf{b}^* - \mathbf{b}_0) = \epsilon^2\}$.

Now the problem has been reduced to minimize the quadratic form $(\mathbf{b}^{**}+\mathbf{b}^*-2\mathbf{b})'\,\mathbf{D}\,(\mathbf{b}^{**}+\mathbf{b}^*-2\mathbf{b}) = (\tilde{\mathbf{b}}^*-\mathbf{g})'\,\mathbf{D}\,(\tilde{\mathbf{b}}^*-\mathbf{g})$ with $\tilde{\mathbf{b}}^* = \mathbf{b}^* - \mathbf{b}$ over the sphere $(\mathbf{b}^* - \mathbf{b}_0)'\,(\mathbf{b}^* - \mathbf{b}_0) \leq \epsilon^2$. But this is just the same problem of finding the extreme bounds on the sphere, i.e.

$$\min\, (\tilde{\mathbf{b}}^* - \mathbf{g})'\,\mathbf{D}\,(\tilde{\mathbf{b}}^* - \mathbf{g}) \qquad \text{s.t.} \qquad (\tilde{\mathbf{b}}^* - \mathbf{b}_0)'\,(\tilde{\mathbf{b}}^* - \mathbf{b}_0) \leq \epsilon^2\,, \tag{37}$$

with $\mathbf{g} = \mathbf{b}^{**} - \mathbf{b}$. Using the quadratic EBA results of Lemma A3 in the Appendix we can express the minimum using $\mathbf{z} = \mathbf{D}(\tilde{\mathbf{b}}^* - \mathbf{g})$ as

$$(\mathbf{x} - \mathbf{b})'\,\mathbf{D}\,(\mathbf{x} - \mathbf{b}) = (\tilde{\mathbf{b}}_0 - \mathbf{g})'\,\mathbf{z} \pm \epsilon\sqrt{\mathbf{z}'\mathbf{z}}\,, \tag{38}$$

or

$$(\tilde{\mathbf{b}}_0 - \mathbf{g})'\,\mathbf{z} \pm \epsilon\sqrt{\mathbf{z}'\mathbf{z}} - \mathbf{z}'\,\mathbf{D}^{-1}\mathbf{z} = 0\,.$$

Making $\mathbf{z}'$ explicit leads to $\mathbf{z}$ which depends only on $\mathbf{b} - \mathbf{b}^{**}$:

$$\mathbf{z} = \mathrm{f}(\tilde{\mathbf{b}}_0,\, \mathbf{g},\, \epsilon,\, \mathbf{D},\, \mathbf{b}) = \mathbf{z}[(\mathbf{b} - \mathbf{b}^{**})] \tag{39}$$

Inserting the inverse transformation $\tilde{\mathbf{b}}^* = \mathbf{D}^{-1}\mathbf{z}[(\mathbf{b}-\mathbf{b}^{**})] + \mathbf{g}$ into (37) we obtain

$$(\mathbf{D}^{-1}\mathbf{z}[\mathbf{g}] + \mathbf{g} - \mathbf{g})'\,\mathbf{D}\,(\mathbf{D}^{-1}\mathbf{z}[\mathbf{g}] + \mathbf{g} - \mathbf{g}),$$

which is the implicit quadratic form (35). ∎

It remains an open problem if (35) can be expressed still further in closed form or could be at least approximated by a simple bounding function.

THEOREM 4.4: [Lower bounded bi-weight variances] *Given the Bayesian normal linear model*

$$\mathbf{y} \sim \mathcal{N}(\mathbf{X}\mathbf{b}, \sigma^2(\mathbf{I}_n + \mathbf{V})), \qquad \mathbf{b} \sim \mathcal{N}(\mathbf{b}^*, \sigma^2(\mathbf{I}_n + \mathbf{H}^*)), \tag{40}$$

then the posterior mean is given by

$$\mathbf{b}^{**} = (\mathbf{G}^{*-1} + \mathbf{H}_V^{-1})^{-1}(\mathbf{G}^{*-1}\mathbf{b}^* + \mathbf{H}_V^{-1}\mathbf{b}_V), \tag{41}$$

with $\mathbf{H}_V = \mathbf{X}'(\mathbf{I}_n + \mathbf{V})^{-1}\mathbf{X}$ *and* $\mathbf{b}_V = \mathbf{H}_V^{-1}\mathbf{X}'(\mathbf{I}_n + \mathbf{V})^{-1}\mathbf{y}$.

We are interested in a global sensitivity analysis for $\mathbf{b}^{**}$ *using the bi-weight class of prior distributions*

$$\begin{aligned}\Gamma_{bi} &= \{\mathbf{b}^*\,\text{fixed}, \mathbf{V} \in \mathcal{M}_+, \mathbf{H}^* \in \mathcal{M}_+\} = \\ &= \{\mathbf{b}^*\,\text{fixed}, \Sigma > \mathbf{I}_n, \mathbf{G}^* > \mathbf{I}_n;\ \Sigma \text{ and } \mathbf{G}^*, \text{ pos. def.}\}\end{aligned} \tag{42}$$

where $\mathcal{M}_+$ *is the set of positive definite matrices and* $\Sigma = \mathbf{I}_n + \mathbf{V}$, $\mathbf{G}^* = \mathbf{I}_n + \mathbf{H}^*$.

a) *For the special configuration* $\mathbf{X}'\mathbf{X} = \mathrm{mr}\mathbf{I}_n$ *and* $\Sigma_1 = \mathbf{I}_n/\mathrm{r}$, *the posterior mean* $\mathbf{b}^{**}$ *is constrained to lie in the (Pascal) limaçon* "LIMA" *where the boundary is given by*

$$\begin{aligned}\text{LIMA} &= \text{hull}\{\mathbf{z} + \mathbf{b}_L \mid \mathbf{z}, \mathbf{b}_L \in \mathcal{B}\}, \\ \mathcal{B} &= \{\mathbf{z} \in \mathcal{R}^K \mid 2(1+\mathrm{m})\mathbf{z}'\mathbf{z} - 2\kappa\sqrt{\mathbf{z}'\mathbf{z}} + 2\mathbf{z}'(\mathbf{b}_L - \mathbf{b}^*) = 0\},\end{aligned} \tag{43}$$

with $\kappa^2 = \|\mathbf{b} - \mathbf{b}^*\|$, *and* $\mathbf{b}_L = (\mathbf{X}'\Sigma_L^{-1}\mathbf{X})^{-1}\mathbf{X}'\Sigma_L^{-1}\mathbf{y}$.

b) *For general matrices* $\mathbf{X}'\mathbf{X}$ *and* Σ_1 *the posterior mean* $\mathbf{b}^{**}$, *the boundary of the set of posterior means, is given by the roots of the "implicit quadratic form" of Lemma* 4.3

$$\mathbf{z}[(\mathbf{b} - \mathbf{b}^{**})]\,\mathbf{D}^{-1}\,\mathbf{z}[(\mathbf{b} - \mathbf{b}^{**})] = 0. \tag{44}$$

PROOF: Rewrite the posterior mean as

$$\mathbf{b}^{**} = (\mathbf{I}_n + \mathbf{H}^*\mathbf{H}_V^{-1})^{-1}(\mathbf{b}^* + \mathbf{H}^*\mathbf{H}_V^{-1}\mathbf{b}_V),$$

and note that

$$\begin{aligned}\mathbf{H}^*\mathbf{H}_V^{-1} &= (\mathbf{I}_n + \mathbf{H}^*)\left(\mathbf{X}'(\mathbf{I}_n + \mathbf{V})^{-1}\mathbf{X}\right)^{-1} = \\ &= (\mathbf{I}_n + \mathbf{H}^*)\left(\mathbf{X}'\mathbf{X} - \mathbf{X}'(\mathbf{I}_n + \mathbf{V}^{-1})^{-1}\mathbf{X}\right) = \\ &= \mathbf{X}'\mathbf{X} + \mathbf{V}^*,\end{aligned} \tag{45}$$

where $\mathbf{V}^* = \mathbf{H}^*\mathbf{H}_V^{-1} - \mathbf{X}'\mathbf{X}$ is some arbitrary positive definite matrix depending on both, $\mathbf{H}^*$ and $\mathbf{V}$. Since $\mathbf{b}_V$ is bounded by Theorem 2.1 by an ellipsoid and $\mathbf{H}^*\mathbf{H}_V^{-1} > \mathbf{X}'\mathbf{X}$ is bounded from below, then we can apply Theorem 3.2 and Lemma 4.3.

5. CONCLUSIONS

This paper has extended the robust Bayesian approach in a ANOVA framework to a class of priors which also allows for uncertainties in the data covariance matrix. The approach can be described as joint sensitivity with respect to the covariance matrices in the first and the second stage of a normal Bayesian framework. It also opens new possibilities for heteroscedasticity diagnostics in linear Bayesian models. The following extended classes of prior distributions have been investigated:

- Fixing the prior mean and letting the prior covariance matrix vary over all positive definite matrices produces the well known feasible ellipsoid.
- By varying the prior means over a circle we can extend the feasible ellipsoid for special cases to a limaçon.
- By bounding and restricting the data matrix we could find the extreme bounds for a combined classical and Bayesian robustness analysis.

It still remains as an open question, if other approaches or different classes of prior distributions could lead to even more smaller sets of posterior means. At least for simple ANOVA set ups like the cross-over design model, we could show that an extended Bayesian robustness approach which includes heteroscedasticity diagnostics is possible.
More research in this direction will show if further classes of priors could produce similar closed form geometric shapes and analytical interpretable and also useful results.

APPENDIX: AUXILIARY LEMMAS

Lemma A.1: [Extreme Bounds Analysis (EBA)] *The extreme bounds of the linear combination* $\psi'\mathbf{b}$ *evaluated on the ellipsoid* $(\mathbf{b}-\mathbf{f})'\mathbf{A}^{-1}(\mathbf{b}-\mathbf{f}) = c$ *are given by*

$$\mathbf{b} = \mathbf{f} \pm \mathbf{A}\psi/\sqrt{c\psi'\mathbf{A}\psi}$$

for $\mathbf{A} = \mathbf{A}'$ *and of full rank or*

$$\psi'(\mathbf{b}-\mathbf{f}) = \pm\sqrt{c\psi'\mathbf{A}\psi}\,.$$

where ψ *is an arbitrary fixed vector of interest for the parameter* $\mathbf{b}$. (Typically ψ is an unit vector, since then the bounds of each coefficient can be calculated.)

PROOF: We express the problem as a Lagrange optimization problem:

$$\mathcal{L}(\mathbf{b},\lambda) = \psi'\mathbf{b} - \lambda\left[(\mathbf{b}-\mathbf{f})'\mathbf{A}^{-1}(\mathbf{b}-\mathbf{f}) - c\right],$$

and the first derivative is set to zero:

$$\frac{\partial\mathcal{L}}{\partial\mathbf{b}} = \psi - \lambda\mathbf{A}^{-1}(\mathbf{b}-\mathbf{f}) = 0\,,$$

leading to

$$(\mathbf{b}-\mathbf{f}) = \mathbf{A}\psi/\lambda\,. \qquad (46)$$

Now multiply by $(\mathbf{b}-\mathbf{f})'\mathbf{A}^{-1}$ to obtain

$$(\mathbf{b}-\mathbf{f})'\mathbf{A}^{-1}(\mathbf{b}-\mathbf{f}) = (\mathbf{b}-\mathbf{f})'\mathbf{A}^{-1}\psi/\lambda$$

Substitute (46) again to obtain

$$c = \psi'\mathbf{A}\psi/\lambda^2$$

or

$$\lambda = \pm\sqrt{c \,/\, \psi' A \psi}\,.$$

Now substituting gives

$$\mathbf{b} - \mathbf{f} = \pm \mathbf{A}\,\psi \sqrt{c \,/\, \psi' A \psi}$$

or

$$\psi' \mathbf{b} = \psi' \mathbf{f} \pm \psi' A \psi \sqrt{c \,/\, \psi' A \psi}, = \psi' \mathbf{f} \pm \sqrt{c\, \psi' A \psi}\,.$$

■

Theorem A.2: [Quadratic extreme bound analysis (QEBA)] *The extreme bounds of the ellipsoid*

$$(\mathbf{x} - \mathbf{b})' \mathbf{H} (\mathbf{x} - \mathbf{b})$$

subject to the constraint that $\mathbf{x}$ *lies on the sphere*

$$(\mathbf{x} - \mathbf{a})' (\mathbf{x} - \mathbf{a}) \leq \epsilon^2$$

are given by all the $\mathbf{x}$ *which satisfy the following equation:*

$$(\mathbf{x} - \mathbf{b})' \mathbf{H} (\mathbf{x} - \mathbf{b}) = (\mathbf{a} - \mathbf{b})' \mathbf{z} \pm \epsilon \sqrt{\mathbf{z}'\mathbf{z}}), \tag{47}$$

where $\mathbf{z} = \mathbf{H}(\mathbf{x} - \mathbf{b})$. *Or, given the inverse transformation* $\mathbf{x} = \mathbf{H}^{-1}\mathbf{z} + \mathbf{b}$, *the solution is given by the set*

$$\{ (\mathbf{H}^{-1}\mathbf{z} + \mathbf{b} \in \mathcal{R}^{K} \mid \pm \epsilon \sqrt{\mathbf{z}'\mathbf{z}}) + \mathbf{z}'(\mathbf{a} - \mathbf{b}) - \mathbf{z}' \mathbf{H}^{-1} \mathbf{z} = 0 \}.$$

PROOF: The Lagrangian function is

$$\mathcal{L}(\mathbf{x}, \lambda) = (\mathbf{x} - \mathbf{b})' \mathbf{H} (\mathbf{x} - \mathbf{b}) - \lambda [(\mathbf{x} - \mathbf{a})' (\mathbf{x} - \mathbf{a}) - \epsilon^2]$$

and the derivative is set to zero

$$\frac{\partial \mathcal{L}}{\partial \mathbf{x}} = 2\mathbf{H}(\mathbf{x} - \mathbf{b}) - 2\lambda (\mathbf{x} - \mathbf{a}) = 0,$$

which yields

$$\mathbf{H}(\mathbf{x} - \mathbf{b}) = \lambda (\mathbf{x} - \mathbf{a})\,. \tag{48}$$

Now it follows that

$$(\mathbf{x} - \mathbf{a})' \mathbf{H} (\mathbf{x} - \mathbf{b}) = \lambda (\mathbf{x} - \mathbf{a})' (\mathbf{x} - \mathbf{a}) = \lambda \epsilon^2,$$

giving $\lambda = (\mathbf{x} - \mathbf{a})' \mathbf{H} (\mathbf{x} - \mathbf{b}) / \epsilon^2$. Substituting λ into (48) gives

$$\epsilon^2 \mathbf{H}(\mathbf{x} - \mathbf{b}) = (\mathbf{x} - \mathbf{a})' \mathbf{H} (\mathbf{x} - \mathbf{b}) (\mathbf{x} - \mathbf{a}),$$

and multiplying by $(\mathbf{x} - \mathbf{b})' \mathbf{H}$ gives a "quartic curve", i.e. a quadratic of quadratic forms:

$$\epsilon^2 (\mathbf{x} - \mathbf{b})' \mathbf{H}^2 (\mathbf{x} - \mathbf{b}) = [(\mathbf{x} - \mathbf{a})' \mathbf{H} (\mathbf{x} - \mathbf{b})]^2,$$

or the roots are

$$(\mathbf{x} - \mathbf{a})' \mathbf{H} (\mathbf{x} - \mathbf{b}) = \pm \epsilon \sqrt{(\mathbf{x} - \mathbf{b})' \mathbf{H}^2 (\mathbf{x} - \mathbf{b})}\,.$$

Now $\mathbf{x} - \mathbf{a} = \mathbf{x} - \mathbf{b} + \mathbf{b} - \mathbf{a} = \mathbf{z} - (\mathbf{a} - \mathbf{b})$ and therefore

$$\mathbf{z}' \mathbf{H}^{-1} \mathbf{z} - (\mathbf{a} - \mathbf{b})' \mathbf{z} = \pm \epsilon \sqrt{\mathbf{z}' \mathbf{z}}$$

which is (47). ■

REFERENCES

[1] BERGER, J.O. (1984) *The Robust Bayesian Viewpoint.* In: *Robustness in Bayesian Statistics, [J. Kadane (ed.)]*, North-Holland, 63-144.

[2] DASGUPTA, A. & STUDDEN, W.J. (1988) *Robust Bayesian Analysis and Optimal Experimental Designs in Normal Linear Models with many Parameters - I.* Purdue University, Report 88-14.

[3] GRIZZLE, J.E. (1965) The Two-Period Cross-over Design and its Use in Clinical Trials. *Biometrics*, **21**, 467-480.

[4] GRIZZLE, J.E. (1974) Corrigenda. *Biometrics* , **30**, 727.

[5] GRIEVE, A.P. (1989) *Crossover versus Parallel Designs.* In: *Statistical Methodology in Pharmaceutical Sciences, [D. Berry (ed.)]*, Marcel Dekker, 239-270.

[6] GRIEVE, A.P. (1985) A Bayesian Analysis of the Two-Period Crossover Design for Clinical Trials. *Biometrics*, **41**, 979-990.

[7] LAWRENCE, D. (1972) *A Catalog of Special Plane Curve.* Dover.

[8] LEAMER, E.E. (1978) *Specification Searches.* John Wiley and Sons.

[9] LEAMER, E.E. (1982) Sets of Posterior Means with Bounded Variance Priors, *Econometrica*, **50**, 725-736.

[10] POLASEK, W. (1984) *Multivariate Regression Systems: Estimation and Sensitivity Analysis for Two-Dimensional Data.* In *Robustness in Bayesian Statistics, [J. Kadane (ed.)]*, North-Holland, 229-309.

[11] POLASEK, W. & POETZELBERGER, K. (1992) *Robust Bayesian Analysis for Simple ANOVA Models.* To appear in *Proceedings of Milan Conference.*

[12] SMITH, A.F.M. (1973) Bayes Estimates in One-way and Two-way Models. *Biometrika*, **60**, 319-329.

Recent Adv. in Stat. and Prob., pp. 219-232
J. Pérez Vilaplana and M. L. Puri (Eds)

Ambiguity, Imprecision and Sensitivity in Decision Theory

D. RÍOS INSÚA
Departamento de Inteligencia Artificial, Facultad de Informática, Universidad Politécnica de Madrid, Campus de Montegacedo, E-28660 Boadilla del Monte, (Madrid), España.

Abstract – Expected Utility Theory has been the cornerstone for decision making under uncertainty for almost fifty years. Numerous experiments have challenged its descriptive and normative validity leading recently to several alternative theories. This paper reviews some of these theories from the perspective of imprecision and ambiguity in judgements, providing new explanations and theories, together with a framework for sensitivity analysis in decision making.

1. INTRODUCTION

For almost fifty years, Expected Utility Theory (EU), Von Neumann and Morgenstern (1945), and Subjective Expected Utility Theory (SEU), Savage (1954), have had a main role in Decision Making under Uncertainty, having, therefore, preeminence in fields such as Economic Analysis, Decision Analysis and (Bayesian) Statistics. These theories provide a rational framework in which any decision making problem should be solved: given a certain structure of the problem and assuming certain rationality properties on the decision maker's (DM) judgements, it suggests measuring his beliefs by means of a probability distribution, his preferences and risk attitudes by means of a utility function and proposing the alternative with maximum expected utility.

These principles have been viewed as the paradigm of reasonable behaviour under uncertainty. However, some experience with applications and, specially, many laboratory experiments have shed doubts about their validity. A first reaction, from a normative point of view, would be that these show only errors of those subjects; the principles can be used as pointers to these errors, leading eventually to corrections. Several experiments report that this is not the case: many well informed people refuse correcting their choices. Paradoxes are seen as challenges to the principles, leading to new theories of decision making allowing for these inconsistencies. A third position would be to encourage additional work on the implementation of normative rules.

This paper analyses the three possibilities from the point of view of imprecision and ambiguity of the DM's judgements. We start by reviewing EU theory, some of its criticisms and explanations and some alternative theories. Then, we provide a more robust approach to decision making which may account for imprecision and might explain some of the criticisms; this theory coupled with an error model given in the last section, provides foundations for a more general approach to sensitivity analysis in Bayesian Decision Theory. From our point of view, this is the most sensible approach to decision making under uncertainty currently available.

2. EXPECTED UTILITY THEORY AND CRITICISMS

This section reviews the foundations of expected utility theory, some of its criticisms and possible counterarguments.

2.1. Foundations of Expected Utility Theory

We summarise some key results in EU theory. These will allow us to introduce the notation, and set up a basis for comparison with models introduced afterwards, specially those of section 4. Detailed expositions may be seen in Fishburn (1982) or French (1986). EU theory deals with decision making under risk. The key result is provided by Von Neumann-Morgestern axioms:

Theorem 1 *Let C be a space of consequences and $\mathcal{P}$ be the set of simple probability measures over C . Let $\preceq$ be a binary relation on $\mathcal{P}$ modelling the subject's preferences. Then, the three conditions*

- *U1. $(\mathcal{P}, \preceq)$ is a weak order,*
- *U2. For $\alpha \in (0,1), p,q,r \in \mathcal{P}$, $p \preceq q \Longrightarrow \alpha p + (1-\alpha)r \preceq \alpha q + (1-\alpha)r$,*
- *U3. For $p,q,r \in \mathcal{P}$ such that $p \preceq q \preceq r$, there is $\alpha \in [0,1]$ such that $\alpha p + (1-\alpha)r \sim q$.*

are equivalent to the existence of a(n affine unique) real function u on C such that

$$p \preceq q \iff \int u dp \leq \int u dq.$$

SEU theory deals with decision making under uncertainty. There are several paths leading to it; one is that of Anscombe and Aumann (1963). The following version is adapted from Schmeidler (1989):

Theorem 2 *Let S be the set of states with an algebra Σ, C be the space of consequences, $\mathcal{P}$ be the set of simple probability measures over C and $\mathcal{F}$ be the set of $\Sigma-$measurable functions from S to $\mathcal{P}$. Let $\preceq$ be a binary relation on $\mathcal{F}$ modelling the subject's preferences. Then, the conditions*

- *A1. $(\mathcal{F}, \preceq)$ is a weak order.*
- *A2. For $\alpha \in (0,1), f,g,h \in \mathcal{F}$, $f \preceq g \Longrightarrow \alpha f + (1-\alpha)h \preceq \alpha g + (1-\alpha)h$.*
- *A3. For $f,g,h \in \mathcal{F}$ such that $f \preceq g \preceq h$, there is $\alpha \in [0,1]$ such that $\alpha f + (1-\alpha)h \sim g$*
- *A4. For $f,g \in \mathcal{F} : f(s) \preceq g(s), \forall s \in S$ it is $f \preceq g$.*
- *A5. $\exists f,g \in \mathcal{F} : f \prec g$*

imply the existence of a(n affine unique) real function u on $\mathcal{P}$ and a unique finitely additive probability measure p on Σ such that

$$f \preceq g \iff \int u(f(.))dp \leq \int u(g(.))dq.$$

In spite of being mormatively appealing, descriptively quite accurate and prescriptively very practical, these principles have been challenged repeatedly. We shall analyse two challenges due to experiments, Allais and Ellsberg paradoxes, and one coming from applications, the issue of imprecision in judgements.

2.2. Allais paradox

One of the earliest challenges to EU theory is due to Allais (1953). His experiment presents two choice situations:

a) Choose one of these options
   (a) Win 1 M ptas for sure
   (b) Win 5 M ptas with probability .1, 1 M with probability .89, win 0 ptas with probability .01
b) Choose one of these options
   (a) Win 1 M ptas with probability .11, 0 ptas with probability .89
   (b) Win 5 M ptas with probability .1, 0 ptas with probability .9

Some subjects prefer (a) to (b) in the first case and (b) to (a) in the second one, in contradiction with EU theory.
Note that Allais experiment could be presented as follows

a) Choose one of these options
   (a) Win 1 M ptas with probability .11 and 1M ptas with probability .89
   (b) With probability .11, win a gamble with probability 10/11 of winning 5 M ptas and 1/11 of winning 0 ptas, and 1 M ptas with probability .89.
b) Choose one of these options
   (a) Win 1 M ptas with probability .11, 0 ptas with probability .89
   (b) With probability .11, win a gamble with probability 10/11 of winning 5 M ptas and 1/11 of winning 0 ptas, and 0 ptas with probability .89.

Choices between pairs of lotteries should be based on parts of the lotteries which are not common. For example, choice 1 should be based on the preferences between

(a) Win 1 M ptas for sure
(b) Win 5 M ptas with probability 10/11 and 0 ptas with probability 1/11

However, choice 2 should be based on exactly the same preferences. A similar argument was proposed by Savage (1954); in spite of this explanation, some people resist to modify their responses.
Typically, Allais paradox is said to challenge the independence principle in EU theory. But see the discussion in section 4 and in Seidenfeld (1988).

2.3. Ellsberg Paradox

Ellsberg (1961) proposed a problem concernig choices between lotteries with consequences of known probabilities and ones with consequences with partially known probabilities. One version is the following:

a) Choose one of these options
   (a) Win 100 ptas with probability 1/3 and 0 ptas with probability 2/3
   (b) Win 100 ptas with probability p and 0 ptas with probability 1-p
b) Choose one of these options
   (a) Win 100 ptas with probability $1 - p$ and 0 ptas with probability p
   (b) Win 100 ptas with probability 2/3 and 0 ptas with probability 1/3

p is not known for lotteries (1b) and (2a). Many people prefer (a) to (b) in case 1 and (b) to (a) in case 2. This is in contradiction with SEU. Raiffa (1961) provides a randomisation argument to illustrate that these individuals should find these lotteries equally desirable. However, as with Allais paradox, the argument does not convince all subjects.
The choice pattern indicated above violates Savage sure thing principle or Anscombe Aumann independence principle and, consequently, no probability measure may explain these preferences.

2.4. Imprecision and Ambiguity

Several facts motivate the recent interest in imprecision and ambiguity in Decision Theory, which actually starts with the classical risk versus uncertainty distinction.
First, there is the advent of Ellsberg type experiments in which there are comparisons between unambiguous and ambiguous (where probabilities are only partially known) lotteries. Ambiguity refers there to non determinacy of the random mechanism generating outcomes.
Second, there is a growing interest in the areas of Expert Systems and Artificial Intelligence on the topic of representation of uncertainty. Some of this work has looked at EU models and rejected them on the grounds that no judgemental input can be as precise as demanded by those. For example, if the probability of a certain event A has to be assessed, we may ask the decision maker to state a region on a probability wheel, such that he finds equally desirable the bets

(a) Win 100 ptas if A happens and 0 otherwise

(b) Win 100 ptas if spin stops at stated region and 0 otherwise

Very often the DM finds difficult to provide such region with precision.
Another possible justification, suggested in Nau (1990), is an *uncertainty principle for psychological observations*, justified by the many mode biases, heuristics and framing effects pervading psychological measurements.
To sum up, and whatever the causes are we typically end up with imprecision in probabilities and utilities, whereas the foundations of EU announce precision. This challenges the weak order assumption in EU axioms.

3. SOME ALTERNATIVE THEORIES

Section 2 presented some challenges to EU theory. A first reaction is to believe that this behaviour is irrational; we mentioned procedures to explain subjects where and how their errors were. These should react correcting their mistakes; however, this has not happened always. As a consequence, several theories have been proposed trying to solve the paradoxes. Some are normative, others are descriptive. We shall provide a review of the ideas underlying some of them.

3.1. Descriptive Theories

We shall describe briefly a number of theories which accomodate, to a certain extent, the paradoxes. Being descriptive, at most their aim would be to predict unaided choices. However, when the purpose is to aid a decision maker in conforming to some guidelines, the models provide few clues for the development of methods for decision aiding. On the other hand, some of these methods have been the basis for more formal approaches.

Non probabilistically processed probabilities Several theories assume that probabilities are processed non probabilistically by decision makers. For example, Einhorn and Hogarth (1985) suggest an anchoring and adjustment strategy for assessing probabilities: an initial assessment p_A is adjusted to reflect the ambiguity in the situation by means of the formula

$$S(p_A) = (1-\theta)p_A + \theta(1-p_A^\beta)$$

with $\theta \in [0,1]$, controlling the amount of ambiguity perceived in the situation, and $\beta \geq 0$, expressing the attitude towards ambiguity. This model may describe paradoxical behaviour in Ellsberg urns.

Probabilistically processed probabilities Other theories assume that probabilities are processed probabilistically via Bayes' theorem. An example is Viscusi's (1989) prospective reference theory, in which individuals treat stated experimental probabilities as imperfect information. For example, a lottery $L = (x_1, p_1; ...; x_n, p_n)$ is processed so that the lottery is ranked according to $\sum P(p_i)u(x_i)$, with $P(p_i) = (\gamma/n + \psi p_i)/(\gamma + \psi)$, which corresponds to procesig prior probabilities of x_i equal to $1/n$, via Bayes' theorem, if there has been a fraction p_i out of Ψ trials giving x_i under a Dirichlet model. This theory predicts Allais and Ellsberg paradoxes.

Families of probabilities with additional criteria Several resolutions model ambiguity by means of a class of probability distributions together with an additional criteria. For example, Ellsberg (1961) provides as solution of his paradox a weighted combination of the minimum expected utility and the estimated expected utility.

Modified utilities A different type of theories consider modified utilities, instead of perturbated probabilities. One such is due to Sarin and Winkler (1992). It deals with lotteries $L = (x, E; y, E^c)$; instead of utilities $u(x)$, they use modified utilities $v(u(x)|u(y)) = v(x|y)$, which try to value not only x but also the payoff y that might have been received. These are influenced by the degree of ambiguity. Consequently, L is valued according to

$$v(L) = pv(x|y) + (1-p)v(y|x)$$

By means of certain choices of v, one may explain Ellsberg paradox.

Cognitive constraints The above models assume that violations reflect information about the stucture of preferences. Another way to explore violations is to explain them as a consequence of cognitive constraints and experiential limitations. Leland (1990) proposes an approximate expected utility theory according to those lines. Suppose an individual has to choose among lotteries of the form $L = (x_1, p_1; ...; x_n, p_n)$; it is assumed that there are underlying true preferences obeying axioms, so that L is ranked according to $\sum p_i u(x_i)$. However, due to cognitive constraints, imprecise utility and probability perceptions, described by density functions $\Phi(u, x_i), \Psi(P, p_i)$ are generated for each x_i and each p_i and the individual draws specific cognitively meaningful values u', p' from these distributions, according to certain rules, so that L is actually evaluated according to $\sum p_i' u'(x_i)$. Allais paradox can be explained in this way.

3.2. Normative Theories

Alternative axiomatisations provide foundations for other Decision Theories. We consider only some allowing for deviations in Allais and/or Ellsberg experiments. Since the issue is taken with independence principle, three possible approaches are

Dropping the principle The other principles (essentially, weak order and continuity) still imply the existence of an ordinal utility function, which allows a Decision Theory. Machina (1982) provides such a model for decision making under risk. See also Machina and Schmeidler (1992) for decision making under uncertainty. Basically, they provide a general framework which allows the comparison of several theories.

Limiting the scope of the principle The idea would be to accept independence only for certain types of acts. Schmeidler (1989) limits independence to comonotonic acts, in what he calls *comonotonic independence.* Two Anscombe Aumann acts f, g are comonotonic if for no s and t in S, $f(s) \prec f(t)$ and $g(t) \prec g(s)$. Basically, substituting independence by comonotonic independence in theorem 2, we get the representation

$$f \preceq g \iff \int u(f(.))dv \leq \int u(g(.))dv$$

where v is a non additive probability and the integral is of Choquet type.

Modifying the principle Finally, several theories substitute the independence principle by a related condition. For example, Chew (1983) replaces it with the *weak substitution axiom*

- For $p, q \in \mathcal{P}, \alpha \in (0,1)$ such that $p \sim q$, there is $\beta \in (0,1)$ such that $\alpha p + (1-\alpha)r \sim \beta q + (1-\beta)r$ for any $r \in \mathcal{P}$.

in theorem 1 and gets the representation

$$p \preceq q \iff \int u(c)w(c)dp(c) \leq \int u(c)w(c)dq(c),$$

under certain conditions.

4. A MORE ROBUST APPROACH TO DECISION MAKING

In this section, we shall present further insights and a new approach to a model suggested in Rios Insua (1992) which allows us to deal with imprecision in a DM's judgements. The model essentially leads to representing beliefs by means of a class of probability distributions, preferences over consequences by means of a class of utility functions and preferences among alternatives by means of inequalities of expected utilities. As an aside, we shall see how this model may cope with paradoxes.
The first result deals with decision making under risk and holds comparison with theorem 1: we drop completeness from the preference relation and adopt a different continuity axiom. The rest of the framework is, though, exactly the same.

Theorem 3 *Let C be the space of consequences and $\mathcal{P}$ be the set of simple probability measures over C . Let $\preceq$ be a binary relation on $\mathcal{P}$ which models the subject's preferences. Then, the three conditions*

- *U1.* $(\mathcal{P}, \preceq)$ *is a quasi order.*
- *U2. For* $\alpha \in (0,1), p, q, r \in \mathcal{P}$, $p \preceq q \iff \alpha p + (1-\alpha)r \preceq \alpha q + (1-\alpha)r$.
- *U3.* $\alpha p + (1-\alpha)r \preceq \alpha q + (1-\alpha)s, \forall \alpha \in (0,1] \Longrightarrow r \preceq s$.

are equivalent to the existence of a set $\mathcal{U}$ *of real functions* u *on* C *such that*

$$p \preceq q \iff \int u dp \leq \int u dq, \forall u \in \mathcal{U}.$$

For decision making under uncertainty, we adopt the framework of Anscombe–Aumann, limited to the simple case.

Theorem 4 *Let* S *be a set of states with an algebra* Σ, P *be the set of simple probability measures over* C *and* $\mathcal{F}$ *be the set of simple* Σ*-measurable functions from* S *to* P. *Let* $\preceq$ *a binary relation modelling the DM's preferences on* $\mathcal{F}$.
Suppose that

- *P1.* $(\mathcal{F}, \preceq)$ *is a quasi order.*
- *P2. For* $\alpha \in (0,1), f, g, h \in \mathcal{F}$, $f \preceq g \iff \alpha f + (1-\alpha)h \preceq \alpha g + (1-\alpha)h$.
- *P3. For* $f, g, h, l \in \mathcal{F}$, $\alpha f + (1-\alpha)g \preceq \alpha h + (1-\alpha)l, \forall \alpha \in (0,1] \Longrightarrow g \preceq l$.
- *P4.* $\exists p, q \in P$ *such that* $f' \prec g'$, *with* $f'(s) = p, g'(s) = q, \forall s \in S$.
- *P5. For* $f, g \in \mathcal{F}$ *such that* $f(s) \preceq g(s), \forall s \in S$, $f \preceq g$.

Then, defining

$$A \preceq_\ell B \iff qI_A + pI_{A^c} \preceq qI_B + pI_{B^c}$$

for $A, B \in \Sigma$, *where* p, q *are two fixed elements of* P *verifying P4, there is a set* $\mathcal{P}$ *of probability distributions* r *such that*

$$A \preceq_\ell B \iff r(A) \leq r(B), \forall r \in \mathcal{P}$$

If, in addition,

- *P6. For* $f, g \in \mathcal{F}$, $f \preceq g \iff x(f,r) \preceq x(g,r), \forall r \in \mathcal{P}$,

where given $f \in \mathcal{F}$ *and a probability distribution* r *over* S, *we define*

$$x(f,r)(C) = \int f(s)(C) dr(s)$$

with $C \subset C$, *then there is a set* $\mathcal{U}$ *of functions* u *on* P *such that*

$$f \preceq g \iff \int u(f(s)) dr(s) \leq \int u(g(s)) dr(s), \forall r \in \mathcal{P}, \forall u \in \mathcal{U} \quad (1)$$

As announced, this model allows us to deal with the case described in section 2.4. Related results may be seen in Giron and Rios (1980) and Nau (1992).
Let us rehearse now how this model might cope with paradoxes. To start with Allais paradox, assuming that $u(5M) = 1, u(0M) = 0$, choice (a) in case 1 leads to the conclusion

$$u(1M) \; > \; .1/.11 \quad (2)$$

whereas choice (b) in case 2 leads to

$$u(1M) \; < \; .1/.11 \quad (3)$$

Suppose that the DM had imprecise preferences, as in theorem 3, modelled by $u \in \mathcal{U}$. If these constraints imply

$$u(1M) \in [a,b] \tag{4}$$

with $.1/.11 \in [a,b]$, then both choices are consistent with $u \in \mathcal{U}$. However, our model would predict no response in those experiments, a possibility not contemplated in the literature. In some informal experiments we have conducted, some subjects opted for this answer. Another interesting point would be to assess $u(1M)$ directly for individuals violating EU; in our experiments, that value was close to $.1/.11$. Therefore, a description of the paradox compatible with the model is that in the first case, the DM fixes his attention in a function compatible with (1) and, in the second one, with a function compatible with (2).

In Ellsberg example, we have that $p \in [0, 2/3]$ (not assuming additional knowledge). Choice (a) in case 1 leads to $p < 1/3$; choice (b) in case 2 leads to $p > 1/3$. Both choices are compatible with the initial imprecision; however, our model would predict no response; again, in informal experiments allowing for this response, we get some subjects answering it. Therefore, an explanation similar to that of Allais paradox seems in order.

5. MODELING ERRORS

The model in section 4 is expressed by the mapping

$$f \rightarrow \left(\int u(f(s))dr(s), u \in \mathcal{U}, r \in \mathcal{P}\right)$$

We consider now a related case in which we have assessed, additionally, $u' \in \mathcal{U}, r' \in \mathcal{P}$. We view this process as one of measurement. As such it will involve an error which we try to model. Rios Insua (1990) provides a review of ideas developed to deal with this case. Recent experimental results, Hey et al (1992), suggest that EU theory may be not so bad descriptively provided that we couple it with some error theory. This is in consonance with some of the descriptive theories mentioned in section 3.1 and the discussion in section 4.

As a consequence of viewing the elicitation process as one of measurement, we may believe that we get an estimate u', r' of u, r, distorted by an error. Therefore, we might be proposing a suboptimal alternative. Sensitivity analysis is the way to attack this problem. Here we provide an error model that coupled with that of section 4 provides a basis for such task. We simplify the notation, assuming that imprecision about utilities and probabilities is described by parameters $w \in S$, the assessed u', r' correspond to a certain estimate ω and $\Psi_f(w)$ is the expected utility of act f for a given choice of w. See Rios Insua and French (1991) for a detailed description of this representation.

5.1. One Cycle Model

Given that the estimate ω of w is subject to error, we try to find a distribution on w, after having measured ω. Let us decompose $\omega = w + v$, where w is the parameter and v the measurement error. Prior to the elicitation, the DM has beliefs about $w \in S$, measured by a density $\pi_0(w)$, and about v, measured by an error density $f_1(v|w)$. Their joint density is

$$f(v, w) = \pi_0(w) f_1(v|w).$$

Standard computations give

$$\pi_0(w|\omega) \propto \pi_0(w) f_1(\omega - w|w).$$

We may use this distribution to analyse unaided pairwise comparisons of alternatives. Two cases of interest might be choices in laboratory experiments or that in which we have an act f_J presented as a competitor of f_*, which is optimal for ω.
A natural way to do this is via hypothesis testing. We shall concentrate on studying the consistency of comparisons between acts f_* and f_J, that is, we want to test:

- $H_0 : f_J \preceq f_*$
- $H_1 : f_* \prec f_J$

Let us see the expressions of some tests, depending on the loss function considered, with $\Psi_*(w), \Psi_J(w)$ designating the expected utilities of a_* and a_J.

- **0-1 loss** The criteria is: reject H_0 if

$$P_{w|\omega}\{\Psi_*(w) < \Psi_J(w)\} > 1/2,$$

- **Loss of optimality** Suppose

$$\begin{aligned} L(w, a_*) &= \Psi_J(w) - \Psi_*(w) \\ L(w, a_J) &= \Psi_*(w) - \Psi_J(w). \end{aligned}$$

  Then, we reject H_0 if

$$E(\Psi_J - \Psi_*) > E(\Psi_* - \Psi_J),$$

  or

$$E(\Psi_J) > E(\Psi_*).$$

- **Regret loss** Suppose

$$\begin{aligned} L(w, a_*) &= \max(\Psi_*(w), \Psi_J(w)) - \Psi_*(w) \\ &= \begin{cases} 0 & \text{if } \Psi_*(w) \geq \Psi_J(w) \\ \Psi_J(w) - \Psi_*(w) & \text{otherwise} \end{cases} \end{aligned}$$

  Similarly for $L(w, a_J)$. After simple computations, we see that we reject H_0 if

$$E(\Psi_J) > E(\Psi_*).$$

- **Hierarchical loss of optimality** If U is a second order utility, defining

$$\begin{aligned} L(w, a_*) &= U(\Psi_J(w)) - U(\Psi_*(w)) \\ L(w, a_J) &= U(\Psi_*(w)) - U(\Psi_J(w)) \end{aligned}$$

  the criteria is: reject H_0 if

$$E(U(\Psi_J)) > E(U(\Psi_*))$$

The expression of these tests may be further specified in some important cases. Suppose $S \subset \mathbf{R}^n$, the pertinent random variable is ξ and let $\mu = E(w|\omega)$ and $V = var(w|\omega)$. When we use regret loss:

- **Linear case.** Suppose $\Psi_j(w) = c_j'w$. Then, we reject H_0 if

$$(c_J - c_*)'\mu > 0.$$

- **Bilinear case.** Suppose $w = (w_1, w_2)$, $\mu = (\mu_1, \mu_2)$ and

$$\Psi_l(w) = w_1' B_l w_2,$$

with $B_l = (b_{ij}^l)$, $w_1 = (w_i)$, $w_2 = (w_j)$. We reject H_0 if

$$\sum (b_{ij}^J - b_{ij}^*)' v_{ij} + \mu_1'(B_J - B_*)\mu_2 > 0.$$

- **General case.** Try approximate $\Psi_j(w)$ by its second order Taylor expansion around μ

$$\Psi_j(w) \simeq \Psi_j(\mu) + (w-\mu)'\nabla\Psi_j(\mu) + .5(w-\mu)H_j(\mu)(w-\mu),$$

where $H_j(\mu)$ is the Hessian of $\Psi_j(w)$, evaluated at $w = \mu$. We reject H_0 (according to the approximation) if

$$\Psi_J(\mu) - \Psi_*(\mu) + .5\ tr((H_J(\mu) - H_*(\mu))V) > 0.$$

When we use 0-1 loss:

- **Linear case.** We reject H_0 when

$$P_\xi(c_*'w \leq c_J'w) = P_\xi((c_* - c_J)'w \leq 0) = F_{(c_* - c_J)'\xi}(0) > 1/2.$$

- **Bilinear case.** We reject H_0 when

$$F_{\xi_1'(B_* - B_J)\xi_2}(0) > 1/2.$$

- **General case.** We reject H_0 when

$$F_{(\Psi_* - \Psi_J)(\xi)}(0) > 1/2.$$

5.2. Several Cycles Model

Since we conceive of the decision making process cyclically, we may modify the inputs ω, S to the analysis. Let us extend the previous approach to several cycles. We assume first that S remains unchanged and, then, give a solution when S changes.
At the second cycle, the prior distribution over w is

$$\pi_1(w) = \pi_0(w|\omega) \propto \pi_0(w) f_1(\omega - w|w).$$

We measure $\omega' = w + v'$, where v' is the error. The distribution over v' depends possibly on v and w and we represent it with a density $f_2(v'|v, w)$. Then, we find that

$$\pi_1(w|\omega') \propto \pi_1(w) f_2(\omega' - w|\omega - w, w).$$

The formula is extended to several cycles.
When S changes from one cycle to the next, we shall assume that $S_{n+1} \subset S_n$. In this case, it is a consequence of Bayes' theorem that if f was the density over w after the n-th

cycle and we know that the $(n+1)$-th support set is S_{n+1}, our prior density for this cycle is proportional to

$$f(w)I_{S_{n+1}}(w).$$

Summing up, if π_{i-1} represents the prior, S_{i-1} the support, ω_i the measurement and $f_i(.|w, v_1, \ldots, v_{i-1})$ the error density at the i-th cycle, $i = 1, \ldots, n+1$, we have

$$\pi_n(w) \propto \pi_0(w)\Pi_{i=1}^{n-1} f_i(\omega_i - w|w, \omega_1 - w, \ldots, \omega_{i-1} - w)I_{S_n}(w)$$

and

$$\pi_n(w|\omega_n) \propto \pi_0(w)\Pi_{i=1}^{n} f_i(\omega_i - w|w, \omega_1 - w, \ldots, \omega_{i-1} - w)I_{S_n}(w).$$

It may be reasonable to assume a sort of Markovian property for the errors such as

$$f_i(.|w, v_1, \ldots, v_{i-1}) = f_i(v_i|w, v_{i-1})$$

Even the v_i's may be judged independent given w, so that

$$f_i(.|w, v_1, \ldots, v_{i-1}) = f_i(v_i|w)$$

Sometimes we can assume that the v_i's are independent of w

$$f_i(.|w, v_1, \ldots, v_{i-1}) = f_i(v_i).$$

The classic error model assumes that the v_i's are i.i.d. and $S = \mathbf{R}^n$. Then, the formula is

$$\pi_n(w|\omega_n) \propto \pi_0(w)\Pi_{i=1}^{n} f(\omega_i - w).$$

5.3. An Example

Let us consider a particular case when $S \subset \mathbf{R}^n$.
For the distribution of w, since we only know that $w \in S$, we may justify the adoption of a uniform distribution over S appealing to invariance arguments and uniqueness properties of the parameters assessed. Following the physical tradition, we shall assume normal errors: due to a big number of sources of errors, the error distribution may be assumed normal.
The analysis is based on these well-known results:

Lemma 1 *Suppose* $X|\theta \sim N(\theta, \Sigma)$ *and* $\theta \sim N(\mu, \Sigma_0)$. *Then*

$$\int f(x|\theta)\pi(\theta)d\theta \propto \exp(-.5(x-\mu)'(\Sigma + \Sigma_0)^{-1}(x-\mu)).$$

Lemma 2

$$\Pi_{i=1}^{n} \exp(-.5(\theta - \mu_i)'\Sigma_i^{-1}(\theta - \mu_i)) \propto \exp(-.5(\theta - T_n Z_n^{-1})Z_n(\theta - T_n Z_n^{-1}))$$

with

- $Z_n = Z_{n-1} + \Sigma_n^{-1}, \quad Z_0 = O$
- $T_n = T_{n-1} + \mu_n \Sigma_n^{-1}, \quad T_0 = \mathbf{0}$

where O is a matrix of zeroes.
Consider the case in which $v_i \sim N(\mu_i, \Sigma_i)$, then

$$\pi_n(w|\omega_n) \propto I_{S_n}(w)\Pi_{i=1}^n \exp(-.5(\omega_i - w - \mu_i)'\Sigma_i^{-1}(\omega_i - w - \mu_i)) =$$
$$= I_{S_n}(w)\Pi_{i=1}^n \exp(-.5(w - (\omega_i - \mu_i))'\Sigma_i^{-1}(w - (\omega_i - \mu_i))) \propto$$
$$\propto I_{S_n}(w)\exp(-.5(w - Z_n^{-1}T_n)'Z_n(w - Z_n^{-1}T_n))$$

with Z_n and T_n as in lemma 2, with $(\omega_n - \mu_n)$ in place of μ_n. Similar results hold in other cases.
The main problem here is computational. We shall study the simple case in which

$$\pi_n(w|\omega_n) \propto I_{S_n}(w)\exp(-.5(w - \hat{\mu}_n)'\Sigma_n^{-1}(w - \hat{\mu}_n)),$$

where Σ_n is diagonal and $S_n = \Pi_{i=1}^t[e_i, f_i]$. In this case, the exact posterior density is

$$\pi_n(w|\omega_n) = I_{S_n}(w)1/(2\pi)^{t/2}\Pi_{i=1}^t \frac{\exp(-.5((w_i - \hat{\mu}_i)/\sigma_i)^2)}{\sigma_i(\Phi((f_i - \hat{\mu}_i)/\sigma_i) - \Phi((e_i - \hat{\mu}_i)/\sigma_i))}.$$

The mean vector is

$$\bar{\mu} = \hat{\mu} + \Sigma_n h,$$

with

$$h_i = \frac{\phi((e_i - \hat{\mu}_i)/\sigma_i) - \phi((f_i - \hat{\mu}_i)/\sigma_i)}{\Phi((f_i - \hat{\mu}_i)/\sigma_i) - \phi((e_i - \hat{\mu}_i)/\sigma_i)}$$

and the matrix V of variances is diagonal with diagonal elements computed easily.
Then, if we use the regret loss:

- **Linear case.** We reject the optimality of a_* if

$$(c_J - c_*)'(\hat{\mu} + \Sigma_n h) > 0.$$

- **Bilinear case.** We reject the optimality of a_* if

$$(\hat{\mu}_1 + \Sigma_{n1}h_1)'(B_J - B_*)(\hat{\mu}_2 + \Sigma_{n2}h_2) > 0.$$

- **General case.** We reject the optimality of a_* if

$$(\Psi_J(\bar{\mu}) - \Psi_*(\bar{\mu})) + .5\Sigma(h_J^{ii}(\bar{\mu}) - h_*^{ii}(\bar{\mu}))v_{ii} > 0.$$

If we use a 0-1 loss, we have to apply numerical integration. Similarly in more complicated cases.

6. CONCLUSIONS

We have introduced a more robust approach to decision making allowing for imprecision in a DM's judgements. The model is essentially coherent since its basic implication is to conduct a family of coherent analyses and base conclusions on their common grounds. We have suggested how this model might explain some paradoxes in experiments, specially if we couple it with the error model proposed.
These two models provide a framework for sensitivity analysis in decision making. The first one allows us to represent the information which the DM is sure about and to discard acts which are definitely worse than others. The second model helps us to analyse possible assessments of utilities and probabilities and eventual pairwise comparisons of acts. Together, and implemented cyclically, provide a sensible, coherent and robust approach to decision making.

Acknowledgments This work has been supported by DGICYT Project PS91-0172 and grants from NATO and UPM. We are grateful to discussions with Sixto Rios and Simon French. A final version was completed while visiting ISDS at Duke University under a MEC grant.

REFERENCES

[1] ANSCOMBE, F. & AUMANN, R. (1963) A Definition of Subjective Probability, *The Annals of Mathematical Statistics*, **34**, 199-205.

[2] ALLAIS, M. (1953) Le Comportement de l'Homme Rational devant le Risque: Critique des Postulats et Axiomes de l'Ecole Americaine, *Econometrica*, **21**, 503-546.

[3] CHEW, S. H. (1983) A Generalization of the Quasilinear Mean with Applications to the Measurement of Income Inequality and Decision Theory Resolving the Allais Paradox, *Econometrica*, **51**, 1065-1092.

[4] EINHORN, H. & HOGARTH, R. (1985) Ambiguity and Uncertainty in Probabilistic Inference, *Psych. Rev.*, **92**, 433-461.

[5] ELLSBERG, D. (1961) Risk, Ambiguity and Savage Axioms, *Quarterly Journal of Economics*, **75**, 643-669.

[6] FISHBURN, P. (1982) *Foundations of Expected Utility*, Kluwer.

[7] FRENCH, S. (1986) *Decision Theory*, Ellis Horwood.

[8] GIRON, F.J. & RIOS, S. (1980) *Quasi Bayesian Behaviour: A More Realistic Approach to Decision Making?*. In: *Bayesian Statistics 3*, [[J.M. Bernardo, M.H. DeGroot, D.V. Lindley and A.F.M. Smith (eds.)], Valencia U.P.

[9] HEY, J., & ORME, C. (1992) *Circles, Triangles and Straight Lines.* Paper presented at FUR VI, Paris.

[10] LELAND, J. (1990) *An Approximate Expected Utility Theory*, Carnegie Mellon University, Dept. of Social and Decision Sciences, Working Paper.

[11] MACHINA, M. (1982) Expected Utility Analysis without the Independence Axiom, *Econometrica*, **50**, 277-323.

[12] MACHINA, M. & SCHMEIDLER, D. (1992) A More Robust Definition of Subjective Probability, *Econometrica*, **60**, 745-780.

[13] NAU, R. (1990) *Indeterminate Probabilities and Utilities on Finite Sets.* Fuqua Working Paper 8929, Duke University.

[14] NAU, R. (1992) Indeterminate Probabilities on Finite Sets, *Annals of Statistics*, **20**, 1737-1767.

[15] RAIFFA, H. (1961) Risk, Ambiguity, and the Savage Axioms: Comment, *Quarterly Journal of Economics*, **75**, 690-694.

[16] RIOS INSUA, D. (1990) *Sensitivity Analysis in Multi-objective Decision Making.* Springer.

[17] RIOS INSUA, D. (1992) Foundations for a Robust Theory of Decision Making: The Simple Case, *TEST*, **1**, 69-78.

[18] RIOS INSUA, D. & FRENCH, S. (1991) A Framework for Sensitivity Analysis in Discrete Multiobjective Decision Making, *EJOR*, **54**, 176-190.

[19] SARIN, R. & WINKLER, R. (1992) Ambiguity and Decision Modeling: A Preference Based Approach, *Journal of Risk and Uncertainty*, **5**, 389-406.

[20] SAVAGE, L. (1954) *The Foundations of Statistics.* John Wiley and Sons.

[21] SCHMEIDLER, D. (1989) Subjective Probability and Expected Utility without Additivity, *Econometrica*, **3**, 571-587.

[22] SEIDENFELD, T. (1988) Decision Theory without 'Independence' or without 'Ordering'?, *Economics and Philosophy*, **4**, 287-315.

[23] VISCUSI, W. (1989) Prospective Reference Theory: Toward an Explanation of the Paradoxes, *Jour. Risk Uncer.*, **2**, 235-264.

[24] VON NEUMANN, J. & MORGESTERN, O. (1945) *Theory of Games and Economic Behaviour*, Princeton University Press.

Recent Adv. in Stat. and Prob., pp. 233-243
J. Pérez Vilaplana and M. L. Puri (Eds)

Local and Global Sensitivity Under Some Classes of Priors

F. RUGGERI
CNR-IAMI, Via A.M. Ampère 56, I-20131 Milano, Italy

Abstract – In Bayesian robustness, we can distinguish between global and local sensitivity analysis. The former is concerned with computing bounds on posterior quantities as the prior varies over a large class; the latter is concerned with quantifying the effect of small changes in the elicited prior on the posterior quantities. Different techniques have been developed to perform such analyses and some of them are briefly presented here. This paper is mainly interested in both global and local sensitivity analysis for classes of prior probability measures specified by means of the concentration function. The effects of small changes either in the likelihood function or in the loss function are then discussed, allowing for a more general approach to local sensitivity.

1. INTRODUCTION

In Bayesian analysis, a single prior probability measure is usually chosen from among those compatible with the prior knowledge, but the choice could heavily affect the inference about the parameters.

According to robust Bayesian methodology, the uncertainty in the priors can be modeled by using a class Γ of probability measures, while the ranges of posterior functions of interest (e.g. probability of sets, mean, variance and loss function) are usually studied as the prior varies in Γ. This range gives a criterion for discriminating among the priors. A small range means that the results are not meaningfully different, ensuring robustness with respect to the prior probability measure. Depending on the problem, different classes Γ of probability measures may be considered. Furthermore, the class Γ should be easily specified, large enough to contain any plausible prior and small enough to exclude unnatural priors, and involve relatively simple computations.

The study of bounds on posterior quantities over a class of priors is usually referred to as global sensitivity analysis. Such an approach has received much attention; see Berger (1984, 1985, 1990), Wasserman (1992) and the references contained therein.

Recently, Bayesian robustness studies have considered also local sensitivity analyses, in which posterior effects of infinitesimal departures from an elicited prior are studied. Extending work by Diaconis and Freedman (1986), Ruggeri and Wasserman (1993) analysed some ε–contamination classes of priors, considering the norm of the Fréchet derivative of the posterior expectation of a function with respect to the prior over the class of all signed measures with zero mass. Otherwise, the analysis is performed considering the derivatives, with respect to a parameter ε, of the bounds on the posterior expectation over

classes Γ_ϵ shrinking towards a given prior probability measure. Ruggeri and Wasserman (1991a) considered the density bounded and the density ratio classes, while this paper deals with the total variation class and provides a more general framework in which local sensitivity analysis is performed. Based on results by Fortini and Ruggeri (1992) on the possibility of defining neighbourhoods of a given probability measure through the concentration function, defined by Cifarelli and Regazzini (1987), classes of priors are obtained by bounding the probability of any measurable subset.
In Section 2, the main results about the concentration function are presented. In Section 3, global sensitivity analysis will be briefly reviewed while Section 4 will present the local analysis. Some concluding remarks are contained in Section 5 about infinitesimal changes either in the likelihood or in the loss function.

2. CONCENTRATION FUNCTION

Cifarelli and Regazzini (1987) defined the concentration function (c.f.), generalising the classical notion of Lorenz concentration curve (also known as the Lorenz-Gini curve), which is well described in Marshall and Olkin (1979, pag.5). The c.f. allows for the comparison of two probability measures Q and P defined on the same measurable space $(\Omega, \mathcal{F})$, looking for subsets where Q is much more concentrated than P (and vice versa). Such a concentration can be expressed by means of a nondecreasing, continuous and convex function $\varphi : [0,1] \to [0,1]$ (the c.f.) and its distance from the straight line connecting the points $(0,0)$ and $(1,1)$, corresponding to the case $Q = P$.
According to the Radon-Nikodym theorem, there is a unique partition $\{N, N^C\} \subset \mathcal{F}$ of Ω and a nonnegative function h on N^C such that $Q(E) = \int_{E\cap N^C} h(\omega)P(d\omega) + Q_s(E \cap N)$, $\forall E \in \mathcal{F}$, $P(N) = 0$, $Q_s(N) = Q_s(\Omega)$, where $Q_a(\cdot) = \int_{\cdot\cap N^C} h(\omega)P(d\omega)$ and Q_s denote the absolutely continuous and the singular part of Q w.r.t. P, respectively. Set $h(\omega) = \infty, \forall\omega \in N$, and define $H(y) = P(\{\omega \in \Omega : h(\omega) \leq y\})$, $c(x) = \inf\{y \in \Re : H(y) \geq x\}$. Finally, let $L(x) = \{\omega \in \Omega : h(\omega) \leq c(x)\}$ and $L^-(x) = \{\omega \in \Omega : h(\omega) < c(x)\}$.

Definition 1 *The function $\varphi : [0,1] \to [0,1]$ is said to be the concentration function of Q w.r.t. P if $\varphi(x) = Q_a(L^-(x)) + c(x)\{x - H(c(x)^-)\}$ for $x \in (0,1)$, $\varphi(0) = 0$ and $\varphi(1) = Q_a(\Omega)$.*

When the dependence on Q is to be emphasized, the c.f. of Q w.r.t. P will be denoted $\varphi_Q(x)$. Furthermore, $\varphi(x)$ is a nondecreasing, continuous and convex function such that $\varphi(x) = \int_0^{c(x)} \{x - H(t)\}\, dt = \int_0^x c(t)dt$ and $\varphi(x) \equiv 0 \Leftrightarrow Q \perp P, \varphi(x) = x\ \forall x \in [0,1] \Leftrightarrow Q = P$.
Under P non-atomic, Theorem 2.2 in Cifarelli and Regazzini (1987) becomes

Theorem 1 *Given any $x \in [0,1]$, there exists B_x such that $P(B_x) = x$ and $\varphi(x) = Q_a(B_x) = \min\{Q(A) : A \in \mathcal{F}$ and $P(A) \geq x\}$.*

It follows that the c.f. provides bounds on the probability of any measurable subset A, because $\varphi(x) \leq Q(A) \leq 1 - \varphi(1-x)$ if $P(A) = x$. Given a nondecreasing, convex function γ, with $\gamma(0) = 0$ and $\gamma(1) = 1$, Theorem 1 allows to express γ-neighbourhoods of a non-atomic measure P, i.e. $K_\gamma = \{Q : \varphi_Q(x) \geq \gamma(x), \forall x \in [0,1]\}$.
As proved in Fortini and Ruggeri (1992), every probability measure in K_γ can be represented as a mixture of probability measures in $E_\gamma = \{Q : \varphi_Q(x) = \gamma(x), \forall x \in [0,1]\}$ and bounds on expectations are computed using the next Theorems.

Theorem 2 *Let f and g be real-valued functions on Ω such that $\int_\Omega |f(\omega)|Q(d\omega) < \infty$ and $0 < \int_\Omega g(\omega)Q(d\omega) < \infty$ for any $Q \in K_\gamma$. Then*

$$\sup_{Q\in K_\gamma} \frac{\int_\Omega f(\omega)Q(d\omega)}{\int_\Omega g(\omega)Q(d\omega)} = \sup_{Q\in E_\gamma} \frac{\int_\Omega f(\omega)Q(d\omega)}{\int_\Omega g(\omega)Q(d\omega)}.$$

The same result holds with "sup" replaced by "inf".

Theorem 3 *Let $H_f(y) = P\{\omega \in \Omega : f(\omega) \le y\}$, $c_f(x) = \inf\{y : H_f(y) \ge x\}$. Then*

$$\sup_{Q\in K_\gamma} \int_\Omega f(\omega)Q(d\omega) = \int_0^1 c_f(x)c(x)\,dx$$

$$\inf_{Q\in K_\gamma} \int_\Omega f(\omega)Q(d\omega) = -\int_0^1 c_{-f}(x)c(x)\,dx,$$

where $c(x) = \gamma'(x)$ a.e.

The last Theorem is well suited for computing prior bounds but it can be applied also to find bounds on posterior expectations, using the linearisation technique presented by Lavine (1988, 1991a, 1991b).

The next Corollary provides bounds on the prior and posterior probability of any subset A, $Q(A)$ and $Q(A|x)$ respectively.

Corollary 1

$$\sup_{Q\in K_\gamma} Q(A) = 1 - \gamma(P(A^C)) \text{ and } \inf_{Q\in K_\gamma} Q(A) = \gamma(P(A));$$

$$\sup_{Q\in K_\gamma} Q(A|x) = \left\{ 1 + \frac{-\int_0^{P(A^C)} c_{-l_1}(x)c(x)dx}{\int_{P(A^C)}^1 c_{l_2}(x)c(x)dx} \right\}^{-1}$$

and

$$\inf_{Q\in K_\gamma} Q(A|x) = \left\{ 1 + \frac{\int_{P(A)}^1 c_{l_1}(x)c(x)dx}{-\int_0^{P(A)} c_{-l_2}(x)c(x)dx} \right\}^{-1},$$

where I_A is the indicator function of the subset A, $l_x(\omega)$ is the likelihood function and, for any $\omega \in \Omega$, $l_1(\omega) = l_x(\omega)I_{A^C}(\omega)$ and $l_2(\omega) = l_x(\omega)I_A(\omega)$.

PROOF. Bounds on prior probabilities follow from Theorem 3, taking $f \equiv I_A$. Then,

$$H_f(y) = \begin{cases} 0 & y < 0 \\ P(A^C) & 0 \le y < 1 \\ 1 & y \ge 1 \end{cases} \quad \text{and} \quad c_f(x) = \begin{cases} 0 & 0 \le x \le P(A^C) \\ 1 & P(A^C) < x \le 1 \end{cases};$$

$$H_{-f}(y) = \begin{cases} 0 & y < -1 \\ P(A) & -1 \le y < 0 \\ 1 & y \ge 0 \end{cases} \quad \text{and} \quad c_{-f}(x) = \begin{cases} -1 & 0 \le x \le P(A) \\ 0 & P(A) < x \le 1 \end{cases}.$$

About the posterior probabilities, it is simple to show (in the supremum case) that

$$\sup_{Q\in\Gamma} Q(A|x) \le \left\{ 1 + \frac{\inf_{Q\in\Gamma} \int_{A^C} l_x(\omega)Q(d\omega)}{\sup_{Q\in\Gamma} \int_A l_x(\omega)Q(d\omega)} \right\}^{-1}. \tag{1}$$

We need to prove that there exists a measure $\hat{Q} \in E_\gamma$ which gives both infimum and supremum in the right hand side of (1), so that the inequality in (1) becomes an equality. Observe that $c_{-l_1}(x) = 0$ for $P(A^C) \leq x \leq 1$ and $c_{l_2} = 0$ for $0 \leq x < P(A^C)$. Then, $\inf_{Q \in K_\gamma} \int_{A^C} l_x(\omega)Q(d\omega) = -\int_0^{P(A^C)} c_{-l_1}(x)c(x)dx$ and $\sup_{Q \in K_\gamma} \int_A l_x(\omega)Q(d\omega) = \int_{P(A^C)}^1 c_{l_2}(x)c(x)dx$.

Like in the proof of Theorem 3 (see Fortini and Ruggeri, 1992), we can find the maximising $\hat{Q} \in E_\gamma$ which gives the smallest possible mass to A^C (corresponding to $L(P(A^C)) = A^C$) and, besides, less mass where the likelihood is higher (in A^C) or smaller (in A). ∎

3. GLOBAL SENSITIVITY ANALYSIS

Most of Bayesian robustness research deals with global sensitivity analysis: the elicitation process ends with specifying just some features (e.g. quantiles) of the prior measure so that all the measures sharing them are considered. Given a class Γ of prior measures, then ranges of prior and posterior functionals are the object of study. Berger (1990) gives a broad review of the most important classes: conjugate, approximately specified moments, neighbourhood (mainly, ε–contaminations), sub-sigma field (mainly, quantile class) and density ratio class (DeRobertis and Hartigan, 1981).

One of the most important contribution is due to Lavine (1991a, 1991b) who defined the density bounded class. Among the other recent contributions, we just mention the γ–neighbourhoods, described in Section 2, and the classes specified either by bounds on the probability of a set (Ruggeri, 1991a, 1991b and 1992) or by the generalised moments conditions (Betrò, Męczarski and Ruggeri, 1992). Global sensitivity techniques will be presented for this last class.

Let X be a random variable on a dominated statistical space denoted by $(\mathcal{X}, \mathcal{F}_\mathcal{X}, \{P_\omega, \omega \in (\Omega, \mathcal{F})\})$; given a sample x from X, the experimental evidence about ω will be expressed by the likelihood function $l_x(\omega)$, which we assume $\mathcal{F}_\mathcal{X} \otimes \mathcal{F}$-measurable. Let $\mathcal{P}$ denote the space of probability measures on the parameter space $(\Omega, \mathcal{F})$, Ω being a Polish space.

Suppose that the elicitation process enables us to specify just some constraints on the generalised moments of the prior measure, i.e. we consider compatible with our prior knowledge all the priors in

$$\Gamma = \{P \in \mathcal{P} : \int_\Omega H_i(\omega)P(d\omega) \leq \alpha_i,\ i = 1, ..., n\}$$

where H_i are P-integrable functions and α_i, $i = 1, ..., n$, are fixed real numbers. Global sensitivity analysis is interested in the range of the posterior functional

$$T_g(P) = \frac{\int_\Omega g(\omega)l_x(\omega)P(d\omega)}{\int_\Omega l_x(\omega)P(d\omega)}, \tag{2}$$

where g is a given function on Ω such that T_g is well defined.

It often turns out that extrema of posterior functionals are attained at a smaller class, typically, the extreme points of the original class Γ, as in Betrò, Męczarski and Ruggeri (1992) who proved, under generalised moments conditions, that

$$\sup_{P \in \Gamma} T_g(P) = \sup_{(\omega, \mathbf{p}) \in T} \frac{\sum_{i=1}^{n+1} g(\omega_i)l_x(\omega_i)p_i}{\sum_{i=1}^{n+1} l_x(\omega_i)p_i},$$

where $\omega = (\omega_1, ..., \omega_{n+1})'$, $\mathbf{p} = (p_1, ..., p_{n+1})'$, such that $\sum_{i=1}^{n+1} p_i = 1$ and

$$T(\subset \Omega^{n+1} \times [0,1]^{n+1}) = \{(\omega, \mathbf{p}) : (H_i(\omega_1), ..., H_i(\omega_{n+1}))\mathbf{p} \leq \alpha_i, \quad i = 1, ..., n\}.$$

Besides, they applied such a result to a new class of priors, defined by conditions on the marginal distributions of data.

4. LOCAL SENSITIVITY ANALYSIS

Suppose that a unique prior P is elicited, but we allow for small changes in the concentration of the probability, due to errors in the elicitation process. Hence, we consider measures which are "functionally close" to P and study the behaviour of prior or posterior functionals, under infinitesimal departures from P.

Recent work on local sensitivity considers either the prior probability measure embedded in an ϵ-contaminated class of priors, computing Fréchet derivatives of posterior functionals (Diaconis and Freedman 1986, Berger 1986, Cuevas and Sanz 1988, Ruggeri and Wasserman 1993, Srinivasan and Truszczynska 1990), or density based classes of priors, computing the derivative of the supremum (and the infimum) of functionals over classes of priors shrinking towards a given prior probability measure (Ruggeri and Wasserman, 1991a).

4.1. Fréchet Derivatives.

Consider a non-atomic measure P and the posterior expectation $T_g(P) = \frac{N_P}{D_P}$, defined in (2), where $N_P = \int_\Omega g(\omega) l_x(\omega) P(d\omega)$ and $D_P = \int_\Omega l_x(\omega) P(d\omega)$.

Fixed $P \in \mathcal{P}$, the posterior expectation is computed for the measures in $\Delta_{\mathcal{M}}(P) = \{\pi; \pi = P + \delta, \delta \in \mathcal{M}\}$ where $\mathcal{M}$ is a subset of Δ, which contains all signed measure δ with $\delta(\Omega) = 0$. Note that Δ is a normed, linear space with norm given by $||\delta|| = d(\delta, 0)$ where $d(P,Q) = \sup_{A \in \mathcal{B}(\Omega)} |P(A) - Q(A)|$ is the total variation metric. We regard T_g as a non-linear operator taking Δ into $\Re$.

Let $\dot{T}_g^P$ be the Frechet derivative of the operator T_g. In other words, $\dot{T}_g^P$ is a linear map on Δ that satisfies $\dot{T}_g^P(\delta) = T_g(P + \delta) - T_g(P) + o(||\delta||)$. We regard $\dot{T}_g^P$ as a measure of how a small change in P affects the posterior expectation.

The norm of $\dot{T}_g^P$ over $\mathcal{M}$ is defined by $||\dot{T}_g^P||_{\mathcal{M}} = \sup_{\delta \in \mathcal{M}} |\dot{T}_g^P| / ||\delta||$ and it is assumed as a measure of the sensitivity of the posterior expectation as the prior varies in $\mathcal{M}$. Let $\Delta_{\mathcal{P}}$ be the set of all signed measures $\delta = \varepsilon(Q - P)$, where $Q \in \mathcal{P}$ and $\varepsilon \in (0, 1]$; considering only discrete Q's, then the set will be denoted $\Delta_{\hat{\mathcal{P}}}$. Under such choices of $\mathcal{M}$, it follows that $\Delta_{\mathcal{M}}(P) = \{\pi; \pi = (1 - \varepsilon)P + \varepsilon Q\}$ is a an ε-contamination class of priors, with Q either any probability measure or any discrete one. Some results in Ruggeri and Wasserman (1993) are now presented.

Theorem 4 $\dot{T}_g^P(\delta) = \{D_P\}^{-1}(N_\delta - \rho D_\delta)$ *where* $\rho = T_g(P)$.

Define $h(\omega) = l_x(\omega)(g(\omega) - \rho)$, $\overline{h} = \sup_{\omega \in \Omega} h(\omega)$, $\underline{h} = \inf_{\omega \in \Omega} h(\omega)$.

Theorem 5

$$||\dot{T}_g^P||_{\Delta_{\mathcal{P}}} = \{D_P\}^{-1}\{\overline{h} - \underline{h}\} \qquad \text{and} \qquad ||\dot{T}_g^P||_{\Delta_{\hat{\mathcal{P}}}} = \{D_P\}^{-1} \max\{\overline{h}, -\underline{h}\}.$$

It is worth remarking that, in most Bayesian robustness calculations, the restriction to discrete distributions does not affect the result so it is interesting to see that the norm of the derivative changes.

4.2. Derivatives of Extrema.

Fréchet derivatives are appealing from a mathematical point of view, since their norms are very meaningful in a functional analytic framework. As discussed in Ruggeri and Wasserman (1993), Fréchet derivatives are not well suited for analysing local sensitivity in some restricted classes (e.g. ε–contaminations under arbitrary classes of contaminating measures). Derivatives of suprema (and infima) of functionals over classes shrinking to P are a very intuitive tool and rather easy to compute. Ruggeri and Wasserman (1993, 1991a) considered, respectively, the ε–contaminations and special cases of density bounded classes, defined by Lavine (1991a, 1991b), and density ratio classes, due to DeRobertis and Hartigan (1981).

Let $E(g)$ and $E(g|x)$ denote, respectively, the prior and posterior expectations of a function g and $\overline{E}_\varepsilon(g)$ and $\overline{E}_\varepsilon(g|x)$ be their maximum values over a class Γ_ε ($\underline{E}$ denotes minima). Some reasons for justifying the interest in prior quantities are given in Wasserman (1992); we just mention that they are useful to check if the elicitation process has lead to reasonable priors.

While the global sensitivity analysis is mainly interested in computing ranges of functionals, it seems interesting to perform the local analysis considering the ranges of the derivatives of suprema and infima of functionals. Modifying slightly the definition in Ruggeri and Wasserman (1991a), consider the following

Definition 2 *Consider a family $\{K_\varepsilon\}$, $\varepsilon \geq \varepsilon_0$, of neighbourhoods of a probability measure P such that $K_{\varepsilon_0} = \{P\}$ and $K_{\varepsilon_1} \supset K_{\varepsilon_2} \Longleftrightarrow \varepsilon_1 > \varepsilon_2$. Given a measurable function g, we define the prior local sensitivity for the prior expectation $E(g)$:*

$$\rho(g) = D(g) - C(g),$$

when there exist the derivatives

$$C(g) = \left\{\frac{\partial \underline{E}_\varepsilon(g)}{\partial \varepsilon}\right\}_{\varepsilon=\varepsilon_0} \quad \text{and} \quad D(g) = \left\{\frac{\partial \overline{E}_\varepsilon(g)}{\partial \varepsilon}\right\}_{\varepsilon=\varepsilon_0}.$$

The posterior local sensitivity $\rho^*(g) = D^*(g) - C^*(g)$ is defined similarly.

Local sensitivity can be defined also for the sequence $\{K_{\gamma_\varepsilon}\}$ of γ–neighbourhoods of a probability measure P, provided that $\gamma_\varepsilon(x) \uparrow x$, for any $x \in [0,1]$, as $\varepsilon \to \varepsilon_0$. When the conditions for the differentiability of an integral w.r.t. a parameter are fulfilled (see, e.g., Letta, 1976, Theorem 5.4.6 and Remark 5.4.8), then Theorem 3 and Corollary 1 can be applied to compute the prior local sensitivity for any measurable function g.

Theorem 6 *Consider a sequence $\{K_{\gamma_\varepsilon}\}$, $\varepsilon_0 \leq \varepsilon \leq \beta$, of γ–neighbourhoods of P defined as before and suppose that the corresponding $c_\varepsilon(x)$ are differentiable in $U = [\varepsilon_0, \alpha]$, $\alpha \leq \beta$, and there exists an integrable function k, defined over $[0,1]$, such that $\sup_{\varepsilon \in U} |c_\varepsilon'(x)| \leq k(x)$, for any $x \in [0,1]$. Denote $c_{\varepsilon_0}'(x) = \left\{\frac{\partial c_\varepsilon(x)}{\partial \varepsilon}\right\}_{\varepsilon=\varepsilon_0}$. Given a real-valued function g, it follows that*

$$\rho(g) = \int_0^1 \{c_g(x) + c_{-g}(x)\}\, c_{\varepsilon_0}'(x)dx.$$

Let $g \equiv I_A$, then it follows that

$$\rho(g) = \int_{P(A^C)}^{1} c'_{\varepsilon_0}(x)dx - \int_0^{P(A)} c'_{\varepsilon_0}(x)dx.$$

The posterior local sensitivity for $g \equiv I_A$ is given by $\rho^*(g) = D^*(g) - C^*(g)$ where $D^*(g)$ and $C^*(g)$ are, respectively,

$$\frac{\int_{P(A^C)}^{1} c_{l_2}(x)\,dx \int_0^{P(A^C)} c_{-l_1}(x)c'_{\varepsilon_0}(x)\,dx}{\left[\int_{P(A^C)}^{1} c_{l_2}(x)\,dx - \int_0^{P(A^C)} c_{-l_1}(x)\,dx\right]^2} - \frac{\int_0^{P(A^C)} c_{-l_1}(x)\,dx \int_{P(A^C)}^{1} c_{l_2}(x)c'_{\varepsilon_0}(x)\,dx}{\left[\int_{P(A^C)}^{1} c_{l_2}(x)\,dx - \int_0^{P(A^C)} c_{-l_1}(x)\,dx\right]^2}$$

and

$$\frac{\int_0^{P(A)} c_{-l_2}(x)\,dx \int_{P(A)}^{1} c_{l_1}(x)c'_{\varepsilon_0}(x)\,dx}{\left[\int_{P(A)}^{1} c_{l_1}(x)\,dx - \int_0^{P(A)} c_{-l_2}(x)\,dx\right]^2} - \frac{\int_{P(A)}^{1} c_{l_1}(x)\,dx \int_0^{P(A)} c_{-l_2}(x)c'_{\varepsilon_0}(x)\,dx}{\left[\int_{P(A)}^{1} c_{l_1}(x)\,dx - \int_0^{P(A)} c_{-l_2}(x)\,dx\right]^2}.$$

PROOF. It is straightforward, from Theorem 3 and Corollary 1. ∎

As a consequence of the previous Theorem, the same results in Ruggeri and Wasserman (1991a) can be obtained.

TOTAL VARIATION NEIGHBOURHOOD. A class Γ^T is said to be a total variation neighbourhood of a probability measure P if it contains all the probability measures Q that satisfy $\sup_{A\in\mathcal{F}} |Q(A) - P(A)| \le \varepsilon$, given a fixed $\varepsilon \in [0, 1]$. As proved by Fortini and Ruggeri (1993), such a class can be described by means of the c.f. and, since P is non-atomic, prior and posterior expectations are maximised and minimised at probability measures which coincide with P over a subset B_1 such that $P(B_1) = 1 - \varepsilon$, give a total mass ε to B_2 where $P(B_2) = 0$ and vanish elsewhere, confirming previous results due to Wasserman and Kadane (1990).

Before analysing the local sensitivity, we need a lemma which holds under very general conditions.

Lemma 1 *Given a measurable function f and a subset $A \in \mathcal{F}$, suppose that there exists a sequence of subsets $\{K_\varepsilon\}$, with $K_\varepsilon \subseteq A$, such that $P(K_\varepsilon) = \varepsilon$ and $\inf_{\omega\in K_\varepsilon} f(\omega) \uparrow$ $\uparrow \sup_{\omega\in A} f(\omega) = \overline{f}$ as $\varepsilon \downarrow 0$, then $\lim_{\varepsilon\downarrow 0} \frac{\int_{K_\varepsilon} f(\omega)P(d\omega)}{\varepsilon} = \overline{f}$. A similar result holds with "sup" replaced by "inf".*

PROOF. It is straightforward to prove that $\overline{f}$ is an upper bound on the limit. Now we show that this upper bound can be achieved. For any ε, there exists $\delta = \delta(\varepsilon)$ such that $\overline{f} - \delta \le f(\omega) \le \overline{f}$ for any $\omega \in K_\varepsilon$. Since $\delta \downarrow 0$ as $\varepsilon \downarrow 0$, then the proof follows immediately. ∎

Suppose now that the conditions for Lemma 1 are fulfilled; the following Theorems allow us to analyse the local sensitivity for any prior expectation and posterior probabilities, respectively.

Theorem 7 *For any measurable function g, it follows that $D(g) = -C(g)$ and $\rho(g) = 2\{\overline{g} - \underline{g}\}$, where $\overline{g} = \sup_{\omega\in\Omega} g(\omega)$ and $\underline{g} = \inf_{\omega\in\Omega} g(\omega)$. Let $g \equiv I_A$, then it follows that $\rho(g) = 2$.*

PROOF. Given $\varepsilon \in [0,1]$, it follows that $\overline{E}_\varepsilon(g) = \varepsilon\overline{g} + \int_{\overline{K}_\varepsilon} g(\omega)P(d\omega)$ and $\underline{E}_\varepsilon(g) = \varepsilon\underline{g} + \int_{\underline{K}_\varepsilon} g(\omega)P(d\omega)$, where $\overline{K}_\varepsilon$ and $\underline{K}_\varepsilon$, respectively, maximises and minimises $\int_K g(\omega)P(d\omega)$ among all the subsets K such that $P(K) = 1 - \varepsilon$.
$D(g)$ is obtained computing $\lim_{\varepsilon\to 0} \frac{\overline{E}_\varepsilon(g) - E_0(g)}{\varepsilon}$ and using Lemma 1. $C(g)$ is computed similarly. ∎

Theorem 8 *The local sensitivity of the posterior probability is given by*

$$\rho^*(g) = 2\frac{S_A \int_{A^C} l_x(\omega)P(d\omega) + S_{A^C}\int_A l_x(\omega)P(d\omega)}{\left[\int_\Omega l_x(\omega)P(d\omega)\right]^2},$$

where $S_A = \sup_{\omega\in A} l_x(\omega)$.

PROOF. We find only $D^*(g)$ because it can be shown that $C^*(g) = -D^*(g)$. Because of (1) and $\varepsilon \downarrow 0$, then, for a given ε, it follows that $\int_A l_x(\omega)Q(d\omega)$ is maximised by $\varepsilon S_A + \int_A l_x(\omega)P(d\omega)$. Furthermore, $\int_{A^C} l_x(\omega)Q(d\omega)$ is minimised by assigning no mass to a subset B_2, with $P(B_2) = \varepsilon$, where $l_x(\omega)q(d\omega)$ is the highest. Since there exists a probability measure (or a sequence of them) which gives both extrema, the inequality in (1) becomes an equality. The result follows by differentiating such supremum w.r.t. ε. ∎
NEIGHBOURHOOD OF THE UNIFORM DISTRIBUTION. The behaviour of the inferences when considering a neighbourhood of the uniform distribution has been considered in some recent works (e.g. Wasserman and Kadane, 1992). In the next example, we consider a neighbourhood which is given by means of the c.f..

Example 1.

Suppose that a coin is flipped twice and that ω is the probability of getting "Head" in a flip. A uniform distribution P is a possible choice as prior measure on Ω. Such a choice can be modified considering measures which slightly differ from it, choosing the probability measures Q such that

$$\sup_{A\in\mathcal{F}:P(A)=x} |P(A) - Q(A)| \le \varepsilon P(A)(1 - P(A)), \varepsilon \ge 0;$$

it follows that $\varphi(x) \ge (1-\varepsilon)x + \varepsilon x^2 = \gamma(x)$, $\forall x \in [0,1]$.
Hence, we consider the random variable $X \sim Bin(2,\omega)$, with density $f(x|\omega) = \binom{2}{x}\omega^x(1-\omega)^{2-x}$, $\omega \in [0,1]$, $x = 0,1,2$, and the class $K_{\gamma_\varepsilon} = \{Q : \varphi(x) \ge (1-\varepsilon)x + \varepsilon x^2, \forall x \in [0,1]\}$.
Applying the results in Fortini and Ruggeri (1992), it follows that maximisations of $E(f)$ and $E(f|x)$ are made over the set of extremal probability measures E_γ, which are such that $c_\varepsilon(x) = (1-\varepsilon) + 2\varepsilon x$, $x \in [0,1]$.
As pointed out in Berger (1985, §§ 3.5.1, 4.7.2), marginal distributions are sometimes relevant, e.g. to construct the prior or to check assumptions on it and the model. A well-known method to select priors is based on the ML–II approach (see Berger, 1985, §3.5.4). Given a class Γ of probability measures and the observed data x, then $\hat{Q} \in \Gamma$ is said to be a ML–II prior if it maximises, all over Γ, the marginal density $m(x|Q) = \int_\Omega f(x|\omega)Q(d\omega)$.

When either $x = 0$ or $x = 2$, then Theorem 3 implies that the maximum value of the marginal density is $1/3 + \varepsilon/6$, obtained for $H_f(y) = \sqrt{y}$ and $c_f(x) = x^2$; when $x = 1$, then the maximum of the marginal density is $1/3 + \varepsilon/12$, obtained for $H_f(y) = 1 - \sqrt{1-2y}$ and $c_f(x) = x - x^2/2$. Given $\varepsilon = 1$, then the same results as in Fortini and Ruggeri (1993) are found.
Computing now the local sensitivity, it follows that $D(f) = -C(f) = 1/12$ for $x = 1$ and $D(f) = -C(f) = 1/6$ otherwise. As a consequence, we can argue that the marginal density at $x = 1$ is the least sensitive to the above changes in the prior.
It is worth noticing that $D(g) = -C(g)$ holds also when $g \equiv I_A$, since $\left\{\frac{\partial c_\varepsilon(x)}{\partial \varepsilon}\right\}_{\varepsilon=\varepsilon_0} =$ $= 2x - 1$ and

$$\int_{P(A^C)}^{1} (2x-1)dx = -\int_0^{P(A)} (2x-1)dx = P(A)P(A^C).$$

Given $x = 2$, we compute now the local sensitivity for the posterior probability $Q(A|x)$ of the subset $A = [0, 1/3]$, when the prior measure varies in K_{γ_ε}. Applying Theorem 6, it follows that $c'_{\varepsilon_0}(x) = 2x - 1$, $c_{l_1}(x) = x^2 I_{(1/3,1]}(x)$, $c_{l_2}(x) = (x - 2/3)^2 I_{(2/3,1]}(x)$, $c_{-l_1}(x) = -(1-x)^2 I_{[0,2/3]}(x)$ and $c_{-l_2}(x) = -(1/3 - x)^2 I_{[0,1/3]}(x)$. The posterior local sensitivity is then $\rho^*(g) = 214/2187 \approx 0.0979$, with $D^*(g) = -C^*(g) = 107/2187 \approx 0.0489$. It is evident that the probability $Q(A|x)$ is quite insensitive to the small changes allowed here.

5. A GENERAL APPROACH TO SENSITIVITY ANALYSIS

While sensitivity to changes in the prior measure has been extensively studied, little attention has been paid to the uncertainty in the likelihood and the loss function (see Lavine, 1991b, and Pericchi and Perez, 1993, on the former), as discussed by Pericchi and Perez (1993).
While global sensitivity analysis is broadly recognised as useful when considering classes of sampling models and loss functions, we think that the local analysis is well justified also for these classes.
Local analysis about loss functions is justified, as an example, when a squared loss function Δ^2, with $\Delta = \hat{\omega} - \omega$, is chosen to estimate ω with $\hat{\omega}$ but equivalence between the costs of overestimating and underestimating is not completely believed. In this case, it is possible to analyse the behaviour of the estimates when a slightly modified version of a convex loss function, due to Varian (1975), is considered. Take

$$L(\Delta) = \frac{2\left[e^{a\Delta} - a\Delta - 1\right]}{a^2}, a \neq 0,$$

and observe that $L(\Delta) \doteq \Delta^2$ for small values of $|a|$. Small departures from the squared loss function, obtained by using $L(\Delta)$ as a goes far from 0, allow the study of the effects on the estimate, when removing symmetry in the costs.
Local sensitivity analysis to changes in the likelihood has been firstly considered by Cuevas and Sanz (1988) who studied the behaviour of the posterior measure, while Ruggeri and Wasserman (1991b) analysed any posterior functionals, under different departures from a specified sampling model. We present here just some results of them.
Given a non-atomic probability measure P, consider, as space of likelihood functions, the set $\mathcal{L}_P^+$ of all the nonnegative functions in the space $\mathcal{L}_P$ of P-integrable functions

$l : \Omega \to \mathcal{R}$, endowed with the norm $||l||_P = \int |l| dP$. For a likelihood function $l \in \mathcal{L}_P$, let $T_g(l)$ be the posterior expectation of a bounded measurable function g, such that

$$T_g(l) \equiv \frac{\int g(\omega) l(\omega) P(d\omega)}{\int l(\omega) P(d\omega)} = \frac{N_l}{D_l}.$$

We consider T_g as a non-linear functional over $\mathcal{L}_P$ and let $\dot{T}_g^l$ be its Fréchet derivative, which we regard as a measure of how a small change in l affects the posterior expectation. Fixed $l \in \mathcal{L}_P^+$, in the next $\Delta_{\mathcal{L}}$ will be the set of all the P-integrable functions λ_ϵ of the form $\lambda_\epsilon = (1-\epsilon)l + \epsilon L$ where L varies in $\mathcal{L}_P^+$ and $\epsilon \in (0, 1]$. Let $||\dot{T}_g^l||$ be the norm of $\dot{T}_g^l$ over all of $\mathcal{L}_P$, which is used as a measure of the sensitivity of the posterior expectation to the likelihood function.

Theorem 9 $\dot{T}_g^l(\lambda) = \{D_l\}^{-1}(N_\lambda - \rho D_\lambda)$, *where* $\rho = T_g(l)$.

Define $k(\omega) = g(\omega) - \rho$, $\overline{k} = \sup_{\omega \in \Omega} k(\omega)$ and $\underline{k} = \inf_{\omega \in \Omega} k(\omega)$.

Theorem 10 $||\dot{T}_g^l|| = \{D_l\}^{-1} \max\{\overline{k}, -\underline{k}\}$.

Acknowledgements. The author thanks David Rios Insua and three referees for their valuable comments.

REFERENCES

[1] BERGER, J. (1984). The Robust Bayesian Viewpoint (with discussion). In: *Robustness of Bayesian Analyses.* [J. Kadane (ed.)]. North-Holland, Amsterdam.

[2] BERGER, J. (1985). *Statistical Decision Theory and Bayesian Analysis.* (2nd Edition). Springer-Verlag, New York.

[3] BERGER, J. (1990). Robust Bayesian Analysis: Sensitivity to the Prior. *J. Statist. Plann. Inference*, **25**, 303-328.

[4] BETRÒ B., MĘCZARSKI, M. & RUGGERI, F. (1992). *Robust Bayesian Analysis under Generalized Moments Conditions.* Quaderno IAMI 92.1, Milano: CNR-IAMI.

[5] CIFARELLI, D.M. & REGAZZINI, E. (1987). On a General Definition of Concentration Function. *Sankhya B*, **49**, 307-319.

[6] CUEVAS, A. & SANZ, P. (1988). On Differentiability Properties of Bayes Operators. In *Bayesian Statistics 3.* [J.M. Bernardo, M.H. DeGroot, D.V. Lindley and A.F.M. Smith (eds.)]. Oxford University Press, Oxford.

[7] DEROBERTIS, L. & HARTIGAN, J. (1981). Bayesian Inference Using Intervals of Measures. *Ann. Statist.*, **9**, 235-244.

[8] DIACONIS, P. & FREEDMAN, D. (1986). On the Consistency of Bayes Estimates. *Ann. Statist.*, **14**, 1-67.

[9] FORTINI, S. & RUGGERI, F. (1992). *On Defining Neighbourhoods of Measures through the Concentration Function.* Quaderno IAMI 92.3, Milano, CNR-IAMI.

[10] FORTINI, S. & RUGGERI, F. (1993). Concentration Functions and Bayesian Robustness. *J. Statist. Plann. Inference*, (To appear).

[11] LAVINE, M. (1988) *Prior Influence in Bayesian Statistics.* Discussion paper 88-06, Institute of Statistics and Decision Sciences, Duke University.

[12] LAVINE, M. (1991a). Sensitivity in Bayesian Statistics: The Prior and the Likelihood. *J. Amer. Statist. Assoc.*, **86**, 396-399.

[13] LAVINE, M. (1991b). An Approach to Robust Bayesian Analysis for Multidimensional Spaces. *J. Amer. Statist. Assoc.*, **86**, 400-403.

[14] LETTA, G. (1976). *Teoria Elementare dell'Integrazione.* Boringhieri, Torino.

[15] MARSHALL, A.W. & OLKIN, I. (1979). *Inequalities: Theory of Majorization and Its Applications.* Academic Press, New York.

[16] PERICCHI, L.R. & PEREZ, M. (1993). Posterior Robustness with More than One Sampling Model. *J. Statist. Plann. Inference*, (To appear).

[17] RUGGERI, F. (1991a). Robust Bayesian Analysis Given a Lower Bound on the Probability of a Set. *Commun. Statist.-Theory Meth.*, **20(5&6)**, 1881-1891.

[18] RUGGERI, F. (1991b). *Robust Bayesian Analysis Given Bounds on the Probability of a Set.* Quaderno IAMI 90.16, Milano, CNR-IAMI.

[19] RUGGERI, F. (1992). Bounds on the Prior Probability of a Set and Robust Bayesian Analysis. *Theory of Probability and Its Applications*, **37 (2)**.

[20] RUGGERI, F. & WASSERMAN, L. (1991a). *Density Based Classes of Priors: Infinitesimal Properties and Approximations.* Technical Report #528, Department of Statistics, Carnegie Mellon University.

[21] RUGGERI, F. & WASSERMAN, L. (1991b). *Infinitesimal Sensitivity of Likelihood Function.* Unpublished manuscript.

[22] RUGGERI, F. & WASSERMAN, L. (1993). Infinitesimal Sensitivity of Posterior Distributions.*Canad. J. Statist.*, (To appear).

[23] SRINIVASAN, C. & TRUSZCZYNSKA, H. (1990). *On the Ranges of Posterior Quantities.* Technical Report #294, Department of Statistics, University of Kentucky.

[24] VARIAN, H.R. (1975). A Bayesian Approach to Real Estate Assessment. In: *Studies in Bayesian Econometrics and Statistics in Honor of Leonard J. Savage.* [S.E. Fienberg and A. Zellner, (eds.)]. North-Holland, Amsterdam.

[25] WASSERMAN, L. (1992). Recent Methodological Advances in Robust Bayesian Inference. In: *Bayesian Statistics 4.* [J.M. Bernardo, J.O. Berger, A.P. Dawid and A.F.M. Smith, (eds.)]. Oxford University Press, Oxford.

[26] WASSERMAN, L. & KADANE, J. (1990). Bayes' Theorem for Choquet Capacities. *Ann. Statist.*, **18**, 1328-1339.

[27] WASSERMAN, L. & KADANE, J. (1992). Symmetric upper Probabilities. *Ann. Statist.*, **20**, 1720-1736.

Stochastic Processes

Recent Adv. in Stat. and Prob., pp. 247-254
J. Pérez Vilaplana and M. L. Puri (Eds)

Strong Limit Theorems for Empirical Processes *

E. CSÁKI
Mathematical Institute of the Hungarian Academy of Sciences, P.O.B. 127, H-1364 Budapest, Hungary

Abstract – Let $X_1, X_2, \ldots$ be i.i.d. random variables each having uniform distribution on $(0,1)$. Put

$$F_n(x) = n^{-1} \sum_{i=1}^{n} I(X_i \leq x) \quad (0 \leq x \leq 1).$$

We study the empirical process $\alpha_n(x) = n^{1/2}(F_n(x) - x)$. Glivenko–well known for the weighted empirical process $\tau(x)\alpha_n(x)$. For the lim inf we show that

$$\liminf_{n\to\infty} \sup_{0<x<1} (\tau(x)|\alpha_n(x)|)(\log\log n)^{1/2} = \pi 8^{-1/2} \left(\int_0^1 \tau^2(x)\,dx\right)^{1/2} \qquad \text{a.s.}$$

under certain conditions on $\tau(x)$. Furthermore, for the standardized empirical process it holds

$$\liminf_{n\to\infty} \sup_{\delta_n<x<1-\delta_n} \frac{\left|\alpha_n(x)\right|}{\left(x(1-x)\right)^{1/2}} \tilde{\beta}_n = c \qquad \text{a.s.,}$$

where $\tilde{\beta}_n$ is a sequence of norming constants depending upon δ_n, and c is a constant.

1. INTRODUCTION

Let $X_1, X_2, \ldots$ be i.i.d. random variables with common distribution function $F(x)$. The empirical distribution function $F_n(x)$ is defined by

$$F_n(x) = n^{-1} \sum_{i=1}^{n} I\{X_i \leq x\}, \tag{1.1}$$

where $I\{A\}$ denotes the indicator of the event A. We study the empirical process

$$\alpha_n(x) = n^{1/2}(F_n(x) - F(x)), \qquad -\infty < x < \infty. \tag{1.2}$$

The basic strong limit result is the Glivenko–Cantelli theorem:

$$\lim_{n\to\infty} \sup_{-\infty<x<\infty} |F_n(x) - F(x)| = 0 \qquad \text{a.s.} \tag{1.3}$$

*Research supported by Hungarian National Foundation for Scientific Research, Grant no. 1905

The law of the iterated logarithm was proved by Chung (1949) and Smirnov (1944):

$$\limsup_{n\to\infty} \frac{\sup_{-\infty<x<\infty} |\alpha_n(x)|}{(\log\log n)^{1/2}} = \sup_{-\infty<x<\infty} (2F(x)(1-F(x)))^{1/2} \quad \text{a.s.} \tag{1.4}$$

In the sequel we assume that $F(x) = x$, $(0 \le x \le 1)$,, i.e. the basic distribution is uniform on (0,1). This is no loss of generality as long as $F(x)$ is assumed to be continuous for every x.

Let $\tau(x)$, $(0 < x < 1)$, be a weight function. In this paper we are concerned with strong limit theorems for the weighted empirical process $\tau(x)\alpha_n(x)$. For $\tau(x)$ we assume the following conditions:

$$\tau(x) \ge 0 \tag{Ai}$$

$$\tau(x) \quad \text{is piecewise continuous} \tag{Aii}$$

$$\tau(x) \quad \text{is nondecreasing in} \quad (0,\delta_1], \quad \text{nonincreasing in} \quad [\delta_2, 1) \tag{Aiii}$$

$$\text{and bounded in} \quad [\delta_1, \delta_2] \quad \text{for some} \quad 0 \le \delta_1 \le \delta_2 \le 1.$$

Lai (1974) and Wellner (1977) proved the Glivenko–Cantelli-type theorem:

Theorem A. *Assume* (Ai), (Aii), (Aiii) *and*

$$\int_0^1 \tau(x)\,dx < \infty. \tag{1.5}$$

Then

$$\lim_{n\to\infty} \sup_{0<x<1} \tau(x)|F_n(x) - x| = 0 \quad \text{a.s.} \tag{1.6}$$

The law of the iterated logarithm (LIL) for the weighted empirical process was proved by James (1975):

Theorem B. *Assume* (Ai), (Aii), (Aiii) *and*

$$\int_0^1 \frac{\tau^2(x)}{\log\log(1/(x(1-x)))}\,dx < \infty. \tag{1.7}$$

Then

$$\limsup_{n\to\infty} \frac{\sup_{0<x<1}(\tau(x)|\alpha_n(x)|)}{(\log\log n)^{1/2}} = \sup_{0<x<1} (2\tau(x)x(1-x))^{1/2} \quad \text{a.s.} \tag{1.8}$$

James (1975) proved also a functional LIL (see also Finkelstein, 1971) and a converse of (1.8) in the sense that if the integral in (1.7) diverges, then the lim sup in (1.8) is equal to ∞.

For further lim sup results we refer to Csáki (1984), Shorack and Wellner (1986) and Einmahl (1987). For recent general results in the case when (1.7) is not satisfied, see Einmahl and Mason (1989), Norvaiša (1990) and Einmahl (1992).

On the other hand, lim inf results are known in particular cases only. Kuelbs (1979) and Mogulskii (1979) proved that

$$\liminf_{n\to\infty} \sup_{0<x<1} |\alpha_n(x)|(\log\log n)^{1/2} = \pi 8^{-1/2} \quad \text{a.s.} \tag{1.9}$$

and this was extended by Shi (1991) to

$$\liminf_{n\to\infty} \sup_{0<x<1} (x^m|\alpha_n(x)|)(\log\log n)^{1/2} = \pi 8^{-1/2}(2m+1)^{-1/2} \quad \text{a.s.} \tag{1.10}$$

for $m \geq 0$.
On the other hand, it was shown in Csáki (1982) that

$$\liminf_{n\to\infty} \sup_{0<x<1} \frac{|\alpha_n(x)|}{(x(1-x))^{1/2}}(2\log\log n)^{-1/2} = 1 \qquad \text{a.s.} \tag{1.11}$$

In Section 2 we extend (1.10) to more general weight functions. In Section 3 we study the lim inf problem for the standardized empirical process and extend (1.11) by considering supremum on intervals $(\delta_n, 1-\delta_n)$. The corresponding lim sup problem was studied in Csáki (1977).
We prove first analogue results for the Kiefer process and then use KMT strong invariance principle to show our results for the empirical process.
A Kiefer process $K(x,y)$ is a two parameter Gaussian process with

$$\mathbf{E}K(x,y) = 0 \tag{1.12}$$

$$\mathbf{E}K(x_1,y_1)K(x_2,y_2) = (x_1 \wedge x_2 - x_1x_2)(y_1 \wedge y_2). \tag{1.13}$$

For fixed x, $K(x,y)(x(1-x))^{-1/2}$ is a standard Wiener process in y and for fixed y, $K(x,y)y^{-1/2}$ is a Brownian bridge in x. The strong invariance principle of Komlós, Major and Tusnády (1975) says that on a suitably rich probability space one can construct a sequence $\{X_n\}_{n=1}^{\infty}$ of i.i.d. uniform $(0,1)$ random variables together with a Kiefer process $K(x,y)$ such that

$$\sup_{0\leq x\leq 1} |n(F_n(x)-x) - K(x,n)| = O(\log^2 n) \quad \text{a.s.} \tag{1.14}$$

as $n \to \infty$.
In fact, we use two simple consequences of this result:
(i) If $\tau(x) \geq 0$, $(0<x<1)$, is bounded, then

$$\sup_{0\leq x\leq 1} (\tau(x)|n^{1/2}\alpha_n(x) - K(x,n)|) = O(\log^2 n) \qquad \text{a.s.} \tag{1.15}$$

as $n \to \infty$.
(ii)

$$\sup_{\delta_n<x<1-\delta_n} \left| \frac{\alpha_n(x)}{(x(1-x))^{1/2}} - \frac{K(x,n)}{(nx(1-x))^{1/2}} \right| = O\left(\frac{\log^2 n}{(n\delta_n)^{1/2}}\right) \qquad \text{a.s.} \tag{1.16}$$

as $n \to \infty$.
For these and related results see also Csörgő and Révész (1981).

2. WEIGHTED EMPIRICAL PROCESS

Here we prove

Theorem 2.1. *Assume* (Ai), (Aii), A(iii) *with* $\delta_1 = 0$, $\delta_2 = 1$. *Then*

$$\liminf_{n\to\infty} \sup_{0<x<1} (\tau(x)|\alpha_n(x)|)(\log\log n)^{1/2} = \pi 8^{-1/2}\left(\int_0^1 \tau^2(x)\,dx\right)^{1/2} \qquad \text{a.s.} \tag{2.1}$$

PROOF: First we prove an analogue result for the Kiefer process, i.e.

$$\liminf_{n\to\infty} \sup_{0<x<1} \left(\tau(x)\left|K(x,n)\right|\right)\left(n^{-1}\log\log n\right)^{1/2} = \pi\, 8^{-1/2}\left(\int_0^1 \tau^2(x)\,dx\right)^{1/2} \qquad \text{a.s.} \tag{2.2}$$

Since $K(x,n)/n^{1/2}$ is a Brownian bridge, it follows from Mogulskii (1974) that

$$\log \mathbf{P}\left(\sup_{0<x<1}(\tau(x)|K(x,n)|) < zn^{1/2}\right) \sim -\frac{\pi^2}{8z^2}\int_0^1 \tau^2(x)\,dx \qquad \text{as } z \to 0 \tag{2.3}$$

and we obtain (2.2) from Theorem 2 of Erickson (1980). Now (2.1) follows from (2.2) via KMT strong invariance (1.15). ■

3. STANDARDIZED EMPIRICAL PROCESS

Now let $\{\delta_n\}_{n=1}^{\infty}$ be a non-increasing sequence, $0 \le \delta_n < 1/2$. Introduce the following notation:

$$\rho_n = \frac{\log((1-\delta_n)/\delta_n)}{\log\log n} \tag{3.1}$$

$$\tilde{\rho}_n = \min(\rho_n, \log n) \tag{3.2}$$

$$\beta_n = (\log(1+\rho_n))^{-1/2} \tag{3.3}$$

$$\tilde{\beta}_n = (\log(1+\tilde{\rho}_n))^{-1/2} \tag{3.4}$$

Theorem 3.1.

$$\liminf_{n\to\infty} \sup_{\delta_n<x<1-\delta_n} \frac{|\alpha_n(x)|}{(x(1-x))^{1/2}}\tilde{\beta}_n = c \qquad \text{a.s.} \tag{3.5}$$

with certain constant c. If

$$\lim_{n\to\infty} \rho_n = \rho \tag{3.6}$$

exists ($0 \le \rho \le \infty$), *then* $c = c(\rho)$ *in (3.5) with*

$$c(0) = \pi/2, \tag{3.7}$$

$$c(\rho) = \gamma(1/\rho)(\log(1+\rho))^{-1/2}, \qquad 0<\rho<\infty \tag{3.8}$$

$$c(\infty) = 2^{1/2}, \tag{3.9}$$

where $\gamma(\lambda)$ *is the smallest positive root* $z = \gamma(\lambda)$ *of the confluent hypergeometric function*

$${}_1F_1\left(-\frac{\lambda}{2},\frac{1}{2},\frac{z^2}{2}\right) = \sum_{m=0}^{\infty} \frac{\left(-\frac{\lambda}{2}\right)_m \left(\frac{z^2}{2}\right)^m}{\left(\frac{1}{2}\right)_m m!} = \sum_{m=0}^{\infty} \frac{(2z^2)^m}{(2m)!}\left(m-1-\frac{\lambda}{2}\right)\cdots\left(-\frac{\lambda}{2}\right). \tag{3.10}$$

First we prove analogue results for the Kiefer process. Now assume that $\delta_n \ne 0$.

Theorem 3.2.

$$\liminf_{n\to\infty} \sup_{\delta_n<x<1-\delta_n} \frac{|K(x,n)|}{(nx(1-x))^{1/2}}\beta_n = c \qquad \text{a.s.} \tag{3.11}$$

with the same c as in (3.5). *If, moreover* (3.6) *holds, then* $c = c(\rho)$ *with* (3.7), (3.8) *and* (3.9).

PROOF: (3.11) was in fact proved in Csáki (1981). Here we prove the more precise results under the condition (3.6). For the case $\rho = \infty$ see also Csáki (1990). The proof is based on sharp estimates of the probability

$$\mathbf{P}_\delta(z) = \mathbf{P}\left(\sup_{\delta<x<1-\delta} \frac{|K(x,n)|}{(nx(1-x))^{1/2}} < z\right) \qquad (0<\delta<1/2). \tag{3.12}$$

Since $K(x,n)/n^{1/2}$ is a Brownian bridge in x, by substituting

$$x = \frac{e^{2s}}{1+e^{2s}},$$

we get that

$$\mathbf{P}_\delta(z) = \mathbf{P}\left(\sup_{0<s<\log\frac{1-\delta}{\delta}} |U(s)| < z\right), \tag{3.13}$$

where $U(t)$ is an Ornstein–Uhlenbeck process, i.e. a Gaussian process with

$$\mathbf{E}U(t) = 0 \tag{3.14}$$

$$\mathbf{E}U(t)U(s) = e^{-|t-s|} \tag{3.15}$$

for $-\infty < t < \infty$, $-\infty < s < \infty$.
Using eigenfunction expansion we obtain (cf. Newell, 1962):

$$\mathbf{P}(\sup_{0<s<t} |U(s)| < z) = \frac{1}{(2\pi)^{1/2}} \sum_{k=1}^{\infty} e^{-\lambda_k t} \left(\int_{-z}^{z} \psi_k(x) e^{-x^2/2}\, dx\right)^2, \tag{3.16}$$

where $\lambda_1 \le \lambda_2 \le \ldots$ and $\psi_1(x), \psi_2(x), \ldots$, resp. are the eigenvalues and normed eigenfunctions, resp. of the Sturm–Liouville problem

$$\psi''(x) - x\psi'(x) = -\lambda\psi(x), \qquad \psi(-z) = \psi(z) = 0. \tag{3.17}$$

Here λ_i and ψ_i depend also on z and it is known that $\{\psi_1, \psi_2, \ldots\}$ form an orthonormal sequence with respect to the weight function $\exp(-x^2/2)$. Let $\lambda(z) = \lambda_1$ be the smallest eigenvalue of the problem (3.17). We need the following properties:

Lemma 3.1.

i) $\lambda(z) > 0$ *is a strictly decreasing continuous function of* z *on* $(0, \infty)$.

ii) *If* $\gamma(\lambda)$ *denotes the continuous inverse of* $z \to \lambda(z)$, *then* $z = \gamma(\lambda)$ *is the smallest positive zero of the confluent hypergeometric function* (3.10). $\gamma(\lambda) > 0$ *is a decreasing continuous function of* λ.

iii) $$\lambda(z) \sim \frac{2z(2\pi)^{1/2}}{e}^{-z^2/2} \qquad \text{as } z \to \infty$$

iv) $$\lambda(z) \sim \frac{\pi^2}{4z^2} \qquad \text{as } z \to 0$$

v) $$\lambda_{k+1} \ge \lambda_1 + k, \qquad k = 1, 2, \ldots$$

PROOF: For (*i*) and (*ii*) we refer to Perkins (1983) and (*iii*) is proved in Newell (1962). To show (*iv*) we rewrite the problem (3.17) by substituting $y(x) = z^{1/2}\psi(zx)$, as

$$y'' - xz^2 y' + \mu y = 0, \qquad y(-1) = y(1) = 0, \tag{3.18}$$

where $\mu = \lambda z^2$. When $z \to 0$, by perturbation theory (cf. Kamke, 1956) we have $\lim_{z\to 0} \mu_1(z) = \mu_1(0) = \pi^2/4$.

To show (v), we conclude from Sturm–Liouville theory (cf. Kamke, 1956) that $\psi_{k+1}(x)$ has k roots in $(-z, z)$, $\psi'_{k+1}(x)$ has $k-1$ roots in $(-z, z)$, $\ldots \psi^{(k)}_{k+1}(x)$ has 2 roots in $(-z, z)$. Denote these roots by $-z_0, z_0$. It is seen from (3.17) that $\eta(x) = \psi^{(k)}_{k+1}(x)$ satisfies

$$\eta'' - x\eta' + (\lambda_{k+1} - k)\eta = 0, \qquad \eta(-z_0) = \eta(z_0) = 0. \tag{3.19}$$

It follows that $\lambda(z_0) = \lambda_{k+1} - k$. But from (i) we get $\lambda_1 = \lambda(z) \le \lambda(z_0) = \lambda_{k+1} - k$, i.e. (v) is proved. ∎

The following estimation holds true:

Lemma 2.2. *For $z > 0$, $t > 0$ we have*

$$\frac{1}{(2\pi)^{1/2}} e^{-\lambda(z)t} \left(\int_{-z}^{z} \psi_1(x) e^{-x^2/2}\, dx\right)^2 \le \mathbf{P}(\sup_{0<s<t} |U(s)| < z) \le \frac{e^{-\lambda(z)t}}{1 - e^{-t}}. \tag{3.20}$$

PROOF The lower estimation in (3.20) is obvious from (3.16). The upper estimation follows from the inequality

$$\frac{1}{(2\pi)^{1/2}} \left(\int_{-z}^{z} \psi_k(x) e^{-x^2/2}\, dx\right)^2 \le \frac{1}{(2\pi)^{1/2}} \int_{-z}^{z} \psi_k^2(x) e^{-x^2/2}\, dx \int_{-z}^{z} e^{-x^2/2}\, dx \le 1$$

and from Lemma 2.1 (v).

By putting $t = \log((1 - \delta_n)/\delta_n)$ and $z = az_n$ with constant a and

$$z_n = (2 \log \rho_n)^{1/2} \qquad \text{if } \rho = \infty \tag{3.21}$$

$$z_n = \gamma(1/\rho) \qquad \text{if } 0 < \rho < \infty \tag{3.22}$$

$$z_n = \pi \rho_n^{1/2}/2 \qquad \text{if } \rho = 0 \tag{3.23}$$

we get as $n \to \infty$,

$$\log \mathbf{P}_{\delta_n}(az_n) \sim -\frac{2a}{\pi^{1/2}} (\log \rho_n)^{1/2} \rho_n^{1-a^2} \log\log n \qquad \rho = \infty \tag{3.24}$$

$$\log \mathbf{P}_{\delta_n}(az_n) \sim -\rho\lambda(a\gamma(1/\rho)) \log\log n \qquad 0 < \rho < \infty \tag{3.25}$$

$$\log \mathbf{P}_{\delta_n}(az_n) \sim -\frac{1}{a^2} \log\log n \qquad \rho = 0 \tag{3.26}$$

From these asymptotic relations we can prove the statements of Theorem 3.2 by using the procedure employed by Mogulskii (1979). The details are omitted.

Now we are ready to prove Theorem 3.1 under the assumption (3.6). Without this assumption (3.5) was proved in Csáki (1981). If $0 \le \rho < \infty$ then the statement immediately follows from Theorem 3.2 via KMT invariance principle (1.16). The same holds true if $\rho = \infty$ but $\delta_n \ge n^{-1}(\log n)^4$ for n large enough. Put

$$\delta'_n = \max(\delta_n, n^{-1}(\log n)^4) \tag{3.27}$$

$$\rho'_n = \frac{\log((1 - \delta'_n)/\delta'_n)}{\log\log n} \tag{3.28}$$

$$\beta'_n = (\log(1 + \rho'_n))^{-1/2}. \tag{3.29}$$

Then it follows that

$$\liminf_{n\to\infty} \sup_{\delta'_n < x < 1-\delta'_n} \frac{|\alpha_n(x)|}{(x(1-x))^{1/2}} \beta'_n = 2^{1/2} \qquad \text{a.s.} \tag{3.30}$$

Since $\beta_n' \sim \tilde{\beta}_n \quad (n \to \infty)$, we have also

$$\liminf_{n\to\infty} \sup_{\delta_n<x<1-\delta_n} \frac{|\alpha_n(x)|}{(x(1-x))^{1/2}}\tilde{\beta}_n \geq 2^{1/2} \qquad \text{a.s.} \tag{3.31}$$

Now assume that there exists a subsequence $\{n_k\}_{k=1}^{\infty}$ such that

$$\delta_{n_k} < n_k^{-1}(\log n_k)^4. \tag{3.32}$$

Then it follows from Jaeschke (1979) that

$$\sup_{\delta_{n_k}<x<1-\delta_{n_k}} \frac{|\alpha_{n_k}(x)|}{(x(1-x))^{1/2}}(\log\log n_k)^{-1/2} \to 2^{1/2} \qquad \text{as } k\to\infty \tag{3.33}$$

in probability and hence

$$\liminf_{k\to\infty} \sup_{\delta_{n_k}<x<1-\delta_{n_k}} \frac{|\alpha_{n_k}(x)|}{(x(1-x)^{1/2}}(\log\log n_k)^{-1/2} \leq 2^{1/2} \qquad \text{a.s.} \tag{3.34}$$

and since $\tilde{\beta}_{n_k} \sim (\log\log n_k)^{-1/2}$ $(k \to \infty)$ we have also

$$\liminf_{n\to\infty} \sup_{\delta_n<x<1-\delta_n} \frac{|\alpha_n(x)|}{(x(1-x))^{1/2}}\tilde{\beta}_n \leq 2^{1/2} \qquad \text{a.s.} \tag{3.35}$$

which together with (3.31) proves Theorem 3.1. ∎

REFERENCES

[1] CHUNG, K.L (1949). An Estimate Concerning the Kolmogorov Limit Distributions, *Trans. Amer. Math. Soc.* **67**, 36–50.

[2] CSÁKI, E (1977). The Law of the Iterated Logarithm for Normalized Empirical Distribution Function, *Z. Wahrsch. verw. Gebiete* **38**, 147–167.

[3] CSAKI, E (1981). *A Note on the Lower Limit of the Standardized Empirical Distribution Function*, Technical Report no. 81–19, Purdue University.

[4] CSÁKI, E (1982). On the Standardized Empirical Distribution Function, *Coll. Math. Soc. J. Bolyai*, **32**, *Nonparametric Statistical Inference*, 123–138, North–Holland, Amsterdam.

[5] CSÁKI, E (1984). Empirical Distribution Function, *Handbook of Statistics, Vol.4, Nonparametric Methods*, 405–430, North–Holland, Amsterdam.

[6] CSÁKI, E (1990). Lim inf Results for the Standardized Empirical Distribution Function, *Probab. Theory and Math. Statist.*, Vol. 1, pp. 238-244 (B. Grigelionis et al., eds.) VSP/Mokslas.

[7] CSÖRGŐ, M. AND RÉVÉSZ, P (1981). *Strong Approximations in Probability and Statistics*, Academic Press, New York.

[8] EINMAHL, J.H.J (1987). *Multivariate Empirical Processes*, CWI Tract, **32**, Mathematisch Centrum, Amsterdam.

[9] EINMAHL, J.H.J (1992). The a.s. Behaviour of the Weighted Empirical Process and the LIL for the Weighted Tail Empirical Process, *Ann. Probab.* **20**, 681–695.

[10] EINMAHL J.H.J. AND MASON, D.M (1989) A Unifying Approach to Functional Laws of the Iterated Logarithm and Glivenko–Cantelli Theorems for Weighted Empirical Processes, *Asymptotic Statistics, 4-th Prague Symp.*, Prague, Czech., 1988, 215–227.

[11] ERICKSON, K.B (1980). Rates of Escape of Infinite Dimensional Brownian Motion, *Ann. Probab.*, **8**, 325–338.

[12] FINKELSTEIN, H (1971). The Law of the Iterated Logarithm for Empirical Distributions, *Ann. Math. Statist.*, **42**, 607–615.

[13] JAESCHKE, D (1979). The Asymptotic Distribution of the Supremum of the Standardized Empirical Distribution Function on Subintervals, *Ann. Statist.*, **7**, 108–115.

[14] JAMES, B.R (1975). A Functional Law of the Iterated Logarithm for Weighted Empirical Distribution, *Ann. Probab.*, **3**, 762–772.

[15] KAMKE, E (1956). *Differentialgleichungen. Lösungsmethoden und Lösungen I. Gewöhnliche Differentialgleichungen*, Akademische Verlag, Leipzig.

[16] KOMLÓS, J., MAJOR, P. AND TUSNÁDY, G. (1975). An Approximation of Partial Sums of Independent r.v.'s and the Sample d.f. I., *Z. Wahrsch. verw. Gebiete*, **32**, 111–131.

[17] KUELBS, J (1979). Rates of Growth for Banach Space Valued Independent Increment Processes, *Lecture Notes in Math.*, No. **709**, 151–169, Springer, New York.

[18] LAI, T.L (1974). Convergence Rates in the Strong Law of Large Numbers for Random Variables Taking Values in Banach Spaces, *Bull. Inst. Math. Acad. Sinica*, **2**, 67–85.

[19] MOGULSKII, A.A (1974). Small Deviations in Space of Trajectories, *Theor. Probab. Appl.*, **19**, 726–736.

[20] MOGULSKII, A.A (1979). On the Law of the Iterated Logarithm in Chung's Form for Functional Spaces, *Theor. Probab. Appl.*, **24**, 405–413.

[21] NEWELL, G.F (1962). Asymptotic Extreme Value Distributions for One Dimensional Diffusion Processes, *J. Math. Mech.*, **11**, 481–496.

[22] NORVAIŠA, R (1990). The Law of the Iterated Logarithm for Random Variables with Heavy Tails and Empirical Processes, *Liet. Matem. Rink.*, **30**, 337–343.

[23] PERKINS, E (1983). On the Hausdorff Dimension of the Brownian Slow Points, *Z. Wahrsch. verw. Gebiete*, **64**, 369–399.

[24] SHI, Z (1991). A Generalization of the Chung–Mogul'skii Law of the Iterated Logarithm, *J. Multivariate Anal.*, **37**, 269–278.

[25] SHORACK, G. AND WELLNER, J (1986). *Empirical Processes with Applications to Statistics*, John Wiley and Sons, New York.

[26] SMIRNOV, N.V (1944). Approximate Laws of Distribution of Random Variables from Empirical Data, *Usp. Mat. Nauk*, **10**, 179–206 (in Russian).

[27] WELLNER, J.A (1977). A Glivenko–Cantelli Theorem and Strong Laws of Large Numbers for Functions of Order Statistics, *Ann. Statist.*, **5**, 473–480. [Correction: ibid., **6** (1978), 1394.]

Recent Adv. in Stat. and Prob., pp. 255-261
J. Pérez Vilaplana and M. L. Puri (Eds)

Bahadur-Kiefer Representations on the Tails

M. CSÖRGŐ[1] and L. HORVÁTH[2]
[1] Department of Mathematics and Statistics, Carleton University, Ottawa, Ontario, Canada K1S 5B6
[2] Department of Mathematics, University of Utah, Salt Lake City, Utah 84112, U.S.A

Abstract – We obtain three limit theorems for the deviation between empirical and quantile processes on the tails.

Let $\{U_i,\ 1 \le i < \infty\}$ be independent random variables, uniformly distributed on $(0,1)$. The order statistics of $U_1, \ldots, U_n$ are denoted by $0 = U_{0,n} \le U_{1,n} \le U_{2,n} \le \ldots \le U_{n,n} \le U_{n+1,n} = 1$. The uniform empirical distribution and quantile functions are respectively defined by

$$E_n(t) = \frac{1}{n} \sum_{1 \le i \le n} I\{U_i \le t\}, \quad 0 \le t \le 1,$$

and

$$U_n(t) = \begin{cases} U_{k,n}, & \frac{k-1}{n} \le t < \frac{k}{n}, \quad 1 \le k \le n, \\ Y_{n,n}, & t = 1 \end{cases}$$

while the uniform empirical and quantile processes e_n and u_n by

$$e_n(t) = n^{1/2}(E_n(t) - t), \qquad 0 \le t \le 1,$$

and

$$u_n(t) = n^{1/2}(t - U_n(t)), \qquad 0 \le t \le 1.$$

Starting with Bahadur (1966) and Kiefer (1967, 1970), several papers have been devoted to the study of the difference $\Delta_n(t) = e_n(t) - u_n(t)$. We summarize two important results on Bahadur-Kiefer representation in the following theorem (cf. Kiefer (1967, 1970)).

Theorem A. *For any* $0 < t_0 < 1$ *we have*

$$\lim_{n\to\infty} P\left\{n^{1/4}|\Delta_n(t_0)| \le x\right\} = \frac{2}{(t_0(1-t_0))^{1/2}} \int_0^\infty \left(2\Phi\left(\frac{x}{\sqrt{u}}\right) - 1\right) \phi\left(\frac{u}{(t_0(1-t_0))^{1/2}}\right) du,$$

where Φ *is the standard normal distribution function and* $\Phi' = \phi$.

We have also

$$n^{1/4}(\log n)^{-1/2} \sup_{0<t<1} |\Delta_n(t)| \xrightarrow{\mathcal{D}} \left(\sup_{0<t<1} |B(t)|\right)^{1/2},$$

where $\{B(t),\ 0 \le t \le 1\}$ *is a Brownian bridge.*

In this note we obtain tail versions of Theorem A.

Theorem 1. *If*

$$k = k(n) \to \infty \quad \text{and} \quad k(n)/n \to 0 \quad (n \to \infty), \tag{1}$$

then we have

$$\lim_{n\to\infty} P\left\{\frac{n^{1/2}}{k^{1/4}}|\Delta_n(k/n)| \le x\right\} = 2\int_0^\infty \left(2\Phi\left(\frac{x}{\sqrt{u}}\right) - 1\right)\phi(u)du \tag{2}$$

and

$$\frac{n^{1/2}}{k^{1/4}}(\log k)^{-1/2} \sup_{0<t\le k/n} |\Delta_n(t)| \xrightarrow{\mathcal{D}} \left(\sup_{0<t<1}|W(t)|\right)^{1/2}, \tag{3}$$

as $n \to \infty$, where $\{W(t),\ 0 \le t \le 1\}$ is a Wiener process.

Proof. First we note that

$$E_n(t) - t = E_n(t) - U_n(E_n(t)) + (U_n(E_n(t)) - t). \tag{4}$$

It follows from the definitions of E_n and U_n that

$$E_n(t) = i/n, \qquad \text{if } U_{i,n} \le t < U_{i+1,n} \tag{5}$$

and, therefore, we have

$$U_n(E_n(t)) = U_{i+1,n}, \qquad \text{if } U_{i,n} \le t < U_{i+1,n}. \tag{6}$$

Thus we conclude

$$\begin{aligned} \sup_{0\le t\le k/n} |U_n(E_n(t)) - t| &\le \max_{0\le i\le nE_n(k/n)} \sup_{U_{i,n}\le t\le U_{i+1,n}} |U_n(E_n(t)) - t| \\ &\le \max_{0\le i\le nE_n(k/n)} |U_{i+1,n} - U_{i,n}|. \end{aligned} \tag{7}$$

Since

$$E_n(k/n)/(k/n) \xrightarrow{\mathrm{P}} 1 \tag{8}$$

and

$$\max_{0\le i\le r(n)} |U_{i+1,n} - U_{i,n}| = \mathcal{O}_P(\log r(n)) \quad (r(n) \to \infty), \tag{9}$$

(7) yields

$$\sup_{0\le t\le k/n} |U_n(E_n(t)) - t| = \mathcal{O}_P(n^{-1}\log k). \tag{10}$$

Csörgő *et al.* (1986) constructed a sequence of Brownian bridges $\{B_n(t),\ 0 \le t \le 1\}$ such that

$$\sup_{0\le t\le r/n} |u_n(t) - B_n(t)| = \mathcal{O}_P(n^{-1/2}\log r) \tag{11}$$

for all $1 < r \le n$. Using the exponential partial sums representation of uniform order statistics we can easily establish

$$nk^{-1/2}(E_n(k/n) - k/n) = \mathcal{O}_P(1). \tag{12}$$

For any Brownian bridge $\{B(t),\ 0 \le t \le 1\}$ we can find a Wiener process $\{W(t),\ 0 \le t \le 1\}$ such that

$$B(t) = W(t) - tW(1). \tag{13}$$

By (13) we have

$$h^{-1/2}(n) \sup_{0\le s\le h(n)} \left|B\left(\frac{k}{n}+s\right) - B\left(\frac{k}{n}\right)\right| \xrightarrow{\mathcal{D}} \sup_{0\le t\le 1} |W(t)|, \tag{14}$$

as $h(n) \to 0$. Putting together (10), (11), (12) and (14), we get

$$\begin{aligned} |e_n(k/n) - B_n(k/n)| &\le |e_n(k/n) - u_n(E_n(k/n))| + |u_n(E_n(k/n)) - \\ &\quad -B_n(E_n(k/n))| + |B_n(E_n(k/n)) - B_n(k/n)| = \\ &= \mathcal{O}_P\left(n^{-1/2}k^{1/4}\right). \end{aligned} \tag{15}$$

Using (4) again, we obtain

$$n^{1/2}k^{-1/4}|\Delta_n(k/n) = n^{1/2}k^{-1/4}|u_n(k/n) - u_n(E_n(k/n))| + \mathcal{O}_P(k^{-1/4}\log k). \tag{16}$$

Now (11), (14), (15) and (16) imply

$$n^{1/2}k^{-1/4}|\Delta_n(k/n)| = n^{1/2}k^{-1/4}\left|B_n\left(\frac{k}{n}\right) - B_n\left(\frac{k}{n} + n^{-1/2}B_n\left(\frac{k}{n}\right)\right)\right| + \mathcal{O}_P(k^{-1/8}). \tag{17}$$

The distribution of B_n does not depend on n and, therefore, (2) is proven if we show

$$\lim_{n\to\infty} P\left\{n^{1/2}k^{-1/4}\left|B(\frac{k}{n}) - B\left(\frac{k}{n} + n^{-1/2}B(\frac{k}{n})\right)\right| \le x\right\} = 2\int_0^\infty \left(2\Phi\left(\frac{x}{\sqrt{u}}\right) - 1\right)\phi(u)\,du. \tag{18}$$

We recall (2.15). By (13) we have

$$\begin{aligned} n^{1/2}k^{-1/4}&\left|B\left(\frac{k}{n}\right) - B\left(\frac{k}{n} + n^{-1/2}B\left(\frac{k}{n}\right)\right)\right| \stackrel{\mathcal{D}}{=} \\ &\stackrel{\mathcal{D}}{=} n^{1/2}k^{-1/4}\left|W\left(\frac{k}{n}\right) - W\left(\frac{k}{n} + n^{-1/2}\left(W\left(\frac{k}{n}\right) - \frac{k}{n}W(1)\right)\right) + \right. \\ &\quad \left. + n^{-1/2}\left(W\left(\frac{k}{n}\right) - \frac{k}{n}W(1)\right)W(1)\right| = \\ &= n^{1/2}k^{-1/4}\left|W(\frac{k}{n}) - W\left(\frac{k}{n} + n^{-1/2}W\left(\frac{k}{n}\right)\right)\right| + \mathcal{O}_P\left(\left(\frac{k}{n}\right)^{1/4}\right) + \mathcal{O}_P\left(n^{-1/4}\left(\frac{k}{n}\right)^{1/4}\right). \end{aligned} \tag{19}$$

Using again the scale-transformation of W we get

$$n^{1/2}k^{-1/4}\left|W\left(\frac{k}{n}\right) - W\left(\frac{k}{n} + n^{-1/2}W\left(\frac{k}{n}\right)\right)\right| \stackrel{\mathcal{D}}{=} k^{1/4}\left|W(1) - W\left(1 + k^{-1/2}W(1)\right)\right|.$$

The joint distribution of $u - W(1 + k^{-1/2}u)$ and $W(1)$ is bivariate normal for each u and hence we obtain

$$\begin{aligned} &\lim_{k\to\infty} P\left\{k^{1/4}\left|W(1) - W\left(1 + k^{-1/2}W(1)\right)\right| \le x\right\} = \\ &= \lim_{k\to\infty}\int_{-\infty}^{\infty} P\left\{k^{1/4}\left|W(1) - W\left(1 + k^{-1/2}W(1)\right)\right| \le x \mid W(1) = u\right\}\phi(u)\,du \\ &= \lim_{k\to\infty}\int_{-\infty}^{\infty} P\left\{k^{1/4}\left|u - W(1 + k^{-1/2}u)\right| \le x \mid W(1) = u\right\}\phi(u)\,du \\ &= \int_{-\infty}^{\infty}\left(2\Phi\left(\frac{x}{\sqrt{|u|}}\right) - 1\right)\phi(u)\,du. \end{aligned}$$

This also completes the proof of (18).

Next we prove (3). By the modulus of continuity of B and W (cf., e.g., Csörgő and Révész (1981)) we have

$$(2h(n)\log k/(nh(n)))^{-1/2} \sup_{0\le t\le k/n} \sup_{0\le s\le h(n)} |W(t+s)-W(t)| \xrightarrow{\mathrm{P}} 1, \tag{21}$$

and

$$(2h(n)\log k/(nh(n)))^{-1/2} \sup_{0\le t\le k/n} \sup_{0\le s\le h(n)} |B(t+s)-B(t)| \xrightarrow{\mathrm{P}} 1 \tag{22}$$

as $h(n)\to 0$ $(n\to\infty)$. Similarly to (15) we can establish

$$\sup_{0\le t\le k/n}\Big|e_n(t)-B_n(t)\Big| \le \sup_{0\le t\le k/n}\Big|e_n(t)-u_n\big(E_n(t)\big)\Big| + + \sup_{0\le t\le k/n}\Big|u_n\big(E_n(t)\big) - \tag{23}$$

$$-B_n\big(E_n(t)\big)\Big| + \sup_{0\le t\le k/n}\Big|B_n\big(E_n(t)\big)-B_n(t)\Big| =$$

$$= \mathcal{O}_P\big(n^{-1/2}\log k\big) + \mathcal{O}_P\big(k^{1/4}n^{-1/2}(\log k)^{1/2}\big).$$

Applying (4), (7) and (10) we get

$$\sup_{0\le t\le k/n}|\Delta_n(t)| = \sup_{0\le t\le k/n}|u_n(t)-u_n(E_n(t))| + \mathcal{O}_P(n^{-1/2}\log k). \tag{24}$$

Now we use (11), (21), (22) and (24), and obtain

$$n^{1/2}k^{-1/4}(\log k)^{-1/2}\sup_{0\le t\le k/n}\Big|\Delta_n(t)\Big| = \tag{25}$$

$$= n^{1/2}k^{-1/4}(\log k)^{-1/2}\sup_{0\le t\le k/n}\Big|B_n(t)-B_n\big(t+n^{-1/2}B_n(t)\big)\Big| + \mathcal{O}_P\big(k^{-1/8}(\log k)^{3/4}\big).$$

Similarly to (19), (21) and (22) yield

$$\sup_{0\le t\le k/n}\Big|B_n(t)-B_n\big(t+n^{-1/2}B_n(t)\big)\Big| \stackrel{\mathcal{D}}{=} \tag{26}$$

$$\stackrel{\mathcal{D}}{=} \sup_{0\le t\le k/n}\Big|W(t)-W\Big(t+n^{-1/2}\big(W(t)-tW(1)\big)\Big)+n^{-1/2}\big(W(t)-tW(1)\big)W(1)\Big| =$$

$$= \sup_{0\le t\le k/n}\Big|W(t)-W\big(t+n^{-1/2}W(t)\big)\Big| + \mathcal{O}_P\big(n^{-1/2}(k/n)^{1/2}\big) + \mathcal{O}_P\big(k^{1/2}n^{-3/4}(\log n)^{1/2}\big).$$

Using again the scale-transformation of the Wiener process we get

$$n^{1/2}k^{-1/4}\sup_{0\le t\le k/n}\Big|W(t)-W\big(t+n^{-1/2}W(t)\big)\Big| \stackrel{\mathcal{D}}{=} k^{-1/4}\sup_{0\le t\le k}\Big|W(t)-W\big(t+W(t)\big)\Big|. \tag{27}$$

Theorem 1B of Deheuvels and Mason (1990) implies

$$k^{-1/4}(\log k)^{-1/2}\sup_{0\le t\le k}|W(t)-W(t+W(t))| \xrightarrow{\mathcal{D}} \Big(\sup_{0\le t\le 1}|W(t)|\Big)^{1/2}, \tag{28}$$

as $k\to\infty$, which completes the proof of (3).

Theorem 1 remains valid for non-uniform random samples. Let $X_1, X_2, \ldots, X_n$ be i.i.d.r.v.'s with distribution function F. The quantile function, i.e. the inverse of F, is denoted by Q. Let

$$t_F = \lim_{t \downarrow 0} Q(t).$$

We assume that

C1. F is continuous on its support, and there is a constant t_F^* such that $t_F < t_F^*$ and F is twice differentiable on $(t_F, t_F^*]$ and $f(x) > 0$ for all $t_F < x \le t_F^*$,

C2.
$$\limsup_{x \downarrow t_F} F(x)(1 - F(x)) \left| \frac{f^1(x)}{f^2(x)} \right| < \infty,$$

where $F' = f$ denotes the density of F.

The order statistics of $X_1, \ldots, X_n$ are denoted by $X_{1,n} \le \ldots \le X_{n,n}$ and the empirical distribution and quantile functions are defined by

$$F_n(x) = \frac{1}{n} \sum_{1 \le i \le n} I\{X_i \le x\}, \qquad -\infty < x < \infty,$$

and

$$Q_n(t) = \begin{cases} X_{k,n}, & \dfrac{k-1}{n} \le t < \dfrac{k}{n}, \quad 1 \le k \le n, \\ X_{n,n}, & t = 1. \end{cases}$$

Next we define the general empirical and quantile processes

$$\alpha_n(x) = n^{1/2}(F_n(x) - F(x))$$

and

$$\rho_n(t) = n^{1/2} f(Q(t)) \left(Q(t) - Q_n(t)\right), \qquad 0 < t < 1.$$

Now we define the difference Δ_n^* between α_n and ρ_n by

$$\Delta_n^*(t) = \alpha_n(Q(t)) - \rho_n(t), \qquad 0 < t < 1,$$

and note in passing that $\alpha_n(Q(t)) = e_n(t)$. We prove two theorems.

Theorem 2. *If* (1) *and* $C1$, $C2$ *hold, then we have*

$$\lim_{n \to \infty} P\left\{ \frac{n^{1/2}}{k^{1/4}} |\Delta_n^*(k/n)| \le x \right\} = 2 \int_0^\infty \left(2\Phi\left(\frac{x}{\sqrt{u}}\right) - 1 \right) \phi(u) du \tag{29}$$

and

$$\frac{n^{1/2}}{k^{1/4}} (\log k)^{-1/2} \sup_{1/(n+1) \le t \le k/n} |\Delta_n^*(t)| \xrightarrow{\mathcal{D}} \left(\sup_{0 \le t \le 1} |W(t)| \right)^{1/2}, \tag{30}$$

as $n \to \infty$.

Assuming one more regularity condition on f, we can study $\Delta_n^*(t)$ on the interval $(0, k/n]$ as well. Namely, for the sake of doing this, we assume also that

C3. f is non-decreasing on $(t_F, t_F^*]$.

Theorem 3. *If* (1) *and* $C1, C2, C3$ *hold, then, as* $n \to \infty$ *we have*

$$\frac{n^{1/2}}{k^{1/4}}(\log k)^{-1/2} \sup_{0<t\le k/n} |\Delta_n^*(t)| \xrightarrow{\mathcal{D}} \left(\sup_{0\le t\le 1} |W(t)|\right)^{1/2}. \tag{31}$$

PROOF OF THEOREM 2. Let $U_i = F(X_i)$. Then $U_1, \dots, U_n$ are i.i.d.r.v.'s, uniformly distributed on (0.1).

The corresponding uniform quantile process is $u_n(t)$. By a two-term Taylor expansion we get

$$\rho_n(s) = u_n(s) - \frac{1}{2}n^{-1/2}\frac{u_n^2(s)}{s(1-s)}\left\{\frac{s(1-s)}{\theta_n(s)\big(1-\theta_n(s)\big)}\right\} \times \tag{32}$$

$$\times\left\{\theta_n(s)\big(1-\theta_n(s)\big)\frac{f'\big(Q(\theta_n(s))\big)}{f^2\big(Q(\theta_n(s))\big)}\right\}\frac{f\big(Q(s)\big)}{f\big(Q(\theta_n(s))\big)},$$

where

$$\min(U_n(s), s) < \theta_n(s) < \max(U_n(s), s). \tag{33}$$

By Wellner (1978), we have

$$\sup_{1/(n+1)\le s\le n/(n+1)} \frac{s(1-s)}{\theta_n(s)(1-\theta_n(s))} = \mathcal{O}_P(1). \tag{34}$$

Using C2 and Lemma 4.5.2 of Csörgő and Révész (1981), we obtain

$$\sup_{1/(n+1)\le s\le \delta} \frac{f(Q(s))}{f(Q(\theta_n(s)))} = \mathcal{O}_P(1), \tag{35}$$

where $\delta = F(t_F^*)$. Theorem 2.6 in Csörgő and Horváth (1990) implies

$$\sup_{1/(n+1)\le t\le k/n} |u_n(t)|/(t(1-t))^{1/2} = \mathcal{O}_P((\log\log k)^{1/2}). \tag{36}$$

Putting together (32) – (36), we obtain

$$\sup_{1/(n+1)\le t\le k/n} |\rho_n(t) - u_n(t)| = \mathcal{O}_P(n^{-1/2}\log\log k). \tag{37}$$

Now the first part of Theorem 2 follows immediately from (2) and (37).

It is well-known that $nU_{1,n}$ is asymptotically an exponential r.v. Thus we have

$$\sup_{0<t\le 1/(n+1)} |u_n(t)| \le n^{1/2}U_{1,n} + n^{1/2}/(n+1) = \mathcal{O}_P(n^{-1/2}). \tag{38}$$

Also, $nE_n(1/(n+1))$ has a Poisson limit distribution, and thus we get

$$\sup_{0<t\le 1/(n+1)} |e_n(t)| \le n^{1/2}E_n(1/(n+1)) + n^{1/2}/(n+1) = \mathcal{O}_P(n^{-1/2}). \tag{39}$$

We proved

$$\frac{n^{1/2}}{k^{1/4}}(\log k)^{-1/2} \sup_{0<t\le 1/(n+1)} |\Delta_n(t)| = \mathcal{O}_P(k^{-1/4}(\log k)^{-1/2}). \tag{40}$$

Now (3), (37) and (40) yield (30).

Proof of Theorem 3. We can assume without loss of generality that

$$\sup_{t_F<t\le t_F^*} F(x)(1-F(x))\left|\frac{f'(x)}{f^2(x)}\right| \le \gamma, \tag{41}$$

where $\gamma > 1$. We consider $\rho_n(t)$ on $(0, 1/(n+1))$. If $U_{1,n} < t$, then by C.3 we have

$$|\rho_n(t)| = n^{1/2}\int_t^{U_{1,n}} \frac{f(Q(t))}{f(Q(s))}ds \le |u_n(t)|. \tag{42}$$

If $U_{1,n} < t \le \dfrac{1}{n+1}$, then Lemma 4.5.2 of Csörgő and Révész (1981) implies

$$|\rho_n(t)| = n^{1/2}\int_{U_{1,n}}^t \frac{f(Q(t))}{f(Q(s))}ds \le n^{1/2}\int_{U_{1,n}}^t \left(\frac{t(1-s)}{s(1-t)}\right)^\gamma ds \le \tag{43}$$

$$\le 2^\gamma n^{1/2}\int_{U_{1,n}}^t (t/s)^\gamma ds \le \frac{2^\gamma}{\gamma-1} n^{1/2}\, t^\gamma\, U_{1,n}^{1-\gamma} \le \mathcal{O}_P(n^{-1/2}).$$

Hence we obtain

$$\sup_{0<t<\frac{1}{n+1}} |\rho_n(t)| = \mathcal{O}_P(n^{-1/2}). \tag{44}$$

Collecting together (3), (37)-(40) and (44) we get Theorem 3.

Acknowledgement. This research was supported in part by an NSERC Canada grant of Miklós Csörgő at Carleton University.

REFERENCES

[1] R.R. BAHADUR (1966). A Note on Quantiles in Large Samples. *Ann. Math. Statist.*, **37**, 577–580.

[2] M. Csörgő, S. Csörgő, L. Horváth and D.M. Mason (1986). Weighted Empirical and Quantile Processes, *Ann. Probab.*, **14**, 31–85.

[3] M. Csörgő and L. Horváth (1990). On the Distributions of the Supremum of Weighted Quantile Processes, *Studia Sci. Math. Hung.*, **15**, 353–375.

[4] M. Csörgő and P. Révész (1981). *Strong Approximations in Probability and Statistics.* Academic, New York.

[5] P. DEHEUVELS AND D.M. MASON (1990). Bahadur–Kiefer–Type Processes, *Ann. Probab.*, **18**, 669–697.

[6] J. KIEFER (1967). On Bahadur's Representation of Sample Quantiles, *Ann. Math. Statist.*, **38**,1323–1342.

[7] J. KIEFER (1970). *Deviations Between the Sample Quantile Process and the Sample D.F.* In: (M.L. Puri, ed.), *Nonparametric Techniques in Statistical Inference.* Cambridge University Press, London, 299–319.

[8] J.A. WELLNER (1978). Limit theorems for the Ratio of the Empirical Distribution Function to the True Distribution Function, *Z. Wahrscheinlichkeitstheorie verw. Gebiete*, **45**, 73–88.

Recent Adv. in Stat. and Prob., pp. 263-267
J. Pérez Vilaplana and M. L. Puri (Eds)

Strong Theorems for Random Walks and Its Local Time

A. FÖLDES
Department of Mathematics, CUNY–College of Staten Island, St. George Campus, 130 Stuyvesant Place, Staten Island, NY-10301, U.S.A.

Abstract – In 1960 Erdös and Taylor proved an almost sure logarithmic average type result for the return times to the origin of the simple symmetric random walk. Since then many authors dealt with similar questions. In this paper a couple of consequences of this result is discussed concerning the local time of the random walk.

Let $X_1, X_2, \ldots$ be a sequence of i.i.d. r.v.-s., and put $S_0 = 0$, $S_n = X_1 + X_2 + \cdots + X_n, n = 1, 2, \ldots$, $P(X_n = 1) = P(X_n = -1) = \frac{1}{2}$. S_n is the so called simple symmetric random walk. Its two-dimensional counterpart is $S_n^* = X_1^* + X_2^* + \cdots + X_n^*$, where X_i^* is a sequence of i.i.d. random vectors with

$$P(X_1^* = e_i) = P(X_1^* = -e_i) = \frac{1}{4} \qquad (i = 1, 2)$$

where e_1 and e_2 are the orthogonal unit vectors in Z_2.

In their classical paper Erdős and Taylor proved the following two results:

Theorem (Erdős-Taylor 1960 Th.5). *Suppose that ρ_k denotes the suffix of the k-th return to the origin, then*

(i) if the random walk is on the line

$$\lim_{n\to\infty} \frac{1}{\log n} \sum_{k=1}^{n} \frac{1}{\sqrt{\rho_k}} = \sqrt{\frac{2}{\pi}} \qquad a.s. \tag{1}$$

(ii) if the random walk is in the plane, then

$$\lim_{n\to\infty} \frac{1}{\log n} \sum_{k=1}^{n} \frac{1}{\log \rho_k} = \frac{1}{\pi} \qquad a.s. \tag{2}$$

The almost sure logarithmic average type results got much attention, especially in the last few years. The connection of (1) and the almost sure central limit theorem was pointed out by Csáki, Földes, Révész (1992). In this note a pair of consequences of the above results will be discussed.

Results (1) and (2) can be reformulated as follows:

$$\lim_{n\to\infty} \frac{1}{\log n} \sum_{k=1}^{n} \frac{I(S_k = 0)}{\sqrt{k}} = \sqrt{\frac{1}{2\pi}} \qquad a.s. \tag{3}$$

$$\lim_{n\to\infty} \frac{1}{\log\log n} \sum_{k=2}^{n} \frac{I(S_k^* = 0)}{\log k} = \frac{1}{\pi} \qquad a.s. \tag{4}$$

Define the local time of the walk on the line by

$$\xi(x,n) = \#\{k;\ 0 < k \le n,\ S_k = x\}$$

Then, as it was pointed out in Csáki-Földes-Révész

$$\sum_{i=1}^{n} \frac{\xi(0,i)}{i^{3/2}} = \sum_{i=1}^{n} \frac{\sum_{k=1}^{i} I(S_k = 0)}{i^{3/2}} = \sum_{k=1}^{n} I(S_k = 0) \sum_{i=k}^{n} \frac{1}{i^{3/2}} \approx 2 \sum_{k=1}^{n} \frac{I(S_k = 0)}{\sqrt{k}}. \tag{5}$$

Thus

$$\lim_{n\to\infty} \frac{1}{\log n} \sum_{i=1}^{n} \frac{\xi(0,i)}{i^{3/2}} = \sqrt{\frac{2}{\pi}} \qquad a.s. \tag{6}$$

This raises the question of a similar result on the plane. Define the local time of the walk on plane by

$$\xi^*(x,n) = \#\{k;\ 0 < k \le n,\ S_k^* = x\}$$

then we have

Theorem 1.

$$\lim_{n\to\infty} \frac{1}{\log\log n} \sum_{k=2}^{n} \frac{\xi^*(0,k)}{k \log^2 k} = \frac{1}{\pi} \qquad a.s. \tag{7}$$

PROOF: Denote the sum in (7) by ζ_n. Then clearly,

$$\zeta_n = \frac{1}{\log\log n} \sum_{k=2}^{n} \frac{\sum_{j=2}^{k} I(S_j^* = 0)}{k \log^2 k} = \frac{1}{\log\log n} \sum_{j=2}^{n} I(S_j^* = 0) \sum_{k=j}^{n} \frac{1}{k \log^2 k}$$

Define

$$\eta_n = \frac{1}{\log\log n} \sum_{j=2}^{n} I(S_j^* = 0) \sum_{k=j}^{\infty} \frac{1}{k \log^2 k}$$

It is easy to see that,

$$\eta_n = \frac{1}{\log\log n} \sum_{j=2}^{n} \frac{I(S_j^* = 0)}{\log j} + o_n(1)$$

Thus by (4)

$$\lim_{n\to\infty} \eta_n = \frac{1}{\pi} \qquad a.s. \tag{8}$$

Moreover,

$$\zeta_n - \eta_n = \frac{1}{\log\log n} \sum_{j=2}^{n} I(S_j^* = 0) \sum_{k=n}^{\infty} \frac{1}{k(\log k)^2} \tag{9}$$

Observe now that,

$$\zeta_n - \eta_n = \frac{1}{\log\log n} \sum_{j=2}^{n} I(S_j^* = 0) \frac{1}{\log n} + o_n(1)$$

Hence,

$$\zeta_n - \eta_n = \frac{1}{\log n \log\log n} \xi^*(0,n) + o_n(1) \tag{10}$$

But from Erdős-Taylor(1960) we also know that,

$$(1+\epsilon)\pi^{-1}\log n \log_3 n \quad \in UUC\,\xi^*(0,n)$$

the function $a(n)$ belongs to the upper-upper class, UUC of $\{Y(n)\}$ if for almost all $\omega \in \Omega$ there exists an $n_0 = n_0(\omega) > 0$ such that $Y(n) < a(n)$ if $n > n_0$ [see also in Révész (1990), p.202]. Consequently, the almost sure limit of ζ_n and η_n coincide, which together with (8) implies the theorem.

Returning to the walk on the line, in Csáki-Földes-Révész (1992), (6) was generalized for additive functionals as well [see also Csáki et al.(1991)]. More precisely, consider

$$A_n = \sum_{i=1}^{n} f(S_i)$$

where $f(\ )$ is a real valued function on integers. Then A_n can be represented as:

$$A_n = \sum_{x=-\infty}^{\infty} f(x)\,\xi(x,n) \tag{11}$$

Assuming that,

$$\sum_{x=-\infty}^{\infty} |f(x)| < \infty \tag{12}$$

and denoting

$$\overline{f} = \sum_{x=-\infty}^{\infty} f(x)$$

it was pointed out that if $\overline{f} \neq 0$ then

$$\lim_{n\to\infty} \frac{1}{\log n} \sum_{i=1}^{n} \frac{A_i}{i^{3/2}} = \frac{2\sqrt{2}}{\sqrt{\pi}}\overline{f} \qquad a.s. \tag{13}$$

In fact, it was proved for a large class of random walks. However, the analogue question when $\overline{f} = 0$ remained open. One of the simplest cases, when $f(1) = 1, f(0) = -1, f(x) = 0$ otherwise, suggests that we have to find the almost sure limit of

$$\frac{1}{\log n} \sum_{k=1}^{n} \frac{\xi(1,k) - \xi(0,k)}{k^{5/4}}$$

as roughly speaking, $\xi(1,k) - \xi(0,k)$ is of order $k^{1/4}$.

Theorem 2. *Suppose that $f(x), x \in \mathbf{Z}$, has finite support and $\overline{f} = 0$. Then for A_n given by (11),*

$$\lim_{n\to\infty} \frac{1}{\log n} \sum_{k=1}^{n} \frac{A_n}{n^{5/4}} = 0 \qquad a.s. \tag{14}$$

For the sake of simplicity, we give the proof for the above-mentioned logarithmic average of $\xi(1,k) - \xi(0,k)$ only, that is, we prove that,

$$\lim_{n\to\infty} \frac{1}{\log n} \sum_{k=1}^{n} \frac{\xi(1,k) - \xi(0,k)}{k^{5/4}} = 0, \qquad a.s. \tag{15}$$

and indicate how to proceed toward (14). The proof of (15) is very simple, based on a recent result of Móri.

Theorem (Móri, 1992) . *Let $Y_1, Y_2, \ldots$ be arbitrary random variables, with finite variances. Suppose that there exists a positive non-increasing function on the positive numbers, and a positive integer m, such that*

$$\int_1^{\infty} h(z) \frac{\log_m(z)}{z \log z}\, dz < \infty \tag{16}$$

and

$$E(Y_i Y_j) \leq h(\tfrac{j}{i}) \qquad for\ all\ 1 \leq i \leq j \tag{17}$$

Then

$$\lim_{n\to\infty} \frac{1}{\log n} \sum_{i=1}^{n} \frac{1}{i} Y_i = 0 \qquad a.s. \tag{18}$$

where $\log_m(x)$ is the m-times iterated logarithm of x.

The second key element of the proof is the following:

Lemma (Földes-Révész, 1991). *Let*

$$I(i,l) = \begin{cases} 1 & if\ \ S_l = i \quad or \quad i+1 \\ 0 & otherwise \end{cases}$$

and

$$D(i,n) = \xi(i+1,n) - \xi(i,n) - \sum_{l=0}^{n} I(i,l) X_{l+1}$$

then

$$|D(i,n)| \leq 2$$

Proof of Theorem 2: It is not hard to see, that according to the previous lemma, to get (15) enough to show that the sequence,

$$Y_i = \frac{\sum_{l=0}^{i} I(0,l) X_{l+1}}{i^{1/4}} \qquad i = 1, 2, \ldots$$

satisfies the conditions of Móri's Theorem. Observe, that for $i \leq j$,

$$E(Y_i Y_j) = \frac{l}{i^{1/4} j^{1/4}} E\Big(\Big(\sum_{l=0}^{i} I(0,l) X_{l+1}\Big)\Big(\sum_{k=0}^{j} I(0,k) X_{k+1}\Big)\Big) = \frac{l}{i^{1/4} j^{1/4}} \sum_{l=0}^{i} E\big(I(0,l)\big) =$$

$$= \frac{1}{(ij)^{1/4}} \sum_{l=0}^{i} P(S_l = 0 \text{ or } S_l = 1) \leq \frac{1}{(ij)^{1/4}} \left(1 + \sum_{l=1}^{i} \frac{c}{\sqrt{l}}\right) \leq c_1 \sqrt[4]{\frac{i}{j}}.$$

The $i = j$ case implies that Var Y_i are finite. Moreover, condition (17) holds with $h(x) = x^{-1/4}$, which satisfies condition (16) [with arbitrary m] implying (15).

To prove now the theorem, one has to observe first, that with the notation

$$U(l,i,n) = \sum_{j=1}^{i} \sum_{k=0}^{n} I(l+j-1,k)\, X_{k+1} \tag{19}$$

we have,

$$|\xi(l+i,n) - \xi(l,n) - U(l,i,n)| \leq 2i \tag{20}$$

Moreover, if $\overline{f} = 0$ and f has finite support then,

$$A_n = \sum_x f(x)\,\xi(x,n) = \sum_{j=1}^{K} f(x_j)\big(\xi(x_j,n) - \xi(x_1,n)\big) \qquad [x_1 < x_2 < ..x_K]$$

Consequently it is enough to prove that

$$\lim_{n\to\infty}; \frac{1}{\log n} \sum_{s=1}^{n} \frac{U(x_1, x_j - x_1, s)}{s^{\frac{5}{4}}} = 0 \qquad a.s. \tag{21}$$

To prove (21), observe, that by (19) it is enough to handle the logarithmic averages of

$$Y_i^j(t) = \frac{\sum_{k=1}^{n} I(y_r + t - 1, k)\, X_{k+1}}{n^{1/4}}$$

for $j = 1, 2, \dots x_r - y_r$. Finally, to prove that

$$\lim_{n\to\infty} \frac{1}{\log n} \sum_{k=1}^{n} \frac{Y_i^j(t)}{k} = 0 \qquad a.s.$$

one has to repeat the proof of (15) which we omit.

Remark. The condition of finite support emerged already in Dobrushins (1955) paper, but it was eliminated in the strong approximation result [(Csáki-Csörgő-Földes-Révész, (1991)]. It would be interesting to know, whether it can be relaxed here, too.

REFERENCES

[1] Csáki, E., Csörgő, M., Földes, A. and Révész, P. (1991). Strong Approximations of Additive Functionals (To appear in *Theoretical Probabi-lity*).

[2] Csáki, E., Földes, A. and Révész, P. (1992). *On Almost Sure Local and Global Central Limit Theorems.* Preprint.

[3] Dobrushin, R.L. (1955). Two Limit Theorems for the Simplest Random Walk on a Line, *Uspehi Mat. Nauk (N.S.)*, **10**, 139-146 (In Russian).

[4] Erdős, P. and Taylor, S.J. (1960). Some Problems Concerning the Structure of Random Walk Paths, *Acta Math.Acad.Sci.Hung.*, **11**, 137-162.

[5] Földes, A. and Révész, P. (1991). *Quadratic Variation of the Local Time of a Random Walk.* Preprint.

[6] Móri, T.F. (1992). *On the Strong Law of Large Numbers for Logarithmically Weighted Sums.* Preprint.

[7] Révész, P. (1990). *Random Walk in Random and Non-Random Environments.* World Scientific Publishing, Singapore.

Recent Adv. in Stat. and Prob., pp. 269-276
J. Pérez Vilaplana and M. L. Puri (Eds)

L_2–Rate of Clustering for Some Gaussian Processes

K. GRILL
Institut für Statistik und Wahrscheinlichkeitstheorie, Technische Universität Wien, Wiedner Hauptstraße 8–10, A–1040 Wien, Austria

Abstract – We investigate the upper limiting behaviour of the L_2-distance of the normalized trajectories $W(T.)/\sqrt{2T\log\log T}$ of a Wiener process from Strassen's class. We show that the right limiting rate is $(\log\log T)^{-2/3}$, which is the same as for the sup-norm which was investigated in a previous paper by the author. In addition, we are able to identify the limiting constant. The method employed can be used to obtain similar results for more general Gaussian processes.

1. INTRODUCTION

Let $(W(t), t \geq 0)$ be a standard Wiener process defined on the probability space $(\Omega, \mathcal{F}, \mathbf{P})$. Furthermore, let

$$\mathcal{S} = \{f(t) = \int_0^t g(u)du : t \in [0,1], \int_0^1 (g(u))^2 du \leq 1\}.$$

Strassen's[9] law of the iterated logarithm states that as $T \to \infty$,

$$\eta_T(t) = \frac{W(Tt)}{\sqrt{2T\log\log T}}, \quad t \in [0,1]$$

is relatively compact with respect to the sup-norm and has set of limit points $\mathcal{S}$. In the present paper, we are going to investigate the upper limiting behaviour of

$$d(T) = \inf\{\ \|\eta_T - f\|_2 : f \in \mathcal{S}\ \}. \tag{1}$$

The most natural norm to study for this distance is of course the sup-norm. This question has been studied by Bolthausen [1],Grill [4,5], and Goodman and Kuelbs [2,3]. The latter also give an upper bound to the limiting rate in (1). They also give a lower bound to the limitig rate with respect to the square norm, using a technique similar to the one developed in the present paper.
In [5], the author proves that the right limiting rate for the sup-norm is $(\log\log T)^{-2/3}$. As the L_2-norm is majorized by the sup-norm, we see that this result gives also an upper bound for the L_2-rate of convergence. In the present paper, it will be shown that $(\log\log T)^{-2/3}$ is the right rate in this case, too. In addition, the "nicety" of the L_2-norm will enable us to calculate the limiting constant not only for the Wiener process, but also for some related Gaussian processes.

In the following, we shall use a somewhat different setting, we may consider independent copies $W_n(t)$ of the Wiener process and define

$$Y_N(t) = W_N(t)/\sqrt{2\log N}.$$

In this case, too, $Y_N(t)$ will converge to Strassen's class. It seems worth noting that the first one to prove this convergence (in the case of the sup norm) was Lai [7]. Also, results by Kuelbs [6] and Ledoux and Talagrand [8] imply this convergence for more general norms under very general assumptions.
In addition, a standard rescaling argument (cf. Goodman and Kuelbs [2]) will yield the result in the original setting by replacing $\log N$ by $\log\log T$. Besides, the Wiener process in this setting may be replaced by some other Gaussian process, or even by some Gaussian random variable in a separable Hilbert space, with Strassen 's class replaced by the unit ball of the reproducing kernel Hilbert space. In priciple, the method presented in this paper is applicable in these cases, too, though we may be at a loss in trying to carry out the calculations.

2. THE WIENER PROCESS

Theorem 1 *Let $(W_N(t))$ be a sequence of Wiener processes, and*

$$Y_N(t) = W_N(t)/\sqrt{2\log N}.$$

Then

$$\limsup(\log N)^{2/3}\inf_{f\in S}\|Y_N - f\|_2 \le 3^{1/2}2^{-3/2}\pi^{-1/3}.$$

If, in addition, $W_N(t)$ is a sequence of independent Wiener processes, then

$$\limsup(\log N)^{2/3}\inf_{f\in S}\|Y_N - f\|_2 \ge 3^{1/2}2^{-3/2}\pi^{-1/3}.$$

Our main tool will be the Karhunen-Loeve expansion for the Wiener process:
If we let

$$K(s,t) = s\wedge t$$

denote the correlation function of the Wiener process, then the integral operator

$$K(f)(s) = \int_0^1 K(s,t)f(t)dt$$

has eigenvalues

$$\lambda_n = \left(\frac{(2n+1)\pi}{2}\right)^{-2}$$

and corresponding normed eigenfunctions

$$\phi_n(x) = \sqrt{2}\sin\frac{(2n+1)\pi x}{2}.$$

Karhunen-loeve theory now states that we can write

$$W(t) = \sum_{n=0}^{\infty}\sqrt{\lambda_n}X_n\phi_n(t),$$

where X_n is a sequence of i.i.d. standard normal random variables. In addition, for any function f from Strassen's class, we have

$$f(t) = \sum_{n=0}^{\infty} \sqrt{\lambda_n} f_n \phi_n(t)$$

with

$$\sum_{n=0}^{\infty} f_n^2 = 1.$$

Now, if we try to find the function from Strassen's class that has minimal L_2-distance from $W(t)$, this amounts to minimizing

$$\sum_{n=0}^{\infty} \lambda_n (X_n - f_n)^2$$

under the constraint

$$\sum_{n=0}^{\infty} f_n^2 = 2 \log N. \tag{2}$$

This can be achieved by the usual Lagrange method: We differentiate

$$\sum_{n=0}^{\infty} \lambda_n (X_n - f_n)^2 + \lambda \sum_{n=0}^{\infty} f_n^2$$

with respect to f_n, and set the derivative equal to zero, obtaining

$$f_n = \frac{\lambda_n X_n}{\lambda + \lambda_n}.$$

Of course, λ has to be chosen in such a way as to satisfy the constraint (2). So, our task is to pin down the joint distribution of

$$d(\lambda) = \sum_{n=0}^{\infty} \lambda_n \left(\frac{X_n \lambda}{\lambda + \lambda_n} \right)^2$$

and

$$I(\lambda) = \sum_{n=0}^{\infty} \left(\frac{X_n \lambda_n}{\lambda + \lambda_n} \right)^2 .$$

To this end, let us consider the moment generating function

$$\Psi(x,t) = \mathbf{E} \exp(xI + td).$$

We get

$$\Psi(x,t) = \prod_{n=0}^{\infty} \left(1 - \frac{2x\lambda_n^2 + 2t\lambda^2\lambda_n}{(\lambda + \lambda_n)^2} \right)^{-1/2} .$$

using the equation

$$\prod_{n=0}^{\infty} (1 + x\lambda_n) = \cosh \sqrt{x},$$

we get

$$\Psi(x,t) = \frac{\cosh \sqrt{\mu}}{\sqrt{\cosh \sqrt{\rho_1} \cosh \sqrt{\rho_2}}},$$

where we have put

$$\mu = 1/\lambda$$

and $\rho_{1,2}$ are the roots of the equation

$$\rho^2 - 2(\mu - t)\rho + \mu^2(1 - 2x) = 0.$$

now, we may use Markov's inequality to obtain an upper estimate for the probability that both I and d exceed some given values:

$$\mathbf{P}(I > I_0, d > d_0) \le \exp(-xI_0 - td_0)\Psi(x, t),$$

which holds for all positive x and t. In order to use this inequality, it seems more convenient to rewrite it in terms of ρ_1 and ρ_2. From the quadratic equation it follows that

$$t = \mu - \frac{\rho_1 + \rho_2}{2}$$

and

$$x = \frac{1}{2} - \frac{\rho_1\rho_2}{2\mu^2}.$$

Thus, our estimate becomes

$$\mathbf{P}(I > I_0, d > d_0) \le \tag{3}$$

$$\le \exp\left(-\frac{I_0}{2} + \frac{\rho_1\rho_2}{2\mu^2}I_0 - \mu d_0 + \frac{\rho_1}{2}d_0 + \frac{\rho_2}{2}d_0 + \log\cosh\sqrt{\mu} - \frac{1}{2}\log\cosh\sqrt{\rho_1} - \frac{1}{2}\log\cosh\sqrt{\rho_2}\right).$$

If we try to minimize the upper bound in equation (3), it turns out that we have to choose ρ_1 close to $-\pi^2/4$, and if we look at the same time for the value of μ that makes this expression a maximum, we have to choose $\rho_2 = \mu$ in such a way that

$$d_0 = -\frac{\rho_1}{\mu^2}I_0 + \frac{1}{2\sqrt{\mu}}\tanh\sqrt{\mu}. \tag{4}$$

As we are mainly interested in the case where μ is large, we have

$$\tanh\mu \approx 1$$

and expressing d_0 by equation (4) we get

$$\begin{aligned}
&\mathbf{P}\big(I > I_0, d > d_0\big) \le \\
&\le \exp\left(-I_0/2 + \frac{\rho_1}{\mu}I_0 - \frac{\sqrt{\mu}}{4}\tanh\sqrt{\mu} + \frac{1}{2}\log\cosh\sqrt{\mu} - \frac{1}{2}\log\cosh\sqrt{\rho_1}\right) = \\
&= \exp\left(-I_0/2 + \frac{\rho_1}{\mu}I_0 + \frac{\sqrt{\mu}}{4} + o(\sqrt{\mu})\right)
\end{aligned}$$

At this point, we let $W_N(t)$ be a sequence of Wiener processes, and define

$$\begin{aligned}
I_{0N} &= 2\log N \\
\mu_N &= (\pi^2 I_N)^{2/3}(1 + \varepsilon)
\end{aligned}$$

and by equation (4),

$$d_{0N} = \frac{\pi^2}{4\mu_N^2} I_{0N} + \frac{1}{\sqrt{2}\mu_N}.$$

Furthermore, let d_N and I_N be the analogues of I and d calculated for W_N. From the above, it follows that for any $\mu = 1/\lambda$ the inequality

$$\mathbf{P}(I_N(\lambda) > I_{0N}, d_N(\lambda) > d_{0N}) \leq \exp(-\log N - C(\varepsilon)(\log N)^{1/3})$$

holds. We need, however, an estimate for the probability

$$\mathbf{P}(\exists\lambda : I_N(\lambda) > I_{0N}, d_N(\lambda) > d_{0N}) \tag{5}$$

To this end, we choose values

$$\mu_{n,m} = (\log N)^{1/2} + m(logN)^{-3/2}$$

and let

$$\lambda_{n,m} = 1/\mu_{n,m}$$

and

$$M = (\log N)^3.$$

Using a few more applications of Markov's inequality , we obtain that the probabilities

$$\begin{aligned}
&\mathbf{P}\Big(I_N(\lambda_{N,0}) > I_{0N}\Big),\\
&\mathbf{P}\Big(d_N(\lambda_{N,M}) > d_{0N}\Big),\\
&\mathbf{P}\Big(\exists\lambda \in [\lambda_{N,m}, \lambda_{N,m+1}] : |I_N(\lambda) - I_N(\lambda_{N,m}| > 1\Big),\\
&\mathbf{P}\Big(\exists\lambda \in [\lambda_{N,m}, \lambda_{N,m+1}] : |d_N(\lambda) - d_N(\lambda_{N,m}| > (\log N)^{-2/3}\Big)
\end{aligned}$$

can all be overestimated by $exp(-\log N - c(\varepsilon)(\log N)^{1/3}$. This finally yields the same estimate for the probability (5). From this point, an application of the Borel-Cantelli lemma and some algebra give us the first part of Theorem 1. For the lower part, we now assume that $W_N(t)$ is a sequence of independent Wiener processes. We want to get a lower estimate for the probabilities in question. We let

$$\begin{aligned}
I_0 &= 2\log N\\
\rho_2 &= \mu = (\pi^2 I_N)^{2/3}(1+\varepsilon),\\
\lambda &= 1/\mu,
\end{aligned}$$

and define $\rho_1 \approx -\pi^2/4$ by the equation

$$\frac{1}{2\sqrt{-\rho_1}} \tan\sqrt{-\rho_1} = \frac{I_0}{\mu} + d_0,$$

where

$$d_0 = \frac{\pi^2}{4\mu_N^2} I_0 + \frac{1}{\sqrt{\mu_N}}.$$

Now, let

$$\begin{aligned}
\eta &= \varepsilon/8,\\
n_0 &= 0,\\
n_1 &= 1\\
n_{j+1} &= \lceil (1+\eta)n_j \rceil.
\end{aligned}$$

We have

$$d(\lambda) = \sum_{n=0}^{\infty} \lambda_n \left(\frac{X_n \lambda}{\lambda + \lambda_n}\right)^2 \geq \sum_{j=0}^{J} \inf_{n_j \leq n < n_{j+1}} \lambda_n \left(\frac{\lambda}{\lambda + \lambda_n}\right)^2 Z_j$$

with

$$Z_j = \sum_{n_j \leq n < n_{j+1}} X_n^2$$

and

$$I(\lambda) \geq \sum_{j=0}^{J} Z_j \inf_{n_j \leq n < n_{j+1}} \left(\frac{\lambda_n}{\lambda + \lambda_n}\right)^2.$$

We choose J in such a way that $n_J = \sqrt{\mu}/\eta$. Now, let

$$z_0 = \frac{1 + \rho_2 \lambda_0}{1 + \rho_1 \lambda_0} + 2\eta\sqrt{\mu},$$

and for $1 \leq j \leq J$

$$z_j = \sum_{n_j \leq n < n_{j+1}} \frac{1 + \rho_2 \lambda_n}{1 + \rho_1 \lambda_n}.$$

If $Z_j > z_j$for all $j = 0, \ldots, J$ then it is readily obtained that

$$I(\lambda) > 2 \log N$$

and

$$d(\lambda > d_0(1 + \varepsilon)^{-1}.$$

Now, Z_j has a χ^2-distribution with $\nu_j = n_{j+1} - n_j$ degrees of freedom. We use the following lower estimate

$$\mathbf{P}\left(\chi_n^2 > x\right) \geq c\frac{\sqrt{n}}{x}\left(\frac{ex}{n}\right)^{n/2} \exp\left(-\frac{x}{2}\right),$$

which holds for $x \geq n$. After some calculation, we obtain the lower bound

$$\mathbf{P}\left(I(\lambda) > 2 \log N,\, d(\lambda) > d_0\right) \geq \prod_{j=0}^{J} \mathbf{P}\left(Z_j > z_j\right) \geq \exp\left(-\, logN + \varepsilon\sqrt{\mu}/8\right).$$

and the second part of Theorem 1 follows by another application of the Borel-Cantelli lemma.

3. MORE GENERAL PROCESSES

The methods developed in the preceding section are fairly general. In fact, closer inspection shows that whenever we are able to calculate the eigenvalues of the associated linear operator, we can follow through along the lines of the above proof. Basically, we only need the value of the smallest eigenvalue and their asymptotic behaviour. In this section, we shall look at two simple examples, namely the d-dimensional Wiener process and the Wiener sheet.

For the d-dimensional Wiener process, the eigenvalues and eigenfunctions are the same as for the one-dimensional case. We only have to replace the standard normal random variables X_n by vectors of d independent standard normal random variables. Thus the moment generating function $\Psi(x, t)$ has to be replaced by its d-th power, and along the lines of the proof of Theorem 1 we get

Theorem 2 *Let $(W_N(t)$ be a sequence of d-dimensional Wiener processes,*

$$\mathcal{S}_d = \left\{ f \in \mathbb{R}^d : f(x) = \int_0^x g(u)\,du : \int_0^1 g^2(u)\,du = 1 \right\},$$

and $Y_N(t) = W_N(t)/\sqrt{2\log N}$. Then

$$\limsup\,(\log N)^{2/3} \inf_{f\in\mathcal{S}_d} \|Y_N - f\| \le d^{2/3}\,3^{1/2}\,2^{-3/2}\,\pi^{-1/3}.$$

If, in addition, $W_N(t)$ is a sequence of independent d–dimensional Wiener processes, then

$$\limsup\,(\log N)^{2/3} \inf_{f\in\mathcal{S}_\Gamma} \|Y_N - f\| \ge d^{2/3}\,3^{1/2}\,2^{-3/2}\,\pi^{-1/3}.$$

Now, let us turn to the Wiener sheet. This is defined as a Gaussian process with parameter space $[0,1]^d$ with expectation zero and covariance

$$\mathbf{E}(W(s)W(t)) = (s_1 \wedge t_1)\ldots(s_d \wedge t_d).$$

The limiting set in this case is

$$\mathcal{S} = \left\{ f(s) = \int_0^{s_1}\cdots\int_0^{s_d} g(t)dt_1\ldots dt_d : \int_0^1\cdots\int_0^1 g^2(t)dt_1\ldots dt_d = 1 \right\}.$$

In this case, for $n \in \mathbb{N}^d$, we have the eigenvalues

$$\Lambda_n = \lambda_{n_1}\ldots\lambda_{n_d}$$

with corresponding eigenfunctions

$$\Phi_n(s) = \phi_{n_1}(s_1)\ldots\phi_{n_d}(s_d).$$

this yields

$$\Psi(x,t) \approx \frac{\exp\left(\sqrt{\mu}\left(\frac{\log\sqrt{\mu}}{\pi}\right)^{d-1}\right)}{\sqrt{\left(\rho_1 + \left(\frac{\pi^2}{4}\right)^d\right)\exp\left(\sqrt{\rho_1}\left(\frac{\log\sqrt{\rho_1}}{\pi}\right)^{d-1}\right)}}.$$

The final result is

Theorem 3 *Let $(W_N(t)$ be a sequence of Wiener sheets, and*

$$Y_N(t) = W_N(t)/\sqrt{2\log N}.$$

Then

$$\limsup \frac{(\log N)^{2/3}\inf_{f\in\mathcal{S}}\|Y_N - f\|}{(\log\log N)^{2(d-1)/3}} \le 3^{1/2}\,2^{-3/2}\,\pi^{-1/3}\left(\frac{2}{9\pi^3}\right)^{2(d-1)/3}.$$

If, in addition, $W_N(t)$ is a sequence of independent Wiener processes, then

$$\limsup \frac{(\log N)^{2/3}\inf_{f\in\mathcal{S}}\|Y_N - f\|}{(\log\log N)^{2(d-1)/3}} \ge 3^{1/2}\,2^{-3/2}\,\pi^{-1/3}\left(\frac{2}{9\pi^3}\right)^{2(d-1)/3}.$$

REFERENCES

[1] BOLTHAUSEN, E. (1978). On the Speed of Convergence in Strassens Law of the Iterated Logarithm. *Ann. Probab.*, **6**, 668–672.

[2] GOODMAN, V. AND KUELBS, J. (1991a). Rate of Clustering in Strassens LIL for Brownian Motion. *J. Theoretical Probab.*, **5**, 265–310.

[3] GOODMAN, V. AND KUELBS, J. (1991b). Rate of Clustering for Some Gaussian Self-similar Processes. *Probab. Th. Rel. Fields*, **88**, 47–75.

[4] GRILL, K. (1987). On the Rate of Convergence in Strassens Law of the Iterated Logarithm. *Probab. Th. Rel. Fields*, **74**, 583–589.

[5] GRILL, K. (1992). Exact Rate of Convergence in Strassens Law of the Iterated Logarithm. *J. Theoretical Probab.*, **5**, 197–204.

[6] KUELBS, J. (1976). A Strong Convergence Theorem for Banach Space Valued Random variables. *Ann. Probab.*, **4**, 744–771.

[7] LAI, T.L. (1974). Reproducing Kernel Hilbert Spaces and the Law of the Iterated Logarithm for Gaussian Processes. *Z. Wahrscheinlichkeitstheorie verw. Gebiete*, **29**, 7–19.

[8] LEDOUX, M. AND TALAGRAND, M. (1988). Characterization of the Law of the Iterated Logarithm in Banach Spaces. *Ann. Probab.*, **16**, 1242–1268.

[9] STRASSEN, V. (1964) An Invariance Principle for the Law of the Iterated Iogarithm. *Z. Wahrscheinlichkeitstheorie verw. Gebiete*, **3**, 211–226.

Recent Adv. in Stat. and Prob., pp. 277-284
J. Pérez Vilaplana and M. L. Puri (Eds)

On the Effects of Noise in Systems Involving Products of Random Matrices

C.C. HEYDE
Department of Statistics, Columbia University, New York, NY 10027, USA; and School of Mathematical Sciences, IAS, Australian, National University, GPO Box 4, Canberra, ACT 2601, Australia

Abstract – This paper is concerned with the separate effects of multiplicative and additive noise on the long term behaviour of simple stochastic vector difference equation systems. The effects of noise on the growth rate (maximum Lyapunov exponent) and indeed on the whole Lyapunov spectrum are discussed. Application to the delayed flowering of biennial plants and to the formulation of harvesting strategies for fish populations are considered.

1. INTRODUCTION

In this paper we shall be concerned with models of the form

$$Y_{n+1} = X_{n+1} Y_n + \epsilon_{n+1} \tag{1}$$

where $\{Y_n\}$ is a sequence of vectors of dimension k, $\{X_n\}$ a stationary ergodic sequence of $k \times k$ random matrices with finite mean matrix and the additive noise vectors $\{\epsilon_n\}$ satisfy $E\epsilon_n = 0$ for each n. We distinguish noise of two kinds in this model; *multiplicative* and *additive*. The multiplicative noise can be seen more clearly by rewriting the model in the form

$$Y_{n+1} = [EX + (X_{n+1} - EX)]Y_n + \epsilon_{n+1} \ .$$

We shall focus on the context of growth and decline rather than that of asymptotic stationary. In this setting we shall observe that the multiplicative noise usually plays a dominant role in determining the growth rate. The additive noise will not, in many cases, influence the growth rate and indeed it may stabilize the process.

2. THE ADDITIVE NOISE

In this section we shall informally examine the effect of the additive noise in (1) on the asymptotics of the system.

Suppose that there is a past history sequence of σ-fields $\{\mathcal{F}_n\}$ for which $E(\epsilon_{n+1}|\mathcal{F}_n) = 0$ a.s. for each n. Then, $E(Y_{n+1}|\mathcal{F}_n) = X_{n+1}Y_n$ a.s. and, if the matrices X_n are nonsingular,

$$\{W_n = X_1^{-1} \dots X_n^{-1} Y_n, \mathcal{F}_n, n \geq 1\}$$

is a martingale. The equation (1) translates into

$$W_{n+1} = W_n + X_1^{-1} \dots X_{n+1}^{-1} \epsilon_{n+1} ,$$

so that

$$W_n = W_0 + X_1^{-1} \sum_{k=1}^{n} X_1^{-1} \dots X_k^{-1} \epsilon_k .$$

Thus, if

$$\sum_1^\infty X_1^{-1} \dots X_k^{-1} \epsilon_k \text{ converges a.s.}$$

we have $W_n \xrightarrow{a.s.} W$ and $Y_n \sim X_n \dots X_1 W$ a.s. on the set $\{W > 0\}$. That is, the asymptotics of Y_n come from those of the product $X_n \dots X_1$.

This crude analysis does oversimplify the subtlety of the problem, for example it may well happen that $P(W = 0) = 1$, but the argument can be formalized in many important cases.

As a prelude to discussing more subtle behaviour we need to introduce a suitable measure of growth. The *Lyapunov exponent* of a sequence of k dimensional random vectors $\{\xi_n\}$ is

$$\lambda = \limsup_{n\to\infty} n^{-1} \log ||\xi_n||$$

where $||\cdot||$ is the Euclidean norm in $\mathcal{R}^k$.

For a system of type (1) where the additive noise component is absent it is possible, under modest additional conditions, to obtain a striking characterization of its asymptotic behaviour. If $E \log_+ ||X_1^{-1}|| < \infty, E||X_1||$, the *multiplicative ergodic theorem of Oseledec* tells us that there are $r (1 \le r \le k)$ real numbers $\lambda_1 < \lambda_2 < \dots < \lambda_r$ and subspaces $E_1, E_2, \dots, E_r$ of $\mathcal{R}^k$ with dimension $\dim E_i = k_i, \sum_{i=1}^r k_i = k$, such that

$$\mathcal{R}^k = E_1 \oplus E_2 \oplus \dots \oplus E_r$$

and the exact Lyapunov exponent of the system $Y_{n+1} = X_{n+1} Y_n$ starting from $Y_0 \in E_i$ is λ_i, i.e.

$$\lambda_i = \lim_{n\to\infty} n^{-1} \log ||Y_n(Y_0)||$$

uniformly for all $Y_0 \in E_i$. The subspaces E_i are *random*. The λ's form the *Lyapunov spectrum.*

This beautiful and delicate result (Oseledec (1968)) does, however, mask the essential probabilistic behaviour of the system for there is an earlier ergodic theorem of Furstenberg and Kesten (1960) which gives

$$\lim_{n\to\infty} n^{-1} \log ||Y_n(Y_0)|| = \lambda_r \quad a.s.$$

independently of the initial population vector. Thus, the largest Lyapunov exponent has a very special role.

It is then an important question as to the effect of introducing additive noise to the system. Our heuristics have suggested that it will mostly not change the dominant Lyapunov exponent. However, a number of interesting effects are possible and one is that the introduction of additive noise may stabilize the system by making the Lyapunov spectrum collapse down to a single point $\max(0, \lambda_r)$. This would mean that *all* the sample paths would have the same asymptotic behaviour when suitably normalized.

Some sufficient conditions to ensure that there is just a single point in the Lyapunov spectrum have been provided by Du, Nhung and Dong (1991). These are, in addition to the Oseledec theorem conditions, that

i) $\{\epsilon_n\}$ are independent, independent of $\{X_n\}$, square integrable and, for some n_0 the scalar product $< \alpha, \epsilon_{n_0} >$ has a continuous distribution for any $0 \neq \alpha \epsilon \mathcal{R}^k$,

ii) $\lim sup_{n \to \infty} n^{-1} \log E||\epsilon_n||^2 \leq 0$ (second moments do not increase too fast),

iii) $P(n^{-1} \log ||\epsilon_n|| > -\eta \; i.o.) = 1$ for any $\eta > 0$ (the noise is not too weak).

3. THE MULTIPLICATIVE NOISE

Our principal concern in this paper is the dominant effects on the asymptotics of the system. Having provided a heuristic discussion to suggest that the additive noise is often unimportant we shall now concentrate on the multiplicative noise by considering the system (1) in which the additive component is absent.

This system has been extensively used in demographic studies, often in a purely deterministic setting, and it is of interest and importance to contrast the stochastic and deterministic formulations. In a demographic setting $\{Y_n\}$ would represent, say, population vectors giving numbers in age classes and $\{X_n\}$ would be non-negative matrices of age specific fertility and mortality rates. Multiplicative noise can be thought of as environmental stochasticity and this is generally regarded as dominating the effect of demographic stochasticity (additive noise).

In a deterministic setting we have $X_n \equiv X$ and $Y_n = X^n Y_0$. Using the Perron-Frobenius theory, the asymptotic behaviour of the system is determined by the largest eigenvalue of X; suppose it is ρ. Then, for any $0 \neq c \in \mathcal{R}^k$,

$$\lim_{n \to \infty} n^{-1} \log c' Y_n = \log \rho \; .$$

Now let us consider the mean of the stochastic system. Suppose that

$$E(X_{n+1} | Y_n, Y_{n-1}, \ldots) = X \quad a.s.$$

Then, $EY_n = X^n EY_0$. The stochastic mean behaves like the deterministic system and

$$\lim_{n \to \infty} n^{-1} \log c' EY_n = \log \rho \; .$$

However, $\log \rho$ is *not* the stochastic growth rate parameter (largest Lyapunov exponent). Assuming that $E \log_+ ||X_1|| < \infty$, the Furstenberg-Kesten ergodic theorem gives

$$\lim_{n \to \infty} n^{-1} \log c' Y_n = \log \mu \; a.s., \tag{2}$$

say, where $\log \mu < \log \rho$ (and μ does not depend on Y_0 or c). The inequality is strict except in trivial cases.

The scalar case $k = 1$ reveals what underlies this difference. Taking logs in $Y_n = X_n X_{n-1} \ldots X_0 Y_0$, we have

$$\log Y_n = \sum_{k=1}^{n} \log X_k + \log Y_0$$

and an application of the ordinary ergodic theorem, together with Jensen's inequality, gives

$$n^{-1} \log Y_n \xrightarrow{a.s.} E \log X_1 < \log EX_1 = \lim_{n \to \infty} n^{-1} \log EY_n \; .$$

The result (2) is very useful, and under modest additional conditions it can be supplied with an associated central limit theorem from which confidence intervals for the growth rate parameter can be provided (e.g. Heyde and Cohen (1985)). Nevertheless, there are problems with the crudeness of the result (2) which gives, for example, $n^{-1}\log Y_{i,n} \xrightarrow{a.s.} \log\mu$ for each entry $Y_{i,n}$ in $Y_n = (Y_{1,n},\ldots,Y_{k,n})'$. It is highly advantageous when asymptotic results can be obtained for $Y_{i,n}$ rather than $\log Y_{i,n}$. However, such results require much more detailed assumptions on the underlying model – for example those of a branching process.

The growth rate parameter (largest Lyapunov exponent) $\log\mu$ being of fundamental importance, its calculation in particular cases is a major issue.

Define $U_n = Y_n/|Y_n|$, where

$$|Y_n| = Y_{1,n} + \ldots + Y_{k,n} \ .$$

In a demographic context $\{U_n\}$ is called the *age structure* and the defining relation (1) for the system can be rewritten as

$$U_{n+1} = X_{n+1}U_n/|X_{n+1}U_n| \ .$$

Then, under mild additional conditions ($\{X_n\}$ follows a countable state space Markov chain is sufficient), the joint distribution (X_n, U_n) converges to an equilibrium distribution, that of (X, U) say, where

$$U \stackrel{d}{=} XU/|XU| \ .$$

Suppose that the probability measures corresponding to X, U are $\alpha(X), \beta(U)$ respectively. Then,

$$\log\mu = \iint \log|XU|\,d\alpha(X)\,d\beta(U) = E\log|XU| \ .$$

This result of Furstenberg and Kesten (1960) provides a formal answer to the problem but explicit calculations have only been successful in a small number of special cases. See for example Tuljapurkar (1990).

4. SOME EXAMPLES

We shall now give two simple examples where the calculation of the largest Lyapunov exponent is vital, and troublesome. The examples are of cases which can usefully be modelled with matrices $\{X_n\}$ having a single random element. The most variable part of the system is taken as random and the other quantities are treated as constants.

(1) *Biennial Life Strategy* [Roerdink (1988)]

Strictly biennial plants flower only in the second year of their life and then they die. However, many biennial species actually exhibit delayed flowering – individuals may flower in the third year or even later.

At face value, delayed flowering seems to be bad strategy for a species. For one thing, more seeds need to be produced to attain the same rate of increase.

It turns out that randomly varying environment have a vital role here. Environmental fluctuations are large and the average rate of increase of the population is low.

To model this example we take t as representing the time in years (taken just prior to the flowering season) and let $Y_{1,t}$ and $Y_{2,t}$ be, respectively the numbers of individuals one year old at t and older than one year at t. Write s for the survival rate per year of individuals older than one year $(0 < s < 1)$, f for the fraction of individuals older than one year that flower in a given year ($0 \leq f \leq 1$) and ϕ_t for the number of offspring per flowering plan in year t. The most variable quantity is ϕ_t and we shall take the ϕ_t to be i.i.d. with finite mean $E\phi$.

Assuming density independent population growth we then obtain the system of difference equations

$$\begin{bmatrix} Y_{1,t+1} \\ Y_{2,t+1} \end{bmatrix} = \begin{bmatrix} 0 & f\phi_t \\ s & (1-f)s \end{bmatrix} \begin{bmatrix} Y_{1,t} \\ Y_{2,t} \end{bmatrix} .$$

In the usual notation, the population mean

$$EY_t = (EX)^t EY_0 \sim \left(\rho(f)\right)^t EY_0$$

as $t \to \infty$ where $\rho(f)$ is the largest eigenvalue of EX:

$$\rho(f) = \frac{1}{2}\left[s(1-f) + \left(s^2(1-f)^2 + 4fE\phi\right)^{\frac{1}{2}}\right] .$$

(This makes sense for the growth of the total population even when $f = 0$ or 1.) Further,

$$\frac{\partial \rho(f)}{\partial f} > 0 \qquad \text{if} \qquad E\phi > s$$

(and $0 < s < 1$) so in all cases of biological releverence the average population grows at its largest rate when $f = 1$. Thus, in a deterministic setting it is not profitable to delay flowering. It turns out, however, that this is *not* the case when random environments are considered.

In the random environment case we need to identify the growth rate $\log \mu(f)$ as a function of f. Instead of the age structure it is useful here to consider

$$r_t(f) = Y_{1,t}/Y_{2,t}$$

and note that the r's satisfy the continued fraction form

$$r_{t+1}(f) = f\phi_t/[sr_t(f) + (1-f)s] .$$

Since there is a stationary limiting age structure, $r_t(f) \xrightarrow{d} r(f)$ where

$$r(f) \stackrel{d}{=} f\phi/[sr(f) + (1-f)s] \tag{3}$$

Then, it is easily seen that

$$\log \mu(f) = E \log[sr(f) + (1-f)s] . \tag{4}$$

If ϕ has an absolutely continuous distribution then (3) gives an integral equation for the density function of $r(f)$. It can be solved if ϕ has a gamma distribution and this was done by Roerdink (1988). Even then, (4) remains rather inaccessible. Roerdink calculated $\log \mu(f)$ numerically in three cases. In one case there was a clear maximum for $f^* < 1$ and in the other cases the maximum, although less than one, was very close to one. He was, however, also able to show for averages over two time periods, that

$$\left(\frac{\partial}{\partial f} \log \mu(f)\right)_{f=1} = -\infty .$$

It turns out, furthermore, that this result continues to hold in general, indicating that delayed flowering is always a preferable strategy in a random environment.

For simple models such as this one, the arguments of Section 2 can easily be applied to check the circumstances under which an additive noise can safely be disregarded in considering $\log \mu$.

(2) Harvesting

Here we consider a population model for some fish species. The population contains newborn, immature and adult individuals. Once fish reach an adult state at which they are large enough to be caught we assume that there is constant probability of survival from year to year. Let β be the average fertility of an adult, while p_n, p_i, p_a are, respectively, the survival probabilities from year to year of newborns, immatures and adults. Then, if we wish to harvest a constant fraction f of the adult stock each year we have the deterministic system

$$\begin{bmatrix} Y_{1,t+1} \\ Y_{2,t+1} \\ Y_{3,t+1} \end{bmatrix} = \begin{bmatrix} 0 & 0 & (1-f)\beta \\ p_n & 0 & 0 \\ 0 & p_i & (1-f)p_a \end{bmatrix} \begin{bmatrix} Y_{1,t+1} \\ Y_{2,t+1} \\ Y_{3,t+1} \end{bmatrix}$$

where $Y_{1,t}, Y_{2,t}, Y_{3,t}$ are, respectively, the numbers of newborns, immatures and adults at the end of year t.

Now the maximum eigenvalue of the matrix of fertility and mortality rates is approximately

$$\rho = 1 + \frac{(1-f)(p_a + \beta p_n p_i) - 1}{\left(3 - 2(1-f)p_a\right)}$$

and for f to be small enough that $\rho > 1$ we must have

$$f < 1 - (p_a + \beta p_n p_i)^{-1} .$$

Furthermore, we know that a corresponding stochastic formulation will require a more conservative harvesting policy to ensure a positive growth rate.

Suppose now that the adult fertility is random and replace β by β_t, the $\{\beta_t\}$ being i.i.d. We need to find a bound on f to ensure that $\log \mu(f) > 1$.

Let

$$\begin{aligned} r_t(f) &= Y_{2,t}/Y_{1,t} \\ s_t(f) &= Y_{3,t}/Y_{1,t} . \end{aligned}$$

The system equations give

$$r_{t+1}(f) = \frac{p_n}{(1-f)\,\beta_t s_t(f)},$$

$$s_{t+1}(f) = \frac{p_i r_t(f) + (1-f)\,p_a s_t(f)}{(1-f)\,\beta_t s_t(f)},$$

and since there is a stationary limiting age structure, $r_t(f) \xrightarrow{d} r(f), s_t(f) \xrightarrow{d} s(f)$, where

$$r(f) \stackrel{d}{=} \frac{p_n}{(1-f)\,\beta s(f)},$$

$$s(f) \stackrel{d}{=} \frac{p_i r(f) + (1-f)\,p_a s(f)}{(1-f)\,\beta s(f)} .$$

Also,

$$\log \mu(f) = E \log[(1-f)\beta s(f)] \ .$$

Despite this deceptively simple form there is no straightforward way to find the maximum f for which $E[\log(1-f)\beta s(f)] > 1$.

5. ASYMPTOTICS FOR DIFFERENCE EQUATIONS

The basic difference equation which underlies the delayed flowering example is

$$Y_{2,t+1} = (1-f)sY_{2,t} + sf\phi_{t-1}Y_{2,t-1} \ , \tag{5}$$

while that for the harvesting example is

$$Y_{3,t+1} = p_a(1-f)Y_{3,t} + p_i p_n(1-f)\beta_{t-1}Y_{3,t-1} \ . \tag{6}$$

Taking $\xi_t = c^{-t}Y_{2,t}$ with c the positive root of

$$c^2 = (1-f)sc + sfE\phi_1$$

in the former case and $\eta_t = d^{-t}Y_{3,t}$ with d the positive root of

$$d^3 = p_a(1-f)d^2 + p_i p_n(1-f)E\beta_1$$

in the latter case, the equations (5), (6) can be rewritten in the forms

$$\xi_{t+1} = a_1\xi_t + a_2\xi_{t-1} + \epsilon_{t-1}^{(\xi)} \ , \tag{7}$$

where $a_1 + a_2 = 1, E(\epsilon_{t+1}^{(\xi)}|\mathcal{F}_{t-1}) = 0$ a.s., and

$$\eta_{t+1} = b_1\eta_t + b_3\eta_{t-2} + \epsilon_{t+1}^{(\eta)} \ , \tag{8}$$

where $b_1 + b_3 = 1, E(\epsilon_{t+1}^{(\eta)}|\mathcal{F}_{t-2}) = 0$ a.s.

The equations (7) and (8) are close in form to those discussed by McQueen (1973). His setting was of a system

$$X_{n+1} = a_1X_n + a_2X_{n-1} + \ldots + a_kX_{n-k+1} + \xi_{n+1} \tag{9}$$

where $E(\xi_{n+1}|\mathcal{F}_n) = 0$ a.s., $\sum_{i=1}^k a_i = 1$, and he obtained conditions under which the system is essentially driven by the corresponding homogeneous difference equation and the noise, now in an additive form, is asymptotically irrelevant. Indeed, let $r_i, 1 \le i \le k$, be the roots of the characteristic equation

$$r^k - (a_1r^{k-1} + a_2r^{k-2} + \ldots + a_k) = 0$$

corresponding to the homogeneous difference equation

$$x_{n+1} = a_1x_n + a_2x_{n-1} + \ldots + a_kx_{n-k+1} \ . \tag{10}$$

Suppose that $\sup_n E|X_n| < \infty$ and $|r_i| \le 1$ for all i with those roots for which $|r_i| = 1$ being simple. Then, on a suitable expanded version of the probability space there exists a sequence of random variables $\{Y_n\}$ satisfying (10) such that $X_n - Y_n \xrightarrow{a.s.} 0$ as $n \to \infty$. However, even when these moment and root conditions are satisfied for (7) and (8), as can easily be achieved, the asymptotic behaviour is fundamentally different from (9). Systems such as (7) and (8) are intrinsically higher dimensional and do not stabilize a.s. as $n \to \infty$ in the way that the one dimensional system (9) does.

REFERENCES

[1] Du, N.H., Nhung, T.V. & Dong, L.S. (1991). *On the one-point Lyapunov spectrum of ergodic stationary difference systems spectrum of ergodic stationary.* Preprint.

[2] Furstenberg, H. & Kesten, H. (1960). Products of random matrices. *Ann. Math. Statist.*, **31**, 457–469.

[3] Heyde, C.C. & Cohen, J.E. (1985). Confidence intervals for demographic projections based on products of random matrices. *Theor. Popul. Biol.*, **27**, 120–153.

[4] McQueen, J.B. (1973). A linear extension of the martingale convergence theorem. *Ann. Probab.*, **1**, 263–271.

[5] Oseledec, V.I. (1968). A multiplicative ergodic theorem: Lyapunov characteristic numbers for dynamical systems. *Trans. Moscow. Math. Soc.*, **19**, 197–231.

[6] Roerdink, J.B.T.M. (1988). The biennial life strategy in a random environment. *J. Math. Biol.*, **26**, 199–215.

[7] Tuljapurkav, S. (1990). *Population Dynamics in Variable Environments.* Springer Lecture Notes in Biomathematics No. 85. Springer, New York.

Recent Adv. in Stat. and Prob., pp. 285-291
J. Pérez Vilaplana and M. L. Puri (Eds)

Asymptotic Normality of the Spectral Density Estimators for Periodically Correlated Stochastic Processes

J. LEŚKOW
Department of Statistics and Applied Probability, University of California, Santa Barbara, CA 93106, U.S.A.

Abstract – A stochastic process $\{X(t); t \in \mathbf{R}\}$ is periodically correlated (PC) when its first and second moments exhibit periodicity . Such nonstationary processes are widely applied in signal processing [Gardner (1991)], oceanography [Dragan and Yavorskii (1982)] or in climatology [Brier *et al* (1983)]. It is known that when the PC process $\{X(t);\ t \in \mathbf{R}\}$ is strongly harmonizable, its covariance function may be represented by a countable family of distributions $\{G_k\}$. In the event that distributions $\{G_k\}$ have densities $\{g_k\}$ we can consider the problem of estimating distributions of the PC process $\{X(t);\ t \in \mathbf{R}\}$ in terms of estimating corresponding spectral densities $\{g_k\}$. Following previous results of the author the estimators of the densities $\{g_k\}$ of the PC process will be presented here. It will be also demonstrated that those estimators are asymptotically normal if the considered process has an appropriate φ mixing property.

1. INTRODUCTION

Let $\{X(t); t \in \mathbf{R}\}$ be a stochastic process which is nonstationary in a periodic way, *i.e.* its mean and covariance functions are periodic. In the sequel, such a process will be called *periodically correlated* (PC). Examples of such processes are frequently encountered in signal processing .

Take for instance $S(t)$ to be stationary noise and $f(t)$ to be a periodic, deterministic and unknown function . Then

$$X(t) = f(t) \cdot S(t)$$

is a nonstationary PC process and there is a need to nonparametrically estimate f from the data. The data may also be affected by the superposition of two periodic trends with different periods, *i.e.*

$$X(t) = Y(t)\cos(2\pi f_1 t + \varphi_1) + Z(t)\cos(2\pi f_2 t + \varphi_2)\ , \tag{1}$$

where $Y(t)$ and $Z(t)$ are independent, stationary processes . If the ratio f_1/f_2 in (1) is a rational number (f_1 and f_2 are commensurate frequencies) then the process $\{X(t); t \in \mathbf{R}\}$ is PC.

Another example of a PC process is

$$X(t) = \sum_{n=-\infty}^{\infty} \sum_{m=1}^{M} \delta_n(m) q_m(t - nT), \tag{2}$$

where $\delta = (\delta_1, ..., \delta_M)$ is a random indicator vector and $q = (q_1, ..., q_M)$ is the finite-energy pulse. In the terminology of Gardner (1986) the process defined in (2) has a cyclic autocorrelation which means exactly that it is PC in our sense.
The interesting question, which will be investigated in the forthcoming paper of Leskow (1992), is a generalization of nonstationary models based on periodicity to models based on almost periodicity (see also Hurd and Leskow (1992)).
The periodic model may also be applied to describing the ocean surface. If $\{X(t); t \in \mathbf{R}\}$ denotes a position of a point on the ocean surface then it is frequently assumed that such a process has a periodic mean and covariance.
Let now $B(t,\tau) = R(t, t+\tau)$, where R is the covariance of $\{X(t);\ t \in \mathbf{R}\}$. For a PC process $\{X(t);\ t \in \mathbf{R}\}$ it is possible to represent the function $B(t,\tau)$ in the form

$$B(t,\tau) = \sum_{k=-\infty}^{\infty} a(k,\tau) \exp(i(2\pi/T)kt) \tag{3}$$

so, basically, the covariance function of the PC process may be represented by a countable family of coefficient functions $\{a(k,\tau)\ :\ \tau \in \mathbf{R}, k - integer\}$. Assume now that the investigated PC process $\{X(t); t \in \mathbf{R}\}$ is harmonizable (see *e.g* Loève (1965)). Then, following Hurd (1989) we can represent this process in the form

$$X(t) = \sum_{k=-\infty}^{\infty} s_k(t) \exp(i(2\pi/T)kt),$$

where s_k are stationary and mutually orthogonal processes and T is the period of the covariance and the mean of the process. Moreover, in such a case the covariance of X may be represented as

$$R(s,t) = \int_{-\infty}^{\infty}\int_{-\infty}^{\infty} \exp(i\gamma_1 s - i\gamma_2 t) r_X(d\gamma_1, d\gamma_2)$$

where the measure r_X on $\mathbf{R}$ x $\mathbf{R}$ is concentrated on lines $\{y = x - (2\pi/T)k; k - integer\}$. Define now G_k as the distribution function corresponding to the restriction of r_X to the $k-th$ line. If such a distribution function G_k has a density g_k and the function $B(t,\tau)$ fulfills certain regularity assumptions then

$$a(k,\tau) = \int_{-\infty}^{\infty} \exp(i\gamma t) g_k(\gamma) d\gamma. \tag{4}$$

Therefore, the problem of estimating the coefficient functions of the PC process $\{X(t);$ $t \in \mathbf{R}\}$ may be resolved by nonparametric estimation of the functions $\{g_k\}$. Following Hurd and Leskow (1992b) we will call functions $\{g_k\}$ *spectral densities* corresponding to the PC process.
The estimator $\hat{g}_A(k,\gamma)$ of the unknown spectral density $g(k,\gamma)$ of the PC process $\{X(t);$ $t \in \mathbf{R}\}$ will be defined as:

$$\hat{g}_A(k,\gamma) = (1/2\pi) \int_{-A}^{A} K(B_A\tau) \hat{a}_A(\lambda_k, \tau) \exp(-i\gamma\tau) d\tau, \tag{5}$$

where

$$\hat{a}_A(k,\tau) = \begin{cases} (1/A) \int_{-A}^{A-\tau} X(t+\tau)X(t) \exp(-i(2\pi kt/T))dt & \text{if } \tau \geq 0, \\ (1/A) \int_{-\tau}^{A} X(t+\tau)X(t) \exp(-i(2\pi kt/T))dt & \text{otherwise.} \end{cases} \tag{6}$$

The covariance averaging kernel $K(\cdot)$ is taken to be even and bounded with bounded support on the real line. Here the sequence B_A is the "window width". We will assume that $B_A \to 0$ but $AB_A^3 \to \infty$.

From the research of Hurd and Leskow (1992a) it is known that the estimator $\hat{a}_A(\lambda_k, \tau)$ is consistent and asymptotically normal provided that the process fulfills certain regularity conditions. The aim of this paper is to show that the estimator defined in (5) is asymptotically normal provided that some additional conditions on the stochastic process $\{X(t);\ t \in \mathbf{R}\}$ are imposed.

2. MAIN RESULT

In the sequel, we will assume that the period T of the PC process $\{X(t);\ t \in \mathbf{R}\}$ is bounded. We need the following conditions:

(A.1) Assume that the PC process $\{X(t);\ t \in \mathbf{R}\}$ is strongly harmonizable, i.e. there exists a spectral process Z such that

$$X(t) = \int_{-\infty}^{\infty} \exp(i\lambda t) dZ(\lambda).$$

(A.2) The stochastic process $\{X(t);\ t \in \mathbf{R}\}$ is φ-mixing with $\varphi(t) = O(t^{-3})$. Moreover, the process $\{X(t);\ t \in \mathbf{R}\}$ has eight bounded moments, *i.e.*

$$\sup_{t \in \mathbf{R}} EX^8(t) \leq M_8 < \infty.$$

Consider also the following process

$$Z(t,\tau) = X(t) \cdot X(t+\tau) - E(X(t) \cdot X(t+\tau)), \tag{7}$$

where τ is the usual "lag" parameter.

For $\underline{\tau} = (\tau_1, \tau_2, \tau_3)$ let $V(t, \underline{\tau})$ correspond to the fourth-order moments of the process $\{X(t);\ t \in \mathbf{R}\}$, *i.e.*

$$V(t, \underline{\tau}) = EX(t)X(t+\tau_1)X(t+\tau_2)X(t+\tau_3). \tag{8}$$

(A.3) The function $V(t, \underline{\tau})$ defined in (8) is periodic in t, i.e. for each $\underline{\tau} \in \mathbf{R}$ we have that $V(t, \underline{\tau}) = V(t+T, \underline{\tau})$ for each $t \in \mathbf{R}$.

(A.4) The nonnegative covariance averaging kernel K has bounded support on the real line, *i.e.* there exists such M that $K(u) > 0$ only when $u \in [-M, M]$. We take $M = 1$ for simplicity . Moreover, the window width B_A is such that $B_A \to 0$ but $AB_A^3 \to \infty$.

(A.5) The process $\{X(t);\ t \in \mathbf{R}\}$ is Holder continuous in L^2 on the real line, *i.e.* there exists a constant H such that for each $t_1, t_2 \in \mathbf{R}$

$$E|X(t_1) - X(t_2)|^2 \leq H|t_1 - t_2|^2 .$$

From the earlier research of Hurd and Leskow (1992a) it is known that under the above assumptions the estimator (1.5) is mean-square consistent.

The central result of this paper is given below.

Theorem 2.1. *Let the assumptions (A.1) through (A.5) be fulfilled. Then the estimator*

$\hat{g}_A(k,\gamma)$ of the spectral density $g(k,\gamma)$ of the PC process $\{X(t);\, t\in\mathbf{R}\}$ is asymptotically normal, i.e. for each integer k and $\gamma\in\mathbf{R}$

$$\left(AB_A^3\right)^{1/2}\left(\hat{g}_A(k,\gamma)-g(k,\gamma)\right)\xrightarrow{d}N_2(0,V_k).$$

Here $\xrightarrow{d}$ denotes the convergence in distribution and the asymptotic variance matrix V_k equals

$$V_k=\begin{bmatrix} v_{11} & v_{12}\\ v_{21} & v_{22}\end{bmatrix},$$

where $v_{ij}=(1/T)\int_0^T P_{ij}(t)dt,\ i,j=1,2$ and

$$P_{ij}(t)=\lim_{A\to\infty}(B_A/2\pi)\int_{-1/B_A}^{1/B_A}\int_{-1}^{1}\int_{-1}^{1}\gamma_{ij}K(u_1)K(u_2)\cdot$$

$$\cdot\, EZ(t,u_1/B_A)Z(t+v,u_2/B_A)\,du_2\,du_1\,dv.$$

Moreover,

$$\begin{aligned}
\gamma_{11}&=\cos\left(\lambda_k(2t+v)\right)\cos\left(-\gamma(u_1+u_2)/B_A\right),\\
\gamma_{12}&=\cos\left(\lambda_k(2t+v)\right)\sin\left(-\gamma(u_1+u_2)/B_A\right),\\
\gamma_{21}&=\sin\left(\lambda_k(2t+v)\right)\cos\left(-\gamma(u_1+u_2)/B_A\right),\\
\gamma_{22}&=\sin\left(\lambda_k(2t+v)\right)\sin\left(-\gamma(u_1+u_2)/B_A\right),\ \textit{where } \lambda_k=(2\pi k/T).
\end{aligned}$$

3. PROOF

Observe first that if

$$\hat{a}_A^*(\lambda_k,\tau)=A^{-1}\int_0^A X(t+\tau)X(t)\exp(-i\lambda_k t)dt \tag{9}$$

then $A^{1/2}(\hat{a}_A(\lambda_k,\tau)-\hat{a}_A^*(\lambda_k,\tau))\to 0$ in probability .
This convergence holds because

$$E[\hat{a}_A(\lambda_k,\tau)-\hat{a}_A^*(\lambda_k,\tau)]^2\le A^{-2}\int_{A-\tau}^{A}E[X(t)\,X(t+\tau)]^2\,dt_1\,dt_2\le A^{-2}\tau\ M_8^{1/4} \tag{10}$$

where M_8 is the bound on the eighth moment (see (A.1)). It is easy to show that the estimator

$$\hat{g}_A^*(k,\gamma)=(1/2\pi)\int_{-A}^{A}K(B_A\tau)\,\hat{a}_A^*(\lambda_k,\tau)\exp\left(-i\gamma\tau\right)d\tau \tag{11}$$

is asymptotically equivalent to $\hat{g}_A(k,\gamma)$. Therefore , in the sequel we will study the asymptotic behaviour of $\hat{g}_A^*(k,\gamma)$ and we will drop the superscript *.
Observe now that the bias term for the studied estimator may be written in the form:

$$(AB_A^3)^{1/2}\left[(1/2\pi)\int_{-A}^{A}K(B_A\tau)E\hat{a}_A(\lambda_k,\tau)\exp(-i\gamma\tau)\,d\tau-\right.$$

$$\left.-(1/2\pi)\int_{-\infty}^{\infty}a(\lambda_k,\tau)\exp(-i\gamma\tau)\,d\tau\right]=$$

$$=(AB_A^3)^{1/2}\left[(1/2\pi)\int_{-A}^{A}K(B_A\tau)E\hat{a}_A(\lambda_k,\tau)-a(\lambda_k,\tau))\exp(-i\gamma\tau)\,d\tau\right]+$$

$$+(AB_A^3)^{1/2}\left[(1/2\pi)\int_{-\infty}^{-A}a(\lambda_k,\tau)\exp(-i\gamma\tau)\,d\tau+(1/2\pi)\int_{A}^{\infty}a(\lambda_k,\tau)\exp(-i\gamma\tau)d\tau\right],$$

where $a(\lambda_k, \tau)$ are the functional coefficients of the shifted covariance function $B(t, \tau)$. Due to the φ-mixing property of the process $\{X(t);\ t \in \mathbf{R}\}$ and similar considerations as in Hurd and Leskow (1992a) we see that the bias tends to zero as A grows.
Therefore, to prove the asymptotic normality for our estimator it suffices to show asymptotic normality of the term $\left(AB_A^3\right)^{1/2}\left(\hat{g}_A(k,\gamma) - E\hat{g}_A(k,\gamma)\right)$. In the subsequent considerations we will fix the value of γ and denote

$$N_A = \left(AB_A^3\right)^{1/2}\left(\hat{g}_A(k,\gamma) - E\hat{g}_A(k,\gamma)\right). \tag{12}$$

Easy calculations show that N_A may be represented in the form

$$N_A = \frac{\left(AB_A^3\right)^{1/2}}{2\pi\ AB_A} \int_0^A U_A(t) \exp(i\lambda_k t)\, dt$$

where

$$U_A(t) = X(t) \cdot Y_A(t) - E(X(t) \cdot Y_A(t))$$

and

$$Y_A(t) = \int_{-1}^{1} K(u)\, X\left(t + \frac{u}{B_A}\right) \exp\left(-i\gamma u/B_A\right) du.$$

Let us first study the asymptotic variance of the sequence N_A. To do that observe that

$$EN_A^2 = (B_A/2\pi A) \iint_{F_A} E\left\{U_A(t)\, U_A(t+v)\right\} \exp\left(i\lambda_k(2t+v)\right) dv\, dt\ +$$
$$+\, (B_A/2\pi A) \iint_{F_A^c} E\left\{U_A(t)\, U_A(t+v)\right\} \exp\left(i\lambda_k(2t+v)\right) dv\, dt,$$

where $v = s - t$ and $F_A = \{(t,v) : 0 \le t \le A \text{ and } A \ge |v| \ge (1/B_A)\}$.
Using similar methods as in the proof of Proposition 2 of Hurd and Leskow (1992a) we see that due to the φ-mixing property of the process $\{X(t);\ t \in \mathbf{R}\}$ the inequality $|E\ U_A(t)U_A(t+v)| \le C\ \ \varphi^{1/2}(|v| - (1/B_A))$ holds if $(t,v) \in F_A$. From the assumption on the speed of convergence to zero of the function φ we see that

$$(1/2\pi A) \iint_{F_A} \varphi^{1/2}\left(|v| - (1/B_A)\right) dv =$$
$$= (1/2\pi) \int_{-A}^{-1/B_A} \varphi^{1/2}\left(-v - (1/B_A)\right) dv + (1/2\pi) \int_{1/B_A}^{A} \varphi^{1/2}\left(v - (1/B_A)\right) dv =$$
$$= (1/\pi) \int_0^{A-1/B_A} \varphi^{1/2}(p)\, dp.$$

Therefore,

$$\left|\frac{B_A}{2\pi A} \iint_{F_A} E\left\{U_A(t)\, U_A(t+v)\right\} \exp\left(i\lambda_k(2t+v)\right) dv\, dt\right| \le \frac{B_A}{\pi} \int_0^\infty \varphi^{1/2}(p)\, dp,$$

so the first term on the RHS of the above equality is of the order $O(B_A)$ and is asymptotically negligible.
Denote the second term as σ_A. Observe that σ_A may be expressed as

$$\sigma_A = \frac{1}{2\pi A} \int_0^A \left[B_A \int_{-1/B_A}^{1/B_A} E\left\{U_A(t)\, U_A(t+v)\right\} \exp\left(i\lambda_k(2t+v)\right) dv\right] dt.$$

Using standard results from the theory of periodic functions we can see that σ_A is asymptotically equal to V_k.
To show the asymptotic distribution result we will apply the central limit theorem of Rootzén and Leadbetter (1990, Theorem 3.5) for mixing arrays to the sequences $Re(N_A)$ and $Im(N_A)$, where Re and Im denote, respectively, the real and the imaginary part of a complex number and N_A that was defined in (12) . We will give the proof for the real part and then apply the classical Cramér-Wold device.
In our context the mixing array will be defined by:

$$\zeta_A(I) = (1/2\pi) \cdot \left(AB_A^3\right)^{1/2} \cdot \left(AB_A\right)^{-1} \Re\left(\int_{A\cdot I} U_A(t)\exp(i\lambda_k t)\,dt\right), \qquad (13)$$

where I is the semiclosed interval in (0,1] [See Leadbetter and Rootzén (1990)].
Now we will prove that the array $\zeta_A(I)$ is a strongly mixing when the process $\{X(t); t \in \mathbf{R}\}$ is φ-mixing and apply Theorem 3.5 of Leadbetter and Rootzén (1990). For the convenience of the reader we recall below the definition of a strongly mixing families. Notations follow these of Leadbetter and Rootzén (1990).
Let $\{\zeta_A;\ A > 0\}$ be a family of additive random functions . Let also $F_{s,t}^A = \sigma\{\zeta_A(u,v] : s \le u < v \le t\}$ and let $\alpha_{A,\epsilon A} = \sup\{|P(B_1 \cap B_2) - P(B_1)P(B_2)| : B_1 \in F_{0,s}^A\ ,\ B_2 \in F_{s+\epsilon,1}^A\ ;\ 0 \le s \le 1-\epsilon\}$. The system $\{\zeta_A;\ A > 0\}$ is called strongly mixing if for each fixed $\epsilon > 0$ we have that

$$\alpha_{A,\epsilon A} \to 0 \qquad \text{for A} \to \infty.$$

Due to (13) it is quite easy to show that in our case $\alpha_{A,\epsilon A} \le \varphi(\epsilon A - (2/B_A))$. Since for each fixed $\epsilon > 0$, $\epsilon A - (2/B_A) \to \infty$([See assumption (A.4)], using Theorem 3.5 of Leadbetter and Rootzén we can establish that $\Re(N_A)$ tends in distribution to $N(0, v_{11})$, where v_{11} was defined in Theorem 2.1.
Applying now the same arguments to the imaginary part of N_A and using the classical Cramér–Wold device we get the conclusion of the Theorem 2.1.

REFERENCES

[1] Amerio, L. and Prouse, G. (1971). *Almost Periodic Functions and Functional Equations.* Van Nostrand Publishing Company, New York.

[2] Besicovitch, A.S. (1932). *Almost Periodic Functions.* London, Cambridge University Press.

[3] Billingsley, P. (1968), *Convergence of Probability Measures.* John Wiley and Sons, New York.

[4] Dehay, D. (1990). Nonlinear Analysis for Almost Periodically Correlated Strongly Harmonizable Processes. *Proceedings of the World Congress of the Bernoulli Society,* Uppsala, Sweden.

[5] Dragan, Ya. P. and Yavorskii, I.N. (1982). *The Rhytmics of the Sea and the Underwater Acoustic Signals.* Naukovah Doomkah, Kiev. (In Russian)

[6] Gardner, W. (1991). Exploitation of Spectral Redundancy in Cyclostationary Signals. *IEEE Signal Processing Magazine,* **8**, 2.

[7] Gladyshev, E.G. (1961). Periodically Correlated Random Se-quen-ces. *Sov. Math.,* **2**, 385–88.

[8] Hurd, H.L. (1989a). Nonparamteric Time Series Analysis for Periodically Correlated Stochastic Processes. *IEEE Trans. Inf. Th.,* **35**, 350 – 359.

[9] HURD, H.L. (1989b). Representation of Strongly Harmonizable Periodically Correlated Processes and their Covariances. *J. Multiv. Anal.*, **29**, 53 – 67.

[10] HURD, H.L. AND GERR, N.L. (1991). Graphical Methods for Determining the Presence of Periodic Correlation. *Journal of Time Series Analysis*, **12**, 4.

[11] HURD, H.L. AND LESKOW, J. (1992a). Estimation of the Fourier Coefficient Functions and their Spectral Densities for φ–mixing almost Periodically Correlated Processes. *Statistics and Probability Letters*, **15**, No 4. (to appear)

[12] HURD H.L. AND LESKOW, J. (1992b). Strongly Consistent and Asymptotically Normal Estimation of the Covariance for almost Periodically Correlated Processes. *Statistics and Decisions*, **10**. (to appear)

[13] LESKOW, J. (1992). *Spectral Estimation for almost Periodically Correlated Processes.* In preparation.

[14] LEADBETTER, M.R. AND ROOTZÉN, H. (1990). *On Central Limit Theory for Families of Strongly Mixing Additive Random Functions.* Technical Report No 295, Center for Stochastic Processes, University of North Carolina, Chapel Hill.

[15] OGURA, H. (1971). Spectral Representation of a Periodic Nonstationary Random Processes. *IEEE Trans. Inform. Theory*, **17**, 143 – 49.

[16] SAKAI, H. (1991), On the Spectral Density Matrix of a Periodic ARMA Process. *Journal of Time Series Analysis*, **12**, 73 – 83.

Recent Adv. in Stat. and Prob., pp. 293-310
J. Pérez Vilaplana and M. L. Puri (Eds)

Covering Large Domains by a Wiener Process

P. RÉVÉSZ
Institut für Statistik und Wahrscheinlichkeitstheorie, Technische Universitat Wien, Wiedner Hauptstraße 8–10, A–1040 Wien, Austria

1. INTRODUCTION

Consider a simple, symmetric random walk on the lattice Z^d. This means that if the moving particle is in $x \in Z^d$ in the moment n then at the moment $n+1$ the particle can move with equal probabilities to any of the $2d$ neighbours of x independently of how the particle achieved x. (The neighbours of an $x \in Z^d$ are those elements of Z^d whose $d-1$ coordinates coincide with the corresponding coordinates of x and one coordinate differs by $+1$ or -1 from the corresponding coordinate of x.) Let $S_n = S(n)$ be the location of the particle after n steps (i.e. in the moment n) and assume that $S_0 = 0$. Equivalently: $S_n = X_1 + X_2 + \cdots + X_n$ $(n = 1, 2, \ldots)$ where $X_1, X_2, \ldots$ is a sequence of independent, identically distributed random vectors with

$$\mathbf{P}\{X_1 = e_i\} = \mathbf{P}\{X_1 = -e_i\} = \frac{1}{2d} \qquad (i = 1, 2, \ldots, d)$$

and $e_1, e_2, \ldots, e_d$ are the orthogonal unit-vectors of Z^d.
The well-known recurrence theorem of Polya ([13]) claims that $\{S_n\}$ is recurrent if $d \le 2$ and transient if $d \ge 3$. That is to say

$$\mathbf{P}\{S_n = 0 \text{ i.o.}\} = \begin{cases} 1 & \text{if} \quad d \le 2, \\ 0 & \text{if} \quad d \ge 3. \end{cases}$$

A trivial consequence of the above theorem is

Theorem 1.1 *For any $x \in Z^d$ we have*

$$\mathbf{P}\{S_n = x \ i.o.\} = \begin{cases} 1 & if \quad d \le 2, \\ 0 & if \quad d \ge 3. \end{cases}$$

We say that the ball

$$Q(r) = \{x \in Z^d, \ \|x\| \le r\}$$

is covered by the random walk $\{S_k\}$ in time n if every point of $Q(r)$ is visited at least once by the random walk during its first n steps, i.e. if for any $x \in Q(r)$ there exists a positive random integer $k = k(x, \omega) \le n$ such that $S_k = x$. Let $R_d(n) = R(n)$ be the largest integer for which $Q(R(n))$ is covered in time n. Then Theorem 1.1 easily implies that

$$\lim_{n \to \infty} R_d(n) = \begin{cases} \infty \text{ a.s.} & \text{if} \quad d \le 2, \\ R_\infty(d) \text{ a.s.} & \text{if} \quad d \ge 3 \end{cases}$$

where $R_\infty(d)$ is a random variable, finite with probability 1.
Erdős and Taylor ([9]) proposed to study the limit behaviour of $R_2(n)$. The first paper devoted to this problem is due to Erdős and Révész ([7]). Since then the following papers appeared on this and related problems [1], [2], [8], [12], [14], [15], [17], [18], [19], [20], furthermore Chapter 22 of the book [16] is dealing with this and similar covering problems. It looks also interesting to study the analogue problems when a Wiener process is considered instead of a simple random walk. Section 7 of the present exposition is devoted to this question.
For later reference we introduce the following notations.

(i)

$$\xi(x,n) = \xi_d(x,n) = \#\{k:\ 0 \le k \le n,\ S_k = x\}, \qquad x \in Z^d,\ n = 0,1,2,\ldots$$

is the local time of $\{S_k\}$.

(ii)

$$\begin{aligned} \rho(0) &= \rho_d(0) = 0, \\ \rho(1) &= \rho_d(1) = \min\{k:\ k > 0,\ S_k = 0\}, \\ \rho(n+1) &= \rho_d(n+1) = \min\{k:\ k > \rho(n),\ S_k = 0\} \quad (n = 1,2,\ldots) \end{aligned}$$

are the returning times to the origin.

(iii) The sequence

$$\{S_{\rho(k)}, S_{\rho(k)+1}, \ldots, S_{\rho(k+1)}\} \qquad (k = 0,1,2,\ldots)$$

is called the $(k+1)$-th excursion of $\{S_k\}$ away from 0.

(iv)

$$\begin{aligned} p(0\,;\ x) &= \mathbf{P}\{\{S_n\} \text{ reaches } x \text{ before returning to } 0\} = \\ &= \mathbf{P}\{\inf\{n:\ n \ge 1,\ S_n = 0\} > \inf\{n:\ n \ge 1, S_n = x\}\} = \\ &= \mathbf{P}\{\xi(x,\rho_1) > 0\}. \end{aligned}$$

In the present exposition I intend to give a summary of the knonw results and to present a few new ones. Most of the proofs appeared in the above mentioned papers are not very simple mostly because of technical reasons. Hence the proofs of the previously appeared theorems will be not presented in details in general. The complete proofs of the new results will presented. Also complete proofs will be given when a simpler proof is available than the known ones.

2. THE CASE $d = 1$.

The limit behaviour of $R_1(n)$ is well-known. We present the corresponding results only to compare them by the multivariate theorems and for the sake of completness.
A simple consequence of a theorem of Csáki and Grill ([6]) (see also [16] p. 45 Consequence 1) is the following:

Theorem 2.1 *Let $f(n)$ be a nondecreasing function. Then*

$$R_1(n) \ge \frac{1}{3} f(n) n^{1/2} \qquad i.o.\ a.s. \tag{2.1}$$

if and only if

$$\sum_{n=1}^{\infty} \frac{f(n)}{n} \exp\left(-\frac{f^2(n)}{2}\right) = \infty. \tag{2.2}$$

Theorem 2.1 describes how big $R_1(n)$ can be. For example it tells us that

$$R_1(n) \geq \frac{1}{3}(2n \log\log n)^{1/2} \quad \text{i.o. a.s.}$$

However, for any $\varepsilon > 0$

$$R_1(n) \leq \frac{1+\varepsilon}{3}(2n \log\log n)^{1/2} \quad \text{a.s.}$$

for all but finitely many n.
A theorem of Hirsch ([11]) answers the question: how small $R_1(n)$ can be. In fact we have

Theorem 2.2 *Let $g(n)$ be a nonincreasing function. Then*

$$R_1(n) \leq g(n) n^{1/2} \qquad \textit{i.o. a.s.} \tag{2.3}$$

if and only if

$$\sum_{n=1}^{\infty} n^{-1} g(n) = \infty. \tag{2.4}$$

As we know $R_1(n)$ is the radius of the longest symmetric interval covered by $\{S_k\}$ in time n. It is natural to ask about the radius of the longest, (not surely symmetric) covered interval. The classical Strassen-theorem ([23]) and a theorem of Csáki ([3]) imply

Theorem 2.3 *Let $\overline{R}_1(n)$ be the radius of the longest covered interval in time n. Then*

$$\limsup_{n\to\infty} \frac{\overline{R}_1(n)}{(2n \log\log n)^{1/2}} = 1/2 \qquad \textit{a.s.} \tag{2.5}$$

Further if $h(n)$ is a nonincreasing function, then

$$\overline{R}_1(n) \leq \frac{1}{2} h(n) n^{1/2} \qquad \textit{i.o. a.s.} \tag{2.6}$$

if and only if

$$\sum_{n=1}^{\infty} \frac{1}{n(h(n))^2} \exp\left(-\frac{\pi^2}{2}\frac{1}{(h(n))^2}\right) = \infty. \tag{2.7}$$

REMARK. Comparing Theorems 2.1, 2.2 and Theorem 2.3 it can be seen that $\overline{R}_1(n)$ can be much bigger than $R_1(n)$.
We say that a moving interval $[a(n), b(n)]$ is *uniformly covered* if

$$\lim_{n\to\infty} \sup_{a(n)\leq x\leq b(n)} \left|\frac{\xi(x,n)}{\xi(0,n)} - 1\right| = 0 \qquad \text{a.s.}$$

We ask about the length of the largest (symmetric) uniformly covered interval. Csáki and Földes ([5]) proved

Theorem 2.4 *Put*

$$g(t) = \frac{t^{1/2}}{(\log t)(\log\log t)^{\rho}}.$$

Then

$$\lim_{n\to\infty} \sup_{|x|\le g(n)} \left|\frac{\xi(x,n)}{\xi(0,n)} - 1\right| \begin{cases} = 0 \ a.s. & if\ \rho > 2, \\ \ge 1 \ a.s. & if\ \rho \le 1. \end{cases}$$

In order to investigate the limit distributions of $R_1(n)$ and $\overline{R}_1(n)$ let

$$M^+(n) = \max_{0\le k\le n} S_k, \qquad M^-(n) = -\min_{0\le k\le n} S_k.$$

Then

$$R_1(n) = \min(M^+(n), M^-(n))$$

and

$$\overline{R}_1(n) = \frac{1}{2}(M^+(n) + M^-(n)).$$

The joint limit distribution of $M^+(n)$ and $M^-(n)$ is known (cf. e.g. Feller (1966)). In fact we have

Theorem 2.5

$$\lim_{n\to\infty} \mathbf{P}\{M^+(n) < a\sqrt{n},\ M^-(n) < b\sqrt{n}\} = \sum_{j=-\infty}^{+\infty} (-1)^j \big(\phi(jc+a) - \phi(jc-b)\big) =$$
$$= \frac{4}{\pi}\sum_{j=0}^{\infty} \frac{1}{2j+1} \exp\left(-\frac{(2j+1)^2\pi^2}{2c^2}\right) \sin\frac{(2j+1)\pi a}{c} = G(a,b)$$

where $\phi(x)$ is the standard normal distribution function, $c = a+b$ and $a > 0, b > 0$.

Having this theorem the limit distributions of $R_1(n)$ and $\overline{R}_1(n)$ can be obtained easily as follows (cf. [10]).

Theorem 2.6 *For any $x > 0$ we have*

$$\lim_{n\to\infty} \mathbf{P}\{R_1(n) < x\sqrt{n}\} = 4\phi(x) - 2 - G(x,x) =$$
$$= 4\phi(x) - 2 - \sum_{j=-\infty}^{+\infty} (-1)^j \left[\phi\big((2j+1)x\big) - \phi\big((2j-1)x\big)\right]$$

and

$$\lim_{n\to\infty} \mathbf{P}\{\overline{R}_1(n) < x\sqrt{n}\} = \int\int_{u+v\le x} d_u d_v G(u,v) =$$
$$= 2\sum_{n=1}^{\infty} (-1)^{n-1} n(2\phi(2xn) - \phi(2x(n+1)) - \phi(2x(n-1))) =$$
$$= 8\sum_{n=0}^{\infty} \exp\left(-\frac{\pi^2(2n+1)^2}{8x^2}\right)\left(\frac{1}{4x^2} + \frac{1}{\pi^2(2n+1)^2}\right).$$

3. COMPLETELY COVERED DISC AROUND THE ORIGIN. THE CASE $d = 2$.

Theorem 3.1 ([16] Theorem 22.1)

$$R_2(n) \le \exp(2(\log n)^{1/2} \log_3 n) \qquad a.s. \tag{3.1}$$

for all but finitely many n.
For any $\varepsilon > 0$

$$R_2(n) \ge \exp((1-\varepsilon)(8\pi)^{-1/2}(\log n \log_3 n)^{1/2}) \qquad i.o.\ a.s. \tag{3.2}$$

For any $C > 0$

$$R_2(n) \le \exp(C(\log n)^{1/2}) \qquad i.o.\ a.s. \tag{3.3}$$

For any $\delta > 0$

$$R_2(n) \ge \exp((\log n)^{1/2}(\log_2 n)^{-1/2-\delta}) \qquad a.s. \tag{3.4}$$

for all but finitely many n.

Here and in what follows $\log_k$ is the k-th iterated of log.
REMARK. (3.2) is slightly stronger than the corresponding statement in [16].
The proofs of (3.2) and (3.4) are based on the following two lemmas.

Lemma 3.1 ([21] pp. 117, 124 and 125)

$$p(0\,;\ \ x) = \frac{\pi + o(1)}{4\log\|x\|} \quad as \quad \|x\| \to \infty. \tag{3.5}$$

Lemma 3.2 ([9] or [16] Theorem 19.1)

$$\lim_{n\to\infty} \mathbf{P}\{\xi(0,n) < x\log n\} = 1 - e^{-\pi x} \tag{3.6}$$

uniformly for $0 \le x < (\log n)^{3/4}$. For any $\varepsilon > 0$ we have

$$\xi(0,n) \le \frac{1+\varepsilon}{\pi}(\log n)(\log_3 n) \qquad a.s. \tag{3.7}$$

for all but finitely many n,

$$\begin{aligned} \xi(0,n) &\ge \frac{1-\varepsilon}{\pi}(\log n)(\log_3 n) && i.o.\ a.s. && (3.8)\\ \xi(0,n) &\le (\log n)(\log_2 n)^{-1} && i.o.\ a.s. && (3.9)\\ \xi(0,n) &\ge (\log n)(\log_2)^{-1-\varepsilon} && a.s. && (3.10) \end{aligned}$$

for all but finitely many n.

A trivial consequence of the above lemma is the following

Lemma 3.3 ([16] Theorems 19.2 and 19.5)

$$\lim_{n\to\infty} \mathbf{P}\left\{\rho(n) < \exp\left(\frac{n}{z}\right)\right\} = \exp(-\pi z)$$

uniformly for $0 < z < n^{3/7}$. For any $\varepsilon > 0$ we have

$$\rho(n) \le \exp(n(\log n)^{1+\varepsilon}) \qquad a.s. \tag{3.11}$$

for all but finitely many n,

$$\rho(n) \geq \exp\left(n(\log n)^{1-\varepsilon}\right) \qquad i.o.\ a.s. \tag{3.12}$$

$$\rho(n) \leq \exp\left((1+\varepsilon)\frac{\pi n}{\log_2 n}\right) \qquad i.o.\ a.s. \tag{3.13}$$

$$\rho(n) \geq \exp\left((1-\varepsilon)\frac{\pi n}{\log_2 n}\right) \qquad a.s. \tag{3.14}$$

for all but finitely many n.

PROOFS OF (3.2) AND (3.4)
Let $x \in Z^2$ and assume that

$$\|x\| \leq \exp\left(\frac{n}{f(n)}\right), \quad f(n) = an^{1/2} \qquad \text{and} \qquad a > \left(\frac{8}{\pi}\right)^{1/2}.$$

Then by Lemma 3.1 the probability, that during the first n excursions x is not visited, is

$$\mathbf{P}\{\xi(x,\rho(n)) = 0\} = \left(1 - \frac{\pi + o(1)}{4\log\|x\|}\right)^n \leq \exp\left(-\frac{\pi + o(1)}{4}f(n)\right).$$

Hence

$$\mathbf{P}\left\{R_2(\rho(n)) < \exp\left(\frac{n}{f(n)}\right)\right\} = \mathbf{P}\left\{\exists x:\ x \in Q\left(\exp\left(\frac{n}{f(n)}\right)\right), \xi(x,\rho(n)) = 0\right\} \leq$$
$$\leq \pi\exp\left(\frac{2n}{f(n)}\right)\exp\left(-\frac{\pi + o(1)}{4}f(n)\right) = \pi\exp\left(-\left(\frac{\pi + o(1)}{4}a - \frac{2}{a}\right)n^{1/2}\right). \tag{3.15}$$

Then by Borel - Cantelli lemma we have

$$R_2(\rho(n)) \geq \exp\left(\frac{n}{f(n)}\right) = \exp(a^{-1}n^{1/2}) \qquad \text{a.s.} \tag{3.16}$$

for all but finitely many n. Let

$$N_1(n) = \exp\left(n(\log n)^{1+\varepsilon}\right) \tag{3.17}$$

and

$$N_2(n) = \exp\left((1+\varepsilon)\frac{\pi n}{\log_2 n}\right). \tag{3.18}$$

By (3.16) and (3.11) resp. (3.13) we get

$$R_2(N_1(n)) \geq R_2(\rho(n)) \geq \exp(a^{-1}n^{1/2}) \qquad \text{a.s.}$$

for all but finitely many n and

$$R_2(N_2(n)) \geq R_2(\rho(n)) \geq \exp(a^{-1}n^{1/2}) \qquad \text{i.o. a.s.}$$

Since by (3.17) resp. (3.18)

$$n \geq \frac{\log N_1(n)}{(\log_2 N_1(n))^{1+2\varepsilon}} \qquad \text{and} \qquad n \geq \frac{1-2\varepsilon}{\pi}(\log N_2(n))(\log_3 N_2(n))$$

(if n is big enough), we have

$$R_2(N_1(n)) \geq \exp(a^{-1}(\log N_1(n))^{1/2}(\log_2 N_1(n))^{-1/2-\varepsilon}) \qquad \text{a.s.} \qquad (3.19)$$

for all but finitely many n and

$$R_2(N_2(n)) \geq \exp\left(a^{-1}\left(\frac{1-2\varepsilon}{\pi}(\log N_2(n))(\log_3 N_2(n))\right)^{1/2}\right) \qquad \text{i.o. a.s.} \qquad (3.20)$$

Clearly (3.20) implies (3.2). In order to see (3.4) let $N_1(n) \leq N < N_1(n+1)$. Then by (3.19)

$$R_2(N) \leq R_2(N_1(n+1)) \leq \exp((\log N)^{1/2}(\log_2 N)^{-1/2-2\varepsilon}) \qquad \text{a.s.}$$

for all but finitely many N. ∎

THE IDEA OF THE PROOF OF (3.1).

Let $x_1, x_2, \dots, x_k$ be distinct points of Z^2 and let

$$m_k(x_1, x_2, \dots, x_k; n) = \mathbf{P}\{\xi(x_1, n) > 0, \xi(x_2, n) > 0, \dots, \xi(x_k, n) > 0\}$$

be the probability of the event that all of $x_1, x_2, \dots, x_k$ are visited before n. Then m_k can be estimated as follows. ∎

Lemma 3.4 ([16] Lemma 22.5) *For any* $0 < q < 1$, $k = 2, 3, \dots$ *we have*

$$m_1(u; n-qn)[\mathcal{M}(qn) + (1-k)m_k(x_1, x_2, \dots, x_k; qn)] \leq$$

$$\leq m_k(x_1, x_2, \dots, x_k; n) \leq \frac{m_1(v,n)\mathcal{M}(n)}{1+(k-1)m_1(v,n)}$$

where $u \in Z^2$ *and* $v \in Z^2$ *are defined by*

$$m_1(v,n) = \max_{1\leq i\leq k} m_1(x_i, n) \quad \text{and} \quad m_1(u,n) = \min_{1\leq i\leq k} m_1(x_i, n).$$

This lemma can be obtained by the exclusion - inclusion principle. Let $N = N(n) \nearrow \infty$ and $k = k(n) \nearrow \infty$ be sequences of positive integers with $N(n) < n^{1/3}$, $k(n) \leq (N(n))^{\varepsilon}$ with some $\varepsilon > 0$. Then for any $n = 1, 2, \dots$ there exists a sequence $x_1 = x_1(n), x_2 = x_2(n), \dots, x_k = x_k(n) \in Z^2$ such that

$$\begin{aligned} N-1 &\leq \|x_i\| \leq N \qquad && (i = 1, 2, \dots, k), \\ N^{1-\varepsilon} &\leq \|x_i - x_j\| \leq N \qquad && (1 \leq i < j \leq k). \end{aligned}$$

Then from Lemma 3.4 by induction we obtain

$$m_k(x_1, x_2, \dots, x_k; n) \leq \exp(-(2-4\varepsilon)\log_3 n). \qquad (3.21)$$

Let

$$\begin{aligned} n_j &= [\exp(e^j)], \\ \tilde{N}(n) &= \exp(2(\log n)^{1/2}\log_3 n), \\ k_{j+1} &= \exp((\log n_{j+1})^{1/2}). \end{aligned}$$

Then by (3.20)

$$\mathbf{P}\{R_2(n_{j+1}) \geq \tilde{N}(n_j)\} \leq \mathbf{P}\{R_2(n_{j+1}) \geq N(n_{j+1})\} \leq$$

$$\leq m_{k_{j+1}}(x_1, x_2, \ldots, x_{k_{j+1}}; n_{j+1}) \leq (j+1)^{-(2-4\varepsilon)}$$

where $x_i = x_i(n_{j+1})$ $(i = 1, 2, \ldots, k_{j+1})$. Hence by the Borel - Cantelli lemma

$$R_2(n_{j+1}) \leq \tilde{N}(n_{j+1}) \qquad \text{a.s.}$$

for all but finitely many j. Let $n_j \leq n < n_{j+1}$. Then

$$R_2(n) \leq R_2(n_{j+1}) \leq \tilde{N}(n_j) \leq \tilde{N}(n).$$

The proof of (3.3) will be presented later. Theorem 3.1 can be considered as a two-dimensional analogue of Theorems 2.1 and 2.2. In case $d = 1$ we have also the limit distribution of $n^{-1/2}R_1(n)$. It is natural to ask about the limit distribution of $R_2(n)$. In fact Kesten formulated the following

Conjecture 1. There exists a distribution function $F(z)$ so that

$$\lim_{n\to\infty} \mathbf{P}\left\{\frac{(\log R(n))^2}{\log n} < z\right\} = F(z).$$

We cannot settle this conjecture, however we can prove the following:

Theorem 3.2 ([16] Theorem 22.2)

$$\begin{aligned}\exp(-8z) \ &\leq \ \liminf_{n\to\infty} \mathbf{P}\left\{\frac{(\log R(n))^2}{\log n} > z\right\} \leq \\ &\leq \ \limsup_{n\to\infty} \mathbf{P}\left\{\frac{(\log R(n))^2}{\log n} > z\right\} \leq \exp\left(-\frac{z}{2}\right). \end{aligned} \tag{3.22}$$

REMARK. Theorem 3.2 is slightly stronger than Theorem 22.2 of [16]. During the preparation of the present paper I was pleased to learn that G. F. Lawler ([12]) proved an even stronger result. In fact his Theorem claims

$$\begin{aligned}\exp(-4z) \ &\leq \ \liminf_{n\to\infty} \mathbf{P}\left\{\frac{(\log R(n))^2}{\log n} > z\right\} \leq \\ &\leq \ \limsup_{n\to\infty} \mathbf{P}\left\{\frac{(\log R(n))^2}{\log n} > z\right\} \leq \exp(-2z).\end{aligned}$$

PROOF OF (3.3).

(3.3) is a trivial consequence of Theorem 3.2. ■

Having Theorem 3.2 it is natural to modify Kesten's conjecture as follows.

Conjecture 2.

$$F(z) = 1 - e^{-\lambda z} \quad \text{with some} \quad 1/2 < \lambda < 8$$

where $F(z)$ is the distribution function in Conjecture 1.

PROOF OF THE LOWER PART OF (3.22).

By Lemma 3.3 and (3.15) for any $a > (8/\pi)^{1/2}$ we have

$$\begin{aligned}&\mathbf{P}\left\{R_2\left(\exp\left(\frac{n}{x}\right)\right) > \exp\left(\frac{\sqrt{n}}{a}\right)\right\} \geq \\ &\geq \ \mathbf{P}\left\{R_2\left(\exp\left(\frac{n}{x}\right)\right) \geq \exp\left(\frac{\sqrt{n}}{a}\right), \rho(n) < \exp\left(\frac{n}{x}\right)\right\} \geq\end{aligned}$$

$$\geq \mathbf{P}\left\{R_2(\rho(n)) \geq \exp\left(\frac{\sqrt{n}}{a}\right), \rho(n) < \exp\left(\frac{n}{x}\right)\right\} \geq$$
$$\geq \mathbf{P}\left\{R_2(\rho(n)) \geq \exp\left(\frac{\sqrt{n}}{a}\right)\right\} - \mathbf{P}\left\{\rho(n) \geq \exp\left(\frac{n}{x}\right)\right\} \geq$$
$$\geq 1 - \pi \exp\left(-\left(\frac{\pi + o(1)}{4}a - \frac{2}{a}\right)n^{1/2}\right) - (1 - \exp(-\pi x)) - \varepsilon \geq$$
$$\geq \exp(-\pi x) - 2\varepsilon$$

for any $\varepsilon > 0$ if n is big enough. Let

$$N = \exp\left(\frac{n}{x}\right), \qquad x = za^2.$$

Then we get

$$\mathbf{P}\{R_2(N) > \exp((z \log N)^{1/2})\} \geq \exp(-\pi a^2 z)$$

i.e. we have the lower-part of (3.22). ■

Proof of the upper part of (3.22).

In this proof we need the following

Lemma 3.5 ([9] or [16] Theorem 19.3)

$$\mathbf{P}\{\xi(x,n) = 0\} = \begin{cases} \dfrac{2\log\|x\|}{\log n}\left(1 + O\left(\dfrac{\log_3\|x\|}{\log\|x\|}\right)\right) & if\ 20 < x < n^{1/3}, \\[2ex] 1 - 2\dfrac{\log(n^{1/2}/\|x\|)}{\log n}\left(1 + O\left(\dfrac{\log_2\left(\frac{\|x\|}{\sqrt{n}}\right)}{\log\left(\frac{\|x\|}{\sqrt{n}}\right)}\right)\right) & if\ n^{1/6} < \|x\| < \dfrac{n^{1/2}}{20}. \end{cases}$$

Clearly in the disc $Q(\exp((z\log n)^{1/2})$ one can find $k = \exp(a(\log n)^{1/2})$ points $x_1, x_2, \ldots, x_k$ such that

$$\|x_i - x_j\| \geq \frac{1}{2}\exp((\sqrt{z} - a)(\log n)^{1/2}) \qquad \text{if} \qquad 1 \leq i < j \leq k$$

if $a < \sqrt{z}$. By Lemmas 3.4 and 3.5 we obtain

$$m_k(x_1, x_2, \ldots, x_k; n) \leq \frac{m_1(v,n)\mathcal{M}(n)}{1 + (k-1)m_1(v,n)}$$

where

$$m_1(v,n) \leq 1 - \frac{2(\sqrt{z} - a)}{(\log n)^{1/2}}.$$

Since for any $0 < A < 1$ and $k = 1, 2, \ldots$

$$\frac{1-A}{1 + (k-1)(1-A)} \leq \frac{1 - A/k}{k} \leq \frac{1}{k}\exp(-A/k),$$

by induction we have

$$m_k(x_1, x_2, \ldots, x_k; n) \leq \exp\left(-2\sqrt{z}\left(1 - \frac{a}{\sqrt{z}}\right)a\right) = \exp\left(-\frac{z}{2}\right)$$

if $a = \sqrt{z}/2$. Hence we have the upper part of (3.22). ■

We also present an analogue of Theorem 2.4.

Theorem 3.3 ([16], Theorem 22.3) *For any* $\varepsilon > 0$ *we have*

$$\lim_{n\to\infty} \sup_{\|x\|\le r_n} \left| \frac{\xi(x,n)}{\xi(0,n)} - 1 \right| = 0 \qquad a.s.$$

where

$$r_n = \exp((\log n)^{1/2}(\log_2 n)^{-1/2-\varepsilon}).$$

4. THE DENSITY OF COVERED POINTS IN LARGER DISCS AROUND THE ORIGIN. THE CASE $d = 2$.

The main message of Section 3 is: $R_2(n)$ is about $\exp((\log n)^{1/2})$. In this Section we investigate the density

$$K(r,n) = \#\{x : \|x\| \le r,\ \xi(x,n) > 0\}\frac{1}{r^2\pi}$$

of the covered points in a disc of radius $r \gg \exp((\log n)^{1/2})$.
The essential message of our next theorem is that if $r \le \exp((\log n)^\alpha)$ $(\alpha < 1)$ then $K(r,n)$ is nearly 1. In fact we have

Theorem 4.1 ([16] Theorem 22.9) *For any* $\varepsilon > 0$

$$\lim_{n\to\infty} K\left(\exp\left(\frac{\log n}{(\log_2 n)^{2+\varepsilon}}\right), n\right) = 1 \qquad a.s.$$

Theorem 4.1 tells us that almost all points of the disc $Q(\exp(\log n)^\alpha)$ will be visited by the random walk $\{S_0, S_1, \dots, S_n\}$ if $1/2 < \alpha < 1$. At the same time by Theorem 3.1 we know that some points of $Q(\exp(\log n)^\alpha)$ will surely not be visited. We might investigate the geometrical properties of the non-visited points. For example: what is the area of the largest non-visited disc within $Q(\exp(\log n)^\alpha)$? (By non-visited disc we mean a disc having only non-visited points.) The following theorem claims that with probability 1 there exists a non-visited disc of radius $\exp((\log n)^\beta)$ within the disc $Q(\exp(\log n)^\alpha)$ for every $\beta < \alpha$ provided that $\alpha > 1/2$. Let

$$Q(u,r) = \{x \in Z^2,\ \|x - u\| \le r\}.$$

Theorem 4.2 ([16] Theorem 22.10) *Let*

$$1/2 < \alpha < 1 \qquad \text{and} \qquad \beta < \alpha.$$

Then there exists a sequence of random vectors $u_1, u_2, \dots$ *such that*

$$Q\big(u_n, \exp((\log n)^\beta)\big) \subset Q\big(0, \exp((\log n)^\alpha)\big) = Q\big(\exp(\log n)^\alpha\big)$$

and

$$\xi(x,n) = 0 \qquad \text{for all} \qquad x \in Q\big(u_n, \exp(\log n)^\beta\big).$$

if n is big enough.

Now we consider the density $K(N,n)$ for even larger N. The case $N = n^\alpha$ will be investigated and for any $0 < \alpha < 1/2$ we prove that $K(n^\alpha, n)$ has a limit distribution. In fact we have

Theorem 4.3

$$\lim_{n\to\infty} \mathbf{P}\{K(n^\alpha, n) < x\} = 1 - (1-x)^{2\alpha/(1-2\alpha)} \qquad (0 \le x \le 1). \tag{4.1}$$

The proof this Theorem is based on the following

Lemma 4.1 ([16] Theorem 22.11) *For any* $s = 1, 2, \ldots$ *we have*

$$\lim_{n\to\infty} \mathbf{E}\Big((K(n^\alpha, n))^s\Big) = (1-2\alpha)^s \prod_{j=1}^{s} \left(1 - \left(1 - \frac{1}{j}\right) 2\alpha\right)^{-1}.$$

PROOF OF THEOREM 4.3.
Lemma 4.1 clearly implies that

$$\lim_{n\to\infty} \mathbf{P}\{K(n^\alpha, n) < x\} = F(x)$$

where $F(x)$ is a distribution function with

$$\int_0^1 x^s dF(x) = (1-2\alpha)^s \prod_{j=1}^{s} \left(1 - \left(1 - \frac{1}{j}\right) 2\alpha\right)^{-1}. \tag{4.2}$$

Hence in order to prove (4.1) it is sufficient to prove that (4.2) holds true with

$$F(x) = 1 - (1-x)^{2\alpha/(1-2\alpha)}.$$

It can be obtained by elementary calculus. ■
A trivial consequence of Theorem 4.3 is

$$\limsup_{n\to\infty} K(n^\alpha, n) = 1 \qquad \text{a.s.} \qquad (0 < \alpha < 1/2) \tag{4.3}$$

and

$$\liminf_{n\to\infty} K(n^\alpha, n) = 0 \qquad \text{a.s.} \qquad (0 < \alpha < 1/2) \tag{4.4}$$

We are also interested in investigating (4.3) and (4.4) when we replace n^α by another sequence. An answer of this question was presented in Theorem 4.1. A complementer of that result is

Theorem 4.4 ([20]) *For any* $0 < \mu < 1/2$

$$\lim_{n\to\infty} K\left(\frac{n^{1/2}}{\exp((\log n)^\mu)}, n\right) = 0 \qquad a.s.$$

(4.3), (4.4), Theorems 4.1 and 4.4 suggest the following two questions.
Question 1. Let $a_n \nearrow \infty$ be a sequence with

$$a_n n^{-\nu} \to 0 \text{ for any } \nu > 0$$

and

$$a_n \exp\left(-\frac{\log n}{(\log\log n)^{2+\varepsilon}}\right) \to \infty \text{ for any } \varepsilon > 0.$$

How can be the properties of $K(a_n, n)$ characterized?

Question 2. Let $b_n \nearrow \infty$ be a sequence with

$$b_n n^{-\nu} \to \infty \text{ for any } 0 < \nu < 1/2$$

and

$$b_n \exp\left(-\frac{n^{1/2}}{\exp((\log n)^{\mu})}\right) \to 0 \text{ for any } 0 < \mu < 1/2.$$

How can be the properties of $K(b_n, n)$ characterized?
Theorem 3.3 told us that the disc

$$Q(\exp((\log n)^{1/2}(\log_2 n)^{-1/2-\varepsilon}))$$

is not only completely covered but also homogeneously covered. The next theorem claims that this fact is true for much bigger discs in some sense. In fact we investigate the numbers of the visited points in the sectors

$$\begin{aligned} \mathcal{A} = \mathcal{A}(\lambda_1, \lambda_2) = \mathcal{A}(\lambda_1, \lambda_2, \alpha, n) = \\ = \{x:\ x = re^{i\varphi} \in Z^2,\ 0 \le r \le n^{\alpha},\ \lambda_1 \le \varphi \le \lambda_2\} \end{aligned}$$

where

$$0 < \alpha < 1/2, \qquad 0 \le \lambda_1 < \lambda_2 \le 2\pi.$$

Theorem 4.5 ([20]) *Let* $0 < \alpha < 1/2,\ 1/2 < \vartheta < 1,\ \varepsilon_n = (\log n)^{-1+\delta}$. *Then*

$$\sup_{\substack{0 \le \lambda_1 < \lambda_2 \le 2\pi \\ \lambda_2 - \lambda_1 \ge \varepsilon_n}} |K(\mathcal{A}, n) - K(\alpha, n)| \to 0 \textit{ in probability as } n \to \infty$$

where

$$K(\mathcal{A}, n) = \#\{x:\ x \in \mathcal{A},\ \xi(x,n) > 0\}\frac{1}{|\mathcal{A}|}$$

and $|\mathcal{A}|$ *is the Lebesgue measure of* $\mathcal{A}$.

5. THE LARGEST COMPLETELY COVERED DISC. THE CASE $d = 2$.

In Section 2 we have seen that the radius $R_1(n)$, of the largest covered, symmetric interval, "in general" is much smaller than $\overline{R}_1(n)$, the radius of the largest covered (not-surely) symmetric interval. In this Section it will be shown that this difference in Z^2 is even larger. In fact in Section 3 we have seen that $R_2(n)$ is about $\exp((\log n)^{1/2})$. In this Section it will be shown that the radius of the largest covered disc is a positive power of n.

We say that the ball

$$Q(u, r) = \{x \in Z^d,\ \|x - u\| \le r\}$$

is covered by the random walk $\{S_k\}$ in time n if every point of $Q(u,r)$ is visited at least once by the random walk during its first n steps, i.e. if for any $x \in Q(u,r)$ we have $\xi(x,n) > 0$. Let $\overline{R}_d(n)$ be the largest integer for which there exists a random $u_n \in Z^d$ such that $Q(u_n, \overline{R}_d(n))$ is covered in time n.
In [18] we proved

Theorem 5.1

$$\overline{R}_2(n) \ge n^{1/50} \qquad a.s.$$

for all but finitely many n.

In fact in [18] we proved a somewhat stronger result, telling that there exists a "nearly" homogeneously covered disc of radius $n^{1/50}$.

Theorem 5.2 *There exists a sequence of random vectors $u = u(n) \in Z^2$ $(n = 1,2,\ldots)$ and an $\varepsilon > 0$ such that*

$$\limsup_{n\to\infty} \sup_{\|x-u\|\le n^{1/50}} \left|\frac{\xi(x,n)}{\xi(u,n)} - 1\right| \le 1-\varepsilon \qquad a.s.$$

and

$$\liminf_{n\to\infty} \frac{\xi(u,n)}{\log^2 n} \ge \varepsilon \qquad a.s.$$

In the present paper we prove

Theorem 5.3 *Let $\chi > \chi_0$ where*

$$\chi_0 = \frac{1}{1+\sqrt{2}} < 0.42.$$

Then

$$\overline{R}_2(n) \le n^{\chi} \qquad a.s.$$

for all but finitely many n.

Theorems 5.1 and 5.3 combined suggest the following
Conjecture 3. There exists a $1/50 < q < \chi_0$ such that

$$\lim_{n\to\infty} \frac{\log \overline{R}_2(n)}{\log n} = q \qquad \text{a.s.}$$

PROOF OF THEOREM 5.3.
Let $0 < \delta < 1/2$, $\alpha > 0$, $\beta > 0$, $\alpha + \beta = 1$, $k = [n^{\alpha(1/2-\delta)}]$ and $u_i = i[n^{\beta(1/2-\delta)}]$ $(i = 0,1,2,\ldots,k)$. Then clearly

$$0 \le u_i \le n^{1/2-\delta},$$
$$u_j - u_i \ge [n^{\beta(1/2-\delta)}] \quad (0 \le i < j \le k).$$

Consider the set

$$A = A(n) = \{x_0, x_1, \ldots, x_k\} \subset Q(0, n^{1/2-\delta})$$

where $x_i = (u_i, 0)$ $(i = 0,1,2,\ldots,k)$. By Lemma 3.5 we obtain

$$m_1(x_j - x_i; n) \le 1 - \beta(1-2\delta) + o(1).$$

Since

$$\frac{1-a}{1+(k-1)(1-a)} = \frac{1}{k}\left(1 - \frac{a}{1+(1-a)(k-1)}\right),$$

by Lemma 3.4 we have

$$\begin{aligned} m_k(x_1, x_2, \ldots, x_k; n) &\le \frac{1-\beta(1-2\delta)+o(1)}{1+(k-1)(1-\beta(1-2\delta)+o(1))}\mathcal{M}(n) = \\ &= \left(1 - \frac{\beta(1-2\delta)+o(1)}{1-\beta(1-2\delta)} \cdot \frac{1}{k}\right)\frac{\mathcal{M}(n)}{k} = \\ &= \exp\left(-\frac{\beta(1-2\delta)+o(1)}{1-\beta(1-2\delta)} \cdot \frac{1}{k-1}\right)\frac{\mathcal{M}(n)}{k} \le \\ &\le \ldots \le \exp\left(-\frac{\beta(1-2\delta)+o(1)}{1-\beta(1-2\delta)}\log k\right) = n^{-K} \end{aligned}$$

where

$$K = K_n = \alpha\left(\frac{1}{2}-\delta\right)\frac{\beta(1-2\delta)}{1-\beta(1-2\delta)} + o(1) = \frac{(1-2\delta)^2}{2}\frac{\beta(1-\beta)}{1-\beta(1-2\delta)} + o(1).$$

Choose

$$0 < \beta = \frac{1-\sqrt{2\delta}}{1-2\delta} < 1 \qquad (0 < \delta < 1/2).$$

Then

$$K = \frac{(1-\sqrt{2\delta})^2}{2} + o(1).$$

Hence we proved

$$\mathbf{P}\{A \text{ is covered }\} = \mathbf{P}\{\xi(x_i,n) > 0,\ i = 0,1,2,\dots,k\} \le n^{-K}.$$

Let

$$B = B(n) = \{(0,u_0),(0,u_1),\dots,(0,u_k)\} \subset Q(0,n^{1/2-\delta}),$$
$$C = C(n) = \{(-u_0,0),(-u_1,0),\dots,(-u_k,0)\} \subset Q(0,n^{1/2-\delta}),$$
$$D = D(n) = \{(0,-u_0),(0,-u_1),\dots,(0,-u_k)\} \subset Q(0,n^{1/2-\delta}).$$

Then

$$\mathbf{P}\{\text{ at least one of } A,B,C,D \text{ is covered }\} \le 4n^{-K}.$$

Observe that if $u \in Q(0,n^{1/2-\delta})$ (or equivalently $0 \in Q(u,n^{1/2-\delta})$) then the disc $Q(u,n^{1/2-\delta})$ contains at least one of the sets A,B,C,D. Hence

$$\mathbf{P}\{\text{at least one of the discs } Q(u,n^{1/2-\delta})\ (u \in Q(0,n^{1/2-\delta})) \text{ is covered }\} \le 4n^{-K}.$$

Similarly one can see for any $v \in Z^2$ that

$$\mathbf{P}\{\text{ at least one of the discs } Q(u,n^{1/2-\delta})\ (u \in Q(v,n^{1/2-\delta})) \text{ is covered }\} \le 4n^{-K}.$$

Choose $\delta < (2+\sqrt{2})^{-2}$ and observe that $K = K_n > 2\delta$ if n is big enough. Let $2\delta < \gamma < K$ and consider the points $S(i[n^{1-\gamma}])$ $(i = 0,1,2,\dots,[n^\gamma])$. Let $\mathbf{Q}_n$ be the set of those discs $Q(u,n^{1/2-\delta})$ for which there exists an i $(i = 0,1,2,\dots,[n^\gamma])$ such that

$$S(i[n^{1-\gamma}]) \in Q(u,n^{1/2-\delta}).$$

Then we have

$$\mathbf{P}\{\text{ at least one element of } \mathbf{Q}_n \text{ is covered }\} \le 4n^\gamma n^{-K}.$$

Let T be a positive integer for which $T(K-\gamma) > 1$ and consider the sequence $n_j = j^T$ $(j = 1,2,\dots)$. Then we obtain: *with probability* 1 *for all* j *big enough all* $Q(u,n_j^{1/2-\delta}) \in \mathbf{Q}_{n_j}$ *will be not covered.* Since $\gamma > 2\delta$ we obtain that for all $i[n^{1-\gamma}] \le j \le (i+1)[n^{1-\gamma}]$ $(i = 0,1,2,\dots,[n^\gamma]-1)$ we have

$$S(j) \in Q(S(i[n^{1-\gamma}]),n^{1/2-\delta}) \qquad \text{a.s.}$$

for all but finitely many n. Consequently if $\delta < (2+\sqrt{2})^{-2}$ then

$$\overline{R}_2(n_j) \le n_j^{1/2-\delta} \qquad \text{a.s.}$$

for all but finitely many j. Since $\overline{R}_2(n)$ is nondecreasing we have the Theorem immediately. ∎

6. THE CASE $d \geq 3$.

As we have seen in the Introduction

$$\lim_{n\to\infty} R_d(n) = R_\infty(d) < \infty \qquad \text{a.s.} \qquad \text{if } d \geq 3.$$

However the results of Section 5 suggest that $\overline{R}_d(n)$ might converge to infinity in case $d \geq 3$ too. This question was studied in [8] and [17] where the following result was proved.

Theorem 6.1 *For any $d \geq 3$ we have*

$$\lim_{n\to\infty} \frac{\log \overline{R}_d(n)}{\log\log n} = \frac{1}{d-2} \qquad a.s.$$

7. COVERING BY A WIENER PROCESS

Studying the analogue properties of a Wiener process we say that the disc

$$D(\alpha) = \{z = (x,y) : \|z\| = (x^2+y^2)^{1/2} \leq \alpha\} \subset \mathbf{R}^2$$

is ρ-covered by $W(\cdot)$ in time T if for every $z \in D(\alpha-\rho)$ there exists a $0 \leq t = t(z,\rho,\omega) \leq T$ such that

$$W(t) \in D(z,\rho) = \{u : \|z-u\| \leq \rho\}.$$

Let $R = R(T,\rho) = R(T,\rho,\omega)$ be the supremum of those r's ($r \geq 0$) for which $D(r)$ is ρ-covered in T. We present the following analogue of (3.1) and (3.4).

Theorem 7.1 ([19]) *For any $\rho > 0$ and $\delta > 0$ there exists a $T_0 = T_0(\rho,\delta,\omega) > 0$ such that*

$$\exp((\log T)^{1/2}(\log_2 T)^{-1/2-\delta}) \leq R(T,\rho) \leq \exp(2(\log T)^{1/2}\log_3 T)$$

for all $T \geq T_0$.

It is also natural to ask whether the above lower estimate of R remains true if ρ is replaced in it by a function $\rho(T) \downarrow 0$ ($T \to \infty$). The following two theorems give answers of this question.

Theorem 7.2 ([19]) *Let*

$$\rho(T) = \rho(T,\gamma) = \exp(-(\log T)^{1/2-\gamma}) \qquad (0 < \gamma < 1/2).$$

Then for any $\delta > 0$ and $0 < \gamma < 1/2$ there exists a $T_0 = T_0(\delta,\gamma,\omega) > 0$ such that

$$R(T,\rho(T)) \geq \exp((\log T)^{1/2}(\log_2 T)^{-1/2-\delta})$$

for all $T \geq T_0$.

Theorem 7.3 ([19]) *Let*

$$\rho_1(T) = \rho_1(T,\gamma) = \exp(-(\log T)^{1/2+\gamma}) \qquad (\gamma > 0).$$

Then there exist a $T_0 = T_0(\gamma,\omega) > 0$ and a $z = z(T,\gamma,\omega) \in D(1)$ such that

$$\mathbf{P}\{W(t) \notin D(z,\rho_1(T)) \textit{ for all } 0 \leq t \leq T \textit{ and } T \geq T_0\} = 1.$$

In order to illuminate the meanings of Theorem 7.2 and 7.3 it is worth while to recall a theorem of Spitzer.

Theorem 7.4 ([22]) *Let $g(t)$ be a positive, nonincreasing function. Then*

$$\|W(T)\| \geq T^{1/2}g(T) \qquad a.s.$$

for any T big enough if and only if

$$\sum_{k=1}^{\infty}(k|\log g(k)|)^{-1} < \infty.$$

Note that the function

$$g(T) = T^{-(\log_2 T)^{1+\beta}}$$

satisfies the above conditions if and only if $\beta > 0$. Hence a Wiener process will meet the disc $D(T^{1/2}g(T))$ in time T i.o. if and only if $\beta \leq 0$. However Theorem 7.3 claims that within the unit circle $D(1)$ there exists a disc of radius $\rho_1(T) \gg T^{1/2}g(T)$, $(\beta = 0)$ not visited by $W(\cdot)$ till T. At the same time Theorem 7.2 tells us that any disc of radius $\rho(T)$ within $D(1)$ or even within

$$D\Big(\exp[(\log T)^{1/2}(\log_2 T)^{-1/2-\delta}]\Big)$$

will be visited by $W(\cdot)$ before T. In order to give the analogues of Theorem 5.1 and 5.3, let $u = (u_1, u_2) \in Z^2$ and define

$$Q(u, N) = \{x = (i,j) : \|x - u\|^2 = (i - u_1)^2 + (j - u_2)^2 \leq N^2\}.$$

Let $r(n)$ be the largest integer for which there exists a random vector $u = u(n) \in Z^2$ such that $Q(u, r(n))$ is covered by the random walk in time n i.e. $\xi(x, n) \geq 1$ for every $x \in Q(u, r(n))$.
We say that the disc $D(\alpha, u) = \{z : \|z - u\| \leq \alpha\}$ is ρ-covered by $W(\cdot)$ in time T if for every $z \in D(\alpha - \rho, u)$ there exists a $0 \leq t = (z, \rho, \omega) \leq T$ such that $W(t) \in D(z, \rho)$.
Let $r(T, \rho)$ be the supremum of those r's ($r \geq 0$) for which there exists a random vector $u = u(T, \omega)$ such that $Q(u, r)$ is ρ-covered by $W(\cdot)$ in time T. On the limit behaviour of $r(T, \rho)$ we present our

Theorem 7.5 ([19]) *Let $\kappa = \overline{\rho} = 0.005$. Then*

$$r(T, T^{-\overline{\rho}}) \geq T^{\kappa} \qquad a.s.$$

if T is big enough.

Unfortunately I cannot characterize the domain of the points of $(\kappa, \overline{\rho})$ for which the above inequality holds true. The choice $\kappa = \overline{\rho} = 0.005$ is just one of the possibilities.
Comparing Theorem 7.5 with Theorem 7.2 and 7.3 we can say that the "largest black hole" is not only much larger than the "black hole" around the origin but it is also "much more black".

Theorem 7.6 ([19]) *For any fixed $\vartheta > 0$ we have*

$$r(T, \vartheta) \leq T^{\chi} \qquad a.s.$$

if T is big enough and $\chi > \sqrt{6} - 2$.

Very likely all the results presented in Sections 3–6 for random walks can be reformulated to Wiener processes. However these reformulations are not trivial at all. Recently Prof. J.F.Le Gall was kind enough to inform me that his student, T. Meyre, proved: for any $\rho > 0$ we have

$$\limsup_{T\to\infty} \frac{(\log R(T,\rho))^2}{\log T(\log_3 T)^2} \geq \frac{1}{32} \qquad \text{a.s.}$$

REFERENCES

[1] Auer, P. (1990). The Circle Homogeneously Covered by Random Walk on Z^2. *Statistics & Probability Letters*, **9**, 403–407.

[2] Auer, P. & Révész, P. (1990). On the Relative Frequency of Points Visited by Random Walk on Z^2. In: *Coll. Math. Soc. J. Bolyai*, **57**: Limit Theorems in Probability and Statistics, [I. Berkes, E. Csáki & P. Révész, (eds.)], 27–33. North-Holland, Amsterdam.

[3] Csáki, E. (1978). On Some Distributions Concerning Maximum and Minimum of a Wiener Process. In: *Coll. Math. Soc. J. Bolyai*, **21**: Analytic Function Methods in Probability Theory. 43–52. North-Holland, Amsterdam.

[4] Csáki, E. (1978). On the Lower Limits of Maxima and Minima of Wiener Process and Partial Sums. *Z. Wahrscheinlichkeitstheorie verw. Gebiete*, **43**, 205–222.

[5] Csáki, E. & Földes, A. (1987). A Note on the Stability of the Local Time of a Wiener Process. *Stochastic Processes and Their Applications*, **25**, 203–213.

[6] Csáki, E. & Grill, K. (1988). On the Large Values of the Wiener Process. *Stochastic Processes and Their Applications*, **27**, 43–56.

[7] Erdos, P. & Rēvész, P. (1988). On the Area of the Circles Covered by a Random Walk. *J. of Multivariate Analysis*, **27**, 169–180.

[8] Erdős, P. & Révész, P. (1992). Three Problems on the Random Walk in Z^d. *Studia Scientiarum Mathematicarum Hungarica.* (To appear).

[9] Erdős, P. & Taylor, S. J. (1960). Some Problems Concerning the Structure of Random Walk Paths. *Acta Math. Acad. Sci. Hungar.*, **11**, 137–162.

[10] Feller, W. (1966). *An Introduction to Probability Theory and its Applications*, Vol. II. John Wiley and Sons, New York.

[11] Hirsch, W. M. (1965). A Strong Law for the Maximum Cummulative Sum of Independent Random Variables. *Comm. Pure Appl. Math.*, **18**, 109–217.

[12] Lawler, G. F. (1991). *On the Covering Time of a Disc by Simple Random Walk in Two Dimensoins.* Technical Report, Duke University.

[13] Pólya, G. (1921). Über eine Aufgabe der Wahrscheinlichkeitsrechnung betreffend die Irrfahrt in Strassennetz. *Math. Ann.*, **84**, 149–160.

[14] Révész, P. (1989). Simple Symmetric Random Walk in Z^d. In: *Almost everywhere Convergence*, [G. A. Edgar and L. Sucheston, (eds.)]. Academic Press, Inc., New York, 369–392.

[15] Révész, P. (1990). Estimates of the Largest Circle Covered by a Random Walk. *The Annals of Probability*, **18**, 1784–1789.

[16] Révesz, P. (1990). *Random Walk in Random and Non-random Environments* World Scientific, Singapore.

[17] Révész, P. (1990). On the Volume of the Spheres Covered by a Random Walk. In: *A Tribute to Paul Erdős.* [A. Baker, B. Bollobás & A. Hajnal., (eds.)]. Cambridge Univ. Press, Cambridge.

[18] Révész, P. (1992). Clusters of a Random Walk on the Plane. *The Annals of Probability.* (To appear).

[19] Révész, P. (1992). Black Holes on the Plane Drawn by a Wiener Process. *Probab. Theory Related Fields.* (To appear).

[20] Revész, P. (1993). A Homogenity Property of the Z^2 Random Walk. *Acta Sci. Math.* (To appear).

[21] Spitzer, F. (1964). *Principles of Random Walk.* Van Nostrand, Princeton, N. J.

[22] Spitzer, F. (1958). Some theorems Concerning 2-dimensional Brownian Motion. *Transactions of the Am. Math. Soc.*, **87**, 187–197.

[23] Strassen, V. (1964). An Invariance Principle for the Law of Iterated Logarithm. *Z. Wahrscheinlichkeitstheorie verw. Gebite*, **3**, 211–226.

Recent Adv. in Stat. and Prob., pp. 311-315
J. Pérez Vilaplana and M. L. Puri (Eds)

On the Existence and Uniqueness for a Stochastic Differential Equation

Tran Hung Thao
Institute of Mathematics, National Center for Scientific Research, P.O. Box 631, Bo Ho, Hanoi, Viet Nam 10.000

Abstract – In this paper, the author studies the existence and uniqueness for a stochastic differential equation arising from the Theory of filtering from point process observations. The solution here is a measure valued Feller process.

1. INTRODUCTION

All processes are supposed to be defined on a complete probability space $(\Omega, \mathcal{F}, P)$ and to satisfy "usual conditions" with a filtration $(\mathcal{F}_t)$. Consider a homogeneous Feller-Markov process X_t taking values in a compact separable Hausdorff space S. Denote by $\mathcal{M}$ (S) the set of all probability measures over S. Then $\mathcal{M}(\mathcal{S})$is also a compact Hausdorff space with the induced topology.
The semigroup P_t, $t \geq 0$ associated with the transition probabilities $P_t(x, E)$ is a Feller semigroup, i.e.

$$P_t f(x) = \int P_t(x, dy) f(y)$$

maps C(S) into itself for all $t \geq 0$ and satisfies

$$\lim_{t \downarrow 0} P_t f(x) = f(x)$$

uniformly in $x \in$ S for all f $\in C(S)$, where $C(S)$ is the space of all real continuous functions over S.
Assume that $(Y_t, t \geq 0)$ is a standard Poisson process, $(h_t, t \geq 0)$ is a $C(S)$ process i.e. $h_t \in C(S)$ for all $t \geq 0$. Set $\mu_t = Y_t - t$. Let σ_0 be an $\mathcal{M}(S)$-valued random variable independent of (μ_t).
An $\mathcal{M}(S)$-valued stochastic process σ_t is called a solution of the following equation

$$\sigma_t(f) = \sigma_0(P_t f) + \int_0^t [\sigma_s(hP_{t-s}f) - \sigma_s(P_{t-s}f)]\, d\mu_s, \tag{1}$$

where

$$\sigma_t(f) = \int f(X_t)\, d\sigma_t \text{ for all } f \in C(S),$$

if σ_t is independent of the σ-field $\sigma(\mu_v - \mu_u; 0 \leq u \leq v \leq t)$ and satisfies this equation. Denote by $\sum_0$ the σ-field generated by σ_0.

2. MAIN RESULT

Theorem. *There exists a unique solution σ_t for (1) for arbitrary initial condition σ_0. Furthermore, this solution σ_t is $\sigma(\mu_s - \mu_0; 0 \le s \le t) \vee \sum_0$-measurable.*

PROOF. The method of Kunita in [2] will be applied to prove the above stated theorem.

a) *The uniqueness of the solution of* (1).

Let σ_t and σ_t' be solutions of (1) with the same initial condition σ_0. Set

$$\rho_t(f) = E(|\sigma_t(f) - \sigma_t'(f)|^2) \tag{2}$$

then

$$\rho_t(f) \le 2E(|\sigma_t(f)|^2 + |\sigma_t'(f)|^2) \le 4 \,\|\, lf \,\|\, l^2, \tag{3}$$

where

$$\|\, lf \,\|\, l \overset{\text{def}}{=} \sup_{x \in S} f(x).$$

On the other hand, by definition (2) of $\rho_t(f)$ we have

$$\rho_t(f) \le E\left[\int_0^t \{\sigma_s(hP_{t-s}f) - \sigma_s'(hP_{t-s}f)\}\, d\mu_s\right]^2 + E\left[\int_0^t \{\sigma_s(P_{t-s}f) - \sigma_s'(P_{t-s}f)\}\, d\mu_s\right]^2$$

$$\le \int_0^t E[\sigma_s(hP_{t-s}f) - \sigma_s'(hP_{t-s}f)]^2\, ds + \int_0^t E[\sigma_s(P_{t-s}f) - \sigma_s'(P_{t-s}f)]^2\, ds.$$

This shows that

$$\rho_t(f) \le \int_0^t [\rho_s(hP_{t-s}f) + \rho_s(P_{t-s}f)]\, ds. \tag{4}$$

We can derive from (3) and (4) that

$$\rho_t(f) \le \int_0^t 4 \,\|\, lhP_{t-s}f \,\|\, l^2\, ds + \int_0^t 4 \,\|\, lf \,\|\, l^2\, ds$$

or in noting that $\|\, lP_{t-s}f \,\|\, l \le \|\, lf \,\|\, l$:

$$\rho_t(f) \le 4 \,\|\, lf \,\|\, l^2 (\,\|\, lh \,\|\, l^2 + 1). \tag{5}$$

Applying (5) to the right hand side of (4) and repeating this procedure n-times we get then

$$\rho_t(f) \le 4 \,\|\, lf \,\|\, l^2 (1 + \,\|\, lh \,\|\, l^2)^n \frac{t^n}{n!}. \tag{6}$$

Let n tend to infinity: the right hand side of (6) ten ds to zero and therefore

$$\forall f \in C(S) \qquad \rho_t(f) = 0.$$

That proves the uniqueness of the solution. ∎

b) *The measurability of the solution.*

Consider the following equation

$$\sigma_t(f) = \sigma_0(P_t f) + \int_0^t [\hat{\sigma}_s(hP_{t-s}f) - \hat{\sigma}_s(P_{t-s}f)]\, d\mu_s, \tag{7}$$

where $\hat{\sigma}_t(f)$ is the truncated of $\sigma_t(f)$ by a finite positive constant K:

$$\begin{aligned} \hat{\sigma}_t(f) &= \sigma_t(f) \quad \text{if } |\sigma_t(f)| \le K, \\ \hat{\sigma}_t(f) &= K \quad \text{if } |\sigma_t(f)| \ge K. \end{aligned} \tag{8}$$

Set

$$\begin{aligned} \sigma_t^{(0)}(f) &= \sigma_0(P_t f), \\ \sigma_t^{(1)}(f) &= \sigma_0(P_t f) + \int_0^t [\hat{\sigma}_s^{(0)}(hP_{t-s}f) - \hat{\sigma}_s^{(0)}(P_{t-s}f)]\, d\mu_s, \\ \sigma_t^{(2)}(f) &= \sigma_0(P_t f) + \int_0^t [\hat{\sigma}_s^{(1)}(hP_{t-s}f) - \hat{\sigma}_s^{(1)}(P_{t-s}f)]\, d\mu_s, \\ &\vdots \\ \sigma_t^{(n)}(f) &= \sigma_0(P_t f) + \int_0^t [\hat{\sigma}_s^{(n-1)}(hP_{t-s}f) - \hat{\sigma}_s^{(n-1)}(P_{t-s}f)]\, d\mu_s, \qquad (9) \\ &\vdots \end{aligned}$$

and

$$\rho_t^{(n)}(f) = E(|\sigma_t^{(n+1)}(f) - \sigma_t^{(n)}(f)|^2). \tag{10}$$

A computational procedure similar to that of the part *a*) yiel ds

$$\rho_t^{(n)}(f) \le \int_0^t [\rho_s^{(n-1)}(hP_{t-s}f) + \rho_s^{(n-1)}(P_{t-s}f)]\, ds, n \ge 1. \tag{11}$$

On the other hand, with $K > \| lh \| l$

$$\begin{aligned} \rho_t^{(0)}(f) &= E(|\sigma_t^{(1)}(f) - \sigma_t^{(0)}(f)|^2 \le \int_0^t E(|\sigma_s^{(0)}(hPt - sf) - \sigma_s^{(0)}(P_{t-s}f)|^2)\, ds \le \\ &\le (K^2+1) \| lf \| l^2 t. \end{aligned} \tag{12}$$

Substituting (12) into the right hand side of (11) we have

$$\rho_t^{(1)}(f) \le (K^2+1) \| lf \| l^2 \frac{t^2}{2!}.$$

By induction

$$\rho_t^{(n)}(f) \le (K^2+1)^n \| lf \| l^2 \frac{t^{n+1}}{(n+1)!}, \qquad n \ge 1. \tag{13}$$

By consequence the series $\sum_{n=0}^{\infty} [\rho_t^{(n)}(f)]^{1/2}$ is convergent. We can see from the following relation

$$\| l\sigma_t^{(m)}(f) - \sigma_t^{(n)}(f) \| l_{L^2(P)} = (E|\sigma_t^{(m)}(f) - \sigma_t^{(n)}(f)|^2)^{1/2} \le \sum_{k=n}^{m-1} [\rho_t^{(k)}(f)]^{1/2}$$

that $\sigma_t^{(n)}(f)$ converges in $L^2(P)$ to a stochastic process $\sigma_t^{\infty}(f)$. Because each $\sigma_t^{(n)}(f)$ is measurable with respect to $\sigma(\mu_v - \mu_u; 0 \le u \le v \le t)$ so $\sigma_t^{\infty}(f)$ is.

Suppose now that σ_t is the $\mathcal{M}$(S)-valued solution of (1). If the constant K is chosen such as $K > \max(\| lf \| l, \| lh \| l)$, then σ_t is also a solution of (7). For f ∈ C(S) we have

$$\| l\sigma_t^{\infty}(f) \| l_{L^2(P)} \leq \| l\sigma_t^{(0)}(f) \| l_{L^2(P)} + \sum_{n=0}^{\infty}[\rho_t^{(n)}(f)]^{1/2} \leq C \| lf \| l, \tag{14}$$

where C is a positive constant depending on t and K only. Set

$$\rho_t^*(f) = E(|\sigma_t(f) - \sigma_t^{\infty}(f)|^2).$$

By virtue of (14) we have

$$\rho_t^*(f) \leq \left[\| l\sigma_t(f) \| l_{L^2(P)} + \| l\sigma_t^{\infty}(f) \| l_{L^2(P)}\right]^2 \leq C' \| lf \| l^2$$

for some C'. Moreover

$$\rho_t^*(f) \leq \int_0^t [\rho_s^*(hP_{t-s}f) + \rho_s^*(P_{t-s}f)]\, ds.$$

By an argument similar to the above we can see that $\rho_t^*(f) = 0$, hence $\sigma_t = \sigma_t^{\infty}$. The measurability of σ_t with respect to $\sigma(\mu_v - \mu_u; 0 \leq u \leq v \leq t) \vee \sum_0$ is thus proved. ■

c) *Existence of the solution.*

It is known from the theory of filtering of a Feller-Markov process X_t from a point process observation Y_t of intensity h_t (see [1] and [3]) that the non-normalized filtering $\sigma_t^*(f), f \in C(S)$ is independent of the σ-field $\sigma(\mu_v - \mu_u; 0 \leq u \leq v \leq t)$ and satisfies Zakai equation

$$\sigma_t^*(f) = \sigma_0^*(P_t f) + \int_0^t [\sigma_s^*(hP_{t-s}f) - \sigma_s^*(P_{t-s}f)]\, d\mu_s, \tag{15}$$

where the non-normalized filtering process $\sigma_t^*(f)$ is defined by

$$\sigma_t^*(f) = E_Q[L_t.f(X_t)/\mathcal{F}_t^Y], f \in C(S),$$

$\mathcal{F}_t^Y$ is the σ-field generated by $(Y_s, s \leq t), L_t = E[\frac{dP}{dQ}/\mathcal{F}_t^Y]$ and Q is a probability under which $\mu_t = Y_t - t$ is a martingale (i.e. Y_t becomes a standard Poisson process) and such that the given probability P is obtained from Q after an absolutely continuous change of measure $Q \to P$.

The non-normalized filtering process σ_t^* is a solution of the equation (15) which formally is similar to (1). This solution can be expressed by a functional

$$\sigma_t^* = \Phi(\sigma_0^*, \mu_s - \mu_0; 0 \leq s \leq t). \tag{16}$$

When Y_t, h_t, μ_t and σ_0 are defined as in the introduction, then the same functional Φ expresses the solution of (1) in terms of σ_0 and $\mu_s - \mu_0, 0 \leq s \leq t$:

$$\sigma_t = \Phi(\sigma_0, \mu_s - \mu_0; 0 \leq s \leq t). \tag{17}$$

Thus the existence of the solution of (1) is proved. ■

3. REMARKS

1. As in [2], we can show that the $\mathcal{M}$(S)-valued solution σ_t of (1) is a Feller-Markov process with the transition probability $\Pi_t(\nu, B)$ defined by

$$\Pi_t(\nu, B) = P(\sigma_t^\nu \in B), \qquad B: \text{ Borel sets of } \mathcal{M}(S),$$

where σ_t^ν is the solution corresponding to the initial condition $\nu \in \mathcal{M}(S)$.

2. The result can be extended for the multidimensional case, where $(Y_t, t \geq 0)$ is an N-vector standard Poisson process, $(h_t, t \geq 0)$ is an N-vector $C(S)$ process, i.e. $h_t = (h_t^1, ..., h_t^N), h_t^i \in C(S)$, $i = 1, ...N$ and μ_t is also an N-vector martingale defined by

$$\mu_t = (\mu_t^1, ..., \mu_t^N), \qquad \mu_t^i = Y_t^i - t, \qquad i = 1, ..., N.$$

Then the corresponding stochastic differential equation is the following

$$\sigma_t(f) = \sigma_0(P_t f) + \int_0^t < \sigma_s(hP_{t-s}f) - \sigma_s(P_{t-s}f), \, d\mu_s >, \tag{18}$$

where the symbol <,> stan ds for the inner product.

REFERENCES

[1] ELLIOTT, R.J. (1988). *filtering and Control for Point Process Observations.* Universität Konstanz, Research Report No 09-86.

[2] KUNITA, H. (1971). Asymptotic Behavior of the Nonlinear Errors of Markov Process. *Multivariate Analysis*, **1**, 4, 365-393.

[3] TRAN HUNG THAO (1992). A Problem of Optimal State Estimation from Discrete Observations. *Revue des Technologies Avancées, Alger*, **6**, 1992.

Recent Adv. in Stat. and Prob., pp. 317-331
J. Pérez Vilaplana and M. L. Puri (Eds)

Nonparametric Inference of a Smooth Distribution Function from Irregular Observations on Time Series

J.A. VILAR FERNÁNDEZ[1] and J.M. VILAR FERNÁNDEZ[2]
[1] Departamento de Matemáticas, Facultad de Informática, Universidad de La Coruña, Campus da Zapateira, E-15071 La Coruña, España
[2] Departamento de Matemáticas, Facultad de Informática, Universidad de La Coruña, Campus da Zapateira, E-15071 La Coruna, España

Abstract – Let $X(t)$ be a stationary continuous-time process. The recursive estimation of the univariate probability distribution function, $F(x)$, given discrete-time samples $X(\tau_1), X(\tau_2), \ldots, X(\tau_n)$, with sampling instants, τ_i irregularly spaced or random (according to renewal process) is studied for process $X(t)$ when the observations satisfy strong mixing conditions. The asymptotic behaviour of a certain class of estimators of $F(x)$, obtained by integrating a kernel estimator of the density, is investigated.

1. INTRODUCTION

Statistical theory has focused considerable attention on the problem of the nonparametric estimation of the several probability curves (density, distribution, regression, hazard, ...) associated with stationary continuous-time processes $X = \{X(t);\ -\infty < t < \infty\}$, on the basis of n observations $X(\tau_1), \ldots, X(\tau_n)$. Generally the sampling instants τ_k are assumed to be deterministic, for instance, according to periodic sampling. This assumption is sometimes incorrect because, in most time series data, the observations are not uniformly recorded over time: small random deviations of the recording device, the own random experimental design, ... There are economic time series in which, for example, the early observations are recorded quarterly, while subsequent ones are recorded monthly. In these situations, the sampling instants may be irregularly spaced or even random.
This paper is concerned with the estimation of the probability distribution function $F(x)$ of a stationary continuous-time process X, when X is observed only at times $T = \{\tau_1, \tau_2, \ldots\}$, where T consists of an increasing sequence of random times.
Under these circumstances the assumption of independence for the variables $X(t)$ is clearly unrealistic and cannot be made. It is therefore necessary to assume that the sample consists of dependent data, and in this case one usual way to model the dependence is to assume some mixing condition. We will consider just one of these mixing dependence structures: the α-mixing condition introduced by Rosenblatt (1.956).
The mean square error (MSE) for two nonparametric recursive estimators with two distinct structures of T is derived : the empirical estimator $F_n(x)$ and an appropriate kernel type estimator, $F^*_{n,\eta}(x)$. In addition, we study the asymptotic contribution of the sampling scheme and its randomness on the performance of the estimators. Finally convergence

rates obtained in this paper and the classical results for regularly spaced observations are compared, and the asymptotic efficiency of both estimators is discussed.

2. THE STOCHASTIC MODEL

Let the continuous-discrete model be defined as the pair $(\mathbf{X}, \mathbf{T})$, where

• The first component, $X = \{X(t), t \in R\}$, is a continuous time process on a probability space $(\Omega, \mathcal{A}, P)$, satisfying:

(1) X is stationary.

(2) X is strongly mixing (α-mixing), and, equivalently, if $\mathcal{F}_a^b$ denote the σ-algebra of events generated by the random variables $\{X(t), a < t < b\}$, $a \leq b$, then for $\tau \geq 0$,

$$\sup\{\,|P(AB) - P(A)\,P(B)\,|,\ A \in \mathcal{F}_{-\infty}^{0},\ B \in \mathcal{F}_{\tau}^{+\infty}\} = \alpha(\tau) \downarrow 0 \qquad \text{as } \tau \uparrow +\infty$$

The value $\alpha(t)$ characterizes the mixing rate and is referred to as the strong mixing coefficient. This notion of asymptotic independence (where the values of X at widely separated instants become independent as the distance that separates them tends to infinity) represent a suitable dependence assumption that is not restrictive. Of course this mixing condition is satisfied by many stationary processes.

• The second component, $T = \{\tau_k;\ k \in N\}$, consists of a strictly increasing sequence of random times. Among the variety of possibilities, we confine ourselves to two specific structures for T:

- *Irregularly Observations* (IO). The processes T is defined by

$$\tau_k = \frac{k}{\beta} + z_k \qquad k = 0, \pm 1, \pm 2, \ldots,\ \beta > 0$$

  where the $z_k^{\prime}s$ are independent identically distributed random variables with a symmetric probability density function $g(t)$ supported on $[-1/2\beta, 1/2\beta]$. Thereby, the *IO* structure is the mixture of a periodic sampling (with a fixed sampling rate β) and a small random deviation (according to $z_k^{\prime}s$).

- *Renewal Process* (RP). The process T is a renewal type process, and equivalently

$$\tau_k = \sum_{i=1}^{k} t_i \qquad k = 1, 2, 3, \ldots \text{ and } \tau_0 \equiv 0$$

  where $\{t_i\}_{i=1}^{+\infty}$ is a sequence of independent identically distributed random variables with a common distribution $G(t)$ on $[0, \infty)$, satisfying

  (1) G is absolutely continuous with density $g(t)$ on $[0, \infty)$.

  (2) $G(0) = 0$.

  (3) $\int_0^{\infty} t\, dG(t) = \beta^{-1} < \infty$.

  In the case of an exponential distribution with $g(t) = \rho e^{-\rho t}$, we call the renewal process a Poisson process of rate ρ.

Although it is not the purpose of this section to make an exhaustive analysis of the renewal theory in continuous time, some important notions will be presented later on.

The first structure (IO) is only partially random: the sampling instants are no longer separated by a fixed distance, although the average distance between consecutive instants is cte. (β^{-1}). However the renewal structure is completely random and now β may be regarded as the chance for a realization of T by time unity as $t \to \infty$ (according to a classical result in renewal theory). Besides fitting many practical situations adequately, both structures turn to a progressive randomness of the sampling data and our goal will be analyze their effect on the asymptotic behaviour of our estimation.

Masry (1983) studies this same model for the nonrecursive estimation of the density $f(x)$, but other authors study different stochastic models; for instance, Stoyanov and Robinson (1991) consider (X,T) with $X(\tau) = a\tau + \sigma T(\tau)$, where T is a standard Wiener process and a and σ are given constants, $a \in R, \sigma > 0$.

Finally, throughout this paper the following hypothesis are assumed:

H1.– *The processes X and T are independent.*

so that the time samples $\{X(\tau_k)\}$ are random variables in the sense that they are measurable with respect to the product σ-algebra generated by the processes X and T. Then, when $T \equiv IO$, the random variables $\{z_k\}$ must be independent of X, while if $T \equiv RP$, H1 is always verified.

3. THE ESTIMATORS

We will consider two *recursive* estimators for $F(x)$. From the practical point of view recursive estimators have a clear advantage over nonrecursive estimators as they can be updated with each additional observation, in contrast to the nonrecursive case where the estimator must be entirely recomputed.

For a sample of size n the natural recursive estimator of a distribution function F is the empirical distribution function F_n which is defined by

$$F_n(x) = n^{-1} \sum_{i=1}^{n} I_{(-\infty,x]}\big(X(\tau_i)\big)$$

where

$$I_{(-\infty,x]}(t) = \begin{cases} 1 & \text{if } t \le x \\ 0 & \text{if } t > x \end{cases}$$

F_n has been studied under dependence conditions by Roussas (1989), Sarda-Vieu (1989) and Izerman-Tran (1990).

An alternative recursive estimator of F can easily be obtained by taking an integrand of a density recursive estimate $f^*_{n,\eta}$. For instance, although more generally, Deheuvels (1974) suggested the kernel type estimator defined from a kernel K and a bandwidth sequence h_n as:

$$f^*_{n,\eta}(x) = \left(\sum_{i=1}^{n} h_i^{\eta}\right)^{-1} \sum_{j=1}^{n} h_j^{\eta} K_j\big(x - X(\tau_j)\big)$$

where $K_j(u) = h_j^{-1}K(h_j^{-1}u)$ and $\eta \in [0,1]$; and K and $\{h_n\}$ verify the following conditions of regularity:

H2.– *K is a real, nonnegative and bounded function on R such that*

$$\int_{-\infty}^{+\infty} K(u)\,du = 1 \qquad \text{and} \qquad K(u) = O(|u|^{-1-\varepsilon}) \qquad \text{for some } \varepsilon > 0$$

H3.–*$\{h_n\}$ is a sequence of positive numbers such that $h_n \to 0$ and $nh_n \to +\infty$ as $n \to +\infty$.*

The corresponding kernel distribution estimator $F^*_{n,\eta}$ is given by

$$F^*_{n,\eta}(x) = \int_{-\infty}^{x} f^*_{n,\eta}(t)\,dt = \left(\sum_{i=1}^{n} h_i^{\eta}\right)^{-1} \sum_{j=1}^{n} h_j^{\eta} K_j^*\big(x - X(\tau_j)\big)$$

where

$$K_j^*(u) = \int_{-\infty}^{u} K_j(s)ds$$

This estimator can be computed recursively by

$$F_{n+1,\eta}^*(x) = A_{n+1}^{-1}\left\{A_n F_{n,\eta}^*(x) + h_{n+1}^{\eta} K_{n+1}^*\left(x - X(\tau_{n+1})\right)\right\}$$

with $A_n = \sum_{i=1}^{n} h_i^{\eta}$, and offers on advantage over F_n as it provides a more smooth estimation.

Azzalini (1981), Faraldo Roca-González Manteiga (1984), Reiss (1981) and Lejeunne-Sarda (1989) have studied a similar nonrecursive type kernel estimator of the distribution function with independent observations. Specifically:

$$F_n^*(x) = \int_{-\infty}^{x} f_n^*(t)dt \qquad \text{where} \qquad f_n^*(t) = \frac{1}{n}\sum_{i=1}^{n} K_i(x - X_i) \tag{1}$$

This same estimator has been considered by Vieu (1991) under the assumption of mixing conditions for X. More recently Vilar (1992) compares the asymptotic behaviour of F_n *(empirical)*, F_n^* *(nonrecursive)* and $F_{n,\eta}^*$ *(recursive)*. The study shows that the loss in bias is compensated by a gain in variance, particularly, in asymptotic terms:

$$\text{Bias}\left(F_{n,\eta}^*(x)\right) > \text{Bias}\left(F_n^*(x)\right) > \text{Bias}\left(F_n(x)\right) = 0$$

$$\text{Var}\left(F_{n,\eta}^*(x)\right) < \text{Var}\left(F_n^*(x)\right) < \text{Var}\left(F_n(x)\right)$$

In order to illustrate these results, we have simulated distinct samples from several time series. The three different estimations of F are compared and displayed in the Figs. 1, 2, 3, and 4.

Next the simulation technique applied in this exploratory study is described.

- The sample size investigated was $n = 100$.
- The Epanechnikov kernel function

$$K(u) = \begin{cases} 3/4(1-u^2) & \text{if } |u| \le 1 \\ 0 & \text{if } |u| > 1 \end{cases}$$

- The bandwidth sequence $h_n = Cn^{-1/5}$, where C is empirically calculated and the parameter $\eta = 0$ were chosen.
- And finally, the A.S.E. *(Average Square Error)* was computed for each probability model and estimator. Concretely:

$$\text{A.S.E.}(F_{EST}) = \frac{1}{m}\sum_{j=1}^{m}\left(F(x_j) - F_{EST}(x_j)\right)^2$$

  where F is the theoretical distribution and F_{EST} is the distribution estimate when F_{EST} is the empirical (F_n), nonrecursive (F_n^*) or recursive ($F_{n,\eta}^*$) estimate.

Following this procedure we have simulated four distinct probability models:

Figure 1. The data X_j are simulated realizations of the autoregressive moving average process: $X_j - 0.7X_{j-1} = a_j - 0.3a_{j-1}$ $(ARMA(1,1))$ where the marginal distribution for X follows a standard Normal law $N(0,1)$.

Figure 2. The data X_j are simulated realizations of two $AR(1)$ processes mixtured giving the same probability both of them, and using the same autoregressive coefficient $\rho = 0.5$.

The Normal laws with parameters $(-1,1)$ and $(1,1)$ were chosen as marginal distributions in each case.

Figure 3. The data X_j are simulated realizations of an $AR(1)$ model with autoregressive coefficient $\rho = 0.5$ where marginal distribution F follows an Negative Exponential law with parameter $\lambda = 1$.

Figure 4. The data X_j are simulated realizations of a 4-dependent process with a marginal law following a $\Gamma(2,1)$ distribution. They were derived as follows:

$$X_j = \sum_{k=j}^{j+3} e_k^2 \qquad \text{where } e_k \text{ are i.i.d. random variables } \mathcal{N}(0,1)$$

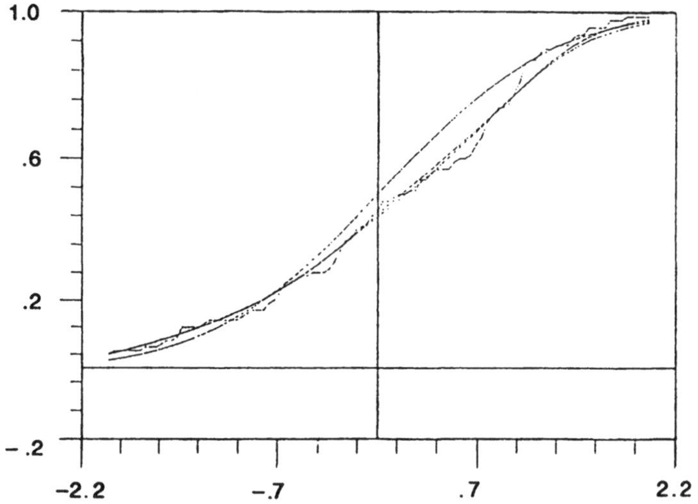

Figure 1. Normal Distribution

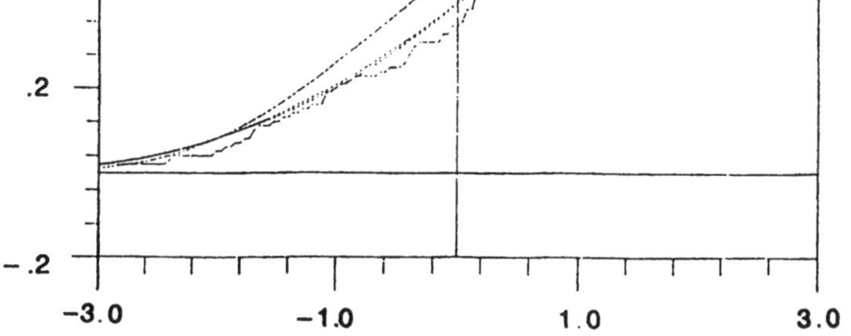

Figure 2. Bimodal Distribution

The $ASE \times 10^3$ obtained for each case are summarized in the next table

Estimate	Fig. 1 N(0,1)	Fig. 2 Bimodal	Fig. 3 E(1)	Fig. 4 Γ(2,1)
Empirical	2.88	4.63	1.97	1.90
Nonrecursive	2.39	3.92	1.26	1.70
Recursive	2.10	3.51	1.12	1.46

Of course this analysis is concerned with observations that are equally spaced. Now, our main goal is to determine if the performance of F_n and $F^*_{n,\eta}$ is the same when the sampling instants are unequally spaced, specifically when $T \equiv IO$ and $T \equiv RP$.

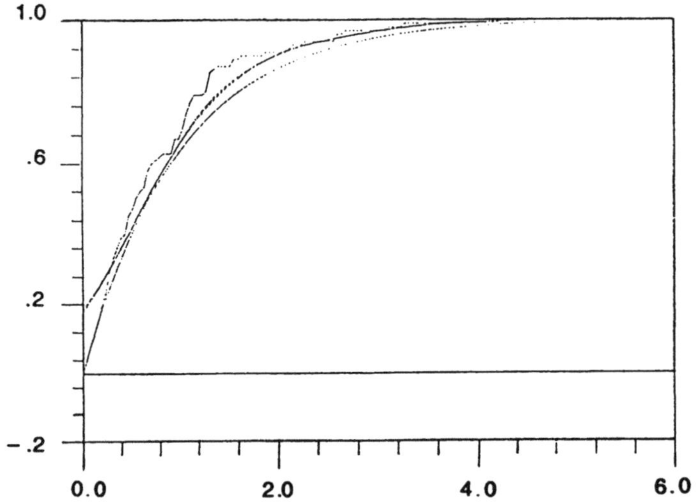

Figure 3. Negative Exponential Distribution

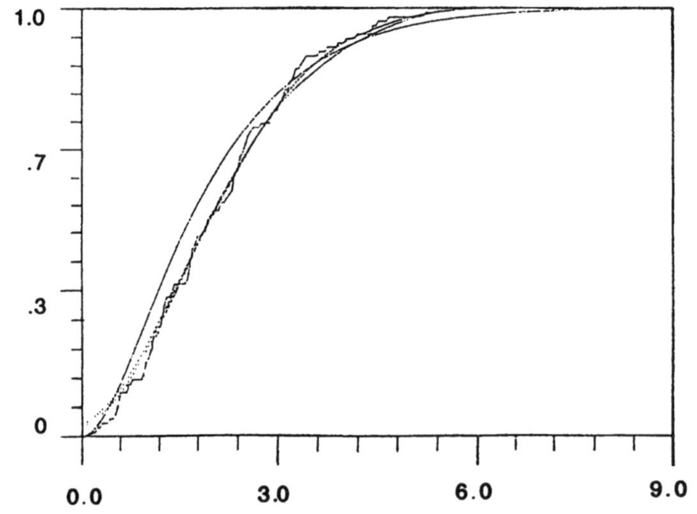

Figure 4. Gamma Distribution

4. THE BIAS.

The assumption **H1** provides that the bias of the estimation does not depend on the particular sampling scheme used nor on the dependence structure type.
Indeed, if $G_k(t)$ and $F_k(x)$ denote the distribution functions of the sampling instant τ_k and the marginal $X(\tau_k)$, respectively, then

$$F_k(x) = \int_{t\in R} P\{X(t) \le x/\tau_k = t\}\, dG_k(t)$$

and by **H1**

$$F_k(x) = \int_{t\in R} F(x)\, dG_k(t) = F(x) \tag{2}$$

Therefore, given **H1** which assumes independence between X and T, the results concerning the bias obtained by Vilar (1992) can be extended to any stochastic model (X,T).
By applying (2) and Toeplitz's Lemma we get the following theorem:

Theorem 1 *Let (X,T) be defined as before with $T \equiv IO$ or $T \equiv RP$. Then under* **H1** *we have:*
(a) $F_n(x)$ is an unbiased estimate of $F(x)$.
(b) Assume that $H2$ and $H3$ hold.
If in addition $\{h_n\}$ satisfies
H4.– $\lim_{n\to\infty} \frac{1}{n}\sum_{i=1}^{n}\left(\frac{h_i}{h_n}\right)^{\eta} = \theta_\eta < \infty$ *with* $\eta \in [0,1]$,
then $F^*_{n,\eta}(x)$ is an asymptotically unbiased estimate of $F(x)$.

It is interesting to remark three observations about this theorem:
1. The theorem is consistent for all sampling schemes.
2. The dependence does not influence on the bias and, therefore, the result coincides with the classical affirmations for independent observations.
3. With respect to bias, $F^*_{n,\eta}$ is always deficient with respect to F_n.
Results on the speed of convergence of the bias can be derived under additional hypothesis like regularity conditions on the underlying density f.

Theorem 2 *Assume that* **H1, H2** *and* **H3** *hold. In addition suppose that*
- *f is r times continuously differentiate at the point x*
- $\int_R |u|^j |K(u)|\, du < \infty$ *for* $j = 1,\ldots,r+1$
- $\int_R g_n(t)\, dt$ *is bounded, where* $g_n(t) = \sup_{z\in R} |f^{(r+1)}(z)|$
- $\sum_{i=1}^{n}\left(\frac{h_i}{h_n}\right)^{\eta+l} \to \theta_{\eta+l} < \infty$ *as* $n \to \infty$, $\eta \in [0,1]$, $l = 1,\ldots,r+1$

Then:

$$E[F^*_{n,\eta}(x)] = F(x) + \sum_{l=1}^{r} \frac{(-h_n)^l}{l!} f^{(l-1)}(x)\frac{\theta_{l+\eta}}{\theta_\eta} K_l + O(h_n^{r+1})$$

Vilar (1992) obtains expressions for the bias of the nonrecursive estimate $F^*_n(x)$ under similar conditions. If both expressions are compared it follows that:

$$\text{Bias}\left[F^*_{n,\eta}(x)\right] - \text{Bias}\left[F^*_n(x)\right] = \sum_{l=1}^{r} \frac{1}{l!} K_l f^{(l-1)}(x)\left[\frac{\theta_{\eta+l}}{\theta_\eta} - 1\right] + O(h_n^{r+1})$$

In order to analyze the value of $\frac{\theta_{\eta+l}}{\theta_\eta}$, the usual choice of the bandwidth is $h_n = Cn^{-\rho}, 0 < \rho < 1, C > 0$. Then $\theta_\eta = \frac{1}{1-\eta\alpha}$ is found. Now, set $r = 2$ and K symmetric ($\Leftrightarrow K_1 = 0$), therefore

$$\text{Bias}\left[F^*_{n,\eta}(x)\right] - \text{Bias}\left[F^*_n(x)\right] = \frac{1}{2}K_2 f'(x)\left[\frac{\theta_{\eta+2}}{\theta_\eta} - 1\right] + O(h_n^2)$$

If we designated $\frac{\theta_{\eta+2}}{\theta_\eta}$ as $H(\eta)$, we have $H(\eta) = \frac{\theta_{\eta+2}}{\theta_\eta} = \frac{1-\eta\rho}{1-\eta\rho-2\rho}$, and it is simply derived that:

1) $H(\eta) > 1, \ \forall \eta \in [0,1]$. Therefore the recursive estimate has a larger bias than a nonrecursive one, although both have the same convergence rate.
2) $H'(\eta) > 0 \ \ \forall \eta \in [0,1]$. Therefore, large values of η result in estimations with a large bias.

5. THE VARIANCE

In this section we establish results related to asymptotic variance of $F_n(x)$ and $F^*_{n,\eta}(x)$ when $T \equiv IO$ and $T \equiv RP$.

Theorem 3 *Let (X,T) be defined as before with $T \equiv IO$ and let us suppose that* **H1** *holds. Then we have:*
(a) If the mixing coefficients $\alpha(t)$ satisfy $\int_0^\infty \alpha(t)\,dt < \infty$, then

$$\text{Var}\big(F_n(x)\big) = \frac{1}{n}\left[F(x)\big(1-F(x)\big) + D_n^{(1)}\right]$$

where $D_n^{(1)}$ appears due to dependence and it is bounded by $2\beta\int_0^\infty \alpha(t)\,dt$.
(b) If K and $\{h_n\}$ verify **H2** *and* **H3** *and, in addition, we assume that the following hypothesis are verified:*
H5.– $\int_0^\infty \alpha(t)^q\,dt < \infty$, *for some* $0 < q < 1$
H6.– $\frac{1}{n}\sum_{i=1}^{n}\left(\frac{h_i}{h_n}\right)^s \to \theta_s < \infty$ *as* $n \to \infty$, $s \in [0,3]$
then:

$$\text{Var}\big(F^*_{n,\eta}(x)\big) \le \frac{1}{n}\left[\frac{\theta_{2\eta}}{\theta_\eta^2}F(x)\big(1-F(x)\big) - h_n\frac{\theta_{2\eta+1}}{\theta_\eta^2}f(x)K_v + 20\frac{\theta_{2\eta}}{\theta_\eta^2}A_n^{1-q}(x)D_n^{(2)} + O(h_n^2)\right]$$

where

$$K_v = 2\int_R uK(u)\,K^*(u)\,du, \quad \text{with} \quad K^*(u) = \int_{-\infty}^{u} K(t)\,dt \tag{3}$$

$$A_n(x) = F(x) - h_n f(x)\frac{2}{1-q}\int_R \left[K^*(z)\right]^{\frac{1+q}{1-q}} K(z)\,dz + O(h_n^2) \tag{4}$$

$$D_n^{(2)} \le \beta\int_0^\infty \alpha(t)^q\,dt$$

Remarks:
[1] First let us state that the bounds for the asymptotic variance of both estimators F_n and $F^*_{n,\eta}$ are written as a sum of terms with clear and specific significance.

Indeed, the first terms: $\left[n^{-1}\frac{\theta_{2\eta}}{\theta_\eta^2}F(x)(1-F(x))\right]$ for $\mathrm{Var}(F_n(x))$ and $\left[n^{-1}\frac{\theta_{2\eta}}{\theta_\eta^2}F(x)\big(1-\right.$ $\left.-F(x)\big) - h_n n^{-1}\frac{\theta_{2\eta-1}}{\theta_\eta^2}f(x)K_v\right]$ for $\mathrm{Var}(F^*_{n,\eta}(x))$, coincide with the classical expressions obtained under the assumption of independent observations.

Obviously, for finite sample size n there will be additional terms due to the mixing dependence of the process X. These are precisely:

$$20n^{-1}\frac{\theta_{2\eta}}{\theta_\eta^2}\beta\int_0^\infty \alpha(t)\,dt \quad \text{for} \quad \mathrm{Var}\big(F_n(x)\big)$$

$$20n^{-1}\frac{\theta_{2\eta}}{\theta_\eta^2}\big[A_n(x)\big]^{1-q}\beta\int_0^\infty \alpha(t)\,dt \quad \text{for} \quad \mathrm{Var}\big(F^*_{n,\eta}(x)\big) \tag{5}$$

as Theorem 3 shows.

[2] Is there some increase in the bound of the asymptotic variance due to the IO structure of the sampling instants? When the data are regularly spaced, Vilar (1992) and Vieu (1991) have established the following asymptotic expressions for the components due to the dependence of $\mathrm{Var}\big(F_n(x)\big)$ and $\mathrm{Var}\big(F^*_n(x)\big)$, where $F^*_n(x)$ is the nonrecursive kernel estimator mentioned in [1]:

$$\frac{2}{n}\int_0^\infty \alpha(t)\,dt \qquad \text{for } \mathrm{Var}\big(F_n(x)\big) \qquad \text{and} \qquad \frac{16}{n}\int_0^\infty \alpha(t)^q\,dt \qquad \text{for } \mathrm{Var}\big(F^*_n(x)\big)$$

If we compare these expressions to those in (5), we seen that, in fact, the IO structure for T contributes to an increase of the variance, which is proportional to β. In fact, the longer β is greater increase in variance. This follows intuitively because if β increases, then the average distance between consecutive instants decreases and the observations $\{X(\tau_k)\}$ become more dependent.

[3] Note that while the convergence rate of the bound is $O(n^{-1})$ for both the equally spaced observations as well as the IO structure; the second dominant term for IO structure is only $O(h_n^{1-q}n^{-1})$ compared to $O(h_n n^{-1})$ for regular observations. Thus, by introducing randomness in the sampling, we have caused the slower convergence to zero of the second term in the asymptotic bound.

[4] Finally, it is also interesting to study the behaviour of the value $G(\eta) = \theta_{2\eta}/\theta_\eta^2$ as $\eta \in [0,1]$. Using the same reasoning for $H(\eta) = \theta_{\eta+2}/\theta_\eta$ as in Section 4, it is easy to show that $G(\eta)$ decreases and $G(\eta) < 1$ for all $\eta \in [0,1]$, when the usual bandwidth $h_n = Cn^{-\rho}$ is used. Therefore the nonrecursive estimate has a larger variance than a recursive one, although both have the same convergence rate. Moreover, the parameter η can be regarded as another smoothing parameter: values of η near 1 result in "smooth" estimates with a small variance, whereas small η yields a small bias.

Next we turn to deriving the asymptotic variance expression for both $F_n(x)$ and $F^*_{n,\eta}(x)$ when the sampling instants $\{\tau_k\}$ are random, constituting a renewal process ($T \equiv RP$) on $[0,\infty)$, as defined in Section 2.

If $G_k(t)$ denotes the distribution function of the kth sampling point τ_k, then $G_k(t)$ is clearly the kth fold convolution of $G(t)$ with itself; and the mean number of sampling instants occurring in the interval $[0,t]$ can be written as

$$H(t) = \sum_{k=1}^{\infty} G_k(t) \qquad t \geq 0 \tag{6}$$

When $G(t)$ is specifically absolutely continuous, with density $g(t)$ then we have

$$H'(t) = \sum_{k=1}^{\infty} g_k(t) := h(t) \qquad t \geq 0 \tag{7}$$

where $h(t)$ is called the *renewal density*. This density is very important in renewal theory and

$$h(t) \to \beta \quad \text{as} \quad t \to \infty \tag{8}$$

is verified, being $\beta^{-1} = E[t_i]$, $\forall i \in N$. Hence we may regard β as the average sampling rate. An exhaustive survey on these results in renewal theory can be found in D.R.Cox (1964), but these notions are sufficient by the proof of following theorem:

Theorem 4 *Let (X,T) be defined as before with $T \equiv RP$ and assume that the hypothesis of Theorem 3 holds. In addition, assume that the renewal density is bounded on $[0,\infty)$. Then*

a)
$$\operatorname{Var}\Big(F_n(x)\Big) = \frac{1}{n}\left[F(x)\Big(1-F(x)\Big) + D_n^{(3)}\right]$$

where due to dependence $D_n^{(3)}$ appears and is bounded by $\int_0^{\infty} h(t)\,\alpha(t)\,dt$.

b)
$$\operatorname{Var}\left(F_{n,\eta}^{*}(x)\right) \leq \frac{1}{n}\left[\frac{\theta_{2\eta}}{\theta_{\eta}^{2}}F(x)\Big(1-F(x)\Big) - h_n\frac{\theta_{2\eta+1}}{\theta_{\eta}^{2}}f(x)K_v + 20\frac{\theta_{2\eta}}{\theta_{\eta}^{2}}A_n^{1-q}(x)D_n^{(4)} + O(h_n^2)\right]$$

where K_v and $A_n(x)$ are expressed in (3) and (4) and

$$D_n^{(4)} \leq \int_0^{\infty} h(t)\alpha^{q}(t)\,dt$$

The conclusions for Theorem 4 (for RP) are similar to those of Theorem 3 (for IO) and will not be repeated here. In fact, the most substantial difference is given by $D_n^{(2)}$ and $D_n^{(4)}$. Note that

$$D_n^{(2)} \leq \beta \int_0^{\infty} [\alpha(t)]^{q}\,dt \qquad \text{and} \qquad D_n^{(4)} \leq \left[\sup_{t\in R} h(t)\right] \int_0^{\infty} [\alpha(t)]^{q}\,dt$$

But by (8), $h(t) \to \beta$ as $n \to \infty$. Moreover, if T is a Poisson process then $H(t) = \beta$ for all $t \in R$, therefore both $D_n^{(2)}$ and $D_n^{(4)}$ have the same bound.

6. SKETCH OF PROOFS

In order to establish the validity of the results stated in this paper we will need some previous lemmas. The first one is an extension of a result commonly used due to Bochner.

Lemma 1 *Suppose that* **H2** *and* **H3** *holds. Then*

$$\lim_{n\to\infty} \int_R K_n^{*}(x-u)f(u)\,du = F(x)$$

at each point $x \in R$.

PROOF:

$$\int_R K_n^*(x-u)f(u)\,du = \int_R \left[\int_{-\infty}^{(x-u)h_n^{-1}} K(v)f(u)\,dv\right] du = \int_R \left[\int_{-\infty}^{x-vh_n} f(u)\,du\right] K(v)\,dv =$$
$$= \int_R F(x-vh_n)K(v)\,dv \to F(x)$$

as $n \to \infty$, by dominated convergence since the integrand is bounded and tends to $K(v)F(x)$ as $n \to \infty$. ∎

We shall also make use of two well known lemmas stated below.

Lemma 2 [Doe, (1973)] *Let $\{X(t); t \in R\}$ be a strong mixing process and let U and V be two random variables that are measurable with respect to the σ-algebras $\mathcal{F}_{-\infty}^{0}$ and $\mathcal{F}_{\tau}^{+\infty}$; $\tau \geq 0$, respectively, and assume $E|U|^{2+\delta} < \infty$, $E|V|^{2+\delta} < \infty$ for some $\delta > 0$. Then*

$$\left|\text{Cov}\{U,V\}\right| \leq 10\left[\alpha(\tau)\right]^{\delta/(2+\delta)}\left\{E|U|^{2+\delta}\,E|V|^{2+\delta}\right\}^{1/(2+\delta)}$$

where $\alpha(\tau)$ is the strong mixing coefficient.

Lemma 3 [Toeplitz] *Let $[a_{n,k}]_{n,k=1}^{\infty}$ be a matrix of real numbers satisfying*

$$\lim_{n\to\infty} a_{n,k} = 0 \qquad \textit{for each } k$$
$$\sum_k |a_{n,k}| \leq C < \infty \qquad \textit{for all } n = 1,2,\ldots$$
$$\lim_{n\to\infty} \sum_k a_{n,k} = A < \infty$$

Let $\{w_n\}$ be a sequence of real numbers converging to a finite limit w as $n \to \infty$. Then

$$\sum_k a_{n,k}w_k \to Aw < \infty \qquad \textit{as } n \to \infty$$

Now the proofs of both Theorems 1 and 2 can be easily derived.

PROOF OF THEOREM 1

Under **H1** the bias does not depend on the particular structure of T. Thus

a) $E[F_n(x)] = n^{-1}\sum_{i=1}^{n} P[X(\tau_i) \leq x] = n^{-1}\sum_{i=1}^{n} F_i(x)$, where $F_i(x)$ denotes the distribution function of $X(\tau_i)$ and by $H1$, $F_i(x) = F(x)$, for all i. Hence $F_n(x)$ is an unbiased estimate of $F(x)$.

b) Under **H1**

$$E[F_{n,\eta}^*(x)] = \left(\sum_{i=1}^{n} h_i^\eta\right)^{-1} \sum_{j=1}^{n} h_j^\eta \nu_j(x)$$

where $\nu_j(x) = \int_R K_j^*(x-u)f(u)\,du$. By Lemma 1, $\nu_j(x) \to F(x)$ as $j \to \infty$, and the result follows from the Lemma 3 and the fact that **H4** is verified. ∎

PROOF OF THEOREM 2

If, in addition, the hypothesis of Theorem 2 are satisfied then

$$E[F_{n,\eta}^*(x)] = \left(\sum_{i=1}^{n} h_i^\eta\right)^{-1} \sum_{j=1}^{n} h_j^\eta \nu_j(x) = \left(\sum_{i=1}^{n} h_i^\eta\right)^{-1} \int_R K(v)F(x-vh_j)\,dv$$

and expanding F in a Taylor series, we have

$$F(x-vh_j) = F(x) + \sum_{l=1}^{r} \frac{(-h_j)^l}{l!} f^{(l-1)}(x)v^l + O(|v|^{r+1})$$

Finally Theorem 2 follows from Lemma 1 and Lemma 3. ■

PROOF OF THEOREM 3

a) Let us denote $I_{(-\infty,x]}(X(\tau_i))$ as A_i. By definition of $\alpha(t)$, we have

$$\begin{aligned}\mathrm{Cov}(A_i, A_{i+k}) &\leq \left|E(A_iA_{i+k}) - E(A_i)E(A_{i+k})\right| = \\ &= \left|P[X(\tau_i)\leq x, X(\tau_{i+k})\leq x] - P[X(\tau_i)\leq x]P[X(\tau_{i+k})\leq x]\right| \leq \\ &\leq \alpha\left[\frac{k+i}{\beta} - \frac{1}{2\beta} - \frac{i}{\beta} - \frac{1}{2\beta}\right] = \alpha\left(\frac{k-1}{\beta}\right) \qquad (9)\end{aligned}$$

Then, by (9), **H1** and the stationarity of the process X we have

$$\begin{aligned}\mathrm{Var}\left[F_n(x)\right] &= n^{-2}\left[\sum_{i=1}^{n}\sum_{i=1}^{n}\mathrm{Var}(A_i) + 2\sum_{k=1}^{n-1}\sum_{i=1}^{n-k}\mathrm{Cov}(A_i, A_{i+k})\right] \leq \\ &\leq n^{-2}\left[nF(x)\big(1-F(x)\big) + 2\sum_{k=1}^{n-1}(n-k)\,\alpha\left(\frac{k-1}{\beta}\right)\right]\end{aligned}$$

Hence, if

$$D_n^{(1)} = 2\sum_{k=1}^{n-1}\left(1-\frac{k}{n}\right)\alpha\left(\frac{k-1}{\beta}\right)$$

we find that

$$D_n^{(1)} \leq 2\sum_{k=0}^{\infty}\alpha\left(\frac{k}{\beta}\right) \leq 2\beta\int_0^{\infty}\alpha(t)\,dt < \infty$$

which completes the proof. ■

b) In order to establish the validity of the second part we proceed as follows:

STEP 1: We define the function $\gamma(t)$ by

$$\gamma(t) = (g*g)(t) = \int_{-\infty}^{+\infty} g(t-y)g(y)\,dy$$

Under the conditions assigned to g, we have that γ is a symmetric density function with support in $[-\beta^{-1},\beta^{-1}]$. Moreover, $\gamma\left(t-\frac{i-j}{\beta}\right)$ is the probability density associated with $z_i - z_j$ for $i \neq$j.

STEP 2:

$$\begin{aligned}\mathrm{Var}[F^*_{n,\eta}(x)] &= \left(\sum_{i=1}^{n}h_i^{\eta}\right)^{-2}\sum_{j=1}^{n}\sum_{l=1}^{n}h_j^{\eta}h_l^{\eta}\mathrm{Cov}\left\{K_j^*\big(x-X(\tau_j)\big), K_l^*\big(x-X(\tau_l)\big)\right\} = \\ &= I_{n,0}(x) + R_n(x) \qquad (10)\end{aligned}$$

where

$$I_{n,0}(x) = \left(\sum_{i=1}^{n}h_i^{\eta}\right)^{-2}\sum_{j=1}^{n}h_j^{2\eta}\mathrm{Var}K_j^*\big(x-X(\tau_j)\big) \qquad (11)$$

$$R_n(x) = \sum_{k=1}^{n}\big(I_{n,k}(x) + I_{n,-k}(x)\big) \qquad (12)$$

$$I_{n,k}(x) = \left(\sum_{i=1}^{n}h_i^{\eta}\right)^{-2}\sum_{j=1}^{n-k}h_j^{\eta}h_{j+k}^{\eta}\mathrm{Cov}\left\{K_j^*\big(x-X(\tau_j)\big), K_{j+k}^*\big(x-X(\tau_j)\big)\right\} \qquad (13)$$

STEP 3: For $k = 0$ we have

$$I_{n,0}(x) = \left(\sum_{i=1}^{n} h_i^{\eta}\right)^{-2} \sum_{j=1}^{n} h_j^{2\eta}\Big\{EK_j^{*^2}\big(x - X(\tau_j)\big) - \Big(EK_j^{*}\big(x - X(\tau_j)\big)\Big)^2\Big\} \tag{14}$$

Put $\mu_j(x) = EK_j^{*^2}\big(x - X(\tau_j)\big)$ and $\nu_j(x) = EK_j^{*}\big(x - X(\tau_j)\big)$ then

$$\mu_j(x) = \int_R M(v)F(x - vh_j)\, dv \tag{15}$$

where $M(x) = 2K(x)\left[\int_{-\infty}^{x} K(t)\, dt\right]$ and $\int_R M(x)\, dx = 1$ is verified. Therefore, by Lemma 1, (11) and (15) we get

$$I_{n,0}(x) = \left(\sum_{i=1}^{n} h_i^{\eta}\right)^{-2} \sum_{j=1}^{n} h_j^{2\eta}\left\{\int_R M(v)F(x - vh_j)\, dv - \nu_j^2(x)\right\} =$$
$$= \left(\sum_{i=1}^{n} h_i^{\eta}\right)^{-2} \sum_{j=1}^{n} h_j^{2\eta}\left\{F(x) + f(x)h_jK_v + O(h_j^2) - \big(F(x) + O(h_j^2)\big)^2\right\}$$

Now, by applying Toeplitz's Lemma and **H6**

$$I_{n,0}(x) = \frac{1}{n}\left[\frac{\theta_{2\eta}}{\theta_\eta^2} F(x)\big(1 - F(x)\big) - h_n \frac{\theta_{2\eta+1}}{\theta_\eta^2} f(x)\, K_v + O(h_n^2)\right] \tag{16}$$

STEP 4: We now show that under assumption **H5** we have

$$R_n(x) \leq \frac{20}{n}\frac{\theta_{2\eta}}{\theta_\eta^2} A_n^{1-q}(x) D_n^{(2)} + O(h_n^2)$$

which together with (16) gives the result. By the stationarity of the process X; $I_{n,k}(x)$ can be written in the form

$$I_{n,k}(x) = \left(\sum_{i=1}^{n} h_i^{\eta}\right)^{-2} \sum_{j=1}^{n-k} h_j^{\eta} h_{j+k}^{\eta} \mathrm{Cov}\Big\{K_j^{*}\big(x - X(0)\big),\, K_{j+k}^{*}\big(x - X(\tau_{j+k} - \tau_j)\big)\Big\}$$

But $\tau_{j+k} - \tau_j = \dfrac{k}{\beta} + (z_{j+k} - z_j)$, and $\gamma\left(t - \dfrac{k}{\beta}\right)$ is the density associated with $z_{j+k} - z_j$, as we pointed out Step 1. Therefore, an estimate of the magnitude of $I_{n,k}(x), k \neq 0$, can be obtained as follows

$$I_{n,k}(x) = \left(\sum_{i=1}^{n} h_i^{\eta}\right)^{-2} \sum_{j=1}^{n-k} h_j^{\eta} h_{j+k}^{\eta} \int_{-\beta^{-1}}^{\beta^{-1}} \mathrm{Cov}\left\{K_j^{*}\big(x - X(0)\big), K_{j+k}^{*}\left(x - X\Big(t + \frac{k}{\beta}\Big)\right)\right\} \gamma(t)\, dt$$

Put $U = K_j^{*}\big(x - X(0)\big)$ and $V = K_{j+k}^{*}\left(x - X\left(t + \dfrac{k}{\beta}\right)\right)$ for every $\delta > 0$ we have

$$A_j(x) := E|U|^{2+\delta} = \int_R t(v)F(x - vh_j)\, dv \tag{17}$$
$$A_{j+k}(x) := E|V|^{2+\delta} = \int_R t(v)F(x - vh_{j+k})\, dv$$

where $t(x) = (2+\delta)\,K(x)\left[\int_{-\infty}^{x} K(z)\,dz\right]^{1+\delta}$ and $\int_R t(x)\,dx = 1$ is verified. Note that $E|U|^{2+\delta} < \infty$, $E|V|^{2+\delta} < \infty$ and $t + (k/\beta) > 0$, for all $k \neq 0$. Therefore, using Lemma 2 [Doe], we find

$$I_{n,k}(x) \le 10\left(\sum_{i=1}^{n} h_i^{\eta}\right)^{-2}\left[\sum_{j=1}^{n-k}\left(h_j^{\eta} A_j(x)^{1/(2+\delta)}\right)\left(h_{j+k}^{\eta} A_{j+k}(x)^{1/(2+\delta)}\right)\right]\times$$
$$\times\int_{-\beta^{-1}}^{\beta^{-1}}\left[\alpha\left(t+\frac{k}{\beta}\right)\right]^{\delta/(2+\delta)}\gamma(t)\,dt \tag{18}$$

The term in square brackets can be bounded by

$$\sum_{j=1}^{n} h_j^{2\eta} A_j(x)^{2/(2+\delta)} \tag{19}$$

which is obtained by applying the Cauchy-Schwartz inequality and some algebra. Now let $\delta = 2q/(1-q)$ with $0 < q < 1$, then by (19) and (18)

$$I_{n,k}(x) \le 10\left(\sum_{i=1}^{n} h_i^{\eta}\right)^{-2}\left(\sum_{j=1}^{n} h_j^{2\eta} A_j(x)^{1-q}\right)\int_{-\beta^{-1}}^{\beta^{-1}}\left[\alpha\left(t+\frac{k}{\beta}\right)\right]^{q}\gamma(t)\,dt \tag{20}$$

Therefore

$$R_n(x) \le \frac{20}{n}\left(\sum_{i=1}^{n} h_i^{\eta}\right)^{-2}\left(\sum_{j=1}^{n} h_j^{2\eta} A_j(x)^{1-q}\right)\sum_{k=1}^{n-1}\int_{-\beta^{-1}}^{\beta^{-1}}\left[\alpha\left(t+\frac{k}{\beta}\right)\right]^{q}\gamma(t)\,dt \tag{21}$$

But since $\alpha(t)$ is non increasing, the last integral on the right side is bounded by $\alpha(\frac{k-1}{\beta})^q$ and therefore the last summand is bounded by

$$\sum_{k=1}^{n-1}\alpha\left(\frac{k-1}{\beta}\right)^{q} \le \sum_{k=1}^{\infty}\alpha\left(\frac{k}{\beta}\right)^{q} = D_n^{(2)} \le \beta\int_0^{\infty}\left[\alpha(t)\right]^{q} dt < \infty$$

by hypothesis **H5**. Finally by applying Toeplitz's Lemma in (21) the proof is established. ■

PROOF OF THEOREM 4

(a) This part can be proven using the same sketch as for Theorem 3(a).
(b) The proof is also similar to that of the Theorem 3(b). Proceeding as in the proof of Theorem 3(b) we write for any x

$$\mathrm{Var}[F_{n,\eta}^{*}(x)] = I_{n,0}(x) + R_n(x)$$

we find that:

$$I_{n,0}(x) = \frac{1}{n}\left[\frac{\theta_{2\eta}}{\theta_\eta^2}F(x)\big(1-F(x)\big) - h_n\frac{\theta_{2\eta+1}}{\theta_\eta^2}f(x)\,K_v + O(h_n^2)\right]$$

Only the estimate for $I_{n,k}(x)$ is different here. Now, for $i > j$, the distribution of $\tau_i - \tau_j := \sum_{k=j+1}^{i} t_k$ is the convolution of $G(t)$ from $k = j+1$ to i, $\left(*_{k=j+1}^{i}\, G(t)\right)$. In fact, since the random variables t_i are independent and identically distributed, we have that

$*_{k=j+1}^{i} G(t)$ is identical to the distribution of $G_{|i-j|}(t)$. Therefore the estimate for $I_{n,k}(x)$ will be

$$I_{n,k}(x) = \left(\sum_{i=1}^{n} h_i^{\eta}\right)^{-2} \sum_{j=1}^{n-k} h_j^{\eta} h_{j+k}^{\eta} \int_0^{+\infty} \mathrm{Cov}\Big\{K_j^*\big(x - X(0)\big),\, K_{j+k}^*\big(x - X(t)\big)\Big\}\, dG_k(t)$$

Applying Doe's Lemma and the Cauchy-Schwartz inequality again we get

$$R_n(x) \leq \frac{20}{n} \left(\sum_{i=1}^{n} h_i^{\eta}\right)^{-2} \left(\sum_{j=1}^{n} h_j^{2\eta} A_j(x)^{1-q}\right) \sum_{k=1}^{n-1} \int_0^{+\infty} [\alpha(t)]^q \, dG_k(t)$$

Since $h(t)$ is bounded on $[0, \infty)$, and by (6) and (7) we have

$$\sum_{k=1}^{n-1} \int_0^{\infty} [\alpha(t)]^q \, dG_k(t) \leq \int_0^{\infty} [\alpha(t)]^q \, d\left[\sum_{k=1}^{\infty} G_k(t)\right] = \int_0^{\infty} [\alpha(t)]^q \, dt =$$

$$= \int_0^{\infty} [\alpha(t)]^q h(t)\, dt = D_n^{(4)} \leq \left[\sup_{t \geq 0} h(t)\right] \int_0^{\infty} [\alpha(t)]^q \, dt < \infty$$

Thus part (b) of the Theorem 4 is finished. ■

REFERENCES

[1] Azzalini, A. (1981). A Note on the Estimation of a Distribution Function and Quantiles by a Kernel Method. *Biometrika*, **68**, 1, 226–228

[2] Cox, D.R. (1962). *Renewal Theory.* Chapman and Hall.

[3] Doe, J. (1973). A Note on Empirical Processes for Strong Mixing Processes. *Ann. Prob.*, **1**, 870–875.

[4] Deheuvels, P. (1974). Conditions Nécessaires et Suffisantes de Convergence Ponctuelle Presque Súre. *C. R. Acad. Sci. Paris Ser. A*, **278**, 1217–1220.

[5] Faraldo Roca, P. & González Manteiga, W. (1984). Obtención del Sesgo, Varianza y Error Cuadrático Medio de una Familia Axiomática de Estimadores No Paramétricos para Funciones de Distribución. *Actas XIV Congreso S.E.I.O., Granada.*

[6] Izenman, A. & Tran, H. (1990). Kernel Estimation of the Survival Function and Hazard Rate Weak Dependence. *Journal of Statistical Planning and Inference*, **24**, 233–247.

[7] Lejeune, P. & Sarda, M. (1989). *Smooth Estimators of Distribution and Density Functions.* Preprint.

[8] Masry, E. (1983). Probability Density Estimation from Sampled Data. *I.E.E.E. Transactions on Information Theory*, **29**, 696–708.

[9] Reiss, R.D. (1981). Nonparametric Estimation of Smooth Distribution Functions. *Scand. J. Statist.*, **8**, 116–119.

[10] Roussas, G. (1989). Some Asymptotic Properties of an Estimate of the Survival Function under Dependence Conditions. *Statistics and Probability Letters*, **8**, 235–243.

[11] Sarda, P. & Vieu, J. (1989). Empirical Distribution Function for Mixing Random Variables. Application in Nonparametric Hazard Estimation. *Statistics*, **20**, 559–571.

[12] Vieu, J. (1991). Quadratic Errors for Nonparametric Estimates under Dependence. *Journal of Multivariate Analysis*, **39**, 2, 324–347.

[13] Vilar, J.M. & Vilar, J.A. (1992). *Estimación No Paramétrica Recursiva de Curvas de Probabilidad.* Preprint.

Queueing Theory and Applications

Recent Adv. in Stat. and Prob., pp. 335-342
J. Pérez Vilaplana and M. L. Puri (Eds)

Performance Evaluation of ATM Network Access Protocols*

C. BLONDIA
Computer Science Department, University of Nijmegen, Toernooiveld, 1, NL-6525 ED Nijmegen, The Netherlands

Abstract – In shared medium communication systems, the medium access control (MAC) protocol arbitrates the access to the physical medium between the different subscribers. In particular, such a protocol is needed when providing network access to B-ISDN subscribers using a passive optical network. Besides the network access delay, the MAC protocol also determines the shape of the traffic at the entrance of the B-ISDN network, and therefore it has a major impact on the overal performance of the system. In this paper we study a MAC protocol which uses a request/permit mechanism. Each subscriber queues the cells ready for transmission and advertises the required bandwidth by means of requests. The available bandwidth is distributed among the subscribers by means of permits. The allocation of these permits is based on the permit distribution algorithm which uses the requests as input, together with the subscriber's agreed peak bit rate.
In order to evaluate different permit distribution algorithms, the system is modeled as a single server queueing system in which the server takes various types of vacations. The input of this system belongs to a general class of discrete-time Markovian arrival procesess which models a variety of traffic sources occuring in B-ISDN. Characteristics of this queueing system can then be translated into performance measures of the access system.

1. INTRODUCTION

Recent progress in the area of optical transmission technology (optical fibers, passive optical splitters and combiners) make it possible for small bussiness and residential customers to share the access resources to the B-ISDN. In particular, when the advantages (low costs) of a Passive Optical Network (PON) with a tree structure is combined with the statistical multiplexing capabilities of the Asynchronous Transfer Mode (ATM), the outcome may be a highly efficient, reliable and cost-effective broadband access facility.
The access to the shared medium is arbitrated by the Medium Access Control (MAC) protocol. Besides the network access delay, the MAC protocol determines the shape of the traffic entering the B-ISDN, and therefore it has a crucial impact on the overal performance of the system. In this paper we study a MAC protocol which uses a request/permit mechanism. Each subscriber queues the cells ready for transmission and advertises its required bandwidth by means of *requests*. The available bandwidth is distributed by

*This work was supported in part by the Commission of the European Communities, under project RACE 2024 (Research and Development on Advanced Communications in Europe) Broadband Access Facilities means of *permits*. The permits are allocated to the different subscribers by means of the *permit distribution algorithm*. This algorithm is a crucial part of the MAC protocol, in particular with respect to

- The required guaranteed bandwidth each subscriber needs in order to ensure a limited interarrival time between successive permit deliveries for a subscriber.
- The way the remaining bandwidth is allocated, i.e. the way permits are distributed (e.g. at the peak rate specified by the subscriber at call set-up).
- The ability to support different types of services (e.g. constant bit rate, variable bit rate, sporadic sources, etc....).

The class of algorithms we investigate uses two parameters to allocate bandwidth : the current number of cells waiting in the subscriber's buffer to be transmitted (specified by the requests), and the peak bit rate specified by the subscriber at call set-up.
In order to evaluate different options for the permit distribution algorithm, the system is modeled by means of the following single server queueing system. The service time is supposed to be constant (i.e. the time needed to transmit a cell). After each service the server takes a vacation, the duration of which depends on the queue size, but which is at least the minimal cell interarrival time corresponding to the agreed peak bit rate. An idle server takes repeated vacations, the length of which is determined by the guaranteed bandwidth. The input of this queueing model belongs to a class of markovian arrival processes which models a variety of traffic sources occuring in B-ISDN (e.g. voice, data, variable bit rate video, etc....). Characteristics of this queueing system can be translated into performance measures of the access system.

2. THE MAC PROTOCOL

The main function of the MAC protocol is to avoid collisions of traffic originating from different subscribers. Moreover it should aim at

- *Efficiency* : the overhead introduced by the MAC protocol should be low.
- *Performance* : the delay and the delay variance introduced by the MAC protocol should be kept within certain bounds defined by the quality of service requirements of the traffic.
- *Fairness* : the traffic of one subscriber should not be subject to more access delay than the traffic of another.

The MAC protocol described in this paper uses a *request/permit* mechanism. The necessary bandwidth a customer needs is claimed by means of a *request*. A request contains the number of cells that have arrived since the previous request was sent (i.e. the number of new arrivals). Requests are sent together with the upstream cells. This implies a lower overhead since the preamble is only needed once. On the other hand, coupling requests with upstream cells does not allow fast reaction on changing traffic situations. Consider for example a subscriber generating traffic according to an on/off source model, and making a transition from the off state to the on state. In the off state the subscriber has nothing to send, and hence he does not request any bandwidth. When he switches to the on state, he cannot request for bandwidth, since he had no upstream cells during the off period. Therefore, each customer needs some *guaranteed bandwidth*, in order to be able to advertise its required bandwidth (by means of requests). This guaranteed bandwidth clearly implies some overhead (in particular when the customer has nothing to send, this bandwidth is wasted). High guaranteed bandwidth implies fast reaction on changing traffic situations but also high overhead, while a low guaranteed bandwidth may cause buffer overflow in the customers premises, unacceptable access delay and an increase of burstiness. A MAC protocol where requests and upstream cells are decoupled, and hence no guaranteed bandwidth is needed, is studied in [3].

A subscriber is informed about the bandwidth it can use by means of *permits.* Permits are sent to the customers preceding the downstream cells. Since for downstream traffic a broadcasting mechanism is applied, the destination of the permit and the destination of the cell are completely decoupled. When a customer receives a permit, it is allowed to send one cell immediately. Remark that permits for multiple cells would involve less overhead, but as cells would enter the network back to back, it would increase the burstiness and add additional delay variation.

The MAC protocol needs an algorithm to distribute the available bandwidth among the active customers, the so called *Permit Distribution Algorithm.* Its input consists of fixed parameters agreed with the network at call set-up (the static part of the traffic characteristics) and the received requests (the dynamic part of the traffic characteristics). The proposed MAC protocol is defined as follows. When a request issued by a customer arrives, the MAC protocol assigns the necessary permits to that customer (the same number as indicated in the request). These permits are then stored in a FIFO queue, together with the permits for the other subscribers, with the additional constraint that a minimal distance between two consecutive permits for the same subscriber is enforced. This minimal distance is determined by the peak rate the subscriber agreed with the network at call set-up. The FIFO queue is called the *permit FIFO queue.*

Hence, the main characteristics of this permit distribution algorithm are :

- The peak rate of each customer is enforced.
- The mechanism by which permits are issued (over all customers) approximates a global FIFO strategy. This FIFO discipline is chosen for fairness reasons and to minimize the access delay variance.
- The available buffer capacity at the customer premises are used to store the cells that wait for transmission (distributed buffering with central control).

Using this protocol, the queueing takes place at the customer premises, while the control of the access medium is achieved by maintaining a permit FIFO queue, by which the transmission instants and the order in which the different subscriber's buffer are emptied, are governed. An implementation of this permit distribution algorithm is described in [3].

3. AN EXAMPLE

In this section we use an example to illustrate the following crucial characteristics of the MAC protocol :

- The reaction speed of the protocol to changing traffic situations (determined by the guaranteed bandwidth).
- The impact of the MAC protocol on the shape of the traffic entering the network (determined by the permit distribution algorithm).

Moreover, we take the round-trip delay into account (i.e. the time a cell needs to go from the subscriber's buffer to the MAC control and back). Consider an access network in which customers are connected to the network by means of links of 150 Mbit/s (effective bandwidth). Assuming a cell length of 53 bytes and a request length of 1 byte, we define a time slot as the time needed to transmit a cell and a request. Clearly, one time slot = 2.88μs. Consider a customer who generates traffic with peak rate 10 Mbit/s and mean rate 1 Mbit/s. Furthermore, this customer is offered a guaranteed bandwidth of 1 Mbit/s (i.e. when the buffer at the customer premises is empty, it is polled every 432μs). Let the round trip delay be 100 μs.

Consider this source when it switches from the off state (i.e. no cells are generated) to a state in which cells are generated at a constant bit rate of 5 Mbit/s (i.e. one cell every 86.4μs). Figure 1 shows the arrival and departure instants of cells. Suppose that the source becomes active just before a polling instant. At this polling instant, the first cell is immediately sent and a request for 0 cells is added (since no cells are waiting in the queue anymore). A new polling cycle is started and at the next polling instant there are already 5 cells waiting. The first cell of these waiting cells is immediately sent and a request for 4 more permits is added. It takes at least one complete round-trip delay before the first permit arrives. Supposing that only this customer is active, permits will now arrive at peak rate (i.e. one permit every 43 μs). Eventually, permits catch up with the incoming traffic, and the last cell of what could be called a busy period is sent with a request for 0 permits. Immediately a polling cycle starts and the previous situation is repeated. From this example we see that

- Due to the low guaranteed bandwidth, a number of cells are accumulated in the customer's buffer before the first cell is sent (in this example 5 cells).
- Cells coming in at a constant bit rate of 5 Mbit/s, are sent to the network in bursts, each at a 10 Mbit/s.

An obvious solution to these problems, namely slow reaction of the MAC protocol and an increase of the burstiness, could be an increase of the guaranteed bandwidth (i.e. shorter polling cycle). However this may lead to unacceptable overhead. A more efficient solution is obtained by adapting the rate at which permits are issued to the rate at which cells arrive at the customer's buffer. Indeed, in this way the MAC protocol does not increase the burstiness of the traffic and introduces less access delay variation. Moreover, it makes busy periods (i.e. periods during which the buffer is not empty) longer, resulting in a decrease of the frequency of polling cycle starts, and hence it introduces less overhead.

From this example we clearly see that an appropriate choice of the permit distribution algorithm leads to less overhead (by decreasing the frequency of ends of busy periods) and less induced burstiness and delay variation.

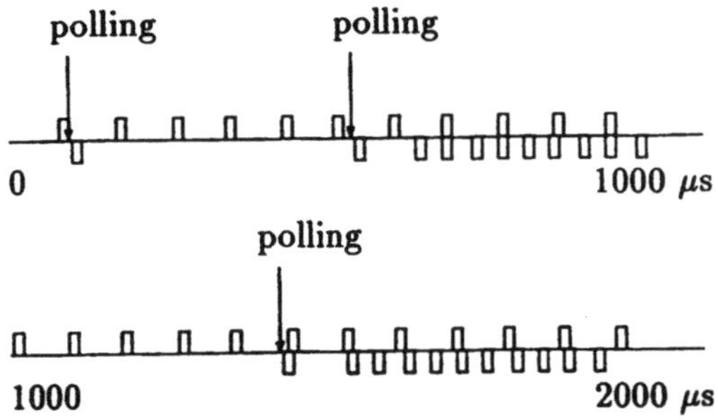

Figure. A Example

In the next section we present a queueing model which can be used to make the appropriate choice of such a permit distribution algorithm.

4. A QUEUEING MODEL FOR THE MAC PROTOCOL

In the previous section we have seen that both the guaranteed bandwidth and the permit distribution algorithm have a major impact on the performance of the access network and on the shape of the traffic entering the public ISDN network. In what follows we describe a queueing model which enables the evaluation of the protocol's performance for variable guaranteed bandwidth values and various permit distribution strategies.

4.1. The Arrival Process

The arrival process belongs to a general class of discrete-time Markovian arrival processes, namely the Discrete-time Batch Markovian Arrival Process (D-BMAP) (see [1]). It is the discrete-time version of the versatile Markovian point process, introduced by Neuts [6] and more recently studied using a more transparant notation by Lucantoni [5]. It includes many well known models for B-ISDN traffic, such as packetized voice and its superposition, high speed data, still pictures, video with and without scene changes and their superposition (see [2], etc... For a detailed discussion we refer to [1]. The analytical tractability of this process when used as input to a queueing system was shown in [2]. For sake of completeness, we recall the definition of a D-BMAP.
Consider a discrete-time Markov chain with $m \times m$ transition matrix $\mathbf{D}$. Assume that at time k the state of this chain is i. With probability $(d_0)_{i,j}$, $1 \leq j \leq m$, there is a transition at instant $k+1$ to state j without arrival, and with probability $(d_n)_{i,j}$, $1 \leq j \leq m$, $n \geq 1$, there is a transition to state j with a batch arrival of size n. The matrix $\mathbf{D_0}$ with elements $(d_0)_{i,j}$, governs transitions without arrivals, while the matrices $\mathbf{D_n}$, $n \geq 1$, with elements $(d_n)_{i,j}$, govern transitions that correspond to arrivals of batches of size n. Hence, the arrival process can be defined as a two dimensional discrete-time Markov process $\{(N(k), J(k)), k \geq 0\}$ on the state space $\{(n,j), n \geq 0, 1 \leq j \leq m\}$, where $\{N(k), k \geq 0\}$ is the counting variable and $\{J(k), k \geq 0\}$ the state of the underlaying Markov chain, also called the phase of the arrival process. We have that $\mathbf{D} = \sum_{n=0}^{\infty} \mathbf{D}_n$.

4.2. The Service Process

The total service mechanism is modeled as a combination of the real service (i.e. the transmission of a cell) and various server vacations (to model the effect of the permit distribution algorithm, the round-trip delay, the guaranteed bandwidth, the multiplexing delay).

i) The Service Time.

The service time is defined to be the constant time to transmit a cell and will be denoted by S.

ii) Permit Distribution.

In view of the example in the previous section, we let the time between consecutive cell transmissions depend on the queue length, with the additional constraint that this inter-transmission time is at least P time slots, P being defined as the number of slots between cells transmitted at peak rate.

If upon departure of a cell there are k cells present in the subscriber's queue, the next permit will be scheduled with an extra delay of V_k time slots, $V_k \geq P$. In what follows we assume that V_k is constant for k fixed, although the analysis holds for V_k a random variable with general distribution. Furthermore, we let V_k be decreasing for k increasing. These general assumptions on V_k allow various possible permit distribution algorithms.

For example, consider a source which can be in three different states, s_i, $i = 1,2,3$. In state s_i, the number of slots between two consecutive cells is d_i. Let $d_i < d_j$ for $i < j$. Denote q the customer's queue length and let T_1 and T_2 be two threshold values such that

- If $q \leq T_1$, then $V_q = d_1$,
- If $T_1 < q \leq T_2$, then $V_q = d_2$,
- If $q > T_2$, then $V_q = d_3$.

iii) The multiplexing time

When a permit is scheduled, it has to compete with all the other permits who are scheduled at the same time slot in the FIFO queue. We let this additional delay be denoted by M.

iv) The round-trip delay

In two boundary cases, the cells suffers from the round-trip delay.

a) Suppose that at a polling instant there are k, $k \geq 2$ cells waiting in the subscriber's queue (see Example in Section 3). The first cell is immediately served (service time S) and the central control of the MAC protocol is informed about the presence of $k-1$ cells. The second cell must wait at least for the round-trip delay R before it can be transmitted.

b) Suppose that the queue contains two cells. If during the interval bounded by the departure instant of the first cell and the start of the service of the second cell, there are k new arrivals, then the first of these new arrivals has to wait at least for the round-trip delay R.

Only in the cases *a)* and *b)* the round-trip delay plays a role.

v) The guaranteed bandwidth delay

When a customer's queue becomes empty, this customer is allowed to send a cell every G time slots (G the number of time slots corresponding to the guaranteed bandwidth). This can be modeled as a server vacation : as soon as the system becomes empty, the server takes a vacation of G slots. If upon return there is at least one cell waiting in the queue it is served immediately, otherwhise the server takes another vacation (i.e. the server takes repeated vacations). When the first cell arrives, it has to wait for the remaining vacation time before it is served.

From the above discussion, we can model the service mechanism as follows.

- A constant service time D time slot (transmission time).
- When the system becomes idle, the server takes repeated vacations, the length of which is constant G time slots.
- After each service, the server takes a vacation. Two cases may occur :

  i) the round-trip delay has to be taken into account (for the conditions see *iv)* above). If k cells are waiting in the buffer, the vacation duration is given by $V_k^{(s)} = \max(R, V_k) + M$, where R is the round-trip delay, V_k the delay imposed by the permit distribution algorithm and M the FIFO multiplexing delay. In this case the vacation is said to be of type r.

ii) the round-trip delay has not to be taken into account. If k cells are waiting in the buffer, the vacation duration is given by $V_k^{(s)} = V_k + M$ (using the same notation as in the previous case). In this case the vacation is said to be of type s.

4.3. The Embedded Markov Chain

Consider a subscriber who generates cells according to a D-BMAP with parameter matrices $\mathbf{D_n}$, $n \geq 0$. The service time of each cell is constant S. When after a service k cells, $k > 0$, are left in the queue, the server takes a vacation. Depending on whether the round-trip delay has to be taken into account or not, the duration is given by $V_k^{(r)}$ or $V_k^{(s)}$. When after a service, the queue is left empty, the server takes repeated vacations, the length of which is G.

Observing the system at cell departure epochs, we may describe the behavior of the system by means of three parameters,

k , $0 \leq k \leq N$, being the number of cells present in the queue at departure time,

l , $l = r, s$, being the type of vacation time after the service,

j , $1 \leq j \leq m$, being the phase of the arrival process.

In this way we obtain a discrete-time Markov chain with state space

$$S = \{(k, l, j), 0 \leq k \leq N, l \in \{r, s\}, 1 \leq j \leq m\}.$$

We introduce the following probability matrices

$[\mathbf{A}_l^{(k)}]_{i,j} = \mathbf{P}$ $\{l$ new arrivals and at the next departure the phase is j given that at the previous departure the phase was i, k customers are present and the vacation after the previous departure was of type $r\}$.

$[\mathbf{B}_l^{(k)}]_{i,j}$ = idem as is defined in a $\mathbf{A}_l^{(k)}$, but with a vacation of type s.

$[\mathbf{C}_l]_{i,j} = \mathbf{P}$ $\{l+1$ new arrivals during the last vacation and the phase is j at the next departure given that the system at the previous departure was empty and that the phase of the arrival process was equal to $i\}$.

The transition matrix of the Markov chain is given by

$$\mathbf{Q} = \begin{pmatrix} \mathbf{C}_0 & \mathbf{C}_1 & 0 & \mathbf{C}_2 & 0 & \mathbf{C}_3 & 0 & \cdots \\ \mathbf{A}_0^{(1)} & \mathbf{A}_1^{(1)} & 0 & \mathbf{A}_2^{(1)} & 0 & \mathbf{A}_3^{(1)} & 0 & \cdots \\ \mathbf{B}_0^{(1)} & \mathbf{B}_1^{(1)} & 0 & \mathbf{B}_2^{(1)} & 0 & \mathbf{B}_3^{(1)} & 0 & \cdots \\ 0 & 0 & \mathbf{A}_0^{(2)} & 0 & \mathbf{A}_1^{(2)} & 0 & \mathbf{A}_2^{(2)} & \cdots \\ 0 & 0 & \mathbf{B}_0^{(2)} & 0 & \mathbf{B}_1^{(2)} & 0 & \mathbf{B}_2^{(2)} & \cdots \\ 0 & 0 & 0 & 0 & \mathbf{A}_0^{(3)} & 0 & \mathbf{A}_1^{(3)} & \cdots \\ 0 & 0 & 0 & 0 & \mathbf{B}_0^{(3)} & 0 & \mathbf{B}_1^{(3)} & \cdots \\ \cdot & \cdot & \cdot & \cdot & \cdot & \cdot & \cdot & \cdots \end{pmatrix}.$$

This chain has a block upper Hessenberg structure. An efficient method to compute the steady state vector is presented in [4]. Let $\overline{\nu}$ be the steady state vector, i.e. $\overline{\nu} = \overline{\nu}\mathbf{Q}$ and $\overline{\nu \mathbf{e}} = 1$. From the steady state vector, we may compute the number of cells present in the

buffer at cell departure instants, for variable G (guaranteed bandwidth) and V_k (permit distribution algorithm). This leads to optimal values for G and V_k.
It is also possible to compute the queue length at polling instants. Let $\overline{\nu}_0/(\overline{\nu}_0\,\overline{e})$ be the conditional distribution when the system is empty. Then the probability that after a polling cycle there are k cells in the queue is given by

$$\frac{\overline{\nu}_0}{\overline{\nu}_0\overline{e}}\,C_k\overline{e}.$$

This is a measure for the reaction speed of the protocol.

REFERENCES

[1] C. Blondia, (1991). A Discrete-time Batch Markovian Arrival Process as B-ISDN Traffic Model.*JORBEL, the Belgian Journal of Operations Research* (To appear).

[2] C. Blondia and O. Casals, (1992). Statistical Multiplexing of VBR sources : A Matrix Analytical Approach, *Performance Evaluation*, **16**, 5-20.

[3] C. Blondia, O. Casals and J. Garcia, (1993). *Performance Evaluation of a MAC Protocol for an ATM Oriented Passive Optical Network*, Proceedings of Telecommunication Systems Conference - Modeling and Analysis, Vanderbilt University, Nashville (Tn), March, pp. 476-488.

[4] G. Latouche, P.A. Jacobs and D.P. Gaver , (1984). Finite Markov Chain Models Skip-free in One Direction, *Naval Research Logistics Quaterly*, **31**, pp. 571-588.

[5] D.M. Lucantoni, (1991). New Results on the Single Server Queue with a Batch Markovian Arrival Process, *Stochastic Models*, **7**, 1991.

[6] M.F. Neuts, (1979). A Versatile Markovian Arrival Process, *J. Appl. Prob.*, **16**, 1979, 764-779.

Recent Adv. in Stat. and Prob., pp. 343-356
J. Pérez Vilaplana and M. L. Puri (Eds)

Approximate Analysis of Statistical Multiplexing of Variable Bit Rate and Periodic Sources *

J. GARCÍA VIDAL[1] and O. CASALS TORRES[2]
[1] Departamento de Arquitectura de Computadores, Universidad Politécnica de Cataluña, Campus Norte, Módulo D6, c/Gran Capitán s/n, E-08034 Barcelona, España
[2] Departamento de Arquitectura de Computadores, Universidad Politécnica de Cataluña, Campus Norte, Módulo D6, c/Gran Capitán s/n, E-08034 Barcelona, España

Abstract – This paper describes a model for the evaluation of the total cell loss probability and of the cell delay variation a periodic flow is subject to when mixed with other variable bit rate connections in an ATM network. The arrival process is modeled by means of a discrete time batch Markovian arrival process (D-BMAP). An exact solution of this model in a realistic situation is not possible due to computing time and memory requirements. An approximate solution is obtained through spectral decomposition and using an asymptotic analysis.

1. INTRODUCTION

The Broadband ISDN is based on the Asynchronous Transfer Mode (ATM) whereby information is transmitted and switched in the form of 53 byte cells. One of the most important aspects of ATM is its flexibility to support constant as well as variable bitrate connections. ATM allows the network to take advantage of the bit rate variations of the individual sources through statistical multiplexing. In this way, these sources share a capacity less than the sum of the individual peak bit rates, achieving a more or less significant multiplexing gain. This multiplexing gain is restricted by the often stringent quality of service requirements of the sources.

In particular, cell loss, cell mean delay and delay variation are the main performance measures that determine this quality of service (QOS). In ATM, bandwidth is dynamically and virtually allocated and the QOS is guaranteed in a probabilistic sense. The dynamic allocation of bandwidth yields variable delay and this can be a problem for services sensitive to delay variation.

When a call is established in an ATM network, a "contract" is established specifying user traffic characteristics and QOS requirements. A policing mechanism is needed to ensure, both for the user and the network, that this contract is not violated. The user traffic characteristics are currently described by the peak cell rate -which is defined as the inverse of the minimum inter-arrival time between consecutive cells of a given connection- so the control reduces to peak cell rate control. Because of cell delay variation (CDV) originated in random queueing delays experienced by successive cells of an ATM connec-

*This work was supported in part by the TIC, Grant no. 0391/89.

tion in multiplexing stages, tolerances have to be introduced in the peak cell rate policing function to accomodate for CDV that may alter the traffic profile offered at the network entry point.

It is very important to evaluate how the introduction of these tolerances can impact the admissible load that can be supported by a network link for meeting a certain QOS. To study that, we consider the so-called Worst Case Traffic (WCT) which is modelled as an on-off periodic source composed of peak rate bursts whose length is the largest compatible with the tolerances introduced in the policing function to accomodate for CDV.

The impact of this WCT on variable bit rate sources is evaluated by using a discrete time queueing system where the arrival process is modeled by means of a discrete time batch Markovian arrival process (D-BMAP). An exact solution of this model in a realistic situation is not possible due to computing time and memory requirements. An approximate solution using spectral decomposition and asymptotic analysis is required.

In recent years, mainly in the context of High Speed Communication Networks, there have been a large number of works trying to quantify the effect of correlated traffic on the performance of statistical multiplexers. It has been shown that the existence of correlations in the traffic can dramatically alter the performance of the system (e.g. [4], [6], [8]). The M+D/D/1 queueing system analysed in [14] and in [5] cannot capture the "background" traffic correlations that are often present in the case of ATM networks. Our basic model consists of a statistical multiplexer in which a periodic source is multiplexed with the traffic generated by other Variable Bit Rate (VBR) sources, which are modelled by the superposition of two-state Markovian sources. This allows us to model traffic behaviours far more complex than those dealt in [5] and [14].

Section 2 of this paper describes the traffic source model. Section 3 is devoted to the analysis method to compute the cell delay variation of a periodic traffic source and the cell loss probability. These results are applied to some numerical examples in Section 4. Finally, conclusions are drawn in Section 5.

2. TRAFFIC MODEL

We consider a statistical multiplexer whose input consists of the superposition of VBR traffic and periodic traffic.

2.1. Traffic Model for VBR Sources

The VBR traffic is modeled as the superposition of N independent on-off sources. Each of these sources can be represented by a Discrete-time Batch Markovian Arrival Process (D-BMAP) [2]. When the source is in state 0 it does not generate any cell, and in state 1, it generates a batch of k cells with probability d_k ($k = 0, 1, \ldots$). The probability that an active, resp. silent, source becomes silent, resp. active, in the next time unit is given by $1 - \alpha$, resp. $1 - \beta$. The probability generating matrix $\mathbf{D}_{ind}(z)$ of each individual source is:

$$\mathbf{D}_{ind}(z) = \begin{pmatrix} \alpha & 1-\alpha \\ d(z)(1-\beta) & d(z)\beta \end{pmatrix}$$

where $d(z) = \sum_{k=0}^{\infty} d_k z^k$.

The eigenvalues of $\mathbf{D}_{ind}(z)$ are:

$$x_{ind}(i,z) = \frac{\alpha + d(z)\,\beta}{2} \pm \sqrt{\left(\frac{\alpha + d(z)\,\beta}{2}\right)^2 - d(z)\,\phi} \qquad i = 0,1$$

with $\phi = \alpha + \beta - 1$. The corresponding right eigenvectors are given by:

$$u_{ind}(0,z) = \begin{pmatrix} \dfrac{x_{ind}(1,z) - \alpha}{x_{ind}(1,z) - x_{ind}(0,z)} \\ \dfrac{-d(z)\,(1-\beta)}{x_{ind}(1,z) - x_{ind}(0,z)} \end{pmatrix} \quad \text{and} \quad u_{ind}(1,z) = \begin{pmatrix} \dfrac{\alpha - x_{ind}(0,z)}{x_{ind}(1,z) - x_{ind}(0,z)} \\ \dfrac{d(z)\,(1-\beta)}{x_{ind}(1,z) - x_{ind}(0,z)} \end{pmatrix}$$

For the left eigenvectors we have:

$$v_{ind}(i,z) = \begin{pmatrix} 1 & \dfrac{x_{ind}(i,z) - \alpha}{d(z)(1-\beta)} \end{pmatrix} \qquad i = 0,1$$

2.1.1. Eigenvalues and eigenvectors of the superposition

When we consider the superposition of N independent two states D-BMAP sources, we obtain a new D-BMAP with 2^N states. The eigenvalues of the new matrix are

$$x_{VBR}(i,z) = x_{ind}^{N-i}(0,z)\,x_{ind}^{i}(1,z).$$

The eigenvectors can be obtained as the Kronecker products of the eigenvectors of the individual sources. To reduce the size of these vectors we aggregate the states that correspond to the same number of active sources, obtaining a new D-BMAP with just $N+1$ states. It can be shown (Appendix B) that the eigenvalues are the same as above while the i-th component of the right eigenvectors is given by:

$$u_{VBR}^{i}(k,z) = \\ = \frac{\sum\limits_{n=max(0,i-N+k)}^{min(k,i)} \binom{k}{n}\binom{N-k}{i-n} u_{ind}^{0}(0,z)^{N-k-i+n}\,u_{ind}^{1}(0,z)^{i-n}}{\binom{N}{i}} u_{ind}^{0}(1,z)^{k-n}\,u_{ind}^{1}(1,z)^{n}$$

and for the i-th component of the left eigenvectors we have :

$$v_{VBR}^{i}(k,z) = \\ = \sum_{n=max(0,i-N+k)}^{min(k,i)} \binom{k}{n}\binom{N-k}{i-n} v_{ind}^{0}(0,z)^{N-k-i+n}\,v_{ind}^{1}(0,z)^{i-n}\,v_{ind}^{0}(1,z)^{k-n}\,v_{ind}^{1}(1,z)^{n}$$

2.2. Traffic Model for Periodic Sources

The periodic traffic consists of a fixed emission sequence which is repeated every T slots. This traffic can be modeled as a D-BMAP whose underlying Markov chain is periodic with T states.

Let b be the total amount of cells generated during a period T. If during the transition from state i to state $(i+1) \bmod T$, r cells are generated, we define $a_i(z)$ as: $a_i(z) = z^r$. The process will be described by the transition matrix:

$$\mathbf{D}_p(z) = \begin{pmatrix} 0 & a_0(z) & 0 & 0 & \cdots & 0 & 0 \\ 0 & 0 & a_1(z) & 0 & \cdots & 0 & 0 \\ \cdots & \cdots & \cdots & \cdots & \cdots & \cdots & \cdots \\ 0 & 0 & 0 & 0 & \cdots & 0 & a_{T-2}(z) \\ a_{T-1}(z) & 0 & 0 & 0 & \cdots & 0 & 0 \end{pmatrix}$$

To obtain the eigenvalues $x_p(k,z)$ of $\mathbf{D}_p(z)$ we must solve:

$$\left|\mathbf{D}_p(z) - x_p(k,z)\mathbf{I}\right| = 0$$

The expression of the determinant is:

$$\begin{aligned} |\mathbf{D}_p(z) - x_p(k,z)\mathbf{I}| &= (-1)^T x_p(k,z) + (-1)^{T-1} a_0(z) \cdots a_{T-1}(z) = \\ &= (-1)^T x_p(k,z)^T + (-1)^{T-1} z^b \end{aligned}$$

so we have T different eigenvalues given by $(k = 0, 1, \cdots, T-1)$:

$$x_p(k,z) = |z|^{\frac{b}{T}} e^{j\frac{b\Omega}{T}} e^{j\frac{2\pi}{T}k}$$

where Ω is the argument of the complex number z between $(-\pi, \pi]$. These expressions are independent of the exact sequence considered and of the length of the period, and depend just on the average traffic of this source. The expression of the i–th component of the right eigenvector $u_p(k,z)$ is:

$$u_p^i(k,z) = u_p^0(k,z) \frac{x_p(k,z)^i}{a_0(z) \cdots a_i(z)}$$

while for the left eigenvectors we have:

$$v_p^i(k,z) = v_p^0(k,z) \frac{a_0(z) \cdots a_i(z)}{x_p(k,z)^i}$$

2.3. Superposition of VBR and Periodic Traffic

The superposition of VBR and periodic traffic is a new D-BMAP described by the $(N+1)*T \times (N+1)*T$ matrix $\mathbf{D}_T(z)$, which is the Kronecker product of the matrices $\mathbf{D}_p(z)$ and $\mathbf{D}_{VBR}(z)$:

$$\mathbf{D}_T(z) = \mathbf{D}_p(z) \otimes \mathbf{D}_{VBR}(z)$$

From that we have that the eigenvalues of $\mathbf{D}_T(z)$ are the product of the eigenvalues of $\mathbf{D}_p(z)$ and $\mathbf{D}_{VBR}(z)$: $x_T(i,k,z) = x_p(i,z) x_{VBR}(k,z)$ and the right and left eigenvectors are:

$$\begin{aligned} u_T(i,k,z) &= u_p(i,z) \otimes u_{VBR}(k,z) \\ v_T(i,k,z) &= v_p(i,z) \otimes v_{VBR}(k,z) \end{aligned}$$

3. ASYMPTOTIC MODEL

Let $\pi_S(n)$ be the vector whose i–th component is the stationary probability of having, just after a service completion, n customers in the queue and the source in state i when the buffer capacity is S. The equations of the steady state queue length distribution are:

$$\pi_S(0)\,\mathbf{D}_{i+1} + \sum_{j=1}^{i+1} \pi_S(j)\mathbf{D}_{i-j+1} = \pi_S(i) \qquad i = 1, 2, \cdots, S$$

$$\pi_S(0)[\,\mathbf{D}_0 + \mathbf{D}_1] + \pi_S(1)\,\mathbf{D}_0 = \pi_S(0)$$

where the matrix $\mathbf{D}_0$ governs transitions that correspond to no arrivals, while the matrices $\mathbf{D}_n$ govern transitions that correspond to arrivals of batches of size n.
These equations are independent of the buffer capacity, so the solution vector is as follows [1]:

$$\pi_S(i) = \pi_S(0)\mathbf{C}_i \tag{1}$$

where the matrix sequence $\mathbf{C}_i$ does not depend on S. In order to get the z–transform of this sequence we consider the infinite buffer case, obtaining:

$$\mathbf{C}(z) = (z-1)\mathbf{D}(z)\big(z\mathbf{I} - \mathbf{D}(z)\big)^{-1}$$

3.1. Spectral Decomposition

Let $\mathbf{V}(z)$ be the matrix

$$\mathbf{V}(z) = \begin{pmatrix} v_T(0,0,z) \\ v_T(0,1,z) \\ \cdots \end{pmatrix}$$

and let $\mathbf{U}(z)$ be given by

$$\mathbf{U}(z) = \big(u_T(0,0,z)u_T(0,1,z)\cdots\big).$$

We choose $u_T(i,k,z)$ and $v_T(i,k,z)$ so that its inner product is 1. Then $\mathbf{D}(z)$ can be expressed as follows

$$\mathbf{D}(z) = \mathbf{U}(z)\mathbf{X}(z)\mathbf{V}(z) = \sum_{i,k} x_T(i,k,z)u_T(i,k,z)v_T(i,k,z),$$

where $\mathbf{X}(z)$ is a diagonal matrix whose elements are the eigenvalues $x_T(i,k,z)$. We deduce that

$$\mathbf{C}(z) = \mathbf{U}(z)\mathbf{Y}(z)\mathbf{V}(z) = \sum_{i,k} y_T(i,k,z)u_T(i,k,z)v_T(i,k,z),$$

where $\mathbf{Y}(z)$ is a diagonal matrix whose elements are defined by

$$y_T(i,k,z) = \frac{(z-1)x_T(i,k,z)}{z - x_T(i,k,z)}.$$

Let θ be the steady state vector of the underlying Markov chain. From (1) we have that

$$\pi_S(0)\sum_{k=0}^{S} \mathbf{C}_k = \theta. \tag{2}$$

The matrix sequence with elements defined by $\mathbf{L}_k = \sum_{i=0}^{k} \mathbf{C}_i$ has $z-$transform

$$\mathbf{L}(z) = \frac{\mathbf{C}(z)}{1-z}.$$

The poles of $\mathbf{L}(z)$ are the roots of the following equations $z = x_T(i,k,z)$ (see Appendix A). For $i = 0, \cdots, T-1$, we have that

$$z = |z|^{\frac{b}{T}} e^{j\frac{b\Omega}{T}} e^{j\frac{2\pi}{T}k} x_{VBR}(k,z), \tag{3}$$

from which we obtain ($l = 0, \cdots, T-b-1$)

$$z = e^{j\frac{2\pi l}{T-b}} x_{VBR}^{\frac{T}{T-b}}(k,z).$$

We call ς_i the poles with module greater or equal to one, and η_i the poles with module smaller than one. It can be shown that, in our case, pole ς_i with minimum module is always real and positive. We call this pole ς_0 and we assume that $\cdots \leq |\eta_1| \leq |\eta_0| \leq$ $\leq 1 < \varsigma_0 < |\varsigma_1| \leq \cdots,$. There are $T-b$ different values of i for which equation (3) has one solution inside the unit circle, while for the other b values we do not have any solution. Then for each value of n, the $z-$transform $x_T(i,n,z)$ has a single pole with module smaller or equal to one for $T-b$ different values of i. Let $\mathbf{Q}(\varsigma_i)$ and $\mathbf{Q}(\eta_i)$ be the residuals associated with the corresponding pole. The elements of the sequence $\mathbf{L}_k$ can be expressed as

$$\mathbf{L}_k = \sum_{i=0}^{m} \mathbf{Q}(\varsigma_i)\varsigma_i^{-k} + \sum_i \mathbf{Q}(\eta_i)\eta_i^{-k} + \mathbf{Q}(1) + \mathbf{R}_k, \tag{4}$$

where $\mathbf{R}_k$ are the elements of a sequence which is of the order $o(\varsigma_m^{-k})$ [1]. A formula for these residuals is given in Appendix A.

3.2. Asymptotic Behaviour of the State Probabilities

Let us consider now the case of an infinite buffer. When the buffer capacity goes to infinity the terms corresponding to the poles with module smaller than one in formula (4) tend to infinity. We obtain the following system of equations (see [9] and [5])

$$\pi_\infty(0) = 1 - \rho_{Tot}$$
$$\pi_\infty(0) u_T(i,k,\eta_j) = 0,$$

where ρ_{Tot} is the global traffic and η_j is the pole associated to the eigenvalue $x_T(i,k,z)$. We can see that there are just $(T-b)*(N+1)$ unknowns to be found, obtaining the following state vectors

$$\pi_\infty(i) = \pi_\infty(0)(1-\varsigma_0)\mathbf{Q}(\varsigma_0)\varsigma_0^{-i} + o(\varsigma_0^{-i}). \tag{5}$$

3.3. Cell Delay Variation

The cell delay variation can be estimated by the $(1-10^{-10})$ quantile of the waiting time (W) distribution scaled by the peak emission period [13]. In our case, the distribution of the waiting time seen by a cell of a periodic traffic is easily derived from equation (5), obtaining

$$p(W=w) = \sum_{h=0}^{w} \sum_{i=0}^{N} \sum_{k=w-h}^{\infty} \frac{p_i(k)\pi_\infty^i(h)}{k+1}$$

where $p_i(k)$ is the probability that i VBR active sources generate k cells.

3.4. Asymptotic Behavior of the Loss Probability

We consider now the case in which the system has a finite capacity of S cells. Let x be a vector with $(N+1)*T$ components. We define x^o as a vector with $(N+1)*(T-b)$ components which is obtained by removing those components of x corresponding to the components for which the vector $\pi_k(0)$ is zero. We also define the $(N+1))*(T-b)\times(N+1)*(T-b)$ matrix X^o as the result of removing the corresponding rows and columns from the $(N+1)*T\times(N+1)*T$ matrix X. From (2) we obtain:

$$\pi_k^o(0)L_k^o = \theta^o.$$

In order to find the asymptotic behavior of the loss probability (Pt) we follow the same steps as in [1]. We have

$$Pt = \frac{\pi_k^o(0)e^o + (\rho_{Tot} - 1)}{\rho_{Tot}}$$

If now we define: $\mathbf{L}_k^{o(a)} = \mathbf{L}_k^{o-1}\det(\mathbf{L}_k^o)$ we obtain:

$$Pt = \frac{\theta^o\mathbf{L}_k^{o(a)}e^o + (\rho_{Tot} - 1)\det(\mathbf{L}_k^o)}{\rho_{Tot}\det(\mathbf{L}_k^o)} \tag{6}$$

and, applying Kramer's rule in (5): $\mathbf{L}_k^{o(a)} = \sum_i \det(\mathbf{L}_k^o[i])$ where matrix $\mathbf{L}_k^o$ is obtained by substituting the i-th row of $\mathbf{L}_k^o$ by the vector θ^o. We define matrix $\mathbf{G}_k^o$ as:

$$\mathbf{G}_k^o = \mathbf{L}_k^o + \frac{1}{\rho_{Tot} - 1}e^o\theta^o$$

and using property P1 in Appendix A of [1]:

$$\det(\mathbf{G}_k^o) = \det(\mathbf{L}_k^o) + \frac{1}{\rho_{Tot} - 1}\sum_i \det\left(\mathbf{G}_k^o[i]\right) \quad \text{and} \quad \sum_i \det\left(\mathbf{L}_k^o[i]\right) = \sum_i \det\left(\mathbf{G}_k^o[i]\right)$$

so that (5) can be expressed as

$$Pt = \frac{(\rho_{Tot} - 1)\det(\mathbf{G}_k^o)}{\rho_{Tot}\det(\mathbf{L}_k^o)}$$

Our goal is now to determine the asymptotic behavior of the determinants of the above formula when k tends to infinity. Let's define matrix $\mathbf{H}_k^o$ as:

$$\mathbf{H}_k^o = \mathbf{Q}^o(\varsigma_0^{-k})\varsigma_0^{-k} + \sum_i \mathbf{Q}^o(\eta_i)\eta_i^{-k}$$

Applying property P2 of Appendix A of [1]: $\det(\mathbf{G}_k^o) = \det(\mathbf{H}_k^o) + o(\varsigma_0^{-k}\eta_1^{-k}\eta_2^{-k}\cdots)$, and

$$\det(\mathbf{L}_k^o) = \det\left(\mathbf{H}_k^o + \frac{1}{1-\rho_{Tot}}e^o\theta^o\right) + o\left(\varsigma_0^{-k}\eta_1^{-k}\eta_2^{-k}\cdots\right)$$

Let us define the matrices

$$\mathbf{U}(z) = [u_T(0,0,z)u_T(0,1,\eta_1)\cdots],$$

$$\mathbf{V}(z) = \begin{pmatrix} v_T(0,0,z) \\ v_T(0,1,\eta_1) \\ \cdots \end{pmatrix}$$

and

$$\mathbf{\Phi}_k = \operatorname{diag}\left(y_T(0,0,\varsigma_0)\varsigma_0^{-k}, y_T(0,1,\eta_1)\eta_1^{-k}, \cdots\right)$$

we have then that matrix $\mathbf{H}_k^o$ can be expressed as

$$\mathbf{H}_k^o = \mathbf{U}^o(\varsigma_0)\mathbf{\Phi}_k\mathbf{V}^o(\varsigma_0)$$

so that we obtain

$$\det(\mathbf{H}_k^o) = \left(\prod_{i,j} y_T(i,j,\eta_l)\eta_l^{-k}\right)\left[\left(y_T(0,0,\varsigma_0)\varsigma_0^{-k}\right) \det\left(\mathbf{U}^o(\varsigma_0)\mathbf{\Phi}_k\mathbf{V}^o(\varsigma_0)\right)\right]$$

where η_l is the pole associated with the eigenvalue $x_T(i,j,z)$ and

$$\det\left(\mathbf{H}_k^o + \frac{1}{1-\rho_{Tot}} e_o\theta_o\right) =$$
$$= \det(\mathbf{H}_k^o) + \left(\prod_{i,j}(y_T i, j, \eta_l)\eta_l^{-k}\right)\left[\frac{1}{1-\rho_{Tot}} \det\left(\mathbf{U}^o(1)\mathbf{V}^o(1)\right) + o(1)\right].$$

Finally

$$Pt = \frac{(1-\rho_{Tot})^2 \det\left(\mathbf{U}^o(\varsigma_0)\mathbf{V}^o(\varsigma_0)\right)}{-\rho_{Tot} \det\left(\mathbf{U}^o(1)\mathbf{V}^o(1)\right)} y_T(0,0,\varsigma_0)\,\varsigma_0^{-k} + o(\varsigma_0^{-k})$$

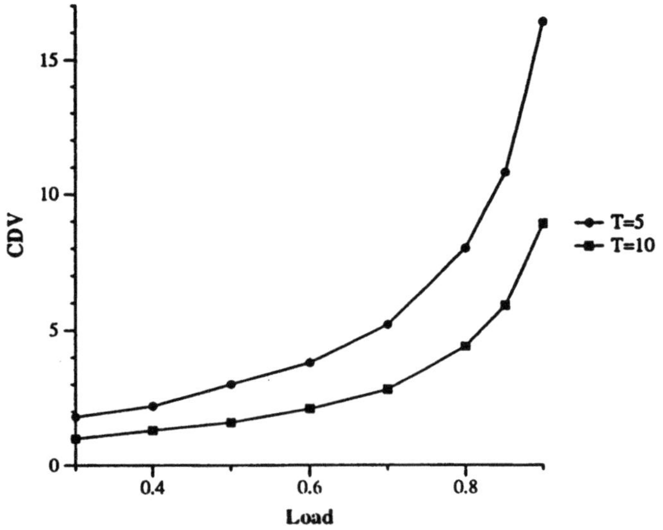

Figure 1. CDV of a CBR source mixed with 15 Bernoulli sources as a function of the total offered load.

4. Numerical Results

In this section we use the model described above to evaluate the CDV a periodic flow is subject to and the cell loss probability, CLP.

4.1. Study of the Cell Delay Variation

The initial time structure of any cell stream is generally altered when passing through a multiplexer. For example, a Constant Bit Rate (CBR) connection (with stringent delay jitter requirements) modeled as a periodic stream of cells, mixed with other VBR connection is subject to jitter. As a consequence of CDV, it may happen that inter-arrival times between consecutive cells of a given connection are less than the peak emission period (clumping effect).

CBR traffic is represented by a periodic traffic with period T, while the VBR traffic is modeled as the superposition of N two-states on-off source with identical parameters: $t_i = t, i = 1, \cdots, N, d_0 = 1 - \frac{1}{t}$ and $d_1 = \frac{1}{t}$. With these assumptions, $p_i(k)$, the probability that i VBR active sources generate k cells is given by:

$$p_i(k) = \binom{i}{k} \left(\frac{1}{t}\right)^k \left(1 - \frac{1}{t}\right)^{i-k}.$$

We compare those results with the ones obtained when the CBR source is mixed with Bernouilli traffic.

The CDV that a Bernouilli traffic introduces on a periodic source increases with the total load (Figure 1). For low traffic load the increase is small but for high loads the slope of the curve becomes much sharper.

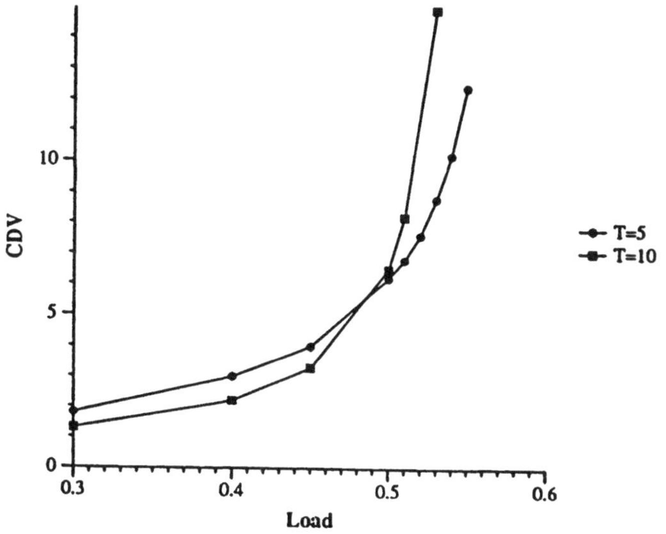

Figure 2. CDV of a CBR source mixed with VBR traffic as a function of the total offered load. The VBR traffic is the superposition of 5 on-off sources with *peak rate/mean rate* = = 2 and sojourn times equal to 10^5 time units.

The degree of CDV depends on the relative rates of the multiplexed streams. When we assume that the superposition of all other streams except the stream whose CDV we wish to evaluate is a superposition of Bernouilli sources, the greatest effect is felt by

the highest rate streams since the number of cells from other streams arriving between successive periodic cells will then be most variable (Figure 1). In that case, for a low rate source, nearly the same number of cells will arrive between any two successive cells, in nearly the same relative positions, and so the delay of each cell will be very similar.
When the CBR connection is mixed with VBR connections modeled as a superposition of on-off sources, the conclusions that can be derived are rather different. The CDV still increases with the load (Figure 2) but much faster and it is not true anymore that, in general, the CDV increases with the CBR rate. The behavior of CDV strongly depends on the characteristics of the multiplexed connections (Figure 2) and, in some cases, the CDV can even increase with the period T. This depends on the burstiness of the VBR sources; in longer periods VBR source state changes are more likely to happen.

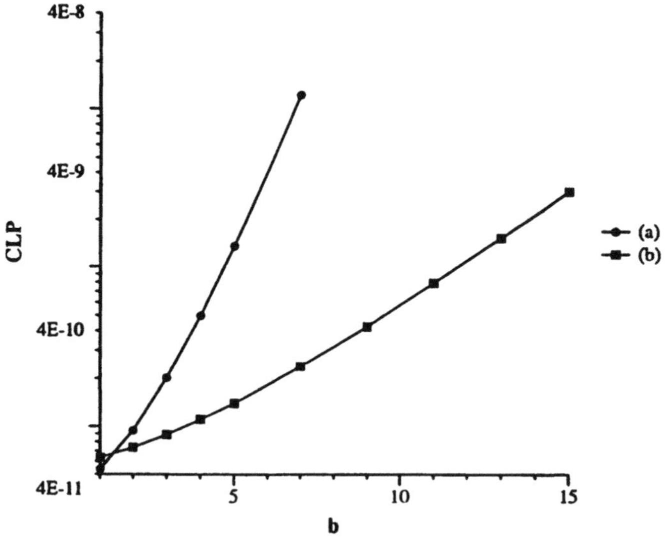

Figure 3. CLP when a WCT source which sends b consecutive cells with $T/b = 5$ is mixed with 15 Bernouilli sources, for different values of the total load and buffer length:
(a) total load of 0.5 and buffer length of 13.
(b) total load of 0.8 with a buffer length of 38.

4.2. Study of the Cell Loss Probability

The necessity of introducing tolerances in the peak cell rate policing function allows a malicious user to send traffic at a higher rate than the negociated peak bit rate. To study the influence of this, we consider the so called Worst Case Traffic (WCT), which is modelled as an on-off periodic source composed of peak rate bursts whose length is fixed and the maximum compatible with the tolerances introduced in the policing function to accomodate for CDV to meet a certain QOS (measured with the CLP parameter). The evaluation is done assuming that there is just one WCT source, so the results are optimistic. But even in this case it can be seen the way in which the QOS suffers from a degradation by the cell clumping (Figures 3 and 4). This degradation is about one

order of magnitude and would be much higher if there were more than one misbehaving customer.
As an example we consider the case of having a superposition of a CBR source, with $T = 5$ with 5 on-off VBR sources with sojourn times equal to 10^5 time units and $peakrate/meanrate = 2$ with a total load of 0.54. According to the CDV obtained and to [13] we need a Leaky Bucket with $N = 13$ credits. This means that a misbehaving customer would be able to send 13 consecutive cells back to back, increasing the CLP of other users.

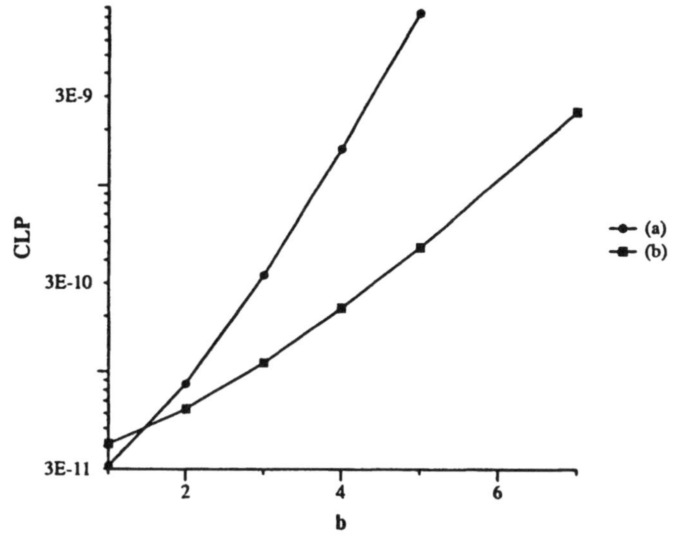

Figure 4. CLP when a WCT source which sends b consecutive cells with $T/b = 5$ is mixed with 5 on-off sources with $peakrate/meanrate = 2$, for different values of the total load and buffer length:
(a) total load of 0.3 and buffer length of 8.
(b) total load of 0.4 with a buffer length of 14.

5. CONCLUSIONS

We have described a model and a solution method, which is a combination of spectral decomposition and asymptotic analysis, leading to an approximate solution for studying the superposition of periodic and VBR traffic.
It is shown that the CDV of a CBR source mixed with VBR sources (modeled as the superposition of on-off sources), increases much faster with the load than when mixed with Bernouilli traffic. Furthermore, in some cases, depending on the burstiness of the VBR traffic, the CDV may increase with the period of the CBR source, contrary with what is observed when modeling the VBR traffic as a Bernouilli process. Clearly, these results are useful when investigating whether special measures have to be taken to guarantee the quality of service (QOS) of the CBR traffic.
It is also very important to obtain a good estimation of the CDV introduced by a mul-

tiplexing stage in order to see if a policing function would be able to control effectively misbehaving sources. The necessity of adding a certain tolerance in the peak cell rate policing function to accomodate for CDV could allow misbehaving customers to degradate the QOS of conforming users.

APPENDIX A

In order to compute the solutions of (3) with module smaller than one, we can use several methods. In [10] the following iterative method is proposed for a similar problem:

$$z_{n+1} = e^{\frac{2\pi l}{T-b}} x_{VBR}^{\frac{T}{T-b}}(k, z_n)$$

A.1. Residuals of L(z)

Let α be a pole of $\mathbf{L}(z)$, that is, α is a root of the equation $z = x(i,z)$ for a given eigenvalue $x(i,z)$ of matrix $\mathbf{D}(z)$. In order to find the residual of this pole we should calculate:

$$\mathbf{Q}(\alpha) = \lim_{z \to \alpha} \left(1 - \frac{z}{\alpha}\right) \mathbf{L}(z) = y(i,\alpha)\, u(i,\alpha)\, v(i,\alpha)$$

with

$$y(i,\alpha) = \lim_{z \to \alpha} \left(1 - \frac{z}{\alpha}\right) \frac{x(i,z)}{x(i,z) - z} \qquad \text{and} \qquad y(i,z) = \frac{1}{1 - \dfrac{d}{dz} x(i,z)}$$

Let ρ be the mean traffic in the queue and let θ be the state probability vector of the underlying Markov chain. In [11] it is shown that:

$$\frac{d}{dz} x(0,1) = \rho$$

and we can choose the vectors $u(0,1)$ and $v(0,1)$ so that:

$$u(0,1) = e, \qquad v(0,1) = \theta$$

and then for the residual in $\alpha = 1$ we have:

$$\mathbf{Q}(1) = \frac{1}{1-\rho} e\theta$$

In [11] it is also shown that:

$$x'(i,z) = v(i,z)\, \mathbf{D}'(z) u(i,z)$$

APPENDIX B

If we consider the superposition of N independent and identical two-state sources, we obtain a Markov chain with 2^N states. Each state can be represented by means of an N–component vector in which the i-th component represents the state of the i–th source. If the k–th state has j active sources, its associate vector is denoted as $e[j,k]$. The number of states with j active sources is given by $\binom{N}{j}$.

Now we consider the aggregation of those states which have the same number of active sources. We obtain a new Markov chain with just $N+1$ states. The state resulting from the aggregation of the states with j active sources will be denoted as j. Let $p(e[j,k]/e[i,n])$ be the one-step transition probability from state $e[i,n]$ to state $e[j,k]$. Analogously we define $p(j/e[i,n])$ to be the probability of going in one step from $e[i,n]$ to any of the states with j active sources, etc.

Let T be a permutation of the elements of an N-component vector. Due to the fact that the sources are identical we have:

$$p\left(e[j,k]/e[i,n]\right) = p\left(T\left(e[j,k]\right)/T\left(e[i,n]\right)\right)$$

and for each $e[j,k]$ and $e[j,h]$,

$$\sum_n p\left(e[j,k]/e[i,n]\right) = \sum_n p\left(e[j,h]/e[i,n]\right)$$

where the summatory in n refers to all the states $e[i,n]$ with the same i. This can be easily shown as follows. Let's consider a permutation T, which transforms the vector $e[j,k]$ into the vector $e[j,h]$. Applying this permutation to the vectors $e[i,n]$ in the first member and considering that the terms in the second summatory are the same as those in the first summatory with the order changed we obtain the equality.
Now it is straightforward to show that

$$\sum_{k,n} p\left(e[j,k]/e[i,n]\right) = \sum_n p\left(j/e[i,n]\right) = \binom{N}{j} \sum_n p\left(e[j,h]/e[i,n]\right)$$

and finally we obtain the following:

$$\sum_n p\left(e[j,k]/e[i,n]\right) = \frac{\binom{N}{i}}{\binom{N}{j}} p\left(j/i\right) \tag{7}$$

Now we are going to find the relationship between the eigenvectors and eigenvalues of the matrices for the original and the aggregated processes. Let $\mathbf{D}(z)$ be the matrix for the original process. The emission of cells does not depends on the future state but just on the actual state, so we have

$$\mathbf{D}(z)^{e[i,n],\, e[j,k]} = d(z,i)\, p\left(e[j,k]/e[i,n]\right)$$

where $d(z,i)$ is a function that just depends on i and z. Let $\mathbf{D}_a(z)$ be the matrix for the aggregated process. We have

$$\mathbf{D}_a(z)^{i,j} = d(z,i)\, p(j/i).$$

Let $H(z)$ be a right eigenvector of the matrix $\mathbf{D}(z)$ associated with the eigenvalue $x(z)$. Then, for each i, n:

$$\sum_j \sum_k d(z,i)\, p\left(e[j,k]/e[i,n]\right) H^{e[j,k]}(z) = x(z)\, H^{e[i,n]}(z).$$

Adding the above equations we obtain:

$$\sum_n \sum_j \sum_k d(z,i)\, p\left(e[j,k]/e[i,n]\right) H^{e[j,k]}(z) = x(z) \sum_n H^{e[i,n]}(z)$$

and changing the order of the summatories:

$$\sum_j \sum_k \sum_n d(z,i)\, p\left(e[j,k]/e[i,n]\right) H^{e[j,k]}(z) = x(z) \sum_n H^{e[i,n]}(z)$$

We use (7), obtaining:

$$\sum_j d(z,i)\, p\left(j/i\right) \frac{H^{e[j,k]}(z)}{\binom{N}{j}} = x(z) \frac{\sum_n H^{e[i,n]}(z)}{\binom{N}{i}}$$

The eigenvalues of both matrices are the same. So if $h(z)$ is a right eigenvector of $\mathbf{D}_a(z)$ associated with the eigenvalue $x(z)$ we obtain

$$h^j(z) = \frac{\sum_k H^{e[j,k]}(z)}{\binom{N}{j}}$$

Using an analogous deduction we have for the left eigenvectors $M(z)$ and $m(z)$ the following:

$$m^j(z) = \sum_k M^{e[j,k]}(z)$$

The derivation of $u_T(i,k,z)$ and $v_T(i,k,z)$ is straightforward.

REFERENCES

[1] A. Baiocchi, (1991). *Asymptotic Behaviour of the Loss Probability of the MMPP/G/1/K Queue*, Int. Rep. n.6/oo2/91, INFOCOM Dept. University "La Sapienza" of Roma.

[2] C. Blondia, (1992). *A Discrete-time Batch Markovian Arrival Process as Traffic Model for B-ISDN*, JORBEL, (submitted).

[3] C. Blondia, (1992). *Performance Evaluation of ATM Network Access Protocols*, IMSI-BAC-4 Proceedings, San Sebastian, Spain, 335–341.

[4] C. Blondia and O. Casals, (1992). *Performance Analysis of Statistical Multiplexing of VBR Sources*, Proceedings INFOCOM '92, Florence, Italy.

[5] P. Brown and A.Simonian, (1988). *Perturbation of a Periodic Flow in a Synchronous Server*, [In PERFORMANCE'87, (P.J. Courtois and G. Latouche Ed.)]. Elsevier Science Publishers.

[6] J. Garcia, O. Casals, (1992). Stochastic Models of Space Priority Mechanisms with Markovian Arrival Processes, *Annals of Operations Research*, **35**, 271–296.

[7] F. Guillemin, P. Boyer and L. Romoeuf,(1992). *The Spacer-controller: Architecture and First Assessments*, Proc. IFIP Workshop on Broadband Communications, Estoril, Portugal.

[8] H. Heffes and D. Lucantoni, (1986). A Markov Modulated Characterization of Packetized Voice and Data Traffic and Related Statistical Multiplexer Performance, *IEEE JSAC*, SAC–4, 6.

[9] I. Ide, (1988). *Superposition of Interrupted Poisson Processes and its Applications to Packetized Voice Multiplexers*, 12th ITC, Torino, Italy.

[10] S.Q. Li, (1990). *A General Solution Technique for Discrete Queueing Analysis of Multimedia Traffic on ATM*, Proceedings of the IEEE GLOBECOM'90, San Diego.

[11] M.F. Neuts, (1981). *Matrix-Geometric Solutions in Stochastic Models. An Algorithmic Approach*, John Hopkins University Press, Baltimore.

[12] E. Wallmeier and T. Worster, (1991). *A Cell Spacing and Policing Device for Multiple Virtual Connections on one ATM Pipe*, Proc. RACE R1022 Workshop on ATM Network Planning and Evolution, London.

[13] G. Niestegge, (1990). The Leaky Bucket Policing Method in ATM Networks, *Int. Journal of Digital and Analog Communication Systems*, **3**, 187–197.

[14] J. Roberts and F. Guillemin, (1992). Jitter in ATM Network and its Impact on Peak Rate Enforcement, *Performance Evaluation*, Special Issue on Modelling of High Speed Telecommunications Systems.

Recent Adv. in Stat. and Prob., pp. 357-369
J. Pérez Vilaplana and M. L. Puri (Eds)

Queues in Tandem with Blocking and Priorities

Ch. LANGARIS
Department of Mathematics, Probability–Statistics and Operational Research Unit, University of Ioannina, GR-45110 Ioannina, Greece

Abstract – In this paper, after a short review of the past work in service systems connected in series with blocking, a two-stage service model without intermediate waiting positions is considered. The system accepts two classes of customers with the first class to have always nonpreemptive priority for service. For such a system and assuming exponential service times and Poisson arrivals we obtain exact formulae for the steady state probabilities and the waiting time of a customer in the system. Finally we discuss the difficulties which arise when the assumption of exponential service times is relaxed and describe briefly the way in which a more general model can be analyzed.

1. INTRODUCTION

A queueing network is a system of service stations arranged in such a way that, between the initial input and the final output, the output of one station or a part of the output becomes immediately the input to another. In general, customers enter the system in a stream having a particular distribution and require service from a number of stations in a prescribed order.

When the waiting space between stations is finite then the blocking phenomenon is observed. Various blocking mechanisms have been considered so far in the literature. We consider here the most commonly used, the type 1 blocking or blocking after service. To explain more precisely the situation, when a customer upon service completion in a node finds the next node, according to his routing, full he cannot proceed further until a service completion in the destination node takes place. In this case the customer is unable to emerge from his service channel and the server is blocked.

An ordered sequence of service stations in series is often said to be "in tandem". Tandem queues with blocking arise frequently in a variety of applications including computer systems, communication networks, flexible manufacturing systems and production lines. In almost all systems of tandem queues with blocking analyzed till now the customers have been considered "similar" as concerning their order for service. However all the above mentioned applications concern in fact priority models in the sense that the arriving customers are usually selected for service according to some "measure of importance" indicated by a priority number. Thus in a communication unit there is a classification of the messages awaiting transmission and also in a computer system the "jobs" are selected for service according to their priority number.

There are many ways to establish a priority rule in a queueing system. The most important of them is the *nonpreemptivepriority* rule in which the low priority customer's service is not interrupted by the arrival of a high priority customer but the high priority stream will take over as soon as the service channel becomes free; and the *preemptivepriority* rule in which the low priority customer's service is always interrupted by the arrival of a high priority customer. In the second case the low priority customer is placed back in the queue and he restarts service, under various policies, when there are no more high priority customers in the system.

The first paper ever published on the equilibrium theory of queueing systems with priority is due to Cobham [7]. He considered both a single channel and a multiple channel queue with r levels of nonpreemptive priorities and gave formulae for the expected waiting times in the system. Preemptive priority models have been studied first by White and Christie [51]. The works of Cobham and White, and Christie were extended and generalized subsequently by many authors under different assumptions regarding the arrival or service characteristics of the system. The interested reader can find a complete bibliography of the earlier works in the subject in Jaiswal [20]. More recently Dalen and Natvig [9] studied the transient waiting times of a GI/M/1 priority model, Nicola [38] considered mixed types of priorities in a single server queue while Kella and Yechiali [24] and Blondia [5] analyzed a M/G/1 multi-queueing system with both infinite and finite capacities and repeated server's vacations. Dshalalow [11] considered priorities and state dependent parameters in a closed system, Kao and Narayanan [22] analyzed a multiprocessor system with preemptive priorities and Stanford [47] studied the interdeparture times of a multi-class $M_i/G_i/1$ queue. Finally Keilson and Servi [23], Epema [12], Simon [46] and Jewkes and Buzacott [21] studied the M/G/1 queueing system with k classes of priorities and feedback.

All the above mentioned works in queues with priorities concern, to our knowledge, systems with a single stage only and not queueing systems in tandem. In this work, after a short review of the past work in tandem queues with blocking we will try to combine these two characteristics (blocking and priority) and to analyze a sequential service model with blocking accepting two classes of customers with the first class to have always priority for service. We will study the model under the assumption of exponential service times. This assumption will be relaxed later and the way in which a more general model can be analyzed will be discussed.

2. TANDEM QUEUES WITH BLOCKING - A SHORT REVIEW

We give here a short review of the past work on tandem queues with blocking focusing our attention particularly on the methodology used to obtain exact results in the case of two stations in tandem. Of course this is not a complete survey of the subject. A comprehensive bibliography of papers on queueing networks with finite capacity queues can be found in Perros [43].

The first work on the model of two stations in tandem is due to Hunt [18]. He allowed either no waiting space or finite space between stages. The arrivals are considered as Poisson random variables while the service times are exponentially distributed. Hunt based his analysis in the Markovian character of the model and by forming differential difference equations he derived expressions for the steady state probabilities and for the maximum possible utilization of the system.

Suzuki [48] extended Hunt's model with no intermediate waiting space by considered arbitrarily distributed service times for both stations. He used the Markov chain formed at the epoches of successive departures from stage 1 to analyze the model while Avi-Itzhak and Yadin [4] studied the same model by observing that in a steady state the first queue of such a system behaves like the M/G/1 queue with the modification that the customer who arrives during an idle period has a service time distribution different from the others. The results in both papers [4] and [48] are steady state results. Prabhu [45] is the first to study the transient behavior of the same model using the concept of Busy Cycle and transformation methods. He derived expressions for the transforms of the virtual waiting time and of the number of customers in the system at time t.

Langaris [31], by connecting the waiting time of the (n+1)-th arriving customer with the corresponding quantity of the nth customer, studied the waiting time process both in the transient and in the steady state for the same model but assuming now a Gamma type input.

Finally Mitchell *et al* [37] are the first who relaxed the assumption of independent service times between stages by assuming that the service times in stages 1 and 2 follow a bivariate distribution with exponential marginals. They studied the model via simulation while Conolly [8] using mathematical methodology derived, for the same model, a functional equation satisfied by the joint probability density function of the customer's waiting times.

In all models considered so far no customers are allowed to wait between stages. Neuts [39] studied the model of two single server stations in series but allowing now k waiting positions before station 2. The arrivals are Poisson and the service times are arbitrarily distributed in the first station and negative exponential in the second. By considering the instants of the successive service completion in stage 1 the author constructed a semi-Markov process and using generating functions arrived at a matrix equation concerning the nth step transition probabilities. Finally Neuts, using the concept of zero-avoiding transition probabilities, gave expressions for the Busy Period. In a second work Neuts [40] generalized this model by considering a wide class of rules to resolve blocking and assuming that the second unit released its customers-when the system is not blocked-according to a stage-dependent death process. This model is studied in terms of Markov renewal branching processes.

Konheim and Reiser [27] studied the same single-server model with finite intermediate positions giving an algorithmic approach to it but now with both service times exponential, while Foster and Perros [13] focused their attention on deriving bounds for the mean blocking time in such a system.

Wont *et al* [52] considered two finite M/M/1 queues in tandem and showed that if the steady state probability vector is partitioned as $\mathbf{X_n} = (\mathbf{P_n}, \mathbf{P_{n+1}})$ then $\mathbf{X_n} = \mathbf{X_o B^n}$ where $\mathbf{B}$ is a suitably defined matrix operator. Using now a number of theorems to locate the eigenvalues of $\mathbf{B}$ they determined the steady state probabilities. Langaris [32] analyzed the same model, but assuming now state dependent arrival and service rates of the birth and death type. He used the formulae obtained to compare the main operational characteristics of the most important state dependent models.

The first study of the two-stage model with multiple service facilities has been made by Makino [36] who assumed that all service times are exponential and the input process is such that the first queue is never empty. Iyama [19] used the imbedded Markov chain technique to extend Makino's results in the case of general service time in the first stage.

A more general two-stage model with multiple servers, finite intermediate space and feedback has been analyzed by Latouche and Neuts [35]. They handled the model using

the Matrix geometric approach while Langaris and Conolly [30] studied the same model (without feedback) using the traditional method of balance equations. In another work Langaris and Conolly [29] analyzed the waiting time of a customer in the same model.
Hillier and Boling [16] are the first who considered a model with more than two stations in tandem. They assumed exponential or Erlang service times, finite buffers and suggested a numerical procedure to calculate the mean output rate and the mean number of customers in the system. Hildebrand [15] extended these results by allowing m single server stations with no waiting space between them and arbitrarily distributed arrivals and service times. Although the model was quite general he succeeded in discovering a Markov process inside it.
Kelly [25], [26] studied the effect of buffer capacity and of reordering the sequence of jobs on the throughput of a model consisting of a series of buffers in tandem while Hillier and So [17] focused their attention on the question of where to place a fixed number of extra servers in a multistage model in order to maximize throughput.
Finally Pinedo and Wolff [44] considered a system of m stations in tandem with zero interstage capacities and equal service times for a customer in all stations and compared the results with the corresponding results in the case of independent service times.

All models so far reviewed are systems of two-and rarely of more than two-queues in tandem. It is obvious that the analysis, by exact methods, of more complicated networks consisting of a large number of queues visited by the customers in a random order and with limited waiting spaces is not an easy task. As a result most of the techniques that have been proposed recently to analyze these networks are used to calculate performance bounds (see Van Dijk [50] and the references therein) or are in the form of numerical methods and approximation. It is obvious that we cannot include all that literature in this short review. For a general description of the most important approximation methods see Gelenbe and Pujolle [14] and also Boxma and Konheim [6], Akyildiz [1], [2], Altiok [3], Perros and Snyder [42], Dallery and Frein [10], Kouvatsos and Xenios [28] and Onvural and Perros [41].

3. A MODEL WITH EXPONENTIAL SERVICE TIMES

Consider a queueing model consisting of two single service points in tandem with no waiting space between them and unlimited waiting space in front of the first. Two classes of customers enter the system: the priority customers (P_1 customers) and the ordinary customers (P_2 customers). Both classes arrive at the first service point according to a Poisson distribution with parameter u_i for the P_i class (i=1,2) and are served according to the exponential distribution with parameters v_1, v_3 for the P_1 and v_2, v_4 for the P_2 customers in the first and second stage respectively. We assume also that P_1 customers have always priority for service in the first queue. Finally we define $u = u_1 + u_2$.
Note that in such a model the customer who, upon service completion at the first service, finds another customer at the second, cannot emerge from the first channel and so that server is unable to serve the next customer. We will describe here without details the essential steps required to analyze the model using its Markovian character. We will focus our attention mainly in the case of nonpreemptive priorities. For a detailed analysis of the same model under a preemptive priority rule see [33].
If we denote by $x_i(t)$ i=1,2 the number of P_i customers who have not yet finished their service in the first queue and by $y_i(t)$ i=1,2 the P_i customers who have finished service in

the first queue but not yet in the second then the states of the system can be described completely by the triple

$$X(t) = (x_1(t), x_2(t), x_3(t))$$

where

$$x_3(t) = \begin{cases} 0 & \text{if} \quad (y_1(t), y_2(t)) = (0,0) \\ 1 & \text{if} \quad (y_1(t), y_2(t)) = (1,0) \\ 2 & \text{if} \quad (y_1(t), y_2(t)) = (0,1) \\ 3 & \text{if} \quad (y_1(t), y_2(t)) = (1,1^*) \\ 4 & \text{if} \quad (y_1(t), y_2(t)) = (0,2) \\ 5 & \text{if} \quad (y_1(t), y_2(t)) = (2,0) \\ 6 & \text{if} \quad (y_1(t), y_2(t)) = (1^*,1) \end{cases}$$

and $*$ is used to denote that the corresponding customer is blocked.
The process $\{X(t) : 0 \leq t < \infty\}$ is an irreducible Markov process and we can define

$$p_{ijk}(t) = Pr[x_1(t) = i, x_2(t) = j, x_3(t) = k] \qquad i,j = 0,1,2,\dots \text{ and } k = 0,1,\dots,6$$
$$\lim_{t\to\infty} p_{ijk}(t) = p_{ijk}$$

where, either $p_{ijk} = 0$ for any (i,j,k) $p_{ijk} > 0$ and $\sum_{ijk} p_{ijk} = 1$ in which case the system settles down to a state of statistical equilibrium.
We have to point out here that in the case of non-preemptive priorities and for $x_3(t) = 0,1,2$ and $i,j \neq 0$ we have to use $p_{ijk}^{(m)}(t)$ and $p_{ijk}^{(m)}$ respectively with the superscript m having the values $m = 1,2$ to denote the class of the customer being served in stage 1 at that time.
If now we assume that the system has reached a state of statistical equilibrium, form the balance equations and define the generating functions

$$P_{ik}(x) = \sum_{j=0}^{\infty} p_{ijk} x^j \qquad P_k(x,y) = \sum_{i=0}^{\infty} P_{ik}(x) y^i \qquad |x| \leq 1 \quad \text{and} \quad |y| \leq 1$$

or $P_{ik}^{(m)}(x)$,$P_k^{(m)}(x,y)$ respectively in the case of non preemptive priorities, then it is easy to obtain in the usual way a set of equations connecting the unknown quantities $P_k(x,y)$ ($P_k^{(m)}(x,y)$ respectively), $k = 0,1,\dots,6$. From this set of equations all the unknowns $P_k(x,y)$ can be obtained as functions of $P_0(x,y)$ while $P_0(x,y)$ is given by

$$P_0(x,y) = \frac{W(x,y)P_{01}(x) - \sum_{i=0}^{2} V_i(x,y) p_{00i}}{D(x,y)} \tag{3.1}$$

where $W(x,y), V_i(x,y)$ are known functions of x and y and

$$D(x,y) = yR_1(x,y)R_3(x,y)R_{13}(x,y) - v_1 v_3 \big(R_1(x,y) + R_3(x,y)\big) \tag{3.2}$$

with

$$R_i(x,y) = u + v_i - u_2 x - u_1 y, \qquad R_{ij}(x,y) = R_i(x,y) + v_j \tag{3.3}$$

In case of non-preemptive priorities $P_0^{(1)}(x,y)$ is given again by (3. 1) but with different functions $W(x,y)$ and $V_i(x,y)$. Thus our problem in both cases is to discover the unknown quantities $P_{01}(x)$ and p_{00i} for $i = 0,1,2$. We state here the following theorem.

Theorem 1 *The function $D(x,y)$, given by (3.2) has for any $|x|<1$ one and only one root in y inside the interval $|y|<1$.*

PROOF: The theorem can be proved by splitting $D(x,y)$ into two polynomials in y,

$$D_1(x,y) = yR_1R_3R_{13}, \quad D_2(x,y) = v_1v_3(R_1+R_3)$$

and comparing the values of $D_1(x,y)$ and $D_2(x,y)$ at the points $y=0$ and $y=1$.
If now we denote by $y^*(x)$ the root of $D(x,y)$ in $|y|<1$ then from (3.1)

$$P_{01}(x) = \frac{V_0^*(x)p_{000} + V_1^*(x)p_{001} + V_2^*(x)p_{002}}{W^*(x)} \tag{3.4}$$

where $W^*(x) = W(x, y^*(x))$, $V_i^*(x) = V_i(x, y^*(x))$ for $i=0,1,2$. Note here that for $x=1$, we have $y^*(1)=1$ and $W^*(1) = V_i^*(1) = 0$.
Now in the case of preemptive priorities one can show (see also [33]) that

Theorem 2 *For any $u<u^*$ the function $W^*(x)$ has a zero, x^* say, in the interval $|x|<1$. The value u^* is the minimal positive root of a known polynomial in u of 6th degree and the inequality $u<u^*$ is the necessary condition for statistical equilibrium.*

In the case of non-preemptive priorities the function $W^*(x)$ is much more complicated and so any attempt to locate the zeros of this function by analytic methods seems almost impossible. Computer results have shown that in this case $W^*(x)$ has similar behavior as in the case of preemptive priorities.

If now x^* is the zero of $W^*(x)$ in $|x|<1$ then from (3.4)

$$V_0^*(x^*)p_{000} + V_1^*(x^*)p_{001} + V_2^*(x^*)p_{002} = 0$$

and so by putting $p_{000}=1$ and using the balance equation

$$up_{000} = v_3p_{001} + v_4p_{002}$$

and the normalizing condition $\sum_k P_k(1,1) = 1$, the generating functions $P_k(x,y)$ can be determined completely.
To study now the waiting time process in our system we assume a FCFS discipline within customers of the same class and also we restrict the model by considering only the case of non-preemptive priorities. For the case of preemptive priorities see [33].
The waiting time process of a high priority customer offers no problems. Thus, if we denote by $f_{ijk}^{(m)}(t)$ the probability density function of the waiting time in stage 1 (including blocking) of an arbitrary P_1 customer who finds upon arrival the system in state (i,j,k) and a P_m (m=1,2) customer being served in the first stage, then using

$$\begin{aligned} w_{ij}(t) &= v_ie^{-v_it}(1-e^{-v_jt}) + v_je^{-v_jt}(1-e^{-v_it}) & i&=1,2 \\ w_{i2}(t) &= v_ie^{-v_it} & j&=3,4 \end{aligned}$$

we obtain easily

$$\begin{aligned} f_{ijk}^{(1)}(t) &= w_{1k+2}(t)^*w_{13}^{(i)}(t) & k&=0,1,2 \\ f_{ijk}^{(2)}(t) &= w_{2k+2}(t)^*w_{14}(t)^*w_{13}^{(i)}(t) & k&=0,1,2 \\ f_{ijk}(t) &= v_ke^{-v_kt*}w_{14}(t)^*w_{13}^{(i)}(t) & k&=3,4 \\ f_{ijk+2}(t) &= v_ke^{-v_kt*}w_{13}^{(i+1)(t)} & k&=3,4 \end{aligned}$$

where $*$ means convolution and $w_{13}^{(i)}(t)$ is the i-fold convolution of $w_{13}(t)$. Thus by taking Laplace transforms and defining the probability density function of the waiting time of a P_1 customer in stage 1, as

$$f(t) = \sum_{ijkm} f_{ijk}^{(m)}(t) p_{ijk}^{(m)}$$

we obtain the Laplace transform $\overline{f}(z)$ of $f(t)$ as

$$\overline{f}(z) = \sum_{k,m} \alpha_{km}(z) P_k^{(m)}(1, \overline{w}_{13}(z)) + \beta(z) \quad \textbf{(3.5)}$$

where $\alpha_{km}(z), \beta(z)$ are known quantities, $\overline{w}_{ij}(z)$ is the Laplace transform of $w_{ij}(t)$ and $P_k^{(m)}(x,y)$ have been found previously.

To analyze now the waiting time of an ordinary customer we need two other random variables, the duration of a Busy Period of P_1 customers in stage 1 and the completion time of an ordinary customer.

Let T_i, $i = 0,1,2$, be a random variable denoting the duration of a Busy Period of priority customers in stage 1, which starts either with no customers in stage 2 ($i = 0$) or with a P_i customer in stage 2 ($i = 1,2$). Let also

$$b_i(t)dt = Pr[t < T_i \leq t + dt]$$
$$\overline{b}_i(z) = \int_0^\infty e^{-zt} b_i(t)dt \qquad \Re(z) \geq 0$$

Then

$$b_i(t) = w_{1i+2}(t)e^{-u_1 t} + w_{1i+2}(t) \sum_{k=1}^{\infty} \frac{(u_1 t)^k}{k!} e^{-u_1 t} * b_1^{(k)}(t) \qquad i = 0,1,2$$

and so

$$\overline{b}_0(z) = \frac{v_1}{z + v_1 + u_1\big(1 - \overline{b}_1(z)\big)} \quad \textbf{(3.6)}$$
$$\overline{b}_i(z) = \overline{w}_{1i+2}\Big(z + u_1\big(1 - \overline{b}_1(z)\big)\Big), \qquad i = 1,2$$

Thus $\overline{b}_1(z)$ is the root in x with smallest absolute value of the equation

$$x = \overline{w}_{13}(z + u_1(1 - x))$$

while $\overline{b}_0(z)$ and $\overline{b}_2(z)$ are known from (3.6).

Completion time of an ordinary customer in stage 1 is the time elapsed from the epoch at which the customer commences his service, until the epoch at which the server is again free to serve the next ordinary customer(if any).

Let now T_{ij}^m be a random variable representing the duration of m consecutive completion times the first of which starts with an i customer in stage 2 and the last one leaves the system with a j customer in stage 2 ($i, j = 0,1,2$).Let also $u_{ij}^m(t), \overline{u}_{ij}^m(z)$ be the probability density function of T_{ij}^m and its Laplace transform respectively. Then

$$u_{i2}^1(t) = e^{-u_1 t} w_{2i+2}(t)$$
$$u_{i1}^1(t) = \sum_{k=1}^{\infty} w_{2i+2}(t) e^{-u_1 t} \frac{(u_1 t)^k}{k!} * b_2(t) * b_1^{(k-1)}(t) \qquad i = 0,1,2$$

and so $\overline{u}_{ij}^1(z)$ can be obtained easily. Now for the m consecutive completion times

$$\overline{u}_{ij}^m(z) = \overline{u}_{i1}^1(z)\overline{u}_{1j}^{m-1}(z) + \overline{u}_{i2}^1(z)\overline{u}_{2j}^{m-1}(z) \qquad i,j=1,2 \qquad m=2,3,\cdots$$

and if we define

$$R_{ij}(z,x) = \sum_{m=1}^{\infty} \overline{u}_{ij}^m(z)x^m \qquad i,j=1,2$$

then we arrive at

$$\overline{u}_{ij}^m(z) = F_{ij}^1(z)\rho_1^{-m}(z) - F_{ij}^2(z)\rho_2^{-m}(z) \qquad i,j=1,2 \tag{3.7}$$

where $F_{ij}^m(z)$ are known functions of z and $\rho_1(z), \rho_2(z)$ are the two roots in x of

$$\left(\overline{u}_{11}^1(z)\overline{u}_{22}^1(z) - \overline{u}_{12}^1(z)\overline{u}_{21}^1(z)\right)x^2 - \left(\overline{u}_{11}^1(z) + \overline{u}_{22}^1(z)\right)x + 1 = 0. \tag{3.8}$$

Let now $g_{ijk}^{(m)}(t)$ be the probability density function of the total time spent in the system by an arbitrary P_2 customer who finds upon arrival the system in state (i,j,k) and a P_m customer being served in the first stage. By using the known densities $b_i(t)$ and $u_{ij}^m(t)$ we can construct now recurrent relations satisfied by $g_{ijk}^{(m)}(t)$. Thus if we define

$$c_m(t) = v_4e^{-v_4t} * \left(w_{23}(t) * u_{11}^m(t) + w_{24}(t) * u_{12}^m(t)\right)$$

then

$$\begin{aligned}
g_{ijk}^{(1)}(t) &= b_k(t)*b_1^{(i-1)}(t)*c_j(t), & k=0,1,2 \quad (3.9)\\
g_{ijk}^{(2)}(t) &= \sum_{m=0}^{\infty} w_{2k+2}(t)\frac{(u_1t)^m}{m!}e^{-u_1t}*b_2(t)*b_1^{(i+m-1)}(t)*c_{j-1}(t), & k=0,1,2\\
g_{ijk}(t) &= \sum_{m=0}^{\infty} v_ke^{-(u_1+v_k)t}\frac{(u_1t)^m}{m!}*b_2(t)*b_1^{(i+m-1)}(t)*c_j(t), & k=3,4\\
g_{ijk}(t) &= \sum_{m=0}^{\infty} v_{k-2}e^{-(u_1+v_{k-2})t}\frac{(u_1t)^m}{m!}*b_1^{(i+m)}(t)*c_j(t), & k=5,6.
\end{aligned}$$

If now we define the probability density function of the waiting time of an ordinary customer in the system by

$$g(t) = \sum_{i,j,k,m} g_{ijk}^{(m)}(t)p_{ijk}^{(m)}$$

then using (3.6) and (3.7) in (3.9) we obtain for the Laplace transform of $g(t)$,

$$\overline{g}(z) = \sum_{k,m} q_{km}^1(z)\,P_k^{(m)}\left(\frac{1}{\rho_1(z)}, \overline{b}_1(z)\right) + \sum_{k,m} q_{km}^2(z)\,P_k^{(m)}\left(\frac{1}{\rho_2(z)}, \overline{b}_1(z)\right) + \gamma(z)$$

where $q_{km}^i(z), \gamma(z)$ are known functions and $P_k^{(m)}(x,y)$ have been found previously. This completes the analysis of the model.

4. GENERAL SERVICE TIMES

Consider now a generalized version of the previously analyzed model in which, the service times of both type of customers in the first stage follow arbitrary distributions with probability density functions (p.d.f.) $b_i(t)$ and cumulative distribution functions (d.f.)

$B_i(t)$, $i = 1,2$, for the $P_i(t)$ class while the service times in the second stage follow the same distribution for both classes with p.d.f. $b(t)$ and d.f. $B(t)$. All three processes are independent, independent from the arrival process and have finite means. The P_1 customers have always a nonpreemptive priority for service. Moreover we assume as before that the priority customers arrive according to the Poisson distribution with parameter u_1 but now the ordinary customers arrive according to a Gamma distribution with parameters m and u_2. Finally the queueing discipline is FCFS for customers of the same class.

In such a model the Markovian property-which previously made the analysis simple-no longer exists and so if one likes to study the system states he has to discover an imbedded Markov chain. By considering, for example, the system at the epoches of successive departures from stage 1, we can recognize easily such a Markov chain.

Here we will discuss briefly the way to study the waiting time of an ordinary customer in such a model. For the complete analysis of a more general system, with different service times for the two classes in the second stage, see [34]. It is known that in all priority models the existence of the high priority customers affects, drastically in many cases, the time spent by an ordinary customer to pass through the system. Thus the analysis of the waiting time process of the low priority customers, in such a model, is of main interest.

Let us begin by defining

$$T_i^* = \max(T_i, T) \qquad i = 1,2$$

where T_i, T represent the service times of a P_i customer in the first and second stage respectively. If $m_i(t)$ is the p.d.f. of T_i^* then it is known that

$$m_i(t) = b_i(t)B(t) + b(t)B_i(t) \qquad i = 1,2.$$

If now $r(x)$ is the p.d.f. of the duration of a Busy period of priority customers in isolation in stage 1, which starts at the time at which a customer commences his service in stage 2 then

$$r(x) = \sum_{k=0}^{\infty} m_1(x)\frac{(u_1 x)^k}{k!}e^{-u_1 x} * r^{(k)}(x)$$

and so the Laplace transform $\overline{r}(z)$ of $r(x)$ is the root of the smallest absolute value of the equation

$$x = \overline{m}_1\big(z + u_1(1-x)\big),$$

where $\overline{m}_i(z)$ is the Laplace transform of $m_i(t)$.

Let finally W_n, N_n be random variables denoting the waiting time (including blocking) in stage 1 of the $n-$th arriving ordinary customer and the number of priority customers left behind upon his departure from the first stage respectively. If

$$f_n(w,k)dw = Pr[w < W_n \leq w + dw, N_n = k]$$

$$g_n(z,x) = \sum_{k=0}^{\infty} x^k \int_0^{\infty} e^{-zw} f_n(w,k)\, dw \qquad \Re(z) \geq 0,\ |x| < 1$$

then by connecting the waiting time of the $(n+1)-$th ordinary customer with the corresponding quantity of the nth ordinary customer and defining the generating function

$$g(z,x,s) = \sum_{n=1}^{\infty} g_n(z,x)s^{n-1} \qquad |s| < 1$$

we finally arrive at

$$g(z,x,s) = \frac{\underline{g}_1(z,x,s) + s\sum\limits_{j=0}^{m-1} \underline{n}_j(z,x,s)\, h_j(u_2,s)}{\left(1-\frac{z}{u_2}\right)^m - s\overline{m}_2\left(z+u_1\left(1-\overline{r}(z)\right)\right)} \tag{4.1}$$

where

$$\underline{g}_1(z,x,s) = \left(\left(1-\frac{z}{u_2}\right)^m - s\overline{m}_2\left(z+u_1\left(1-\overline{r}(z)\right)\right)\right)g_1(z,x) + s\overline{m}_2\left(z+u_1(1-x)\right)g_1\left(z,\overline{r}(z)\right)$$

$$\underline{n}_j(z,x,s) = \left(\left(1-\frac{z}{u_2}\right)^m - s\overline{m}_2\left(z+u_1\left(1-\overline{r}(z)\right)\right)\right)n_j(z,x) + s\overline{m}_2\left(z+u_1(1-x)\right)n_j\left(z,\overline{r}(z)\right)$$

Note here that $g_1(z,x)$ corresponds to the waiting time of the first ordinary customer and can be derived easily, $n_j(z,x)$ are known functions and $h_j(u_2,s)$ are the unknown quantities which must be determined.

Theorem 3 *For $|s| < 1$ or for $|s| \leq 1$ and*

$$\varrho = u_1 E(T_1^*) + \frac{u_2}{m} E(T_2^*) > 1 \tag{4.2}$$

the relation

$$\left(1-\frac{z}{u_2}\right)^m - s\overline{m}_2\left(z+u_1\left(1-\overline{r}(z)\right)\right) \tag{4.3}$$

has in z, m (distinct if $s \neq 0$) zeros, $z = a_{r-1}(s)$ $(r = 1,2,\dots m)$ say, in the region $\Re(z) > 0$. These zeros are given by

$$a_{r-1}(s) = u_2(1 - v_r(s)) \qquad r = 1,2,\dots m$$

where $v_r(s)$ is the only root of

$$x = \epsilon_r \left[s\overline{m}_2\left(u_2(1-x) + u_1\left(1-\overline{r}\left(u_2(1-x)\right)\right)\right)\right]^{1/m}$$

in the unit circle $|x| < 1$ and $\epsilon_r = e^{2\pi i r/m}$.

PROOF: The proof follows the lines of the proof of Lemma 1, p. 82 of [49] and is a consequence of Rouche's theorem.
By putting now these m roots of the denominator in the numerator of (4.1), we obtain $h_j(u_2,s)$ and so finally $g(z,x,s)$ is completely known.
If now

$$f_n(w) = \sum_{k=0}^{\infty} f_n(w,k)$$

$$\lim_{n\to\infty} f_n(w) = f(w)$$

then by using Abel's theorem and observing that for $\varrho > 1$ the roots $z = a_n(1)$ lie inside the region $\Re(z) > 0$ while for $\varrho \leq 1$ the last root becomes $a_{m-1}(1) = 0$ one can prove

Corollary 1 *For $\varrho < 1$ the function $f(w)$ is a proper p.d.f. with Laplace transform*

$$g(z) = \frac{\sum_{j=0}^{m-1} (-1)^j Q_j \underline{n}_j(z,1,1)}{\left(1 - \frac{z}{u_2}\right)^m - \overline{m}_2\left(z + u_1\left(1 - \overline{r}(z)\right)\right)} \tag{4.4}$$

where Q_j are known constants.

Thus from (4.4) all moments of the waiting time in stage 1 of an ordinary customer can be obtained and this completes the analysis.

REFERENCES

[1] AKYILDIZ, I. F. (1988). Mean Value Analysis for Blocking Queueing Networks. *IEEE Trans. Soft Eng.*, **SE-14**, 418-429.

[2] AKYILDIZ, I. F. (1989). Product Form Approximations for Queueing Networks with Multiple Servers and Blocking. *IEEE Trans. Comput.*, **38**, 99-115.

[3] ALTIOK, T. (1989). Approximate Analysis of Queues in Series with Phase-type Service Times and Blocking. *Opns. Res.*, **37**, 601-610.

[4] AVI-ITZHAK, B. AND YADIN, M. (1965). A Sequence of Two Servers with no Intermediate Queue. *Man. Science*, **11**, 553-564.

[5] BLONDIA, C. (1989). A Finite Capacity Multiqueueing System with Priorities and with Repeated Server Vacations. *Queueing Systems*, **5**, 313-330.

[6] BOXMA, O. J. AND KONHEIM, A. G. (1981). Approximate Analysis of Exponential Queueing Systems with Blocking. *Acta Informatica*, **15**, 19-66.

[7] COBHAM, A. (1954). Priority Assignment in Waiting Line Problems. *Opns. Res.*, **2**, 70-76.

[8] CONOLLY, B. W. (1979). *Two Stage Tandem Queue with No Waiting Room between Stages and Correlated Exponential Service in Both Stages.* Research Report 79/2, Chelsea College, University of London.

[9] DALEN, G. AND NATVIG, B. (1980). On the Transient Waiting Times for a GI/M/1 Priority Queue. *J. Appl. Prob.*, **17**, 227-234.

[10] DALLERY, Y. AND FREIN, Y. (1989). A Decomposition Method for the Approximate Analysis of Closed Queueing Networks with Blocking. In: *Proc. First International Workshop on Queueing Networks with Blocking*, [Perros and Altiok (eds.)]. North-Holland, Amsterdam, 193-215.

[11] DSHALALOW, J. (1991). On Single Server Closed Queues with Priorities and State Dependent Parameters. *Queueing Systems*, **8**, 237-254.

[12] EPEMA, D. H. J. (1991). Mean Waiting Times in a General Feedback Queue with Priorities. *Performance Evaluation*, **12**, 45-58.

[13] FOSTER, F. G. AND PERROS H. G. (1980). On the Blocking Process in Queue Networks. *Eur. J. of Oper. Res.*, **5**, 276-283.

[14] GELENBE, H. AND PUJOLLE, M. (1987). *Introduction to Queueing Networks.* John Wiley and Sons. New York.

[15] HILDEBRAND, D. K. (1967). Stability of Finite Queue, Tandem Server Systems. *J. Appl. Prob.*, **4**, 571-583.

[16] HILLIER, F. S. AND BOLING, R. W. (1967). Finite Queues in Series with Exponential or Erlang Service Times - A Numerical Approach. *Opns. Res.*, **15**, 286-303.

[17] HILLIER, F. S. AND SO, K. (1989). The Assignment of Extra Servers to Stations in Tandem Queueing Systems with Small or No Buffers. *Performance Evaluation*, **10**, 219-231.

[18] HUNT, G. C. (1956). Sequential Arrays of Waiting Lines. *Oper. Res.*, **4**, 674-683.

[19] IYAMA, T. (1979). The Efficiency of Two-stage Series Lines. *J. of the Oper. Res. Soc. of Japan*, **22**, 321-337.

[20] JAISWAL, N. K. (1968). *Priority Queues.* Academic Press, New York and London.

[21] JEWKES, E. AND BUZACOTT, J. (1991). Flow Time Distributions in a k Class M/G/1 Priority Feedback Queue. *Queueing Systems*, **8**, 183-202.

[22] KAO, E. AND NARAYANAN, K. (1991). Modeling a Multiprocessor System with Preemptive Priorities. *Management Science*, **37**, 185-197.

[23] KEILSON, J. AND SERVI, L. D. (1988). The Preempt-resume M/G/1 Queue with Feedback and Clocked Schedules. *Performance Evaluation*, **9**, 191-208.

[24] KELLA, O. AND YECHIALI, U. (1988). Priorities in M/G/1 Queue with Server Vacations. *Naval Research Logistics*, **35**, 23-34.

[25] KELLY, F. P. (1982). The Throughput of a Series of Buffers. *Adv. Appl. Prob.*, **14**, 633-653.

[26] KELLY, F. P. (1984). Blocking, Reordering, and the Throughput of a Series of Servers. *Stochastic Processes and their Applications*, **17**, 327-336.

[27] KONHEIM, A. G. AND REISER, M. (1976). A Queueing Model with Finite Waiting Room and Blocking. *Journal of the A. C. M.*, **23**, 328-341.

[28] KOUVATSOS, D. AND XENIOS, N. (1989). MEM for Arbitrary Queueing Networks with Multiple General Servers and Repetitive Service Blocking. *Performance Evaluation*, **10**, 169-195.

[29] LANGARIS, C. AND CONOLLY, B. (1984). On the Waiting Time of a Two-stage Queueing System with Blocking. *J. Appl. Prob.*, **21**, 628-638.

[30] LANGARIS, C. AND CONOLLY, B. (1985). Two Queues in Tandem with Multiple Facilities and Restricted Waiting Space. *Metron*, **XLIII**, 55-66.

[31] LANGARIS, C. (1986). The Waiting Time Process of a Queueing System with Gamma-type Input and Blocking. *J. Appl. Prob.*, **23**, 166-174.

[32] LANGARIS, C. (1989). On a Two-stage Birth and Death Queueing Process. *Opns. Res.*, **37**, 488-497.

[33] LANGARIS, C. (1992) *A Sequential Service System with Preemptive Priority and Blocking.* University of Ioannina, Department of Mathematics, Technical Report 213. (To appear)

[34] LANGARIS, C. (1992) *Waiting Time Analysis of a Two-stage Queueing System with Priorities between Customers.* To appear in *Queueing Systems.*

[35] LATOUCHE, G. AND NEUTS, M. F. (1980). Efficient Algorithmic Solutions to Exponential Tandem Queues with Blocking. *SIAM J. Alg. Disc. Meth.*, **1**, 93-110.

[36] MAKINO, T. (1964). On the Mean Passage Time Concerning some Queueing Problems of the Tandem Type. *J. Oper. Res. Soc. of Japan*, **7**, 17-47.

[37] MITCHELL, C. R. , PAULSON, A. S. AND BESWICK, C. A. (1977). The Effect of Correlated Exponential Service Times on Single Server Tandem Queues. *Naval Res. Log. Quarterly*, **24**, 95-112.

[38] NICOLA, V. (1986). A Single Server Queue with Mixed Types of Interruptions. *Acta Informatica*, **23**, 465-486.

[39] NEUTS, M. F. (1968). Two Queues in Series with a Finite, Intermediate Waiting Room. *J. Appl. Prob.*, **5**, 123-142.

[40] NEUTS, M. F. (1970). Two Queues in Series, Studied in Terms of a Markov Renewal Branching Process. *Adv. Appl. Prob.*, **2**, 110-149.

[41] ONVURAL, R. AND PERROS, H. G. (1988/89). Equivalencies between Open and Closed Queueing Networks with Finite Buffers. *Performance Evaluation*, **9**, 263-269.

[42] PERROS, H. G. AND SNYDER, M. (1988/89). A Computationally Efficient Approximation Algorithm for Analyzing Queueing Networks with Blocking. *Performance Evaluation*, **9**, 217-224.

[43] PERROS, H. G. (1989). A Bibliography of Papers on Queueing Networks with Finite Capacity Queues. *Performance Evaluation*, **10**, 255-260.

[44] PINEDO, M. AND WOLFF, R. (1982). A Comparison between Tandem Queues with Dependent and Independent Service Times. *Opns. Res.*, **30**, 464-489.

[45] PRABHU, N. U. (1967). Transient Behavior of a Tandem Queue. *Management Science*, **13**, 631-639.

[46] SIMON, B. (1984). Priority Queues with Feedback. *Journal of A. C. M.*, **31**, 134-149.

[47] STANFORD, D. (1991). Interdeparture Time Distribution in the Non-preemptive Priority $\Sigma M_i/G_i/1$ Queue. *Performance Evaluation*, **12**, 43-60.

[48] SUZUKI, T. (1964). On a Tandem Queue with Blocking. *J. Oper. Res. Soc. of Japan*, **6**, 137-157.

[49] TAKACS, L. (1962). *Introduction to the Theory of Queues.* Oxford University Press. New York.

[50] VAN DIJK, N. M. (1989). A Simple Throughput Bound for Large Closed Queueing Networks with Finite Capacities. *Performance Evaluation*, **10**, 153-167.

[51] WHITE, H. AND CHRISTIE, L. (1958). Queueing with Preemptive Priorities or with Breakdown. *Opns. Res.*, **6**, 79-90.

[52] WONG,B., GIFFIN, W. AND DISNEY, R. L. (1977). Two finite M/M/1 Queues in Tandem: A Matrix Solution for the Steady State. *Opsearch*, **14**, 1-18.

Recent Adv. in Stat. and Prob., pp. 371-387
J. Pérez Vilaplana and M. L. Puri (Eds)

Assessing the Effect of Bursts of Arrivals on the Characteristics of a Queue*

D. LIU[1] and M. F. NEUTS[2]
[1] Department of Systems Engineering, The Case School of Engineering, Case Western Reserve University, Cleveland, OH 44106-7070, U.S.A.
[2] Department of Systems and Industrial Engineering, College of Engineering and Mines, The University of Arizona, Tucson, AZ 85721, U.S.A.

Abstract – The input process to a single server queue is thinned in a manner which selectively deletes customers who arrive in bursts. The effect of this thinning on the steady-state characteristics of the queue is studied for the $MAP/M/1$ queue and is illustrated by various numerical examples. The same analysis can be applied, but with more involved numerical computations, to the $SM/MAP/1$ model. The purpose of this study is to assess how features of burstiness of the MAP are reflected in the characteristics of a queue to which it serves as input process.

1. INTRODUCTION

In this paper, a procedure is described to quantify the effect of bursts of closely spaced arrivals on the characteristics of a single server queue. The procedure and the class of queues for which it may be studied in an algorithmically tractable manner can be substantially generalized; the essential ideas of such a generalization will be described later in the paper. Initially, we shall consider the $MAP/M/1$ queue, for which the analysis is particularly straightforward.

The $MAP/M/1$ queue is the single server queue with exponential service times of rate μ and with a Markovian arrival process (with single arrivals) as input. The MAP is the point process generated by the transition epochs of a Markov renewal process whose transition probability matrix is of the form

$$H(x) = \int_0^x \exp(D_0 u) du D_1, \quad \text{for} \quad x \geq 0, \tag{1}$$

where D_0 and D_1 are square matrices of order m whose sum $D = D_0 + D_1$ is an *irreducible* infinitesimal generator. The matrix D_0 has negative diagonal elements, nonnegative off-diagonal elements and its inverse exists. The matrix D_1 is nonnegative. Such a Markov renewal process (and the extension allowing for group arrivals) is known as a *Markovian arrival process*. Because of its appealing matrix-analytic properties and ease of algorithmic implementation, it is extensively used in queueing theory. We refer to Asmussen and Ramaswami [1], Blondia [2], Lucantoni [5], Lucantoni, Meier-Hellstern and Neuts [4], Neuts

*This research was supported in part by Grant Nr. DDM-8915235 from the National Science Foundation.

[8] for detailed discussions. Particular Markovian arrival processes are the $PH-$renewal process and the Markov-modulated Poisson process [8] and (independent) *superpositions* of these.
In the $MAP/M/1$ queue, the arrivals occur at the transition epochs of that Markov renewal process. If we denote by θ the stationary probability vector of the generator D, the stationary arrival rate of customers is given by $\lambda^* = \theta D_1\mathbf{e}$. The matrix Laplace-Stieltjes transform $h(s)$ of $H(\cdot)$ is given by

$$h(s) = (sI - D_0)^{-1}D_1. \tag{2}$$

The stochastic matrix $h(0+) = (-D_0)^{-1}D_1$ is not necessarily irreducible; it is indeed quite common that D_1 has some columns of zeros. This has little effect on the analysis to be presented, but must be borne in mind when examining the structure of certain matrices that arise in our discussion. By elementary calculations, it is seen that the matrix $h(0+)$ has a unique invariant probability vector which is given by $\mathbf{h} = (\lambda^*)^{-1}\theta D_1$. The components of the vector $\mathbf{h}$ are the probabilities of the various phases at an arbitrary arrival epoch in the stationary MAP.

1.1. The $MAP/M/1$ Queue

It is well-known that the $MAP/M/1$ queue may be studied as the Markov process with infinitesimal generator Q given by

$$Q = \begin{pmatrix} D_0 & D_1 & 0 & 0 & 0 & 0 & \cdots \\ \mu I & D_0 - \mu I & D_1 & 0 & 0 & 0 & \cdots \\ 0 & \mu I & D_0 - \mu I & D_1 & 0 & 0 & \cdots \\ 0 & 0 & \mu I & D_0 - \mu I & D_1 & 0 & \cdots \\ 0 & 0 & 0 & \mu I & D_0 - \mu I & D_1 & \cdots \\ \vdots & \vdots & \vdots & \vdots & \vdots & \vdots & \ddots \end{pmatrix},$$

and may therefore be studied by the methodology in Neuts [6]. The queue is stable if and only if

$$\rho = \lambda^*\mu^{-1} < 1, \tag{3}$$

and the stationary probability vector $\mathbf{x}$, when partitioned into $m-$vectors $\mathbf{x}_0, \mathbf{x}_1, \ldots$ of the generator Q is given by

$$\mathbf{x}_i = \theta(I - R)R^i, \quad \text{for} \quad i \geq 0, \tag{4}$$

where the $m \times m$ matrix R is the unique nonnegative matrix with Perron eigenvalue less than one, satisfying the equation

$$\mu R^2 + R(D_0 - \mu I) + D_1 = 0. \tag{5}$$

The vector $\mathbf{x}$ is the joint stationary density of the queue length and the phase of the MAP at an arbitrary time epoch. The vector $\mathbf{y} = [\mathbf{y}_0, \mathbf{y}_1, \ldots]$ given by

$$\mathbf{y}_i = \frac{1}{\lambda^*}\theta(I - R)R^iD_1, \quad \text{for} \quad i \geq 0, \tag{6}$$

is the joint stationary density of the queue length and the phase of the MAP just prior to an arbitrary arrival epoch.
By postmultiplying by $\mathbf{e}$ in (5) and using the fact that the matrix $I - R$ is nonsingular, we obtain the equality

$$\mu R\,\mathbf{e} = D_1\mathbf{e}. \tag{7}$$

If we premultiply in that equality by θ, we immediately obtain that

$$\theta R\,\mathbf{e} = \rho, \tag{8}$$

which further implies that the stationary probability that the queue is empty at an arbitrary time is given by $\mathbf{x}_0\mathbf{e} = 1 - \rho$. The stationary probability that an arriving customer finds the system empty is $\mathbf{y}_0\mathbf{e} = 1 - (\lambda^*)^{-1}\theta R D_1\mathbf{e}$. By evaluating the transform $w(s)$ given by

$$w(s) = \sum_{i=0}^{\infty} \theta(I - R)R^i \left(\frac{\mu}{\mu + s}\right)^i \mathbf{e},$$

we routinely obtain the stationary probability distribution $W(\cdot)$ of the waiting time at an arbitrary time epoch. That distribution is explicitly given by

$$W(x) = 1 - \theta R \exp[-\mu(I - R)x]\mathbf{e}, \quad \text{for} \quad x \geq 0. \tag{9}$$

The stationary waiting time distribution $W_A(\cdot)$ of an arbitrary customer is similarly obtained; it is explicitly given by

$$W_A(x) = 1 - \frac{1}{\lambda^*}\theta R \exp[-\mu(I - R)x]D_1\mathbf{e}, \qquad \text{for} \quad x \geq 0. \tag{10}$$

Remark. We emphasize that the nonnegative matrix R is not necessarily irreducible. When the matrix D_1 has zero rows, the corresponding rows of R also vanish. However, the matrix R has a simple eigenvalue η of maximum modulus which is real, satisfies $0 < \eta < 1$, and has a corresponding *positive left eigenvector* $\mathbf{u}$, which may be normalized to be a probability vector. See Neuts [6, Chapter 1.]

1.2. A Thinning Operation

Concurrent with the given MAP input to the queue, we consider an independent Poisson process of rate α. Transitions in the MAP are *labeled* if and only if they are immediately preceded by an event in the Poisson process. Only labeled customers are admitted into the queue. The labeled point process is again an MAP with the parameter matrices of order $2m$:

$$\tilde{D}_0(\alpha) = \begin{pmatrix} D - \alpha I & \alpha I \\ 0 & D_0 \end{pmatrix}, \quad \tilde{D}_1(\alpha) = \begin{pmatrix} 0 & 0 \\ D_1 & 0 \end{pmatrix}.$$

With a very high rate α, most points will be labeled and as α decreases, fewer will. In addition, it is to be expected that the attrition of arrivals will be most pronounced where the original points are "clustered." By varying the rate α and imagining that only labeled arrivals are admitted for service, we obtain a systematic way of studying the effect of bursts of arrivals on the steady-state distributions of the queue. Using a Poisson marking process is the simplest of a number of selective thinning processes which can be used in assessing the effect of bursts of arrivals. See [9] and [10] for generalizations and related considerations. As we shall discuss in what follows, the queue with the labeled MAP as input can be studied essentially in the same manner as the given $MAP/M/1$ queue.

2. PRELIMINARY PROPERTIES

The $2m$–state Markov process governing the labeled MAP has the generator $\tilde{D}(\alpha) = \tilde{D}_0(\alpha) + \tilde{D}_1(\alpha)$, which is seen to be irreducible by inspecting the allowable transitions and using the fact that D is irreducible and D_0 nonsingular. The Markov process cannot stay forever in the sets of states $1, ..., m$ or $1', ..., m'$ and there is at least one link having positive probability leading from each of these sets of states to the other.

It is routinely verified that the stationary probability vector $\tilde{\theta}(\alpha) = [\tilde{\theta}_0(\alpha), \tilde{\theta}_1(\alpha)]$ of the generator $\tilde{D}(\alpha)$ is given by

$$\tilde{\theta}_0(\alpha) = \theta D_1(\alpha I - D_0)^{-1}, \quad \tilde{\theta}_1(\alpha) = \alpha\theta(\alpha I - D_0)^{-1}, \tag{11}$$

so that the arrival rate of labeled customers is given by

$$\lambda^*(\alpha) = \alpha\theta(\alpha I - D_0)^{-1} D_1 \mathbf{e}. \tag{12}$$

The $MAP(\alpha)/M/1$ queue to which only labeled customers are admitted is studied as the QBD–process with the generator $Q(\alpha)$ which has the same form as Q, except that the matrices D_0 and D_1 are replaced by $\tilde{D}_0(\alpha)$ and $\tilde{D}_1(\alpha)$ and that the identity matrix is now of order $2m$. That queue is stable if and only if

$$\rho(\alpha) = \lambda^*(\alpha)\mu^{-1} < 1. \tag{13}$$

It is clearly not necessary that the original $MAP/M/1$ queue be stable. In that case, the $MAP(\alpha)/M/1$ queue will be stable only if the rate α is less than α°, the unique solution of the equation

$$\lambda^*(\alpha) = \mu. \tag{14}$$

Moreover, $p(\alpha) = \rho^{-1}\rho(\alpha) = (\lambda^*)^{-1}\lambda^*(\alpha)$ is the fraction of the original customers who are admitted. In Neuts [10] the inverse function $\chi(p)$ of $p(\alpha)$ for the rescaled MAP with $\lambda^* = 1$ is proposed as the *index of burstiness* of the MAP. In examining the characteristics of the stable $MAP(\alpha)/M/1$ queue for examples of the arrival process, it will be convenient to rescale so that $\lambda^* = \mu = 1$ and after specifying values of the traffic intensity $\rho(\alpha) = p$ to compute various distributions and moments for the queue with $\alpha = \chi(p)$. We shall refer to this case used in our numerical examples as the *normalized model.* Very few cases are tractable in scalar expressions. For the sake of testing computer codes, the simple case of the $M/M/1$ queue is useful. In that case, the corresponding $M(\alpha)/M/1$ queue is the special $PH/M/1$ queue for which the interarrival distribution is the generalized Erlang distribution with Laplace-Stieltjes transform $\lambda(s + \lambda)^{-1}\alpha(s + \alpha)^{-1}$. Elementary explicit solutions for that case are easily obtained.

3. MATRIX-GEOMETRIC SOLUTION OF THE QUEUE WITH LABELED ARRIVALS

The matrix-geometric solution of the stable $MAP(\alpha)/M/1$ queue is formally similar to that of the original $MAP/M/1$ model, but there are important simplifications which follow from the special structure of the matrix $\tilde{D}_1(\alpha)$. Because the first m rows of that matrix are zero, the minimal nonnegative solution to the matrix equation

$$\mu R^2(\alpha) + R(\alpha)[\tilde{D}_0(\alpha) - \mu I] + \tilde{D}_1(\alpha) = 0, \tag{15}$$

is of the form

$$R(\alpha) = \begin{pmatrix} 0 & 0 \\ R_0(\alpha) & R_1(\alpha) \end{pmatrix},$$

where the matrices $R_0(\alpha)$ and $R_1(\alpha)$ are obtained by solving the system of equations

$$\begin{aligned} R_0(\alpha) &= R_1(\alpha)R_0(\alpha)L + M, \\ R_1(\alpha) &= [\mu R_1(\alpha)^2 + \alpha R_0(\alpha)]K, \end{aligned} \tag{16}$$

with the $m \times m$ matrices K, L and M defined by

$$K = (\mu I - D_0)^{-1}, \quad L = \mu[(\alpha + \mu)I - D]^{-1}, \quad M = \mu^{-1}D_1 L.$$

We denote by $\mathbf{x}(\alpha)$ the stationary probability vector of the generator $Q(\alpha)$. Partitioned into $2m$–vectors $\mathbf{x}_i(\alpha)$, that vector is given by

$$\mathbf{x}_i(\alpha) = \tilde{\theta}(\alpha)[I - R(\alpha)]R^i(\alpha), \quad \text{for} \quad i \geq 0. \tag{17}$$

If each vector $\mathbf{x}_i(\alpha)$ is itself partitioned into two m–vectors $\mathbf{x}_{0,i}(\alpha)$ and $\mathbf{x}_{1,i}(\alpha)$, these are given by

$$\begin{aligned} \mathbf{x}_{0,0}(\alpha) &= \tilde{\theta}_0(\alpha) - \tilde{\theta}_1(\alpha)R_0(\alpha), \\ \mathbf{x}_{0,i}(\alpha) &= \tilde{\theta}_1(\alpha)[I - R_1(\alpha)][R_1(\alpha)]^{n-1}R_0(\alpha), \quad \text{for} \quad i \geq 1, \\ \mathbf{x}_{1,i}(\alpha) &= \tilde{\theta}_1(\alpha)[I - R_1(\alpha)][R_1(\alpha)]^{n}, \quad \text{for} \quad i \geq 0. \end{aligned} \tag{18}$$

The expressions for the probability vectors $\mathbf{y}_i(\alpha)$, the joint distribution of the queue length and the phase at the arrival of an arbitrary *labeled* customer may be similarly written in forms convenient for numerical computation. In the numerical examples discussed in Section 6, we provide values of the 95-th percentile of the probability density $\{\mathbf{y}_i(\alpha)\mathbf{e}\}$. If, for a given queueing model, we rather arbitrarily define a "long" queue as one in excess of the queue length experienced by five percent of all admitted customers, then that percentile can serve to compare what long queues are for different models. By applying the equality (7) to the $MAP(\alpha)/M/1$ queue and using the particular form of the matrix $R(\alpha)$, we derive the equality

$$\mu[R_0(\alpha) + R_1(\alpha)]\mathbf{e} = D_1\mathbf{e}. \tag{19}$$

The fraction of idle time in the stationary $MAP(\alpha)/M/1$ queue is given by $1 - \rho(\alpha)$, while the probability that an arriving labeled customer enters service immediately is $\mathbf{y}_0(\alpha)\mathbf{e} = 1 - [\lambda^*(\alpha)]^{-1}\tilde{\theta}_1(\alpha)R_1(\alpha)D_1\mathbf{e}$.

3.1. The Waiting Time Distributions

The special form of the matrix $R(\alpha)$ results in major simplifications in the expressions for the stationary waiting time distributions $W(x;\alpha)$ and $W_A(x;\alpha)$, respectively at an arbitrary time and of an arbitrary arriving customer.

Theorem 1 *The probability distribution $W(x;\alpha)$ is given by*

$$W(x;\alpha) = 1 - \mu^{-1}\tilde{\theta}_1(\alpha)\exp\{-\mu[I - R_1(\alpha)]x\}D_1\mathbf{e}, \quad \textit{for} \quad x \geq 0. \tag{20}$$

The probability distribution $W_A(x;\alpha)$ is given by

$$W_A(x;\alpha) = 1 - [(\lambda^*(\alpha)]^{-1}\tilde{\theta}_1(\alpha)\exp\{-\mu[I - R_1(\alpha)]x\}R_1(\alpha)D_1\mathbf{e}, \tag{21}$$

for $x \geq 0$.

PROOF: If we write $\exp\{-\mu[I-R(\alpha)]x\} = \Phi(x)$, and partition the matrix $\Phi(x)$ into four $m \times m$ blocks $\Phi_{ij}(x), i,j = 0,1$, then clearly $\Phi_{01}(x) = 0$. Routinely solving the differential equations for $\Phi(x)$ obtained by partitioning the matrix, we get that

$$\Phi_{00}(x) = e^{-\mu x}I, \quad \Phi_{11}(x) = \exp\left\{-\mu[I - R_1(\alpha)]x\right\},$$
$$\Phi_{10}(x) = \mu e^{-\mu x}\int_0^x \exp\left[\mu R_1(\alpha)u\right] du \; R_0(\alpha). \tag{22}$$

Substitution into the equations (9) and (10) adapted to the $MAP(\alpha)/M/1$ queue, we obtain the expressions (20) and (21). In the course of the calculation for (21), we use the equality (19). ∎

3.2. What the Waiting Time Distributions Reveal

As we consider examples of the normalized model for decreasing values of $p < 1$, a rapid decrease of the (common) decay parameter of the exponential tails of $W(x;\alpha)$ and $W_A(x;\alpha)$ suggests the presence of intense bursts in the arrival process. If, for a given value of p, the jump at 0 in $W_A(x;\alpha)$ is significantly larger than $W(0+;\alpha) = 1-p$, this is attributable to long intervals in the arrival process during which few arrivals occur. A corresponding slow decrease in $1 - W_A(x;\alpha)$ at high values of p indicates that the arrivals occur in long-lived clusters. These are resistant to the thinning and, therefore, the waiting times of many admitted customers remain long.

4. THE CAUDAL CHARACTERISTIC CURVE

A graph of the Perron-Frobenius eigenvalue $\eta(\alpha)$ of the matrix $R(\alpha)$ is called a *caudal characteristic curve* of the $MAP(\alpha)/M/1$ queue. The significance of that eigenvalue and the interpretation of a graph illustrating its dependence on a parameter of the queue are discussed in Neuts [7]. It is clear that $R(\alpha)$ has an eigenvalue 0 of multiplicity (at least) m and that the other eigenvalues are those of $R_1(\alpha)$. $\eta(\alpha)$ is therefore the P-F eigenvalue of $R_1(\alpha)$. Moreover, as is readily seen from the expressions in (20) and (21), the waiting time distributions have asymptotically exponential tails with the common decay parameter $\mu[1 - \eta(\alpha)]$. By application of the following theorem, the eigenvalue $\eta(\alpha)$ can be evaluated without computing the matrix $R_1(\alpha)$.

Theorem 2 *If the $\phi(z)$ denotes the eigenvalue of largest real part of the matrix $D_0 + z^{-1}D_1$, for $z > 0$, then $\phi(z)$ is a simple eigenvalue which is positive and decreasing for $0 < z < 1$. The maximal eigenvalue of the matrix R is the unique solution of the equation*

$$\phi(z) = \mu(1 - z). \tag{23}$$

The eigenvector $\mathbf{u}$ *of* R *corresponding to* η *is the positive left eigenvector of the matrix* $D_0 + \eta^{-1}D_1$ *corresponding to the eigenvalue* $\phi(\eta)$.

PROOF: The stated properties of the eigenvalue of maximum real part of the irreducible matrix $D_0 + z^{-1}D_1$ are immediate consequences of the Perron-Frobenius theorem. Let η and $\mathbf{u}$ be as defined, then if we premultiply by $\mathbf{u}$ in the equation (5), the resulting equation may be rewritten as

$$\mathbf{u}(D_0 + \eta^{-1}D_1) = \mu(1 - \eta)\mathbf{u}.$$

This shows that η satisfies the equation (23). The existence and uniqueness of a solution in the interval (0,1) to that equation, under the necessary and sufficient condition that the queue is stable, is established for considerably more general models in Neuts [6]. ■

By numerical solution of the equation (23) for the case of the coefficient matrices $\tilde{D}_0(\alpha)$ and $\tilde{D}_1(\alpha)$, for various values of α, a graph of the caudal characteristic curve may be obtained. An examination of the caudal characteristic curve is particularly useful in exploring cases for which the order m of the matrices D_0 and D_1 is large and computation of the matrix R may be onerous. For our examples, which involve only matrices of modest order, we have not computed the caudal characteristic curve.

5. THE MEAN BUSY PERIOD

It is to be anticipated that the duration of the busy period of the $MAP(\alpha)/M/1$ queue will be quite sensitive to burstiness in the input process. Short, intense bursts of arrivals during any busy period will greatly extend its subsequent duration. The removal of such bursts by the proposed thinning will result in a rapid decrease of the mean and the variance of the busy period. Similarly, an arrival process whose most salient features are long, sustained periods with many arrivals will have very long busy periods separated by long idle periods. The lengths of such busy periods will only be slowly reduced by thinnings with a high value of p.

The successive busy periods of the $MAP/M/1$ queue are dependent. That can be exploited to examine, for example, the serial correlations of successive busy periods to gain insight into the macroscopic behavior of the queue. We shall not pursue that subject here, but will derive matrix-analytic expressions for some features of an "arbitrary" busy period. Since the $MAP/M/1$ queue is an instance of a $QBD-$process, the analysis of its busy period is an application of the discussion in Chapter 3 of Neuts [6]. We shall therefore only sketch the essential steps of the derivations.

The Laplace-Stieltjes transform $\hat{G}(s)$ of the semi-Markov matrix $G(\cdot)$ describing the *fundamental period* of the queue is for $s \geq 0$ the minimal nonnegative solution to the equation

$$D_1\hat{G}^2(s) - (sI + \mu I - D_0)\hat{G}(s) + \mu I = 0. \tag{24}$$

For the stable queue, the matrix $G = \hat{G}(0+)$ is a positive stochastic matrix, whose invariant probability vector will be denoted by $\mathbf{g}$.

The probability vector $\mathbf{y}^{\circ}$ of the phase probabilities at the beginning of a busy period in the stationary version of the queue is given by

$$\mathbf{y}^{\circ} = [\lambda^{*} - \theta R D_1 \mathbf{e}]^{-1} \theta (I - R) D_1. \tag{25}$$

The Laplace-Stieltjes transform $\beta(s)$ of the duration of the busy period is therefore given by

$$\beta(s) = \mathbf{y}\hat{G}(s)\mathbf{e}. \tag{26}$$

We shall derive expressions for the first two moments of the length of the busy period. The matrices $-\hat{G}'(0+)$ and $\hat{G}''(0+)$ will be denoted by M_1 and M_2 respectively. Routine differentiations in (24) lead to the equations:

$$D_1 G M_1 + D_1 M_1 G - G - (\mu I - D_0) M_1 = 0, \tag{27}$$

$$D_1 G M_2 + D_1 M_2 G + 2 D_1 M_1^2 - 2M_1 - (\mu I - D_0) M_2 = 0. \tag{28}$$

By standard calculations the vector $M_1\mathbf{e}$ may be expressed as

$$M_1 \mathbf{e} = (\mu I - D - D_1 G)^{-1} \mathbf{e} = (I - G + \mathbf{e}\mathbf{g})[(\mu I - D_1)\mathbf{e}\mathbf{g} - D]^{-1} \mathbf{e}. \tag{29}$$

The existence of the inverses can be verified by an elementary argument and also follows from more general results in Neuts [8]. We note the equality $\mathbf{g}M_1\mathbf{e} = (\mu - \lambda^{*})^{-1}$, which is a useful consequence of (29). The expression

$$M_2 \mathbf{e} = 2[\mu I - D - D_1 G]^{-1} [D_1 M_1 - I] M_1 \mathbf{e}, \tag{30}$$

unavoidably requires the computation of the matrix M_1 by an iterative solution of the linear matrix equation (28). The explicit formula (29) for $M_1\mathbf{e}$ is useful in constructing accuracy checks and stopping criteria. When D_1 has many zero columns (as is the case for the $MAP(\alpha)/M/1$ queue) it is worthwhile to apply the stopping criterion to the matrix $D_1 M_1$ which is all that is needed for use in (30). The mean and variance of the duration of the busy period are, of course, given by $\mu_{BP} = \mathbf{y}^{\circ} M_1 \mathbf{e}$ and $\sigma_{BP}^2 = \mathbf{y}^{\circ} M_2 \mathbf{e} - \mu_{BP}^2$ respectively.

The application of these results to the $MAP(\alpha)/M/1$ queue is straightforward. For matrices of high order, there is considerable computational efficiency in exploiting the special structure of the coefficient matrices $\tilde{D}_0(\alpha)$ and $\tilde{D}_1(\alpha)$. This is readily done by partitioning all matrices to be computed into four $m \times m$ blocks.

6. NUMERICAL EXAMPLES

In this section, we present a set of related numerical examples. These may serve to illustrate how various queue characteristics respond to thinning with reference to Poisson processes of different rates α. In general, such a "look through a queue" can be used to quantify the effect of the burstiness of the input stream.

For each example, the arrival rate λ^{*} and the service rate μ are initially set equal to one, so that the corresponding queue is critical. A number of values of the fraction p of arrivals that will be labeled and admitted for service are specified. p is then also the traffic intensity $\rho(\alpha)$ of the queue with labeled arrivals. The corresponding values of the marking rate α are next computed by solving the equation (12). For each case, the mean virtual waiting time $E(W)$, the mean waiting time $E(W_A)$ of an arbitrary labeled customer at arrival, the mean busy period μ_{BP} and the 95-th percentile $\phi(0.95)$ of the queue length distribution at an arbitrary labeled arrival epoch are listed. We computed $\phi(0.95)$ as the first index for which the probability distribution of the queue length at arrivals exceeds 0.95. These quantities will be listed for a set of examples with varying burstiness characteristics. For purposes of comparison, we first present a table of results for the $M/M/1$ queue (Table 1). We recall that the thinned versions of that queue correspond to a simple $PH/M/1$ queue with a generalized Erlang interarrival time distribution with two phases only.

Table 1.
Queue Characteristics for the Thinned $M/M/1$ Queue.

p	α	$E(\mathbf{W})$	$E(\mathbf{W_A})$	$\mu_{\mathbf{BP}}$	$\phi(0.95)$
0.50	1.00	0.81	0.62	1.62	3
0.60	1.50	1.20	1.00	2.00	4
0.70	2.33	1.89	1.70	2.71	6
0.80	4.00	3.38	3.22	4.24	11
0.90	9.00	8.13	8.03	9.12	25
0.91	10.11	9.20	9.11	10.22	28
0.92	11.50	10.54	10.45	11.60	32
0.93	13.29	12.26	12.19	13.37	37
0.94	15.67	14.57	14.49	15.75	44
0.95	19.00	17.77	17.70	19.07	54
0.96	24.00	22.54	22.48	24.07	68
0.97	32.33	30.42	30.37	32.40	92
0.98	49.00	45.80	45.74	49.07	138
0.99	99.00	89.19	89.10	99.11	268

6.1. The Principal Examples

To construct a generic set of examples, we used an $MMPP$ with 10 phases. The underlying Markov process with generator D cyclically visits its ten phases. The first five exponential phases have a common parameter σ_1 and the remaining five a common parameter σ_2. The values of σ_1 and σ_2 are selected so that, in all cases, the mean cycle time is 50. This corresponds to requiring that

$$\frac{1}{\sigma_1} + \frac{1}{\sigma_2} = 10. \tag{31}$$

During the first five phases, arrivals occur at a high rate λ_1, while during the remaining phases, there is a low arrival rate λ_2. These rates are appropriately scaled to guarantee that $\lambda^* = 1$. In thinking about the examples, it is useful to view λ_2 as the rate of a background Poisson process, superimposed on an $MMPP$ with rate $\lambda_1 - \lambda_2$ in the first five phases and no arrivals during sojourns in the remaining five. The normalized rate λ_2 is also the traffic intensity offered by the Poisson arrivals. Setting the overall arrival rate λ^* equal to one is equivalent to the equation

$$\lambda_2 + (\lambda_1 - \lambda_2)\left(\frac{\sigma_2}{\sigma_1 + \sigma_2}\right) = 1. \tag{32}$$

In selecting the value for $(10\sigma_1)^{-1}$, we fix the ratio κ of the expected duration of the periods with high arrival rate to the mean cycle time. In our examples, we achieve different degrees of burstiness in the arrival process by varying the parameters κ and λ_2. With smaller values of κ, the arrivals during the first five phases are concentrated in shorter intervals, so that the corresponding arrival process is more bursty. With larger values of λ_2, the share of the bursty component process is reduced.

Table 2.
Queue Characteristics for Example 1.

p	α	$E(W)$	$E(W_A)$	μ_{BP}	$\phi(0.95)$
0.50	98.96	8.94	16.89	30.04	42
0.60	149.46	14.45	23.09	42.01	59
0.70	233.63	23.64	32.77	61.37	85
0.80	401.97	41.90	51.37	99.47	139
0.90	906.98	96.20	105.89	212.93	300
0.91	1019.20	108.24	117.94	238.10	336
0.92	1159.48	123.27	132.99	269.56	381
0.93	1339.83	142.59	152.33	309.99	438
0.94	1580.31	168.35	178.09	363.90	515
0.95	1916.98	204.38	214.14	439.36	623
0.96	2421.98	258.42	268.19	552.54	785
0.97	3263.64	348.46	358.23	741.14	1054
0.98	4946.97	528.47	538.08	1118.34	1593
0.99	9996.93	1068.29	1078.08	2249.88	3210

In Example 1, we set $\kappa = 1/101$ and $\lambda_2 = 0$, so that $\sigma_1 = 10.1$, $\sigma_2 = 0.101$, and $\lambda_1 = 101$. This results in the following computed values of the same descriptors of the queue as in Table 2.

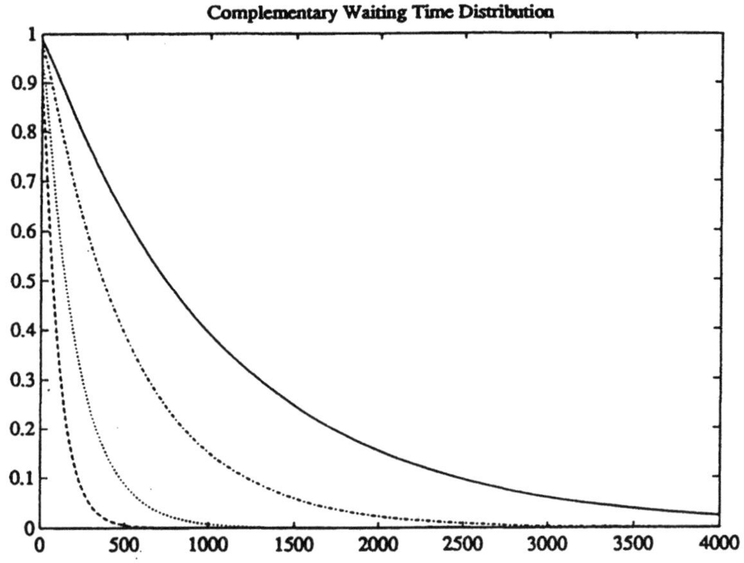

Figure 1a. Graph of the complementary virtual waiting time distribution for the queue of Example 1. In decreasing order, the graphs corresponds to $p = 0.99, 0.98, 0.95$ and 0.90.

Graphs of the complementary waiting time distributions at an arbitrary time and at an arbitrary arrival epoch (of a labeled customer) are displayed in Figures 1(a) and 1(b).

The graphs correspond, in decreasing order, to the traffic intensities 0.99, 0.98. 0.95 and 0.90.

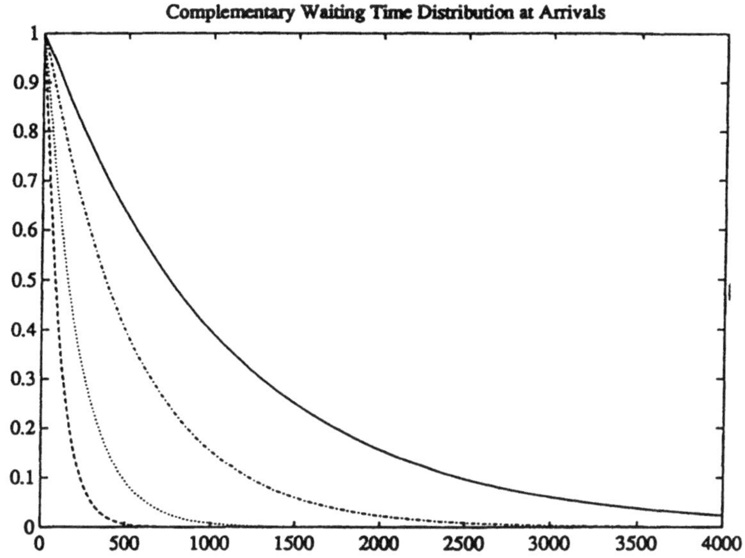

Figure 1b. Graph of the complementary virtual waiting time distribution an arbitrary arrival for the queue of Example 1. In decreasing order, the graphs corresponds to $p = 0.99, 0.98, 0.95$ and 0.90.

Example 2 is obtained from Example 1, by stretching the intervals with the high arrival rate by a factor of 10, while keeping the mean cycle length the same. The corresponding parameter values are $\kappa = 10/101$, $\sigma_1 = 1.01$, $\sigma_2 = 101/910 = 0.110989$, and $\lambda_1 = 10.1$. Computed values of the queue descriptors are listed in Table 3.

Table 3.
Queue Characteristics for Example 2.

p	α	$E(\mathrm{W})$	$E(\mathrm{W_A})$	μ_{BP}	$\phi(0.95)$
0.50	9.90	7.22	13.43	23.35	34
0.60	14.95	11.89	18.82	33.53	48
0.70	23.36	19.69	27.12	49.81	70
0.80	40.19	35.14	42.92	81.67	115
0.90	90.69	81.02	89.03	176.30	251
0.91	101.91	91.19	99.20	179.28	282
0.92	115.94	103.88	111.91	223.50	319
0.93	133.97	120.19	128.24	257.21	368
0.94	158.02	141.93	149.99	302.14	433
0.95	191.68	172.35	180.42	365.03	524
0.96	242.18	217.96	226.05	459.36	661
0.97	326.34	293.96	302.05	616.55	888
0.98	494.65	445.90	454.00	930.91	1343
0.99	999.60	901.50	909.61	1873.92	2708

Example 3 is obtained from Example 1, by stretching the intervals with the high arrival rate by a factor of 50. The corresponding parameter values are $\kappa = 50/101$, $\sigma_1 = 0.202$,

$\sigma_2 = 101/510 = 0.198039$, and $\lambda_1 = 2.02$. Computed values of the queue descriptors are listed in Table 4.

Table 4.
Queue Characteristics for Example 3.

p	α	$E(\mathbf{W})$	$E(\mathbf{W_A})$	$\mu_{\mathbf{BP}}$	$\phi(0.95)$
0.50	1.98	1.99	2.97	4.34	9
0.60	2.99	3.73	5.22	7.54	15
0.70	4.67	6.83	8.76	13.21	24
0.80	8.04	12.99	15.24	24.34	41
0.90	18.14	31.02	33.47	56.81	94
0.91	20.39	35.00	37.46	63.97	105
0.92	23.19	39.96	42.43	72.92	120
0.93	26.80	46.33	48.81	84.41	139
0.94	31.61	54.81	57.31	99.73	164
0.95	38.34	66.67	69.18	121.15	200
0.96	48.44	84.44	86.97	153.28	253
0.97	65.28	114.06	116.58	206.80	341
0.98	98.95	173.23	175.77	313.82	518
0.99	199.96	350.65	353.19	634.81	1050

In Example 4, the case of Example 1 is modified so that the arrivals in the "background" Poisson arrival stream account for half of the traffic intensity. The values of σ_1 and σ_2 are as in Example 1, but the corresponding values of $\lambda_1 = 51$, $\lambda_2 = 0.5$. Values of the queue descriptors are listed in Table 5 and, for comparison with Example 1, the graphs of the corresponding complementary waiting time distributions are given in Figures 2(a) and 2(b).

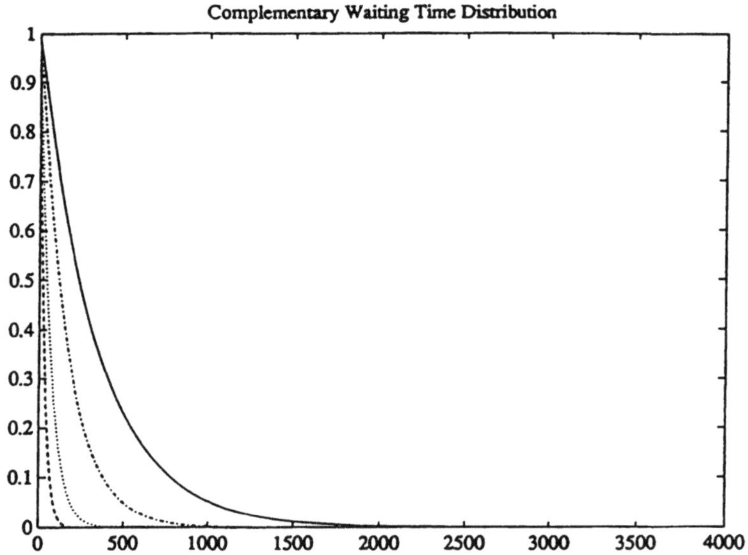

Figure 2a. Graph of the complementary virtual waiting time distribution for the queue of Example 4. In decreasing order, the graphs corresponds to $p = 0.99, 0.98, 0.95$ and 0.90.

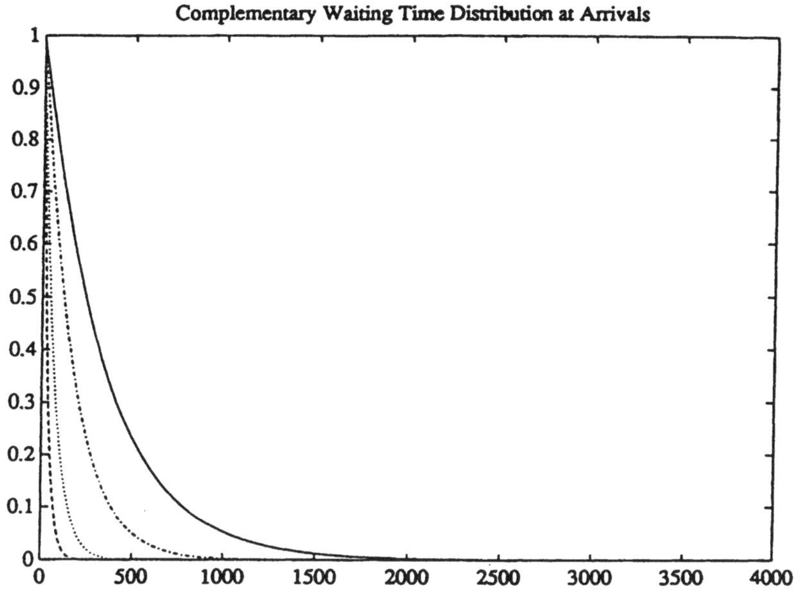

Figure 2b. Graph of the complementary virtual waiting time distribution an arbitrary arrival for the queue of Example 4. In decreasing order, the graphs corresponds to $p =$ 0.99, 0.98, 0.95 and 0.90.

Table 5.
Queue Characteristics for Example 4.

p	α	$E(W)$	$E(W_A)$	μ_{BP}	$\phi(0.95)$
0.50	4.51	1.01	1.02	1.98	4
0.60	14.29	2.07	2.44	2.85	9
0.70	34.92	4.55	5.50	4.40	18
0.80	77.84	10.13	11.66	7.48	35
0.90	207.64	27.34	29.38	16.72	85
0.91	236.52	31.17	33.26	18.78	97
0.92	272.61	35.97	38.10	21.34	111
0.93	319.02	42.13	44.30	24.64	130
0.94	380.92	50.35	52.56	29.04	154
0.95	467.57	61.85	64.11	35.19	188
0.96	597.56	79.10	81.40	44.43	240
0.97	814.22	107.86	110.20	59.82	326
0.98	1247.52	165.36	167.74	90.59	498
0.99	2547.51	337.83	340.25	182.93	1015

In Example 5, the case of Example 2 is modified so that the Poisson arrivals account for half of the traffic intensity. The values of σ_1 and σ_2 are as in Example 2. Correspondingly $\lambda_1 = 5.55$, $\lambda_2 = 0.5$. Numerical results are given in Table 6.

Table 6.
Queue Characteristics for Example 5.

p	α	$E(\mathbf{W})$	$E(\mathbf{W_A})$	$\mu_{\mathbf{BP}}$	$\phi(0.95)$
0.50	1.87	1.14	1.28	1.99	5
0.60	3.28	2.21	2.69	2.82	10
0.70	5.84	4.45	5.35	4.31	17
0.80	11.17	9.36	10.70	7.31	32
0.90	27.46	24.48	26.20	16.34	76
0.91	31.09	27.85	29.60	18.35	86
0.92	35.63	32.06	33.85	20.86	99
0.93	41.47	37.47	39.29	24.08	115
0.94	49.26	44.69	46.54	28.39	136
0.95	60.17	54.79	56.67	34.41	166
0.96	76.54	69.94	71.85	43.44	212
0.97	103.83	95.18	97.13	58.50	287
0.98	158.40	145.67	147.64	88.61	438
0.99	322.14	279.18	299.08	178.96	892

Example 6 is a similar modification of Example 3. The Poisson arrivals again account for half of the traffic intensity and the values of σ_1 and σ_2 are as in Example 3. Correspondingly $\lambda_1 = 1.51$, $\lambda_2 = 0.5$. For this example, we refer to Table 7.

Table 7.
Queue Characteristics for Example 6.

p	α	$E(\mathbf{W})$	$E(\mathbf{W_A})$	$\mu_{\mathbf{BP}}$	$\phi(0.95)$
0.50	1.15	0.91	0.82	1.76	3
0.60	1.76	1.49	1.48	2.33	6
0.70	2.79	2.65	2.78	3.41	9
0.80	4.87	5.26	5.58	5.79	17
0.90	11.14	13.44	13.93	13.15	41
0.91	12.53	15.26	15.77	14.79	47
0.92	14.27	17.54	18.07	16.84	53
0.93	16.51	20.47	21.01	19.48	62
0.94	19.50	24.38	24.93	23.01	74
0.95	23.68	29.84	30.41	27.94	90
0.96	29.96	38.03	38.62	35.34	114
0.97	40.41	51.68	52.28	47.69	155
0.98	61.33	78.97	79.59	72.37	237
0.99	124.08	160.80	161.43	146.43	482

6.2. Comments on the Numerical Results

The values of the marking rate α change in a predictable way across the examples and we note that for high values of p, the marking rate is close to the value $p(1-p)^{-1}$ needed to mark a fraction p of the events in a Poisson process of rate one. With a high marking rate, the arrivals during the periods with the lower arrival rate (if any) will almost certainly be marked, so that the value of the marking rate is most significantly affected by the high arrival rate.

Examples 1 through 6 illustrate very well and in a decreasing manner, the phenomenon of " coalescence of excursions" described in Neuts [7]. Using the results in Table 1 for the familiar $M/M/1$ queue as a reference point, we see that with bursty arrivals, there is a strong likelihood that the input in one burst has not been cleared before another burst arrives. With a significant number in the queue, more bursts are likely to come in before the current work is cleared. This is evident from the values of μ_{BP} which remain much higher than those in Table 1, even when the fraction of admitted customers is low. Comparing the values for μ_{BP} for Examples 1 and 2, we see that this is an important feature of bursty arrivals, since there is only a modest decrease in the mean durations of the busy periods when the individual bursts are spread out even by a factor of ten. With the choice of the parameters in Example 3, the " modulation" in the arrival rate is quite gentle, yet the mean busy periods for the high traffic intensities are five or six times those for comparable cases of the $M/M/1$ queue.

Coalescing excursions also carry the queue lengths to high values. Whereas five percent of admitted customers experience queues in excess of 268 for the $M/M/1$ case with $p = 0.99$, the corresponding values (characteristic of " long" queues) for the first three examples are respectively 3210, 2708 and 1050. With the parameter values in these examples, each burst (total number of arrivals during a period with arrivals at the high rate) contains an average of 49.5 customers. Thinking of the " arbitrary" arrival as being midway in the burst to which it belongs, suggests that in Example 1, that 5% of those arrivals may still find waiting customers who arrived during at least the

$$(3210 - 24.75)(49.5)^{-1} \approx 64 \tag{33}$$

preceding bursts. The corresponding numbers for Examples 2 and 3 are 56 and 21. It is interesting to compare these " rough" estimates to those of Examples 4-6. In these, the average number admitted per cycle (including those of the Poisson background process) is also 50. Using the values of $\phi(0.95)$ in Tables 5-7 corresponding to $p = 0.99$, a similar calculation yields the approximate values 20, 17 and 9, which suggest that with a smoother arrival process, far fewer bursts survive long enough in the queue to contribute to the large queue lengths.

As anticipated, there is only a small relative difference in the expected waiting times at arbitrary times and at admitted arrivals at high traffic intensities. These means are, in fact, related by the formula

$$E(W_A) = \frac{\mu}{\lambda^*(\alpha)} E(W) - \frac{1}{\mu}, \tag{34}$$

The stated formula is an easy consequence of formulas (20) and (21). The response of these mean values to greater burstiness in the arrival stream is similar to that for the μ_{BP}. In fact, as seen in Examples 1-3, for the higher values of p, the expected waiting times are roughly 1/2 of the expected busy period. Interestingly, in Examples 4-6, the mean busy periods for corresponding $p-$values are smaller than the average waiting time. Although confirmation by detailed computations of the distributions of the lengths of busy periods is needed, this suggests that, with the high share of the Poisson background arrivals, there are more short busy periods. Customers served during these have short waiting times. Less frequent, but much longer busy periods account for the large number of customers who experience long waiting times.

A final, general remark is in order. By applying the marking procedure to the critical $MAP/M/1$ queue, the effect of burstiness on the queue and the different impact of selective thinning have been illustrated. When applied to a subcritical queue, say with a higher value of μ, the same relative effect of burstiness on the queue descriptors is seen, though the numerical values are smaller. The principal effect on the queue of bursty arrivals lies in the increased occurrence of coalescing excursions. That effect inherently depends on the speed of service and the importance of bursty arrivals is therefore always determined by the queue to which they serve as input. As noted in Neuts [7], the determination of the regime(s) in which a stable queue exhibits coalescing excursions is an important subject for investigation.

7. EXTENSION TO THE $SM/PH/1$ QUEUE

The analysis discussed in the preceding sections can be formally extended to the $SM/PH/1$ queue. In that queue, the arrivals of customers occur at the transition epochs of an $m-$state irreducible Markov renewal process with transition probability matrix $F(\cdot)$. The service times have a $PH-$distribution with irreducible representation (β, S). As noted in Section 4.2a of Neuts [6], that queueing model has a matrix-geometric analysis which is entirely analogous to that of the $GI/PH/1$ queue.
The arrival process may be thinned exactly as discussed for the MAP and the resulting point process is also generated by the transition epochs of an $m-$state Markov renewal process. In order to specify its transition probability matrix, we define the matrix of mass-functions

$$F^{\circ}(x;\alpha) = \int_0^x e^{-\alpha u} dF(u), \quad \text{for} \quad x \geq 0. \tag{35}$$

The matrix Laplace-Stieltjes transforms of $F(\cdot)$ and $F^{\circ}(\cdot;\alpha)$ will be respectively denoted by $f(s)$ and $f^{\circ}(s;\alpha)$. The transition probability matrix $G(\cdot;\ \alpha)$ of the thinned arrival process then has the Laplace-Stieltjes transform

$$g(s;\alpha) = [I - f^{\circ}(s;\alpha)]^{-1}[f(s) - f^{\circ}(s;\alpha)]. \tag{36}$$

Clearly, the principal challenge in carrying out the same analysis as we have done for the $MAP/M/1$ model for this substantial generalization lies in its algorithmic implementation. Even the evaluation of the transition probability matrix $G(\cdot;\ \alpha)$ requires the solution of a linear matrix integral equation obtained by the inversion of the transform in (36). Also, while efficient computer codes for the solution of the $GI/PH/1$ queue exist, we are unaware of similar implementations for the $SM/PH/1$ queue. If such an effort were contemplated, we would recommend the procedures developed in Lucantoni and Ramaswami [3] and the waiting time analysis in Sengupta [11].

REFERENCES

[1] ASMUSSEN, S. AND RAMASWAMI, V. (1990) Probabilistic Interpretation of some Duality Results for the Matrix Paradigms in Queueing Theory. *Stoch. Mod.*, **6**, 715-733.

[2] BLONDIA, C. (1989) The $N/G/1$ Finite Capacity Queue. *Stoch. Mod.*, **5**, 273-294, .

[3] LUCANTONI, D. M. AND RAMASWAMI, V. (1985) Efficient Algorithms for Solving the Non-Linear Matrix Equations Arising in Phase Type Queues. *Stoch. Models*, **1**, 29-52.

[4] LUCANTONI, D. M., MEIER-HELLSTERN, K. S., AND NEUTS, M. F. (1990) A Single Server Queue with Server Vacations and a Class of Non-Renewal Arrival Processes. *Adv. Appl. Prob.*, **22**, 676-705.

[5] LUCANTONI, D. M. (1991) New Results on the Single Server Queue with a Batch Markovian Arrival Process. *Stoch. Mod.*, **7**, 1-46.

[6] NEUTS, M. F. (1981) *Matrix-Geometric Solutions in Stochastic Models: An Algorithmic Approach.* Baltimore: The Johns Hopkins University Press.

[7] NEUTS, M. F. (1986) The Caudal Characteristic Curve of Queues. *Adv. Appl. Prob.*, **18**, 221-54.

[8] NEUTS, M. F. (1989) *Structured Stochastic Matrices of $M/G/1$ Type and Their Applications.* New York: Marcel Dekker Inc.

[9] NEUTS, M. F. (1990) *The Square Wave Spectrum of a Markov Renewal Process.* In: "Proceedings of a Symposium on Stochastic Processes and their Applications in honor of Professor S. K. Srinivasan", Bombay, 1990.

[10] NEUTS, M. F. (1993) The Burstiness of Point Processes. *Stoch. Mod.*, **9**, 445-466.

[11] SENGUPTA, B. (1989) Markov processes whose Steady State Distribution is Matrix-Exponential with an Application to the $GI/PH/1$ Queue. *Adv. Appl. Prob.*, **21**, 159-180.

Recent Adv. in Stat. and Prob., pp. 389-398
J. Pérez Vilaplana and M. L. Puri (Eds)

Transient Analysis of Finite State Birth and Death Process with Absorbing Boundary States

O.P. SHARMA[1] and N.S.K. NAIR[2]
[1] Department of Mathematics, Indian Institute of Technology, New Delhi, India 110016
[2] Computer Centre, National Institute of Immunology, Shahid Jit singh Marg, New Delhi, India 110067

Abstract – Transient analysis of finite state birth and death process with absorbing barriers is carried out. Closed form expressions for state probabilities and density functions of time of absorption for each case is obtained along with the moments. The results are expressed in terms of eigenvalues of symmetric tridiagonal matrix which can be easily computed.

1. INTRODUCTION

Finite state birth and death process with either one or both absorbing boundary states has very useful applications in queueing and biological models. Researchers have studied this process for certain particular cases. Specifically busy period of the Markovian queues [Sharma and Dass (1989)], population growth models without immigration [Bhat (1982)], epidemic and communicable disease models [Goel and Richter-Dyn (1974)] are some examples. However the transient analysis of the process with absorbing barriers has not received much attention in the literature. An attempt is made in this article to give a unified approach to the problem and obtain closed form expressions for the transient state probabilities. Probability distribution of the time of absorption and its moments are also determined. Some applications of the process are highlighted.
We consider the finite state time homogeneous birth and death process $\{X(t), 0 \leq t < \infty\}$ with state space, $S = \{0, 1, 2, \ldots, N\}$ and λ_n and μ_n, $(0 \leq n \leq N)$, as birth and death parameters. The states 0 and N are boundary states. Based on the nature of λ_n and μ_n at the boundary states, the following three possible cases can arise namely,

i) state 0 is absorbing and state N is reflecting, ie., $\lambda_0 = 0$ and $\mu_N > 0$
ii) state 0 is reflecting and state N is absorbing, ie., $\lambda_0 > 0$ and $\mu_N = 0$
iii) both the boundary states are absorbing, ie., $\lambda_0 = \mu_N = 0$.

Defining,

$$\begin{aligned} p_n(t) &= \Pr\{X(t) = n/X(0) = k\} \\ p_n'(t) &= \frac{d}{dt} p_n(t) \end{aligned} \tag{1}$$

and

$$\psi(n, \theta) = \int_0^\infty e^{-\theta t} p_n(t)\, dt\,, \tag{2}$$

We discuss these cases one by one.

2. PROCESS WITH STATE 0 AS ABSORBING AND STATE N AS REFLECTING

Here the differential difference equations satisfied by the process are,

$$
\begin{aligned}
p_0'(t) &= \mu_1 p_1(t) \\
p_1'(t) &= -(\lambda_1+\mu_1)\, p(t) + \mu_2 p_2(t) \\
p_n'(t) &= \lambda_{n-1}\, p_{n_1}(t) - (\lambda_n+\mu_n) p_n(t) + \mu_{n+1}\, p_{n+1}(t)\,, \quad 2 \le n \le N-1 \\
p_N'(t) &= \lambda_{N-1}\, p_{N-1}(t) - \mu_N\, p_N(t)
\end{aligned} \tag{3}
$$

After taking Laplace transforms and simplification, we get,

$$
\begin{aligned}
\theta\, \psi(0,\theta) - \mu_1\, \psi(1,\theta) &= p_0(0) \\
(\theta+\lambda_1+\mu_1)\, \psi(1,\theta) - \mu_2 \psi(2,\theta) &= p_1(0) \\
-\lambda_{n-1}\, \psi(n-1,\theta) + (\theta+\lambda_n+\mu_n)\, \psi(n,\theta) - \mu_{n+1}\, \psi(n+1,\theta) &= p_n(0), \quad 2 \le n \le N-1 \\
-\lambda_{N-1}\, \psi(N-1,\theta) + (\theta+\mu_N)\, \psi(N,\theta) &= p_N(0)
\end{aligned} \tag{4}
$$

which can be expressed in the matrix form as,

$$\mathbf{A}\, \psi(\theta) = \delta \tag{5}$$

where, $\psi(\theta) = [\,\psi(0,\theta),\ \psi(1,\theta), \ldots, \psi(N,\theta)\,]'$ and $\delta = [\,p_0(0),\, p_1(0), \ldots, p_N(0)\,]'$ are column vectors of order $N+1$, and assuming that the process starts at k,

$$p_k(0) = 1 \qquad \text{and} \qquad p_n(0) = 0,\ \forall n \ne k,$$

and $\mathbf{A}$ is the tridiagonal matrix of order $N+1$ given by,

$$
\mathbf{A} = \begin{bmatrix}
\theta & -\mu_1 & 0 & \ldots & 0 & 0 \\
0 & \lambda_1+\mu_1+\theta & -\mu_2 & \ldots & 0 & 0 \\
0 & -\lambda_1 & \lambda_2+\mu_2+\theta & \ldots & 0 & 0 \\
\vdots & \vdots & \vdots & \vdots & \vdots & \vdots \\
0 & 0 & 0 & \ldots & \lambda_{N-1}+\mu_{N-1}+\theta & -\mu_N \\
0 & 0 & 0 & \ldots & -\lambda_{N-1} & \mu_N+\theta
\end{bmatrix}
$$

Therefore,

$$\psi(\theta) = \mathbf{A}^{-1}\delta \tag{6}$$

where $\mathbf{A}^{-1}$is the inverse of the matrix $\mathbf{A}$ which obviously exists. The matrix $\mathbf{A}$ can be partitioned as,

$$\mathbf{A} = \begin{bmatrix} \mathbf{A}_{11} & \mathbf{A}_{12} \\ \mathbf{0} & \mathbf{A}_{22} \end{bmatrix}$$

where, $\mathbf{A}_{11} = \theta\,,\qquad \mathbf{A}_{12} = [-\mu_1,\, 0, \ldots, 0]_{1\times N}$ and

$$
\mathbf{A}_{22} = \begin{bmatrix}
\lambda_1+\mu_1+\theta & -\mu_2 & \ldots & 0 & 0 \\
-\lambda_1 & \lambda_2+\mu_2+\theta & \ldots & 0 & 0 \\
\vdots & \vdots & \vdots & \vdots & \vdots \\
0 & 0 & \ldots & \lambda_{N-1}+\mu_{N-1}+\theta & -\mu_N \\
0 & 0 & \ldots & -\lambda_{N-1} & \mu_N+\theta
\end{bmatrix}
$$

Then from algebra,

$$\mathbf{A}^{-1} = \begin{bmatrix} \mathbf{A}_{11}^{-1} & \mathbf{A}_{11}^{-1}\mathbf{A}_{12}\mathbf{A}_{22}^{-1} \\ \mathbf{0} & \mathbf{A}_{22}^{-1} \end{bmatrix}$$

and

$$\mathbf{A}_{11} = \theta \quad \longrightarrow \quad \mathbf{A}_{11}^{-1} = 1/\theta$$

It can be easily seen that the diagonal matrix, $\mathbf{D} = \text{diag}\,(d_1,\, d_2,\, \ldots,\, d_N)$ with

$$d_1 = 1 \qquad \text{and} \qquad d_r = \sqrt{\frac{\mu_2\,\mu_3\,\ldots\,\mu_r}{\lambda_1\,\lambda_2\,\ldots,\,\lambda_{r-1}}} \qquad 2 \le r \le N$$

transforms $\mathbf{A}_{22}$ into a symmetrical tridiagonal matrix, say, $\mathbf{B}(\theta)$, given by,

$$\mathbf{B}(\theta) = \mathbf{D}\mathbf{A}_{22}\mathbf{D}^{-1} = \begin{bmatrix} \lambda_1 + \mu_1 + \theta & -\sqrt{\lambda_1\mu_2} & \ldots & 0 & 0 \\ -\sqrt{\lambda_1\,\mu_2} & \lambda_2 + \mu_2 + \theta & \ldots & 0 & 0 \\ \vdots & \vdots & \vdots & \vdots & \vdots \\ 0 & 0 & \ldots & \lambda_{N-1} + \mu_{N-1} + \theta & -\sqrt{\lambda_{N-1}\mu_N} \\ 0 & 0 & \ldots & -\sqrt{\lambda_{N-1}\mu_N} & \mu_N + \theta \end{bmatrix} \qquad (7)$$

Let $P_r(\theta)$ and $Q_r(\theta)$ be the determinants of bottom right and top left $r \times r$ square matrices formed from $\mathbf{B}(\theta)$, these polynomials satisfy the difference equations,

$$P_r(\theta) - (\lambda_{N-r+1} + \mu_{N-r+1} + \theta)\, P_{r-1}(\theta) + \lambda_{N-r+1}\mu_{N-r+1}\, P_{r-2}(\theta) = 0, \quad 2 \le r \le N$$

$$Q_r(\theta) - (\lambda_r + \mu_r + \theta)\, Q_{r-1}(\theta) + \lambda_{r-1}\mu_r\, Q_{r-2}(\theta) = 0, \quad 2 \le r \le N-1 \qquad (8)$$

with initial conditions,

$$P_0(\theta) = Q_0(\theta) = 1$$
$$P_1(\theta) = \mu_N + \theta, \qquad Q_1(\theta) = \lambda_1 + \mu_1 + \theta$$

We also define $P_{-1}(\theta) = \frac{1}{\theta} = Q_{-1}(\theta)$.
Now if $\mathbf{A}_{22}^{-1} = (a^{ij}(\theta))\,, 1 \le i, j \le N$. Then $a^{ij}(\theta)$ can be written in terms of$P_r(\theta)$ and $Q_r(\theta)$, ([4], [5]) as,

$$\begin{aligned} a^{ij}(\theta) &= \mu_{i+1}\,\ldots\,\mu_j\,\frac{P_{N-j}(\theta)\,Q_{i-1}(\theta)}{|\mathbf{B}(\theta)|}\,, && i < j \\ &= \frac{P_{N-i}(\theta)\,Q_{i-1}(\theta)}{|\mathbf{B}(\theta)|}\,, && i = j \\ &= \lambda_j\,\ldots\,\lambda_{i-1}\,\frac{P_{N-i}(\theta)\,Q_{j-1}(\theta)}{|\mathbf{B}(\theta)|}\,, && i > j\,, \quad 1 \le i,j \le N \end{aligned} \qquad (9)$$

and hence,

$$\mathbf{A}^{-1} = \begin{bmatrix} \frac{1}{\theta} & \frac{\mu_1}{\theta}\,a^{11}(\theta) & \frac{\mu_1}{\theta}\,a^{12}(\theta) & \ldots & \frac{\mu_1}{\theta}\,a^{1N}(\theta) \\ \mathbf{0} & & \mathbf{A}_{22}^{-1} & & \end{bmatrix} \qquad (10)$$

Using it in (6) we get, for $k = 0$,

$$\begin{aligned} \psi(n,\theta) &= \tfrac{1}{\theta} && n = 0 \\ &= 0 && 1 \le n \le N \end{aligned}$$

and for $1 \le k \le N$,

$$\begin{aligned}\psi(n,\theta) &= \mu_{n+1}\,\dots\,\mu_k\,\frac{P_{N-k}(\theta)\,Q_{n-1}(\theta)}{|\mathbf{B}(\theta)|}, && 0 \le n < k\\ &= \frac{P_{N-k}(\theta)\,Q_{k-1}(\theta)}{|\mathbf{B}(\theta)|}, && n = k\\ &= \lambda_k\,\dots\,\lambda_{n-1}\,\frac{P_{N-n}(\theta)\,Q_{k-1}(\theta)}{|\mathbf{B}(\theta)|}, && k < n \le N \qquad (11)\end{aligned}$$

Eigenvalues of the matrix $\mathbf{B}(0)$ are real, positive and distinct [3]. Let these be $\alpha_1, \alpha_2, \dots, \alpha_N$, so that we can write,

$$|\mathbf{B}(\theta)| = \prod_{i=1}^{N} (\theta + \alpha_i) \qquad (12)$$

and

$$|\mathbf{B}(0)| = P_N(0) = \mu_1\,\mu_2\,\dots\,\mu_N$$

Using (12) in (11) and making use of partial fractions and simplification we get, for $1 \le k \le N$,

$$\begin{aligned}\psi(n,\theta) &= \frac{\delta_{0n}}{\theta} + \mu_{n+1}\,\dots\,\mu_k\,\sum_{i=1}^{N}\left[\frac{A_{ni}}{\theta} + \alpha_i\right], && 0 \le n < k\\ &= \sum_{i=1}^{N}\left[\frac{A_{ki}}{\theta} + \alpha_i\right], && n = k\\ &= \lambda_k\,\dots\,\lambda_{n-1}\,\sum_{i=1}^{N}\left[\frac{A_{ni}}{\theta} + \alpha_i\right], && k < n \le N \qquad (13)\end{aligned}$$

where

$$A_{ni} = \frac{P_{N-l}(-\alpha_i)\,Q_{q-1}(-\alpha_i)}{\prod\limits_{\substack{j=1\\ j\neq i}}^{N} (\alpha_j - \alpha_i)}, \qquad n = 0, 1, 2, \dots, N$$

where $l = \max(n,k)$ and $q = \min(n,k)$.
Taking inverse Laplace transform, the expression for transient state probabilities is obtained as, for $k = 0$,

$$\begin{aligned}p_0(t) &= 1\\ p_n(t) &= 0, \qquad 1 \le n \le N\end{aligned}$$

and for $1 \le k \le N$,

$$\begin{aligned}p_n(t) &= \delta_{0n} + \mu_{n+1}\,\mu_{n+2}\,\dots\,\mu_k\,\sum_{i=1}^{N} A_{ni}\,e^{-\alpha_i t}, && 0 \le n < k\\ &= \sum_{i=1}^{N} A_{ki}\,e^{-\alpha_i t}, && n = k\\ &= \lambda_k\,\lambda_{k+1}\,\dots\,\lambda_{n-1}\,\sum_{i=1}^{N} A_{ni}\,e^{-\alpha_i t}, && k < n \le N \qquad (14)\end{aligned}$$

Here if we let $t \to \infty$, $p_0(t)$ converges to unity and all other probabilities tend to zero which guarantees the ultimate absorption of the process at the state 0.
Let T be the random variable representing the time of absorption of the process at 0 and let $f(t)$ denotes its probability density function, then $f(t)$ is given by,

$$\begin{aligned} f(t) &= \mu_1 \mu_2 \dots \mu_k \sum_{i=1}^{N} A_{ni} e^{-\alpha_i t}, && 0 \le t < \infty \\ &= 0, && \text{elsewhere} \end{aligned} \tag{15}$$

and if m_r be its r-th moment about the origin, then

$$m_r = \mu_1 \mu_2 \dots \mu_k \overline{|r+1} \sum_{i=1}^{N} A_{li} \alpha_i^{-r-l} \tag{16}$$

As a particular case, take $\lambda_n = \lambda$ and $\mu_n = \mu$. Here, $f(t)$ gives the busy period distribution of the $M/M/1/N$ queue and in this case the transient probabilities of the process and the probability density function of the busy period duration are given by (14) and (15) respectively. These results agrees with the one reported by Sharma and Dass (1989).
Similarly one can study applications of this process in biology. For example, consider the spread of an epidemic in a finite population. Let N be the size of the population in which n individuals are infectives and $N-n$ are susceptibles at time t. Let $X(t)$ be the process representing the number of infectives at time t, μ be the mean rate of infectives being removed and λ be the mean rate of contact between two members of the population which makes a susceptible an infective. Then $X(t)$ is a birth and death process with state 0 as absorbing and state N as reflecting. Birth and death parameters are,

$$\lambda_n = \lambda n(N-n) \qquad \text{and} \qquad \mu_n = n\mu, \quad 0 \le n \le N$$

Here, $p_n(t)$ and $f(t)$ are given again by (14) and (15). In particular the expected size and duration of the epidemic are respectively,

$$\begin{aligned} E[X(t)] &= \sum_{i=1}^{N} e^{-\alpha_i t} \left[\overline{|k+1} \sum_{n=1}^{k-1} \frac{\mu^{k-n}}{\overline{|n}} A_{ni} + \frac{\overline{|N-k+1}}{\overline{|k}} \sum_{n=k}^{N} \frac{\overline{|n+1}}{\overline{|N-n+1}} \lambda^{n-k} B_{ni} \right] \\ m_1 &= \overline{|k+1}\, \mu^k \sum_{i=1}^{N} \bar{A}_i \alpha_i^{-2} \end{aligned}$$

3. PROCESS WITH STATE N AS ABSORBING AND STATE 0 AS REFLECTING

This process can also be dealt with in a similar way. Laplace transform of the differential difference equations takes the matrix form,

$$\mathbf{A}\, \psi(\theta) = \delta \tag{17}$$

where,

$$\mathbf{A} = \begin{bmatrix} \lambda_0 + \theta & -\mu_1 & 0 & \dots & 0 & 0 \\ -\lambda_0 & \lambda_1 + \mu_1 + \theta & -\mu_2 & \dots & 0 & 0 \\ 0 & -\lambda_1 & \lambda_2 + \mu_2 + \theta & \dots & 0 & 0 \\ \vdots & \vdots & \vdots & \vdots & \vdots & \vdots \\ 0 & 0 & 0 & \dots & \lambda_{N-1} + \mu_{N-1} + \theta & 0 \\ 0 & 0 & 0 & \dots & -\lambda_{N-1} & \theta \end{bmatrix}$$

and $\psi(\theta)$ and δ are as defined earlier.
Proceeding in a similar way to Section 2, we finally obtain the transient state probabilities as, for $0 \le k \le N-1$,

$$\begin{aligned} p_n(t) &= \mu_{n+1}\,\mu_{n+2}\,\dots\,\mu_k \sum_{i=1}^{N} A_{ni}\, e^{-\alpha_i t}\,, && 0 \le n < k \\ &= \sum_{i=1}^{N} A_{ki}\, e^{-\alpha_i t}\,, && n = k \\ &= \frac{\delta_{Nn}}{\theta} + \lambda_k\,\lambda_{k+1}\,\dots\,\lambda_{n-1} \sum_{i=1}^{N} A_{ni}\, e^{-\alpha_i t}\,, && k < n \le N \end{aligned} \tag{18}$$

and for $k = N$,

$$\begin{aligned} p_n(t) &= 0\,, && 0 \le n \le N-1 \\ p_n(t) &= 1\,, && n = N \end{aligned}$$

where

$$A_{ni} = \frac{P_{N-l-1}(-\alpha_i)\, Q_q(-\alpha_i)}{\prod\limits_{\substack{j=1 \\ j\neq i}}^{N} (\alpha_j - \alpha_i)}\,, \qquad n = 0, 1, 2, \dots, N$$

where $l = \max(n, k)$ and $q = \min(n, k)$.
Here also if we take $t \to \infty$, $p_N(t)$ converges to unity and all other probabilities tend to zero which guarantees the ultimate absorption of the process at the state N, and probability density function of the time of absorption is given by,

$$\begin{aligned} f(t) &= \lambda_k\,\dots\,\lambda_{N-1} \sum_{i=1}^{N} A_{N-1i}\, e^{-\alpha_i t}\,, && 0 \le t < \infty \\ &= 0\,, && \text{elsewhere} \end{aligned} \tag{19}$$

4. PROCESS WITH BOTH BOUNDARY STATES AS ABSORBING

In this case the matrix representation of Laplace transforms of differential difference equations becomes,

$$\mathbf{A}\,\psi(\theta) = \delta \tag{20}$$

where,

$$\mathbf{A} = \begin{bmatrix} \theta & -\mu_1 & 0 & \dots & 0 & 0 \\ 0 & \lambda_1 + \mu_1 + \theta & -\mu_2 & \dots & 0 & 0 \\ 0 & -\lambda_1 & \lambda_2 + \mu_2 + \theta & \dots & 0 & 0 \\ \vdots & \vdots & \vdots & \vdots & \vdots & \vdots \\ 0 & 0 & 0 & \dots & \lambda_{N-1} + \mu_{N-1} + \theta & 0 \\ 0 & 0 & 0 & \dots & -\lambda_{N-1} & \theta \end{bmatrix}$$

Now we can partition the matrix $\mathbf{A}$ as,

$$\mathbf{A} = \begin{bmatrix} \mathbf{A}_{11} & \mathbf{A}_{12} \\ \mathbf{0} & \mathbf{A}_{22} \end{bmatrix}$$

with, $\mathbf{A}_{11} = \theta$, $\mathbf{A}_{12} = [-\mu_1, 0, \ldots, 0]_{1\times N}$ and $\mathbf{A}_{22}$ is a matrix which can be partitioned again as,

$$\mathbf{A}_{22} = \begin{bmatrix} \mathbf{B}_{11} & \mathbf{0} \\ \mathbf{B}_{21} & \mathbf{B}_{22} \end{bmatrix}$$

where,

$$\mathbf{B}_{22} = \theta\,, \qquad \mathbf{B}_{21} = [0, 0, \ldots, -\lambda_{N-1}]_{1\times(N-1)}$$

and

$$\mathbf{B}_{11} = \begin{bmatrix} \lambda_1+\mu_1+\theta & -\mu_2 & \ldots & 0 & 0 \\ -\lambda_1 & \lambda_2+\mu_2+\theta & \ldots & 0 & 0 \\ \vdots & \vdots & \vdots & \vdots & \vdots \\ 0 & 0 & \ldots & -\lambda_{N-2} & \lambda_{N-1}+\mu_{N-1}+\theta \end{bmatrix}$$

$\mathbf{B}_{11}$is a tridiagonal matrix of order $N-1$ which can be symmetrized so that $\mathbf{B}(\theta)$ is given by,

$$\mathbf{B}(\theta) = \begin{bmatrix} \lambda_1+\mu_1+\theta & -\sqrt{\lambda_1\mu_2} & \ldots & 0 & 0 \\ -\sqrt{\lambda_1\mu_2} & \lambda_2+\mu_2+\theta & \ldots & 0 & 0 \\ \vdots & \vdots & \vdots & \vdots & \vdots \\ 0 & 0 & \ldots & -\sqrt{\lambda_{N-2}\mu_{N-1}} & \lambda_{N-1}+\mu_{N-1}+\theta \end{bmatrix} \tag{21}$$

Again let $P_r(\theta)$ and $Q_r(\theta)$ be the determinants of bottom right and top left $r\times r$ square matrices formed from $\mathbf{B}(\theta)$. So inverse of the matrix $\mathbf{B}_{11}$ can be written in terms of $P_r(\theta)$ and $Q_r(\theta)$ as,

$$\mathbf{B}_{11}^{-1} = \left(b^{ij}(\theta)\right), \qquad 1 \le i,j \le N-1$$

where

$$\begin{aligned} b^{ij}(\theta) &= \mu_{i+1}\ldots\mu_j\,\frac{P_{N-j-1}(\theta)\,Q_{i-1}(\theta)}{|\mathbf{B}(\theta)|}, && i<j \\ &= \frac{P_{N-i-1}(\theta)\,Q_{i-1}(\theta)}{|\mathbf{B}(\theta)|}, && i=j \\ &= \lambda_j\ldots\lambda_{i-1}\,\frac{P_{N-i-1}(\theta)\,Q_{j-1}(\theta)}{|\mathbf{B}(\theta)|}, && i>j,\ 1\le i,j\le N-1 \end{aligned}$$

Therefore

$$\mathbf{A}_{22}^{-1} = \begin{bmatrix} & \mathbf{B}_{11}^{-1} & & & 0 \\ \frac{\lambda_{N-1}}{\theta}\,b^{N-11}(\theta) & \frac{\lambda_{N-1}}{\theta}\,b^{N-12}(\theta) & \ldots & \frac{\lambda_{N-1}}{\theta}\,b^{N-1\,N-1}(\theta) & \frac{1}{\theta} \end{bmatrix}$$

and hence

$$\mathbf{A}^{-1} = \begin{bmatrix} \frac{1}{\theta} & \frac{\mu_1}{\theta}\,b^{11}(\theta) & \frac{\mu_1}{\theta}\,b^{12}(\theta) & \ldots & \frac{\mu_1}{\theta}\,b^{1\,N-1}(\theta) & 0 \\ 0 & & \mathbf{B}_{11}^{-1} & & & 0 \\ 0 & \frac{\lambda_{N-1}}{\theta}\,b^{N-11}(\theta) & \frac{\lambda_{N-1}}{\theta}\,b^{N-12}(\theta) & \ldots & \frac{\lambda_{N-1}}{\theta}\,b^{N-1\,N-1}(\theta) & \frac{1}{\theta} \end{bmatrix}$$

Thus the solution is given by, for $k = 0$,

$$\begin{aligned}\psi(n,\theta) &= \frac{1}{\theta} && n = 0\\ &= 0 && 1 \le n \le N\end{aligned}$$

for $1 \le k \le N-1$,

$$\begin{aligned}\psi(n,\theta) &= \mu_{n+1}\ \dots\ \mu_k \frac{P_{N-k-1}(\theta)\,Q_{n-1}(\theta)}{|\,\mathbf{B}(\theta)\,|}, && 0 \le n < k\\ &= \frac{P_{N-k-1}(\theta)\,Q_{k-1}(\theta)}{|\,\mathbf{B}(\theta)\,|}, && n = k\\ &= \lambda_k\ \dots\ \lambda_{n-1} \frac{P_{N-n-1}(\theta)\,Q_{k-1}(\theta)}{|\,\mathbf{B}(\theta)\,|}, && k < n \le N\end{aligned}$$

and for $k = N$,

$$\begin{aligned}\psi(n,\theta) &= 0 && 0 \le n \le N-1\\ &= \frac{1}{\theta} && n = N\end{aligned} \tag{22}$$

The eigenvalues of the matrix $\mathbf{B}(0)$ are positive and distinct, let these be $\alpha_1, \alpha_2, \dots, \alpha_{N-1}$, then,

$$|\,B(\theta)\,| = \prod_{i=1}^{N-1} (\theta + \alpha_i)$$

and by using partial fractions we write, for $1 \le k \le N-1$,

$$\begin{aligned}\psi(n,\theta) &= \frac{\delta_{0n}\pi_0}{\theta} + \mu_{n+1}\ \dots\ \mu_k \sum_{i=1}^{N-1}\left[\frac{A_{ni}}{\theta} + \alpha_i\right], && 0 \le n < k\\ &= \sum_{i=1}^{N-1}\left[\frac{A_{ki}}{\theta} + \alpha_i\right], && n = k\\ &= \frac{\delta_{Nn}\pi_N}{\theta}\lambda_k\,\lambda_{k+1}\ \dots\ \lambda_{n-1} \sum_{i=1}^{N-1}\left[\frac{A_{ni}}{\theta} + \alpha_i\right], && k < n \le N\end{aligned} \tag{23}$$

where

$$\pi_0 = \mu_1\,\mu_2\ \dots\ \mu_k \frac{P_{N-k-1}(0)}{\prod_{i=1}^{N-1} \alpha_i}, \qquad \pi_N = \lambda_k\,\lambda_{k+1}\ \dots\ \lambda_{N-1} \frac{Q_{k-1}(0)}{\prod_{i=1}^{N-1} \alpha_i},$$

$$A_{ni} = \frac{P_{N-l-1}(-\alpha_i)\,Q_{q-1}(-\alpha_i)}{\prod_{\substack{j=1\\ j\neq i}}^{N-1} (\alpha_j - \alpha_i)}, \qquad n = 0,1,2,\dots,N$$

where $l = \max(n,k)$ and $q = \min(n,k)$.
Now, inverse Laplace transformation gives the transient state probabilities as, for $k = 0$

$$\begin{aligned}p_n(t) &= 1\,, && n = 0\\ &= 0\,, && 1 \le n \le N\end{aligned}$$

and for $1 \le k \le N-1$,

$$\begin{aligned} p_n(t) &= \delta_{0n}\pi_0 + \mu_{n+1}\,\mu_{n+2}\,\cdots\,\mu_k \sum_{i=1}^{N-1} A_{ni}\, e^{-\alpha_i t}\,, && 0 \le n < k \\ &= \sum_{i=1}^{N-1} A_{ki}\, e^{-\alpha_i t}\,, && n = k \\ &= \delta_{Nn}\pi_N + \lambda_k\,\lambda_{k+1}\,\cdots\,\lambda_{n-1} \sum_{i=1}^{N-1} A_{ni}\, e^{-\alpha_i t}\,, && k < n \le N \end{aligned}$$

and for $k = N$

$$\begin{aligned} p_n(t) &= 0\,, && 0 \le n \le N-1 \\ &= 1\,, && n = N \end{aligned} \qquad (24)$$

Thus transient state probabilities are determined explicitly. Here π_0 and π_N can be further simplified and obtained as,

$$\begin{aligned} \pi_n &= \left[1 + \frac{\lambda_{N-1}}{\mu_{N-1}} + \frac{\lambda_{N-1}\lambda_{N-2}}{\mu_{N-1}\mu_{N-2}} + \ldots + \frac{\lambda_{N-1}\lambda_{N-2}\cdots\lambda_{k+1}}{\mu_{N-1}\mu_{N-2}\cdots\mu_{k+1}}\right]\beta^{-1}, && n=0 \\ &= 0, && 1 \le n \le N-1 \\ &= \frac{\lambda_k\lambda_{k+1}\cdots\lambda_{N-1}}{\mu_k\mu_{k+1}\cdots\mu_{N-1}}\left[1 + \frac{\lambda_{k-1}}{\mu_{k-1}} + \frac{\lambda_{k-1}\lambda_{k-2}}{\mu_{k-1}\mu_{k-2}} + \ldots + \frac{\lambda_{k-1}\lambda_{k-2}\cdots\lambda_1}{\mu_{k-1}\mu_{k-2}\cdots\mu_1}\right]\beta^{-1}, && n=N \end{aligned} \qquad (25)$$

with

$$\beta = \left[1 + \frac{\lambda_{N-1}}{\mu_{N-1}} + \frac{\lambda_{N-1}\,\lambda_{N-2}}{\mu_{N-1}\,\mu_{N-2}} + \ldots + \frac{\lambda_{N-1}\,\lambda_{N-2}\,\cdots\,\lambda_1}{\mu_{N-1}\,\mu_{N-2}\,\cdots\,\mu_1}\right]$$

Consider a particular case by taking $\lambda_n = \lambda$ and $\mu_n = \mu$. New $\mathbf{B}(0)$ takes the simple form,

$$\mathbf{B}(0) = (\lambda + \mu)\,\mathbf{I}_{N-1} - \sqrt{\lambda\mu}\,\mathbf{D}_{N-1}$$

where $\mathbf{I}_{N-1}$ is the identity matrix of order $N-1$ and $\mathbf{D}_{N-1}$ is the simple symmetric tridiagonal matrix of order $N-1$ having all diagonal elements as zeros and offdiagonal entries are unity. The eigenvalues of $\mathbf{D}_{N-1}$ are,

$$a_i = 2\cos(i\pi\,/\,N)\,, \qquad i = 1, 2, \ldots, N-1$$

and hence eigenvalues of $\mathbf{B}(0)$ are,

$$\alpha_i = \lambda + \mu - \sqrt{\lambda\mu}\, a_i\,, \qquad i = 1, 2, \ldots, N-1$$

In this case, $P_r(-\alpha_i) = \left(-\sqrt{\lambda\mu}\right)^r \bar{P}_r(-a_i)$ and $Q_r(-\alpha) = \left(-\sqrt{\lambda\mu}\right)^r \bar{Q}_r(-a_i)$ where

$$\bar{P}_r(a_i) = \bar{Q}_r(a_i) = \begin{vmatrix} a_i & 1 & 0 & \ldots & 0 & 0 & 0 \\ 1 & a_i & 1 & \ldots & 0 & 0 & 0 \\ \vdots & \vdots & \vdots & \vdots & \vdots & \vdots & \vdots \\ 0 & 0 & 0 & \ldots & 1 & a_i & 1 \\ 0 & 0 & 0 & \ldots & 0 & 1 & a_i \end{vmatrix}_{(r \times r)}$$

and we obtain the simplified form of constants as

$$A_{ni} = \left(-\sqrt{\lambda\mu}\right)^{n-k} \frac{\bar{P}_{N-k-1}(-a_i)\,\bar{Q}_{n-1}(-a_i)}{\prod\limits_{\substack{j=1\\ j\neq i}}^{N-1} (a_j - a_i)},$$

Also

$$\beta = \frac{1-\rho^N}{1-\rho},$$

$$\pi_0 = \frac{1-\rho^{N-k}}{1-\rho^N} \quad \text{and} \quad \pi_N = \rho^{N-k}\frac{1-\rho^k}{1-\rho^N}.$$

Now in the light of the new constants the transient solution can be written as, for $1 \leq k \leq N-1$,

$$\begin{aligned} p_n(t) &= \delta_{0n}\pi_0 + \mu^{k-n} e^{-(\lambda+\mu)t} \sum_{i=1}^{N-1} A_{ni}\, e^{\sqrt{\lambda\mu}\, a_i t}, && 0 \leq n < k \\ &= e^{-(\lambda+\mu)t} \sum_{i=1}^{N-1} A_{ki}\, e^{\sqrt{\lambda\mu}\, a_i t}, && n = k \\ &= \delta_{Nn}\pi_N + \lambda^{n-k} e^{-(\lambda+\mu)t} \sum_{i=1}^{N-1} A_{ni}\, e^{\sqrt{\lambda\mu}\, a_i t}, && k < n \leq N \end{aligned}$$

An immediate application of this process in biology is the modelling of incubation period. Consider the growth of bacterium in an infectives body. Let there be k bacteria initially introduced into the susceptibles body by contact with an infected individual. Let us assume that each bacteria in the susceptible body has a probability $\lambda\,\delta t$ of dividing in to two bacteria and at the same time $\mu\,\delta t$ of dying in the time δt. Then this process of bacteria growth is a simple birth and death process with, $\lambda_n = n\lambda$ and $\mu_n = n\mu$. In the infected body symptoms appear when the number of bacteria reaches a certain level, say, N which is large enough compared to the initial number, k. Then the incubation period of the disease can be obtained by studying the above birth and death process between absorbing barriers 0 and N.

REFERENCES

[1] Bhat, U. N. (1982). *Elements of Applied Stochastic Processes.* New York: John Wiley and Sons.

[2] Goel, N. S. & Richter-Dyn, N. (1974). *Stochastic Models in Biology.* New York: Academic Press.

[3] Ledermann, W. & Reuter, G. E. H. (1954). Spectral Theory for the Differential Equations of Simple Birth and Death Processes. *Phil. Trans. Roy. Soc. London*, Ser. A, **246**, 321–369.

[4] Lewis, J. W. (1981). Inversion of Tridiagonal Matrices. *Numerische Mathematik*, **38**, 333–345.

[5] Nair, N. S. K. (1992). *Homogeneous Finite Source Birth and Death Process with Applications.* PhD Thesis, Indian Institute of Technology, New Delhi, India.

[6] Sharma, O. P. (1990). *Markovian Queues.* Chichester: Ellis Horwood Ltd.

[7] Sharma, O. P. & Dass, J. (1989). Initial Busy Period Analysis for a Multichannel Markovian Queues. *Optimization*, **20**, 317–323.

Recent Adv. in Stat. and Prob., pp. 399-414
J. Pérez Vilaplana and M. L. Puri (Eds)

Pollaczek Method in Queueing Analysis

R. SYSKI
Department of Mathematics, University of Maryland, College Park, MD 20742, USA

Abstract – Felix Pollaczek's contributions to Queueing Theory are widely acclaimed and frequently quoted, but his methods are much less popular. Basically, his solution for the general system $GI/G/s$ is reduced to the solution of a system of s linear integro-functional equations. Essential use of theory of functions of several complex variables, including contour integration, made his approach hard to follow.

It is the purpose of this paper to present Pollaczek theory in a simplified form and to explain how and why it works. Attention is here concentrated on the case $s > 1$, which differs considerably in difficulty from the well known case $s = 1$.

In addition, the paper discusses extensions obtained by other workers influenced by the Pollaczek method. Relations with other methods are also briefly discussed.

1. INTRODUCTION

The impact of Pollaczek's work on Queueing Theory is well known to every queueing theoretician. Felix Pollaczek (1 Dec. 1892-29 Apr. 1981) whose first papers appeared in the $30's$, was the first to obtain (in 1953) the complete solution for the waiting time in the $GI/G/s$ system with the strict order service, published later in the book form in 1961. His highly individual approach, based on multidimensional contour integrals, was at first received with polite admiration. It was frequently quoted and used for purpose of comparison, but its ideas were far less popular. However, usefulness and power of the Pollaczek method were established later by P. Le Gall, J. W. Cohen and J.H.A. de Smit who extended the method to cover joint distributions of various quantities of interest. This in turn lead to a method based on boundary theory, developed by J.W. Cohen and J.O. Boxma in their recent book [1983].

It should be added that results obtained by the Pollaczek method were confirmed in special cases by results obtained by other simpler methods, specially designed for these cases. Nevertheless, the Pollaczek method still remains the most general theory for $GI/G/$s.

It is impossible in a short paper like this one to give all details of the highly involved derivations in what is now known as the Pollaczek method. Instead, it is more advantageous to characterize the method by providing the outline of its basic ideas, indicating major steps in the argument. This will also help to appreciate significance of extensions made by other authors. An interested reader should consult references for details.

In the paper [44] written in 1965, the present author characterized the Pollaczek method as being *simple but complicated.* It is simple, because basic ideas and procedures are simple; it is complicated, because the technical manipulations (although of standard type) are not

easy. This characterization was confirmed and expanded by P.Le Gall in his comments during discussion of the above paper.
In order to describe in a convincing way the character of the theory, it may be appropriate to quote from a letter of Mr. Pollaczek to the present author, which reads:

> *"Some of the guiding ideas of the theory outlined in my paper may be approximately stated as follows. Under the rule of strict order servicing of an s server system with given initial conditions, all quantities we are interested in, like waiting times, busy periods,etc, are definite functions of a great number of stochastic variables of given distributions, and of some parameters signifying given initial values. Therefore all unknown expressions for probabilities and distribution functions are in fact Lebesgue-Stieltjes integrals, the integrands of which are formed by means of the aforesaid definite functions.*
>
> *In my theory the task of carrying out these integrations is reduced to the problem of resolving one or several systems of s linear non-homogeneous integro-functional equations of a new kind. The circumstance that this theory employs uniquely analytic methods and dispenses with all resources of classical Probability Calculus probably accounts for the fact that hitherto my methods have been employed by nobody save myself."*

As Pollaczek stressed on many occasions, his method belongs to the domain of Complex Variable Theory, and not to Probability Theory. Indeed, his work is essentially analytic in character, with probabilistic aspects remaining in the background. This distinction between analytic and probabilistic methods in theory of queues was mildly criticized by Dr. J. Kingman in discussion of another Pollaczek paper at the meeting in Chapel Hill (in 1964), but Mr. Pollaczek defended it rather forcefully; see [36].
The present paper is based primarily on Pollaczek's works listed in References, especially on his paper presented in Chapel Hill [36] and on his seminar lectures given at the University of Maryland in 1963. An excellent exposition by P.Le Gall [17] and a remarkable thesis of J.H.A. de Smit [8] were very helpful. The author's earlier papers on Pollaczek's work were utilized, but the present paper differs significantly in emphasis.
Attention is here concentrated on the case $s > 1$, and its importance is stressed. Case of $s = 1$ is only briefly mentioned because it was completely treated by J.W. Cohen [3]; see also [44].
Section 2 treats preliminaries, and Section 3 describes basic ideas of the method. Section 4 discusses extensions obtained by other authors indicating the influence of Pollaczek's method on their work. Section 5 discusses related theories.

2. PRELIMINARIES

This Section introduces basic assumptions as well as notation to be used in this paper. Authors who continued Pollaczek work usually employed notation different from his, and even modified some of his assumptions, making comparison difficult for the readers. In order to minimize confusion, this paper uses a new notation. However, presentation of the material follows that of Pollaczek as closely as convenient, but arrangement of lemmas and theorems is that of the present writer.
2.1. The system under consideration is a multiserver queue with general input and general service time, denoted by $GI/G/s$. It consists of s servers and of an infinite waiting room.

The waiting is only possible when all servers are busy, and waiting customers are served in order of their arrival ("strict order service") by available servers ("full availability").
It is supposed that customers arriving to the system are numbered 0, 1, 2, ... (the first customer carrying number 0). Let K_n be the instant of arrival of the n-th customer ($n = 0, 1, 2, \ldots$) with $K_0 = 0$; then, $T_n = K_{n+1} - K_n$ represents interarrival time between the arrival of the n-th and the $(n+1)$-th customers. Let S_n be the service time of the n-th customer ($n \geq 0$). Put also for each $n : Z_n = S_n - T_n$.
Stochastic properties of the system are described by the following assumptions. The processes (T_n) and (S_n) are independent, each consisting of independent identically distributed random variables. Their distributions are assumed to be absolutely continuous and are denoted by:

$$\mathbf{P}(T_n \leq t) = A(t), \qquad \mathbf{P}(S_n \leq t) = B(t), \quad t \geq 0$$

with $A(t) = B(t) = 0$ for $t \leq 0$. The corresponding characteristic functions are defined by:

$$A^*(z) = \mathbf{E}e^{zT_n} = \int_0^\infty e^{zt}\, dA(t), \qquad B^*(z) = \mathbf{E}e^{zS_n} = \int_0^\infty e^{zt}\, dB(t)$$

where z is complex and integrals are assumed to converge for $\Re z$ in an interval around zero. The mean values $\mathbf{E}T_n$ and $\mathbf{E}S_n$ are assumed to be finite (this may replace less restrictive assumption on analyticity of characteristic functions).
Denoting by $G(t), -\infty < t < +\infty$, the distribution of Z_n, its characteristic function is clearly $B^*(z)A^*(-z)$.
Here attention will be restricted to discussion of the waiting time, although Pollaczek considered other variables of interest; see sub-section 3.3. Thus, denote by W_n the waiting time of the n-th customer, $n \geq 0$. The main achievement of Pollaczek's method is the determination of the generating function of characteristic functions of W_n, conditional on initial states. This result will be presented in Section 3 after further preliminaries.
2.2. The first step of analysis describes more precisely the operation of the system. For this purpose, Pollaczek introduced "Markovian parameters" $L_{nv}, (v = 1, \ldots, s)$, such that

$$K_n + L_{n1}, \ldots, K_n + L_{ns},$$

for $n = 0, 1, 2, \ldots$, where $L_{nv}, v = 1, \ldots, s$ represent the amount of work, brought in by customers who arrived before K_n, which the v-th server has to do at time K_n. Note that L_{nv} may be negative meaning that at time K_n the v-th server has been idle during the period of length $-L_{nv}$. In other words, $K_n + L_{nv}$ are instants at which the v-th server finishes serving the last of those among the first $n-1$ customers which it serves.
Actualy, Pollaczek defined $K_n + L_{nv}$ as the last s instants of terminations of service times initiated before the arrival of the n-th customer at K_n(and consequently having indices less than n), taken in arbitrary order; see [33], [36]. P. Le Gall stated in his book [17] that this definition is not sufficient, but should refer to each server individually. In the discussion at the Lisbon conference, Mr. Pollaczek explained that for the strict order service both definitions coincide when waiting time is of interest; see [44]. Later at the Chapel Hill conference, in his reply to discussion, Pollaczek mentioned that the use of L_{nv} is by no means an essential feature of his theory, but is preferable to the use of other variables. The definition adopted here is that used by de Smit in his thesis [8]. In their independent work, Kiefer and Wolfowitz used similar definition; see [15].

It may be useful to comment further on this issue. Let $X_t \equiv X(t)$ denote the number of customers in the system at time t. Then, $X(K_n-)$ denotes the number of customers met by the n-th customer at the instant of his arrival. If this number is less than s, the n-th customer does not wait. If it is larger than s, the subsequent arrivals do not influence waiting time of the n-th customer. (One has here a modified process starting at random time K_n). Eventually, the n-th customer finds himself at the head of the queue with all s servers engaged by s customers who arrived before him. Hence, L_{nv} represent time intervals between the arrival of the n-th customer and the termination of these s services. Therefore, the waiting time W_n of the n-th customer is equal to the smallest of $L_{nv}, (v = 1, 2, \ldots, s)$, if it is positive, and zero in the opposite case. Thus,

$$W_n = \text{ Max } [0, \min (L_{n1}, \ldots, L_{ns})]$$

or, in the Pollaczek notation, $W_n = \min^+(L_{n1}, \ldots, L_{ns})$.
The following convenient notation will be needed later. Arrange L_{nv} in an increasing order:

$$L_{ni_1} \leq L_{ni_2} \leq \ldots \leq L_{ni_s}$$

where $(i_1, \ldots, i_s)$ is the corresponding permutation of $(1, \ldots, s)$, and write for the v-th element of the above sequence:

$$L^+_{ni_v} = \max (0, L_{ni_v}),$$

for $v = 1, \ldots, s$. Next, to simplify notation write

$$W_{nv} = L^+_{ni_v},$$

for $v = 1, \ldots, s$. Hence, one has for $v = 1$ as already noted: $W_{n1} = W_n = L^+_{ni_1}$. Thus, W_{nv} are symbolic waiting times elapsing between instant K_n and the instant when at least v servers become free from services started before time K_n .
The quantities L_{nv} describe completely the influence of customers arriving before the instant K_n on the waiting time of the n-th customer; hence the name "Markovian parameters" with respect to K_n. In particular, for $n = 0, K_0 = 0$ and $L_{01}, \ldots, L_{0s}$ represent the instants at which all s servers will be free from services originated before 0; if all $L_{0v} > 0$, then the first customer (numbered 0-th) will wait. Thus, $L_{01}, \ldots, L_{0s}$ describe the initial conditions, and Pollaczek takes them as given numbers, say $l_{01}, \ldots, l_{0s}$.
The following Lemma expresses this dependence precisely.
Lemma 2.1. *The waiting time W_n of the n-th customer is a definite (although very complicated) function of the $2n$ random variables S_0, T_0, $\ldots, S_{n-1}$, T_{n-1} and of the s initial conditions $l_{01}, \ldots, l_{0s}$; in symbols*

$$W_n = h_n(S_0, T_0, \ldots, S_{n-1}, T_{n-1}; l_{01}, \ldots, l_{0s}).$$

The function h_n is symmetric in parameters l_{0v}, $(v = 1, \ldots, s)$.
The proof follows easily from definitions; see [33], [44]. It is based on a recurrence relation expressing L_{nv} in terms of $L_{(n-1)v}$, and on successive elimination. As this recurrence relation will be needed later, it is worthwhile to state it separately. For this purpose note that the last s services with indices smaller than n will be terminated at instants :

$$K_{n-1} + W_{n-1} + S_{n-1} = K_{n-1} + L^+_{n-1,i_1} + S_{n-1} = L_{ni_1} + K_n,$$

for $v = 1$ and

$$K_{n-1} + L_{n-1,i_v} = L_{ni_v} + K_n,$$

for $v \neq 1$. Hence,

$$L_{ni_1} = L^+_{n-1,i_1} + S_{n-1} - T_{n-1} \tag{RR}$$

for $v = 1$.

$$L_{ni_v} = L_{n-1,i_v} - T_{n-1}$$

for $v \neq 1$. Consequently, quantities $L_{n1}, \ldots, L_{ns}$ are equal, apart from their order, to quantities on the right side of the above recurrence relation. Frequently, one can assume for simplicity that $L_{n1} = L_{ni_1}$.

2.3. The main feature of the Pollaczek method is the use of analytic representations of distributions, especially the distribution of

$$Y^+ = \max(O, Y)$$

where Y is a given random variable with distribution function $U(y)$ and characteristic function $\varphi(z)$. Then, the characteristic function of Y^+ is expressed by a contour integral of the form

$$\mathbf{E}e^{zY^+} = \frac{1}{2\pi i}\int_{-i\infty+\delta}^{i\infty+\delta} \mathbf{E}(e^{-\xi Y})\frac{z}{\xi(z+\xi)}d\xi$$

with $0 < \Re(\xi) < \Re(-z)$ and where the contour of integration is a line parallel to the imaginary axis (traversed upwards) at $\Re z = \delta > 0$, and within the analyticity region of $\mathbf{E}(e^{-\xi Y})$.

In the same fashion, write $a_{(v)} = \min{}_{(v)}(a_1, \ldots, a_s)$ for the v-th element of a sequence $a_1, \ldots, a_s$ of numbers arranged in an increasing order. Thus, $a_{(1)} = \min(a_1, \ldots, a_s)$ and $a_{(s)} = \max(a_1, \ldots, a_s)$. Hence, as before, let

$$a^+_{(v)} = \max\left(O, a_{(v)}\right),$$

for $v = 1, \ldots, s$, and write a^+ when $v = 1$.

Lemma 2.2. *Let* $a^+ = \min^+(a_1, \ldots, a_s)$. *Then,*

$$1 - e^{-qa^+} = \frac{1}{(2\pi i)^s}\int_{C_1}\cdots\int_{C_s} e^{a_1\xi_1+\ldots+a_s\xi_s}\frac{q}{q+\xi_1+\ldots+\xi_s}\frac{d\xi_1}{\xi_1}\cdots\frac{d\xi_s}{\xi_s}$$

where contours $C_1, \ldots, C_s$ *are as above, and* $\Re q > 0, \Re(q + \xi_1 + \ldots + \xi_s) > 0$.

For proof see [33], [35] and also [44]; Pollaczek proved that multiple integrals in Lemma 2.2 are absolutely convergent.

In order to simplify notation concerning multiple integrals in products of complex planes, it is convenient to use operators J_v and N_v defined by Pollaczek [33] and by de Smit [8] :

$$J_v f(\xi) = \frac{1}{2\pi i}\int_{C_v} f(\xi)\,\frac{d\xi_v}{\xi_v} = \frac{1}{2\pi i}\int_{-i\infty+\delta}^{i\infty+\delta} f(\xi)\,\frac{d\xi_v}{\xi_v},$$

$$N_v f(\xi) = f(\xi_1, \ldots, \xi_{v-1}, \xi_{v+1}, \ldots, \xi_s).$$

Here $f(\xi)$ denotes a function of a vector $\xi = (\xi_1, \ldots, \xi_s)$. The path of integration C_v is parallel to, and situated at right of, the imaginary axis at distance $R\xi_v = \delta > 0$, and is transversed from below to above. Note that operators J_v and N_v are associative and

$$J\,f(\xi) \equiv \prod_{v=1}^{s} J_v f(\xi) = J_1 J_2 \ldots J_s\, f(\xi)\,,$$

$$N\,f(\xi) \equiv \prod_{v=1}^{s} N_v f(\xi) = N_1 N_2 \ldots N_s\, f(\xi)\,.$$

Denote by P a collection of j distinct elements from indices $1,\ldots,s$, say $i_1,\ldots,i_j$, and let $P(j)$ be the set of all combinations of size j out of s indices $1,\ldots,s$. Denote by P^c a collection of all indices which do not belong to P. Define:

$$PJ = \prod_{v=1}^{j} J_{i_v} \qquad\qquad P^cJ = \prod_{v=j+1}^{s} J_{i_v}$$

and the analogous expressions for PN and P^cN . Furthermore, the symbol $\sum_{P\in P(j)}$ indicates summation over all $\binom{s}{j}$ elements P of $P(j)$. Write also for the inner product of two vectors ξ and a :

$$\xi a = \xi_1 a_1 + \ldots + \xi_s a_s,$$

and for the sum:

$$S\xi = \xi_1 + \ldots + \xi_s.$$

Finally, in most cases the function f will be symmetric in its variables $\xi_1,\ldots,\xi_s$. If f is not symmetric in some variables, these variables will be separated from symmetric ones by a semi-colon, for example $f(\xi_1,\ldots,\xi_s;\rho)$.
Using the notation introduced here, assertion of Lemma 2.2 takes much simpler form:

$$\bar{e}^{qa^+} = (N - J)\, e^{\xi a}\, q\, (q + S\xi)^{-1}.$$

In Section 3 use will be made of operators:

$$\sum_{P\in P(j)} PJ(P^cN - P^cJ)\, f(P\xi; S\xi)$$

for $j = 0,1,\ldots,s-1$. It is instructive to consider the simplest example, namely $s = 2$. It is clear that one has for $j = 0$:

$$f(0) - Jf(\xi_1 + \xi_2)$$

and for $j = 1$:

$$J_1\, f(\xi_1;\, \xi_1) + J_2\, f(\xi_2;\, \xi_2) - J\, f(\xi_1;\, \xi_1 + \xi_2) - J\, f(\xi_2;\, \xi_1 + \xi_2)$$

where $J = J_1 J_2$.

3. THE MAIN THEOREM

This Section summarizes Pollaczek's basic argument for the determination of the generating function of characteristic functions for the waiting time in the $GI/G/s$ system (with $s > 1$). Presentation uses concepts introduced in Section 2. For complete proof see the original papers [33], [35], [36].

3.1. It is required to find the following generating function :

$$\Phi(q,r) = \sum_{n=0}^{\infty} \Phi_n(-q) r^n \tag{1}$$

where q and r are complex numbers with $\Re q > 0$ and $| r | < 1$, and where $\Phi_n(-q)$ denotes the characteristic function

$$\mathbf{E}\Big(e^{-qW_n}\Big| l_{o1}, \ldots, l_{0s}\Big) \equiv \mathbf{E}\Big(e^{-qW_n}\Big| l_{0v}\Big), \tag{2}$$

conditional on the initial conditions $l_{01}, \ldots, l_{0s}$, regarded as given. Consider vectors $l_0 = (l_{01}, \ldots, l_{0s})$ and $\xi = (\xi_1, \ldots, \xi_s)$, and recall operator notation from sub-section 2.3. The following is the main result.

Theorem 3.1. *The generating function $\Phi(q,r)$, defined above, is given by*

$$\Phi(q,r) = \sum_{j=0}^{s-1} \sum_{P \in P(j)} PJ(P^c N - P^c J)\, e^{\xi l_0}\, V_j(P\xi;\, S\xi), \tag{3}$$

where the functions V_j are given by the system of s linear simultaneous integral equations (for $j = 0, 1, \ldots, s-1$) :

$$\begin{aligned}
&[1 - rA^*(-y)]\, V_j(\xi_1, \ldots, \xi_j; y) - \delta_{j0}\, q(q+y)^{-1} = \\
&\quad = r\,(2\pi i)^{-1} \int_{-i\infty}^{i\infty} [B^*(\alpha) - 1]\, \alpha^{-1}\, [y - (\xi_1 + \ldots + \xi_j) - \alpha]^{-1} M\, d\alpha,
\end{aligned} \tag{4}$$

where

$$\begin{aligned}
M = {} & [y - (\xi_1 + \ldots + \xi_j)]\, A^*(-y)\, V_{j+1}(\xi_1, \ldots, \xi_j, \alpha;\, y) - \\
& - \alpha A^*(-\alpha - \xi_1 - \ldots - \xi_j)\, V_{j+1}(\xi_1, \ldots, \xi_j, \alpha;\, \alpha + \xi_1 + \ldots + \xi_j),
\end{aligned}$$

and the equation (which determines V_s):

$$\sum_{j=0}^{s} \sum V_j(\xi_{i_1}, \ldots, \xi_{i_j};\, y) = 0 \tag{5}$$

where the inner summation has to be extended over all $\binom{s}{j}$ combinations of indices.

Moreover, the functions $V_j(\xi_1, \ldots, \xi_j; y)$ are analytic for $\Re\xi_v > 0, \Re y > 0$, and symmetric in ξ_v. Note that in (3) one has $y = \xi_1 + \ldots + \xi_j$. As already mentioned, it is impossible to reproduce here the whole proof, and the reader should consult references. Only a short outline is given below.

3.2. For simplicity assume that random variables L_{nv}, $(v = 1, \ldots, s)$ are arranged in increasing order, so $W_n = L_{n1}^+$.

i) Replacing in Lemma 2.2 vector $a = a_v$, $(v = 1, \ldots, s)$ by vector $L_n = L_{nv}$, $(v = 1, \ldots, s)$, one obtains a random variable:

$$e^{-qW_n} = 1 - Je^{\xi L_n}\, q\,(q + S\xi)^{-1} \tag{6}$$

which is a symmetric function of L_{nv}.

On the other hand, it follows from Lemma 2.1 and definition of expectation of a function of independent random variables that

$$\mathbf{E}\Big(e^{-qW_n} \mid l_{0v}\Big) = \int_0^\infty dA(t_0) \int_0^\infty dB(s_0) \dots \int_0^\infty dA(t_{n-1}) \int_0^\infty dB(s_{n-1})\, e^{-qw_n} \qquad (7)$$

where $w_n = h(t_0, s_0, \dots, t_{n-1}, s_{n-1}; l_{01}, \dots, l_{0s})$ is the value of the random variable W_n from Lemma 2.1.

ii) Evaluation of these integrals is now performed in stages. First replace L_{nv} in (6) by $L_{n-1,v}$ according to recurrence relation (RR) from Section 2. The inner product becomes:

$$L_n \xi = L^+_{n-1,1}\, \xi_1 + S_{n-1}\, \xi_1 + \sum_{v=2}^{s} L_{n-1,v}\, \xi_v - T_{n-1} \sum_{v=1}^{s} \xi_v.$$

As $L_{n-1,v}$ depend only on random variables $T_0, S_0, \dots, T_{n-2}, S_{n-2}$, but not on T_{n-1} and S_{n-1}, one obtains the conditional expectation

$$\mathbf{E}\Big(e^{-qW_n} \mid L_{n-1,v}\Big) = \int_0^\infty dA(t_{n-1}) \int_0^\infty dB(s_{n-1})\, e^{-qW_n} \qquad (8)$$

which is a symmetric function of $L_{n-1,v}$. Integration in (8) is taken over values of random variables T_{n-1} and S_{n-1}, leading to terms involving corresponding characteristic functions A^* and B^*. Thus, (8) is represented, when random variables are replaced by their values, as a sum of Fourier integrals in the real variables $l_{n-1,v}$.

Repeating the process successively for $k = 1, 2, \dots, n$, one finds that

$$\mathbf{E}\Big(e^{-qW_n} \mid l_{n-k,v}\Big) = \int_0^\infty dA(t_{n-k}) \int_0^\infty dB(s_{n-k})\, \mathbf{E}\Big(e^{-qW_n} \mid l_{n-k+1,v}\Big), \qquad (9)$$

which is a symmetric function of $l_{n-k+1,v}$. Note that (9) reduces to (8) for $k = 1$ because

$$\mathbf{E}\Big(e^{-qW_n} \mid l_{n,v}\Big) = e^{-qw_n},$$

but for $k = n$ it yields the required expression (7).

iii) The crux of the matter is the evaluation of integrals (9). The procedure is ingenious but highly involved; see pages 13–19 in [33] and pages 7–13 in [36]. Here only short outline of these important derivations can be made.

For a given k, (9) will be represented as the sum of $j+1$ terms ($j = 0, 1, \dots, s$) each containing $\binom{s}{j}$ of $(s-j)$–fold Fourier integrals with respect to variables $l_{n-k,v}$, and also a term independent of these variables. The individual Fourier integrals are of the form appearing in (6), with random variables replaced by their values, but involve also auxiliary functions $v_{kj}(\xi_1, \dots, \xi_j;\, y), j = 0, 1, \dots, s$. These functions are symmetric in $\xi_1, \dots, \xi_j$ and satisfy recurrence relations and initial conditions; see formulas (1.39) – (1.42) in [33]. In particular, for $k = n$, the expression for $\mathbf{E}\Big(e^{-qW_n} \mid l_{o,v}\Big)$ has the same form as (3), but with V_j replaced by v_{kj}.

iv) Finally, introducing the generating functions

$$V_j\Big(\xi_1, \dots, \xi_j;\, y\Big) = \sum_{k=0}^{\infty} r^k\, v_{kj}(\xi_1, \dots, \xi_j;\, y),$$

one converts the above mentioned recurrence relations for v_{kj} into relations (4) – (5) for V_j. Similarly, one obtains the formula (3) for the generating function $\Phi(q, r)$. This concludes the outline of proof of Theorem 3.1. Note that in the case when the 0–th customer finds all servers free, then formula (3) reduces to

$$\Phi(q,r) = V_0(0) = \sum_{r=0}^{\infty} r^n \, \mathbf{E}\left(e^{-qW_n} \mid 0\right). \tag{10}$$

3.3. Some further results obtained by Pollaczek can be briefly mentioned here; see [33], [36] for details. In particular, he showed that the system of equations (4) – (5) can be solved by a finite number of steps in the following two cases:

a) $B^*(z)$ is a rational function of z, $A^*(z)$ arbitrary,

b) $B^*(z)$ is a polynomial in e^{bz} (b constant), $A^*(z)$ a rational function of z.

Moreover, Pollaczek considered several special cases, in particular $M/G/s$, including $M/D/s$. For the system $GI/M/s$, he obtained the explicit form of the limiting distribution function

$$W(t) = 1 - A\,e^{-\alpha t},$$

$t \geq 0$, extending Smith's theorem (also found later by D.G.Kendall who used a regeneration points technique).

As shown by Kiefer and Wolfowitz [15], W_n converges in distribution to the random variable W, provided that

$$\rho = \mathbf{E}S/(s\mathbf{E}T) < 1, \tag{11}$$

Hence, the characteristic function $\mathbf{E}\left(e^{-qW}\right)$ of the limiting distribution (independent of initial conditions) can be obtained from (10) by an Abelian theorem and evaluation of residue at $r = 1$. Pollaczek showed also that his technique can be extended to the study of joint distributions like $\mathbf{P}(W_n \leq t,\ W_{n+m} \leq t')$, leading to equations similar to (3).

Of great theoretical importance is his study of the distribution of the successive minima of random variables L_{nv}, denoted by $\mathbf{W}_{nv}$ in sub–section 2.2. Pollaczek determined the generating functions for individual $\mathbf{W}_{nv}$, and showed that the resulting equations have similiar forms to equations (3), (4), (5); see pages 38-45 in [33] and also de Smit [8].

Pollaczek studied queue length and the number of busy servers, but did not consider the virtual waiting time and the busy period. However, he discussed "temporal blocking" and extended his theory to multi–queue system where a customer joins that queue which gives him the least waiting time.

In separate papers, not discussed here, Pollaczek considered systems with no possibility of waiting ("loss systems"); see [32].

4. EXTENSIONS

As already mentioned, Pollaczek's method did not achieve popularity because of its analytical difficulties. Nevertheless, due to efforts of a handful of dedicated workers, its advantages and power became recognized. This section lists major works where the method was applied and extended.

4.1. The first on the list is P. Le Gall who in his book [17], published in 1962, gave the complete account of the theory. In addition, he worked out in detail the case $s = 2$, making thus the theory more accessible. LeGall's own work concerned extension of Pollaczek's method to random servicing. In the book, he treated the case of a single server with regenerative input and the ordinary as well as group service. For $s > 1$, he considered random selection both for exponential and constant service times.

In subsequent papers, Le Gall used Pollaczek's method consistently to solve teletraffic problems; as an illustration see one of his recent contributions [18], where he generalized an earlier study of tandem queues by O.J. Boxma (also a disciple of Pollaczek).

4.2. Pollaczek's method is the basic tool in the treatise by J.W. Cohen on the single server queue [3], whose first edition appeared in 1969. Cohen gave a complete and rigorous description of Pollaczek theory for the $GI/G/1$ system. Moreover, he extended it to cover other quantities like busy period and busy cycle, virtual waiting time, idle period, queue length, and also discussed group arrivals, group service and random service.

In his analysis, Cohen was interested in generating functions, similar to Pollaczek's $\Phi(q,r)$ discussed in Section 3. The associated functional equation involved two unknown functions, and the starting point of determination of a solution was the search for "zero pairs" of a specified kernel. This approach, closely related to the Wiener-Hopf technique, may be interpreted in terms of potential theory, namely as a Riemann–Hilbert problem (which includes the Dirichlet problem as a special case). This observation lead in natural way to treatment of $GI/G/1$ queues as boundary value problems.

Thus, a new fertile field of investigations has been opened, as Cohen and Boxma have shown in their monograph [5]. The method is indeed powerful and recently has been applied by Cohen and his collaborators to study of exit problems in multidimensional random walks.

4.3. The most comprehensive extension of Pollaczek method was made in the thesis by J.H.A. de Smit [8], a student of Cohen. Starting with description of main features of Pollaczek argument, de Smit considered the joint generating function $\Phi_v(q,r)$ of the waiting time W_n and the successive minimum W_{nv} (from Section 2) for $v \neq 1$, and some other random variables.

He first established the integral identity for $s > 1$, which generalized Lemma 2.2 in its operator form stated in sub-section 2.3. Next, he investigated properties of generalized functions v_{kj}, and gave the complete proof of the main theorem, preceded by detailed discussion of the case $s = 2$. His final formula for $\Phi_v(q,r)$ has the same operator form as equation (3) in Section 3.

On this basis, de Smit determined joint distributions of several quantities characterizing the behavior of the system. These included the actual and the virtual waiting times, busy period, busy cycle, total number of customers at instant of arrival and at an arbitrary instant. He mentioned briefly a stationary joint distribution of W_{nv} as $n \to \infty$, using the Kiefer–Wolfowitz result, and included a short description of a variant of the system $GI/G/s$. In the last two chapters of thesis, the results were applied to detailed investigation of systems $GI/M/s$ and $GI/E_m/s$.

4.4. Although many authors referred to Pollaczek work in their papers, only a few attempted to study the method in depth. Notable exceptions, in addition to those quoted above, can be found in the writing of J.H.B. Kemperman and L. Takacs.

In his book, published in 1961, Kemperman [14] recognised that the Pollaczek's solution of $GI/G/1$ system is in fact a contribution to solution of the one-sided absorption problem as well as to the problem of finding the maximum of successive partial sums.

Similarly, Pollaczek's solution for the $GI/G/1$ queue was extended by Takacs [52] in the following manner. Consider the function of the form

$$f(q) = \mathbf{E}\left(Ve^{-qY}\right)$$

where V and Y are complex and real random variables, respectively. Define a transformation $\mathbf{T}$ by

$$\mathbf{T}\{f(q)\} = f^+(q) = \mathbf{E}\left(Ve^{-qY^+}\right).$$

Let $g(q)$ and $\Phi_0(q)$ be two functions with $\mathbf{T}\{\Phi_0(q)\} = \Phi_0(q)$, and define a sequence of functions $\Phi_n(q)$ for $n = 1, 2, \ldots$, by the recurrence relation

$$\Phi_n(q) = \mathbf{T}\{g(q)\,\Phi_{n-1}(q)\}.$$

It is required to find the generating function

$$\Phi(q,r) = \sum_{n=0}^{\infty} \phi_n(q)\, r^n.$$

Takacs showed that $\Phi(q,r)$ satisfies the functional equation

$$\mathbf{T}\{[1 - rg(q)]\,\Phi(q,r)\} = \Phi_0(q),$$

and obtained its explicit solution. This equation and its solution generalize results of Pollaczek for the $GI/G/1$ queue, as discussed in [30]. Moreover, the significance of the kernel $1 - rg(q)$ is clearly visible here. Observe also the symbolic analogy between the case of $s = 1$ and the case of $s > 1$ described in Section 3. For further comments see the next Section 5.

5. RELATED METHODS

This Section provides commentary on the connection between independent works by other researchers and results obtained by Pollaczek. It should be stressed that these authors did not use the Pollaczek's method, making only marginal reference to its existence, yet their results agreed naturally with those of Pollaczek. However, all these results, obtained under different assumptions, are closely interrelated and their comparison is often very delicate. Morover, because of his use of "disheartening integrals", many of Pollaczek's contributions were often not recognized and were rediscovered later. On some occasions, this produced priority arguments.

5.1. The case of a single server $GI/G/1$ was solved by D.V. Lindley [19] in 1952, followed by W.L. Smith [39] in 1953. Pollaczek's theory was ready around 1952 and its complete formulation appeared in the book form [30] in 1957. The present author summarized in [44] Pollaczek's work on the single server queue, and discussed its interpretation from the point of view of modern developments. It suffices therefore to comment briefly on some points of interest.

i) Pollaczek obtained the integral equation for the generating function $\phi(q,r)$ of the characteristic functions of W_n, following analogous argument described here in Sections 2 and 3. He solved the equation explicitly by the use of calculus of residues; see subsection 4.3. He also succeeded in showing existence of the limiting distribution $W(t)$ of the waiting time W_∞ (a limit in distribution of a sequence W_n), and obtained the explicit expressions for the mean and for the probability of no delay under equilibrium conditions.

Recurrence relation (RR) from Section 2 indicates that for the single server, waiting times W_n form a Markov chain with transition probabilities

$$\mathbf{P}(W_{n+1} \le t \mid W_n = s) = G(t - s), \qquad t \ge s$$

where G is the distribution function of the difference $Z_n = S_n - T_n$. The limiting distribution satisfies the Lindley integral equation:

$$W(t) = \int_{-\infty}^{t} W(t-s)\, dG(s), \qquad t \geq 0.$$

This is the equation of the Wiener–Hopf type, and its solution was obtained by W. Smith.
ii) Connections of Wiener–Hopf equations with fluctuation theory of partial sums of independent random variables were discussed by F. Spitzer in his book [41] on random walks. The formula for the characteristic function of maximum of partial sums, known in literature as Spitzer's formula, was earlier deduced by Pollaczek [27]. Extension of both derivations was carried out by L. Takacs [50] later in 1970. It seems proper that this important result should be referred to as Pollaczek–Spitzer formula; see the author's contribution to discussion of Pollaczek paper at the Chapel Hill conference [43].
Similarly, the special case $M/G/1$ was solved directly by Pollaczek in 1930; see also Pollaczek's reply to the comment made by L. Takacs at the Chapel Hill conference [36]. The celebrated formula for the mean waiting time, which was since rediscovered many times, is now known in literature as the Pollaczek–Khintchine formula.
Factorization involved in Wiener–Hopf equations has become an important subject of research. Its use in boundary value problems by Cohen and Boxma has been already discussed. Studies by N. U. Prabhu [38] extend to stochastic processes occurring in various applications.
5.2. Although in principle, extension of interpretation from $s = 1$ to $s > 1$ should be relatively easy, it is manifestly clear that such a task is loaded with difficulties. Increase in dimensionality always leads to greater complexity, but in this case converts the problem into a different and more involved one.
Indeed, one can consider multiple Wiener–Hopf equations, multiple random walks, multiple integrals, but the analysis becomes too complicated and it is hard to get concrete solutions.
First of all the process (W_n) is no longer Markovian, and one must consider the vector Markov process $(L_{n1}, \ldots, L_{ns})$ in order to use Markovian machinery. Indeed, Pollaczek results are analogous to the multidimensional queueing equation of Wiener–Hopf type, developed by Kiefer and Wolfowitz.
These authors considered the process $(W_{n1}, \ldots, W_{ns})$, as in Section 2, with the joint distribution function

$$W_n(t_1, \ldots, t_s) = \mathbf{P}(W_{n1} \leq t_1, \ldots, W_{ns} \leq t_s),$$

and showed that, with $t = (t_1, \ldots, t_s)$, $\lim W_n(t) = W(t)$ exists, as $n \to \infty$, and is the proper distribution function when $\rho < 1$; see (11) in Section 3.
The limiting distribution $W(t)$ is then given by the solution of the integral equation

$$W(t) = \int \sigma\{W(t), x, y\}\, dA(x)\, dB(y),$$

where σ denotes the known function of W, and the integral is extended over the region in s–dimensional space. In general, the integrand contains 2^{s-1} terms, and the solution of the above integral equation is almost an impossible task. For $s = 1$, the equation reduces to Lindley's equation quoted above.The method used by the authors consists of device of "dominating " the given stochastic process by an approximating "lattice process" to which the available theorems on the denumerable Markov chains can be applied.

5.3. It should be pointed out that Pollaczek mathematical interests extended beyond his contributions to queueing problems and probability. Indeed, he contributed also to mathematical analysis and to mathematical physics. His work on successive maxima of partial sums, already mentioned, actually treats a problem in analysis. It was later extended by other writers, especially by Takacs, as already noted.

Another important contribution is his work on orthogonal polynomials and special functions. It seems proper to close this paper with brief comment on a topic which is known in today's literature as Pollaczek polynomials and Pollaczek spaces.

In the late 40's and early 50's, Pollaczek introduced several forms of orthogonal polynomials which are generalizations of classical orthogonal polynomials, but possess certain important properties; see [23] – –[26] and also a summary by A. Erdelyi [11]. Pollaczek polynomials were later investigated by many writers, especially by L. de Branges who introduced Pollaczek spaces of entire functions; see [6], [7]. In recent years orthogonal polynomials find applications which link Markov processes (with independent increments) with Hamiltonian dynamical systems; see [12].

Here, a brief account of Pollaczek polynomials defined on the real line is given for illustrative purposes, [26]. The polynomials $P_n(x)$ involving parameters λ and ϕ, where $\lambda > 0$ and $0 < \phi < \pi$, are defined recursively for $-\infty < x < \infty$ by:

$$P_{-1}(x) = 0, \qquad P_0(x) = 1,$$

$$nP_n(x) - 2\Big[(n-1+\lambda)\cos\phi + x\sin\phi\Big]P_{n-1}(x) + (n-2+2\lambda)P_{n-2}(x) = 0$$

for integers $n \geq 1$. The generating function is now:

$$\sum_{n=0}^{\infty} P_n(x)\, z^n = \left(1 - ze^{\phi}\right)^{-\lambda+ix}\left(1 - ze^{-\phi}\right)^{-\lambda-ix}, \qquad |\, x \,| < 1,$$

and the weight function is :

$$w(x) = \pi^{-1}\left(2\sin\phi\right)^{2\lambda-1} e^{-(\pi-2\phi)x}\left|\Gamma(\lambda+ix)\right|^2.$$

Orthogonality of the system of polynomials $P_n(x)$ is expressed by their scalar product relative to $w(x)$:

$$\int_{-\infty}^{\infty} P_n(x)\, P_m(x)\, w(x)\, dx = a_n\, \delta_{nm}$$

where a_n is the square of the L_2–norm of P_n, relative to the weight function w. These polynomials can also be expressed in terms of hypergeometric series:

$$n!P_n(x) = \left[\Gamma(2\lambda+n)\,/\,\Gamma(2\lambda)\right]e^{in\phi}\,{}_2F_1\left(-n,\, \lambda+ix;\, 1-e^{-2i\phi}\right).$$

Pollaczek polynomials of this form were used by de Branges in his study of Pollaczek spaces [6].

PERSONAL NOTE

This author was privileged to know Mr. and Mrs. Pollaczek personally, since the Teletraffic Congress in the Hague in 1958. They visited Maryland in 1963 where Mr. Pollaczek conducted a seminar on his method [34]. This personal contact, extended through years, was highly beneficial in understanding the man and his work.

Mr. Pollaczek was awarded the John von Neumann Theory Prize (for 1977) by ORSA/TIMS, but as he was unable to travel to America the informal ceremony took place (in July 1977) among close friends in the hospitable home of Mr. and Mrs. Le Gall, in the scenic countryside of Normandy. Wim Cohen presented Mr. Pollaczek with the gold medal on behalf of the societies, and we all offered our congratulations to Mr. and Mrs. Pollaczek [48]. A few years later, it was Cohen's sad task to write the obiturary expressing deep feelings of loss [2].
It is with humility that this paper is dedicated to the memory of the Great Master F. Pollaczek.

REFERENCES

[1] BOXMA, O.J., SYSKI, R. (EDS.) (1988). *Queueing Theory and its Applications, Liber Amicorum for J.W. Cohen.* North-Holland, Amsterdam.

[2] COHEN, J.W. (1981). Obituary: Felix Pollaczek, *J. Appl. Prob.*, **18**, 959-963.[1]

[3] COHEN, J.W. (1982). *The Single Server Queue.* North-Holland, Amsterdam. (Second edition).

[4] COHEN, J.W. (1986). *Some Samples of Modelling.* In [13], 88-108.

[5] COHEN, J.W., BOXMA, O.J. (1983). *Boundary Value Problems in Queueing System Analysis.* North-Holland, Amsterdam.

[6] DE BRANGES, L. (1968). *Hilbert Spaces of Entire Functions.* Prentice Hall, N.J.

[7] DE BRANGES, L., TRUTT, D. (1968). Meixner and Pollaczek Spaces of Entire Functions. *J. Math. Anal. Appl.*, **22**, 12-24, 1968.

[8] DE SMIT, J.H.A. (1971). *Many Server Queueing Systems.* Ph.D. Thesis, Delft University of Technology, 1971.

[9] DE SMIT, J.H.A. (1973). A Simple Analytic Proof of the Pollaczek-Wendel Identity for Ordered Partial Sums. *Ann. Prob.*, **1**, 348-351.

[10] DE SMIT, J.H.A. (1975). On Multi-server Queues. In *Proc.* XX *International Meeting of the Institute of Management Sciences*, [E. Shlifer (ed.)]. Jerusalem Acad. Press, **2**, 555-558.

[11] ERDELYI, A. (1953). *Higher Transcendental Functions.* McGraw Hill, N.Y., **2**, 218-221.

[12] FEINSILVER, P.J. (1978). *Special Functions, Probability Semi-groups, and Hamiltonian Flows.* Lecture Notes in Math. # 696. Springer Verlag, N.Y.

[13] GANI, J. [(ED.).] (1986). *The Craft of Probabilistic Modelling, A Collection of Personal Accounts.* Springer Verlag, N.Y.

[14] KEMPERMAN, J.H.B. (1961). *The Passage Problem for a Stationary Markov Chain.* University of Chicago Press.

[15] KIEFER, J., WOLFOWITZ, J. (1955). On the Theory of Queues with Many Servers. *Trans. Amer. Math. Soc.*, **78**, 1-18.

[16] KIEFER, J., WOLFOWITZ, J. (1956). On the Characteristics of the General Queueing Process with Applications to Random Walk. *Ann. Math. Stat.*, **27**, 147-161.

[17] LE GALL, P. (1962). *Les Systèmes avec ou sans Attente et les Processus Stochastiques.* Dunod, Paris.

[18] LE GALL, P. (1988). *Some Comments on the Work of J.W. Cohen and New Results in the Theory of Queueing Networks.* In [1], 307-324.

[19] LINDLEY, D.V. (1952). The Theory of Queues with a Single Server. *Proc. Cambridge Philos. Soc.*, **48**, 277-289.

[1] A complete list of Pollaczek publications appears in this paper

[20] POLLACZEK, F. (1929–1930). Über eine Aufgabe der Wahrscheinlichkeitstheorie. *Math. Z.*, **32**, 64–100, 729–750.

[21] POLLACZEK, F. (1942). Sur Quelques Lois Asymptotiques de la Théorie de l'Encombrement des Réseaux Téléphoniques. *Ann. Univ. Lyon, (3). A*, **5**, 21–34.

[22] POLLACZEK, F. (1945). Résolution de Certaines Équations Integrales de Deuxième Espèce. *J. Math. Pures et Appl.*, **24**, 73–93.

[23] POLLACZEK, F. (1949). Sur une Généralisation des Polynômes de Legèndre. *C. R. Acad. Sci. Paris*, **228**, 1363–1365.

[24] POLLACZEK, F. (1949). Systèmes de Polynômes Biorthogonaux a Coefficients Reéls. *C. R. Acad. Sci. Paris*, **228**, 1553–1556.

[25] POLLACZEK, F. (1949). Systèmes de Polynômes Biorthogonaux qui Généralisent les Polynômes Ultrasphériques. *C. R. Acad. Sci. Paris*, **228**, 1998–2000.

[26] POLLACZEK, F. (1950). Sur une Famille de Polynômes Orthogonaux qui Contient les Polynômes d'Hermite et de Laguerre comme Cas Limites. *C. R. Acad. Sci. Paris*, **230**, 1563–1565.

[27] POLLACZEK, F. (1952). Fonctions Caractéristiques de Certaines Répartitions Définies au Moyen de la Notion d'Ordre. Applications a la Théorie des Attentes. *C. R. Acad. Sci. Paris*, **234**, 2334–2336.

[28] POLLACZEK, F. (1953). Sur une Généralisation de la Théorie des Attentes. *C. R. Acad. Sci. Paris*, **236**, 578–580.

[29] POLLACZEK, F. (1956). *Sur une Généralisation de Polynômes de Jacobi.* Mem. Sci. Math., **131**. Gauthier–Villars, Paris.

[30] POLLACZEK, F. (1957). *Problèmes Stochastiques Poses par le Phénomène de Formation d'une Queue d'Attente à un Guichet et par des Phénomènes Apparentes.* Mem. Sci. Math., **136**. Gauthier–Villars, Paris.

[31] POLLACZEK, F. (1958). *Sur le Calcul des Probabilités de Délai d'Attente à une Groupe des s Lignes Téléphoniques.* Preprint # 10, ITC, The Hague.

[32] POLLACZEK, F. (1959). Application de la Théorie de Probabilités à des Problèmes Poses par l'Encombrement des Réseaux Téléphoniques. *Ann. Telecom.*, **14**, 165–183.

[33] POLLACZEK, F. (1961). *Théorie Analytique des Problèmes Stochastiques Relatifs à un Groupe de Lignes Téléphoniques avec Dispositif d'Attente.* Mem. Sci. Math., **150**. Gauthier–Villars, Paris.

[34] POLLACZEK, F. (1963). *Analytic Method for the Many Server Problem in Queueing Theory.* Seminar, University of Maryland, (Not published).

[35] POLLACZEK, F. (1963). *Summary of "Théorie Analytique ...".* The Institute for Fluid Dynamics and Applied Mathematics, University of Maryland, Technical Note BN–323.

[36] POLLACZEK, F. (1965). *Concerning an Analytic Method for the Treatment of Queueing Problems.* In: [40], 1–25, (discussion, 25–42).

[37] POLLACZEK, F. (1975). Order Statistics of Partial Sums of Mutually Independent Random Variables. *J. Appl. Prob.*, **12**, 390–395.

[38] PRABHU, N.U. (1972). Wiener–Hopf Factorization for Convolution Semi–groups. *Z. Wahr. ver. Geb.*, **23**, 103–113.

[39] SMITH, W.L. (1953). On the distribution of Queueing times. *Proc. Cambridge Philos. Soc.*, **49**, 449–461.

[40] SMITH, W.L., WILKINSON, W.E., (EDS.) (1965). *Proceedings of a Symposium on Congestion Theory held at Chapel Hill, August 1964.* University of North Carolina Press.

[41] SPITZER, F. (1964). *Principles of Random Walk.* Van Nostrand, N.J..

[42] SYSKI, R. (1963/1965). *Queueing Theory, Pollaczek Method.* Seminar Lectures, University of Maryland. (Not published).

[43] SYSKI, R. (1965). Note on Pollaczek–Spitzer Identity. In: [40], 30–34.

[44] SYSKI, R. (1967). Pollaczek Method in Queueing Theory. In: *Queueing Theory*, [R.Cruon (ed.)]. The English Universities Press, London, 33–60.

[45] SYSKI, R. (1975). Queueing Theory Symposium —- Introduction and Summary. In *Proc. XX International Meeting of the Institute of Management Sciences*, [E. Shlifer (ed.)]. Jerusalem Acad. Press, **2**, 507–508.

[46] SYSKI, R. (1985). Multiserver Queues. In: *Encyclopedia of Statistical Sciences, [S. Kotz, N.L. Johnson (eds.)]*. John Wiley, N.Y., **5**, 727–732.

[47] SYSKI, R. (1986). *Introduction to Congestion Theory*, North–Holland, Amsterdam. (Second edition).

[48] SYSKI, R. (1986). *Markovian Models - An Essay*. In [13], 109–125.

[49] SYSKI, R. (1988). *Markov Processes and Teletraffic*. In: [1], 95–126.

[50] TAKACS, L. (1970). On the Distribution of the Maximum of Sums of Mutually Independent and Identically Distributed Random Variables. *Adv. Appl. Prob.*, **2**, 344–354.

[51] TAKACS,L. (1972). On a Formula of Pollaczek and Spitzer. *Studia Mathematica*, **XLI**, 27–34.

[52] TAKACS, L. (1973). On a Method of Pollaczek, *Stoch. Proc. and Applications*, **1**, 1–9.

[53] TAKACS, L. (1977). An identity for ordered partial sums. *J. Combinatorial Theory, A*, **23**, 364–365.

Recent Adv. in Stat. and Prob., pp. 415-428
J. Pérez Vilaplana and M. L. Puri (Eds)

A Study on a Stochastic System with Multiple MMPP Inputs Subject to Access Function

L. TRALHÃO[1]; J.M.F FERNÁNDES CRAVEIRINHA[2] and J. PAIXÃO[3]
[1] Departamento de Matemática, Universidade de Coimbra, INESC - Núcleo de Coimbra, P-3000 Coimbra, Portugal
[2] Departamento de Engenharia Electrotécnica, Universidade de Coimbra, INESC - Núcleo de Coimbra, P-3000 Coimbra, Portugal
[3] Departamento de Estatística, Investigação Operacional e Computadores, Faculdade de Ciências da Universidade de Lisboa, P-1300 Lisboa, Portugal

Abstract – Problems of performance analysis in telecommunication networks led in the past to the concept of restricted availability systems in which the connection paths may be such that an incoming call may be unsuccessful even when there are still idle circuits in the destination group. On the other hand a number of studies suggest that the MMPP ("Markov Modulated Poisson Process") could be used successfully for modelling certain types of superpositions of complex teletraffic flows. In particular the MMPP is the exact model for the superposition of independent IPPs ("Interrupted Poisson Processes"), proposed by Kuczura, as a rather accurate approximation to overflow traffics, resulting from the overflow of Poisson inputs in loss systems with exponential distribution of the service times (model of great interest in circuit-switched networks with alternative routing). In this paper the authors extend previous work by considering the exact analysis of a system with finite queueing capacity whose inputs are defined from a number of independent MMPPs each being subject to an "access function" which defines the conditional probability of an arrival being rejected for a given number of occupations in the system. Beyond overviewing the calculations of the stationary distribution, waiting times, the overall congestion and the marginal blocking probabilities, experienced by the different MMPP input streams, the analysis of the associated marginal overflow point processes, the accepted process and the calculation of the moments of the sojourn times, are approached. A basic result of this work is the stochastic equivalence between the queue with multiple MMPP inputs, each with a certain access function, and the queue with a single MMPP input and a "multi-dimensional", phase-dependent access function. At a methodological level the models confirm the great potentialities of the matrix analytic methods proposed by Neuts and put in evidence the necessity of seeking to take advantage of the specific properties of the mathematical entities used in the models (namely with respect to the associated infinitesimal generators and systems of linear equations) in order to enhance their algorithmic tractability.

1. INTRODUCTION AND MOTIVATION

Problems of performance analysis in telecommunication networks led in the past to the concept of restricted availability systems in which the connection paths may be such that an incoming call may be unsuccessful even when there are idle circuits (or servers) in the destination group. In classical studies of teletraffic link systems "loss functions" were used to represent in simple mathematical terms the effects of the restricted availability with respect to the arriving calls for service. This function ($\omega(\nu)$) is defined as the conditional probability that a call arriving when there are ν occupied servers is rejected. In particular Wallstrom [25] used this concept for calculating the blocking probability of restricted availability overflow systems arising in teletraffic networks with alternative routing. Although these systems typically did not have queueing facilities, modern technologies may provide systems with limited waiting room (say k queueing positions in a buffer). We also may consider teletraffic systems where decisions regarding the acceptance or rejection of a call are of a probabilistic nature and based on the number of calls already in progress or waiting for service, mechanism which could be also represented by some specific type of loss functions. An example could be the case of "load sharing" schemes of adaptative dynamic routing in digital networks [3] in which calls rejected by a given route are offered to alternative routes according to a set of probabilities which are a function of the states (number of accupations) of the individual groups of servers in the different arcs of the network.

Possible application of this type of model to stochastic service systems of different nature (for example a public attendance system in an organization with correlated arrival intensities and overflow to similar systems) where arriving users may give up entering the system, as a result of a probabilistic decision expressed as an increasing function of the number of users waiting for service at the arrival epochs may also be envisaged. Such system could equally by modelled using appropriate loss functions.

On the other hand a number of studies ([4], [6], [7], [12], [22]) suggest that the MMPP ("Markov Modulated Poisson Process") could be used successfully for modelling certain types of superpositions of complex teletraffic flows. In particular the MMPP is the exact model for the superposition of independent IPPs ("Interrupted Poisson Processes"), proposed by Kuczura [8], as a rather accurate approximation to overflow traffics, resulting from the overflow of Poisson inputs in loss systems with exponential distribution of the service times (model of great interest in circuit-switched networks with alternative routing).

The m-MMPP point process may be defined as a doubly stochastic Poisson Process where the intensity process $\{\lambda(t), t \geq 0\}$ is governed by an ergodic Markov process, with m states, i.e.:

$$\lambda(t) := \lambda_{I(t)}$$

where the R.V. (Random Variable) $I(t)$ indicates the state, at instant t, of an ergodic Markov process. When $I(t) = f(f \in 1, \ldots, m)$ the MMPP is said to be in phase f.

The MMPP is also a particular case of the "Versatile Markovian Point Process" model of Neuts [14] and may also be treated as a particular case of the recently proposed Markovian Arrival Process model, (see [9], [16]), which has been considered for modelling features of data traffic streams, cf. [17].

In [13], a number of results and the analysis of the system MMPP/G/1 were presented. References [6] and [7], show the application of the MMPP to the modelling of traffic flows in data and packetised voice networks, while [12] presents an exact analysis of a full availability queueing system with exponential service times and arrivals consisting of the superposition of a Poisson and multiple independent IPPs processes. The 2-MMPP or SPP ("Switched Poison Process") in particular, was shown to be a useful model for approximating certain processes in data networks, according to the studies by Heffes ([4], [6]) and Rossiter [19].

In a recent study Arvidson [1] considers a model of a loss system with negative exponentially distributed times, being offered a number of 5-MMPPs and Poisson inputs, each being accepted at random with a probability dependent upon the number of busy servers at the call arrival instants. This model was used to obtain an approximation for calculating moments of blocking periods, time-dependent loss probabilities and moments of overflow intervals for different classes of calls in circuit-switched links with priorities.

All these developments led us to approach, in previous works ([23], [24]), the exact analysis of a loss system with a m-MMPP input, a finite queue of capacity k, N servers with negative exponential service times and a loss function $\omega(\nu) := 1 - \alpha_\nu$ as illustrated in Fig. 1.

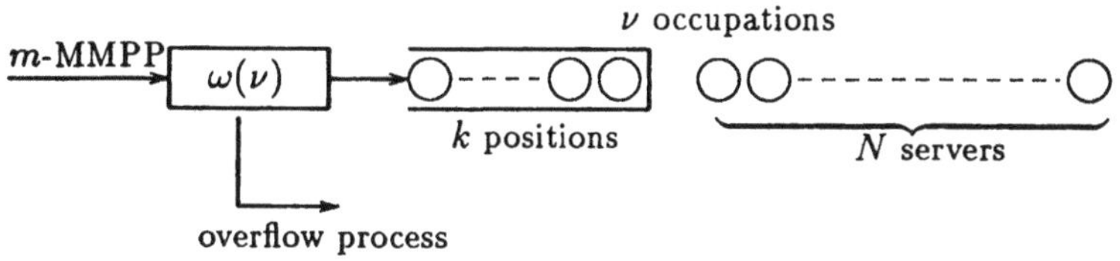

Figure 1

In [24] the authors used the Versatile Markovian Point Process [14] to approach the characterization of key stochastic processes associated with the system of Fig. 1: the process associated with the rejected users (overflow process), the process associated with the accepted users (accepted process) and the departure process. Also the calculations of key parameters for approximating these processes by SPPs, using the methods given in [19] and [4] (this one being only applied to the overflow process) were developed and discussed. The practical interest in the analysis of those processes is quite clear in the context of networks of queues. In particular, for circuit-switched communication networks with alternative routing the calls rejected by a first group of servers (overflow traffic) (namely in the case of restricted availability, as is the case of telephone exchanges) are offered to a second choice group of servers, see for example [10]. Also the calls leaving a queue will contribute to the input of subsequent queues, in the study of networks with tandem queues. Finally the characterization of the process associated with the accepted users (accepted process) may be required to support the analysis of the performance of the system being modeled with the queue, namely in the case of teletraffic switching systems, such as in the study [5].

In this paper the authors extend previous work by considering the exact analysis of a system with finite queueing capacity whose inputs are defined from a number of independent MMPPs each being subject to an "access function" which defines the conditional probability of an arrival being rejected for a given number of occupations in the system, as illustrated in Fig. 2. Beyond reviewing the calculations of the stationary distribution, waiting times, the overall congestion and the marginal blocking probabilities, experienced by the different MMPP input streams, the analysis of the associated marginal overflow point processes, the accepted point process and the calculation of the moments of the sojourn times, are approached.

The organization of the paper is as follows. Section 2 presents the assumptions and basic results of the system stochastic analysis. Section 3 overviews calculations of stationary and waiting time distributions and presents the expressions for the calculation of the moments of the times the system spends in certain states (sojourn times). In Section 4 the global and marginal congestions are derived. The stochastic characterization of the global and marginal overflow point processes, based on the Versatile Markovian Point Process Model [14], is presented in Section 5. Finally Section 6 overviews the analysis of the accepted point process and Section 7 highlights some general conclusions.

2. ASSUMPTIONS AND BASIC RESULTS

The system under study (Fig. 2) includes L independent MMPP inputs, the underlying Markov processes of which have $m_1, \dots, m_L$ states, respectively. These point processes are characterized by the corresponding infinitesimal generators $A_1, \dots, A_L$ and the intensity diagonal matrices $\Lambda_1, \dots, \Lambda_L$, of orders $m_1, \dots, m_L$ – (A_r, Λ_r) sources [6]. The access of each m_r-MMPP input to the queue is controlled by an "access function" $\alpha_r(\nu)$ $(r = 1, \dots, L)$ (which may also be envisaged as a probabilistic "gate") such that $\alpha_r(\nu)$ represents the conditional probability of an arrival of the $MMPP_r$ be accepted, being known that there are ν users in the system; $\alpha_r(\nu) \in]0,1]$, $\nu \in \{0, \dots, N+k-1\}$, and without loss of generality we will assume $\alpha_r(N+k) = 0$. The service times are assumed negative exponentially distributed with mean μ^{-1}.

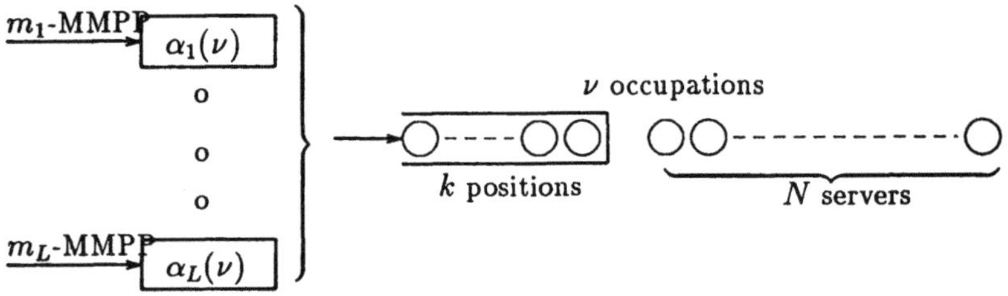

Figure 2

Let us consider the set

$$\mathcal{F} = \{\underline{f} = (f_1, \dots, f_r, \dots, f_L) : f_1 \in [1, m_1] \cap \mathbb{N}, \dots, f_r \in [1, m_r] \cap \mathbb{N}, \dots, f_L \in [1, m_L] \cap \mathbb{N}\} \quad \textbf{(2.1)}$$

with cardinal $\#\mathcal{F} = m := \prod_{r=1}^{L} m_r$.

The analysis of this system is based on the known fact [11] that the superposition of L independent MMPPs is also a MMPP, of order m, whose underlying Markov process has space state $\mathcal{F}$ and, as generator, the $\mathcal{F} \times \mathcal{F}$ matrix A:

$$A = A_1 \oplus \dots \oplus A_L := \bigoplus_{r=1}^{L} A_r := \sum_{r=1}^{L} \left[\left(\bigotimes_{j=1}^{r-1} I_{m_j} \right) \otimes A_r \otimes \left(\bigotimes_{j=r+1}^{L} I_{m_j} \right) \right] \quad \textbf{(2.2)}$$

The associated diagonal matrix of intensities is the $\mathcal{F} \times \mathcal{F}$ matrix Λ:

$$\Lambda = \Lambda_1 \oplus \dots \oplus \Lambda_L := \bigoplus_{r=1}^{L} \Lambda_r \quad \textbf{(2.3)}$$

where $\otimes$ and $\oplus$ indicate the Kronecker product and the Kronecker sum of matrices, respectively, and I_s is the identity matrix of order s.

Proposition. *The system of Fig. 2 is stochastically equivalent to a system of the type of Fig. 1, having as input the m-MMPP superposition of the L MMPPs and having an "access function" $\alpha(\nu, \underline{f})$ dependent not only on the number of users in the system but also on the phase $\underline{f}$ of the superposition MMPP:*

$$\alpha(\nu, \underline{f}) = \frac{\sum_{r=1}^{L} \alpha_r(\nu)\, {}^r\lambda_{f_r}}{\sum_{r=1}^{L} {}^r\lambda_{f_r}} \tag{2.4}$$

where ${}^r\lambda_{f_r}$ is the intensity of the input m_r-MMPP in phase f_r.

PROOF. Let $\underline{\lambda}$ be the vector with indices in $\mathcal{F}$:

$$\underline{\lambda} := (\lambda_{(1,\ldots,1)}, \ldots, \lambda_{\underline{f}}, \ldots, \lambda_{(m_1,\ldots,m_L)}) \qquad \underline{f} \in \mathcal{F} \tag{2.5}$$

where:

$$\lambda_{\underline{f}} = {}^1\lambda_{f_1} + {}^2\lambda_{f_2} + \ldots + {}^L\lambda_{f_L} \tag{2.6}$$

Therefore we have, from (2.3):

$$\Lambda = \operatorname{diag}(\underline{\lambda}) \tag{2.7}$$

We may also state that the Markov process $\{X_t, \geq 0\}$ describing the system under study has state space I'

$$I' = \{(\nu, \underline{f}) : \nu \in [0, N+k] \cap \mathbf{Z},\ \underline{f} \in \mathcal{F}\} \tag{2.8}$$

Let us assume the system is in state $(\nu, \underline{f})$ at instant $t = 0$. The conditional probability of the system being in state $(\nu+1, \underline{f})$ at instant t is:

$$P\{X_t = (\nu+1, \underline{f}) | X_0 = (\nu, \underline{f})\} = [\alpha_1(\nu)^1\lambda_{f_1} + \alpha_2(\nu)^2\lambda_{f_2} + \ldots + \alpha_L(\nu)^L\lambda_{f_L}]t + o(t) \tag{2.9}$$

By defining a "multidimensional" access function $\alpha(\nu, \underline{f})$ according to (2.4) we may write:

$$P\{X_t = (\nu+1, \underline{f}) \,|\, X_0 = (\nu, \underline{f})\} = \alpha(\nu, \underline{f})\, \lambda_{\underline{f}} t + o(t) \tag{2.10}$$

Therefore by introducing the access function $\alpha(\nu, \underline{f})$ in the model of Fig. 1 we obtain a system stochastically equivalent to the original system of Fig. 2. ∎

Let $\underline{\alpha}(\nu)$ be the $\mathcal{F} \times \mathcal{F}$ diagonal matrix:

$$\underline{\alpha}(\nu) := \operatorname{diag}\Big(\alpha(\nu, 1, 1, \ldots, 1), \ldots,\ \alpha(\nu, \underline{f}), \ldots,\ \alpha(\nu, m_1, m_2, \ldots, m_L)\Big)$$

Then, according to (2.4):

$$\underline{\alpha}(\nu) = \left[\bigoplus_{r=1}^{L} \Big(\alpha_r(\nu)\, \Lambda_r\Big)\right] \Lambda^{-1} \tag{2.11}$$

We may proceed using f, element of $\mathcal{G} := [1, m] \cap \mathbf{N}$ or $\underline{f}$, element of $\mathcal{F}$ indifferently. Then we may say that the Markovian stochastic process describing the system has state space:

$$I = \{(\nu, f) : \nu \in [0, N+k] \cap \mathbf{Z},\ f \in \mathcal{G}\} \tag{2.12}$$

Now, according to [24], section 2.3, the Markov process $\{X_t\}$ has infinitesimal generator Q with the following block structure:

$$Q = \begin{bmatrix} A-\underline{\alpha}(0)\Lambda & \underline{\alpha}(0)\Lambda & & & & & \\ \mu I & A-\underline{\alpha}(1)\Lambda-\mu I & \underline{\alpha}(1)\Lambda & & & & \\ & & \cdots & & & & \\ & & N\mu I & A-\underline{\alpha}(N)\Lambda - N\mu I & \underline{\alpha}(N)\Lambda & & \\ & & & N\mu I & A-\underline{\alpha}(N+1)\Lambda - N\mu I & \underline{\alpha}(N+1)\Lambda & \\ & & & & & \cdots & \\ & & & & & N\mu I & A - N\mu I \end{bmatrix} =$$

$$= \mu\Big(\mathrm{diag}(\underline{h})\, F_{N+k+1}\Big) \otimes I_m + \mathrm{diag}\Big(\underline{\alpha}(0)\,\Lambda, \ldots, \underline{\alpha}(\mathrm{N}+\mathrm{k})\,\Lambda\Big)\, \mathrm{H}^{\mathrm{m}}_{\mathrm{N+k+1}} +$$
$$+ I_{N+k+1} \otimes A - \mathrm{diag}\Big(\underline{\alpha}(0)\,\Lambda, \ldots, \underline{\alpha}(N+k)\,\Lambda\Big) - \mu\, \mathrm{diag}(\underline{h}) \otimes I_m \qquad \textbf{(2.13)}$$

where $\underline{h} := (f(0), \ldots, f(N+k))$, with

$$f(\nu) := \begin{cases} \nu & \text{if } \nu \in [0, N] \cap \mathbf{Z} \\ N & \text{if } \nu \in [N, N+k] \cap \mathbf{Z} \end{cases}$$

I_r is the identity matrix of order r, and H_r and F_r are the r order matrices:

$$H_r := \begin{bmatrix} 0 & 1 & & & \\ & 0 & 1 & & \\ & & \ddots & & \\ & & & 0 & 1 \\ & & & & 0 \end{bmatrix} \qquad F_r := \begin{bmatrix} 0 & & & & \\ 1 & 0 & & & \\ & & \ddots & & \\ & & 1 & 0 & \\ & & & 1 & 0 \end{bmatrix}$$

3. SOME PERTINENT DISTRIBUTIONS

3.1. Stationary Distributions

Here we will review the calculation (from Q, [2.13]) of pertinent stationary distributions, using standard Markov theory.

The stationary probability distribution, π, is obtained as the unique solution of the linear system:

$$\begin{cases} \pi\, Q = 0 \\ \pi\, e\; = 1 \end{cases} \qquad \textbf{(3.1)}$$

where e is a column vector $e := [1, 1, \ldots, 1]^T$.

The stationary probability of the number ν of users in the system is:

$$\gamma_\nu = \pi_\nu\, e = \sum_{f=1}^{m} \pi_{\nu f} \qquad \textbf{(3.2)}$$

where $\pi_\nu = [\pi_{\nu f}], (f = 1, \ldots, m)$.

The steady state distribution of the number of simultaneous occupations in the N servers is:

$$\zeta_r = \begin{cases} \gamma_r & \text{if } r \in \{0, 1, \ldots, N-1\} \\ \sum\limits_{\nu=N}^{N+k} \gamma_\nu & \text{if } r = N \end{cases} \qquad \textbf{(3.3)}$$

From (3.3) the moments $\mathbf{E}[X^r]$ of the carried traffic are immediately obtained by considering the random variable X representing the number of simultaneous occupations in the N servers (in equilibrium conditions).

3.2. Waiting Times

Also according to [24] and using the elegant formulation of Neuts [15] we can say that the actual waiting time (i. e. the time an arbitrary accepted user must wait until the beginning of its service) has phase type (PH) distribution with representation (β, T), where:

$$\beta_j = \beta'_{k-j+1} \qquad \forall j \in [1,k] \cap \mathbf{N} \tag{3.4}$$

$$T = \begin{bmatrix} -N\mu & N\mu & & & \\ & -N\mu & N\mu & & \\ & & \cdots & & \\ & & & -N\mu & N\mu \\ & & & & -N\mu \end{bmatrix} = -N\mu(I_k - H_k) \tag{3.5}$$

with:

$$\beta'_r = \begin{cases} \gamma^*_{N+r-1} & \text{if } r \in [1,k] \cap \mathbf{N} \\ \sum\limits_{\nu=0}^{N-1} \gamma^*_\nu & \text{if } r = 0 \end{cases}$$

and $\{\gamma^*_\nu\}$ representing the probability of an accepted user finding ν occupations in the system:

$$\gamma^*_\nu = \frac{\sum\limits_{f=1}^{m} \pi_{(\nu,f)}\, \lambda_f\, \alpha(\nu, f)}{\sum\limits_{\nu=0}^{N+k} \sum\limits_{f=1}^{m} \pi_{(\nu,f)}\, \lambda_f\, \alpha(\nu, f)} \tag{3.6}$$

The noncentral moments of the actual waiting time distribution can now be obtained from the relation [15]:

$$\omega_n := \mathbf{E}[W^n] = (-1)^n\, n!\, (\beta T^{-n} e) \tag{3.7}$$

A recursive scheme for calculating these moments was derived in [24]:

$$\begin{aligned} T^{-n} &= -(N\mu)^{-n} \sum_{\nu=0}^{k-1} \epsilon^{(n)}_\nu\, H^\nu_k := \\ &:= -(N\mu)^{-n} \left(\epsilon^{(n)}_0\, I_k + \epsilon^{(n)}_1\, H^1_k + \ldots + \epsilon^{(n)}_\nu\, H^\nu_k + \ldots + \epsilon^{(n)}_{k-1}\, H^{k-1}_k \right) = \\ &= -(N\mu)^{-n} \begin{bmatrix} \epsilon^{(n)}_0 & \cdots & \epsilon^{(n)}_\nu & \cdots & \epsilon^{(n)}_{k-1} \\ & \ddots & \vdots & \ddots & \vdots \\ & & \epsilon^{(n)}_0 & \cdots & \epsilon^{(n)}_\nu \\ & & & \ddots & \vdots \\ & & & & \epsilon^{(n)}_0 \end{bmatrix} \end{aligned} \tag{3.8}$$

with $\epsilon_\nu^{(n)}$, being given recursively by:

$$\epsilon_\nu^{(n)} = \begin{cases} \sum_{s=0}^{\nu} \epsilon_s^{(n-1)} & \text{if } n > 1 \\ 1 & \text{if } n = 1 \end{cases} \tag{3.9}$$

$\forall \nu \in [0, k-1] \cap \mathbf{Z}$, $n \in \mathbf{N}$.

3.3. Sojourn Times

Let us consider the process $\{Z_t : t \geq 0\}$ with state space $L := \{0, \ldots, N\}$, defined by:

$$Z_t = n \quad \text{iff} \quad X_t \in B_n \tag{3.10}$$

where:

$$B_n := \begin{cases} \{(\nu, f) \in I : \nu = n\} & \text{if } n \in \{0, \ldots, N-1\} \\ \{(\nu, f) \in I : \nu \geq N\} & \text{if } n = N \end{cases} \tag{3.11}$$

Therefore Z_t is in state n when X_t is in the set of states B_n. Here we will determine the steady state moments of the sojourn times of $\{Z_t\}$ in the various states of its space state. For this purpose we shall make use of the theory developed by Rubino *et al.* [21]. Let P be the transition probabilities matrix of the uniformized chain [18] of $\{X_t\}$ i.e.:

$$P = I_{m(N+k)} + \frac{1}{M} Q \tag{3.12}$$

where: $M \geq \max_{i \in I} \{|q_{ii}|\}$. According to $\left[[21], \text{(Corollary 3.3)}\right]$, the rth non-central moment of the sojourn time (T) of $\{X_t\}$ in the set B_n (or, in other words, the sojourn time of Z_t in state n) is given by:

$$\mathbf{E}[T^r] = (-1)^r \, r! \, x_n \, A_n^{-r} \, e_{B_n} \tag{3.13}$$

where e_{B_n} is a column vector of "ones", of order $\# B_n$; A_n is the $B_n \times B_n$ submatrix of Q. According to (3.11), A_n has order m, for $n < N$, and order $m(k+1)$, for $n = N$. For $n < N$:

$$A_n = A - \underline{\alpha}(n) \, \Lambda - f(n) \, \mu \, I_m \tag{3.14}$$

The row vector x_n is given in Corollary 2.7 of the same paper [21]:

$$x_n = \frac{1}{\pi_n (I_{B_n} - P_n) e} \pi_n (I_{B_n} - P_n) \tag{3.15}$$

where π_n is the subvector of π corresponding to the set B_n; I_{B_n} is the $B_n \times B_n$ identity matrix and P_n is the $B_n \times B_n$ submatrix of P. From (3.12):

$$P_n = I_{B_n} + \frac{1}{M} A_n \tag{3.16}$$

Substituting (3.16) in (3.15) we have:

$$x_n = \frac{1}{\pi_n A_n e} \, \pi_n \, A_n \tag{3.17}$$

Introducing now (3.17) in (3.13) we finally obtain:

$$\mathbf{E}[T^r] = (-1)^r \, r! \, \frac{1}{\pi_n A_n e} \, \pi_n \, A_n^{1-r} \, e \tag{3.18}$$

4. GLOBAL AND MARGINAL CONGESTION

The mean blocking and the marginal blocking probabilities (also designated as global and marginal congestions, respectively) are performance parameters of major importance for this type of queues, namely when they model teletraffic systems. These congestions are easily obtained from the stationary distribution π and from the phase-dependent access function (2.4).

4.1. Global Congestion

Let us consider the MMPP, the underlying Markov process of which has infinitesimal generator Q (2.13), and arrival intensity λ_f in every phase (ν, f). This (N+k+1)m-MMPP has an intensity diagonal matrix:

$$\Lambda_\iota = I_{N+k+1} \otimes \Lambda \tag{4.1}$$

This point process is stochastically equivalent to the MMPP superposition of the L input MMPPs, and its event-stationary version is obtained [11] by choosing the initial phase (at $t = 0$) according to the probability measure p:

$$p = (\pi\, \Lambda_\iota\, e)^{-1}\, \pi\, \Lambda_\iota \tag{4.2}$$

The component p_i, $i = (\nu, f)$ of p is the stationary probability of an arrival belonging to phase i, that is, the probability of finding the system in state $i = (\nu, f)$. Therefore the global congestion (or mean blocking probability), that is the probability of an arbitrary arrival being rejected, is given by:

$$B = \sum_{\nu=0}^{N+k} \sum_{f=1}^{m} p_{(\nu,f)} \big(1 - \alpha(\nu, f)\big) \tag{4.3}$$

4.2. Marginal Congestions

Let us now consider the MMPP the underlying Markov process of which has infinitesimal generator Q and having arrival intensity ${}^r\lambda_{f_r}$ in every phase $(\nu, f_1, \ldots, f_r, \ldots, f_L)$. This MMPP has intensity diagonal matrix:

$${}^r\Lambda_\iota = I_{N+k+1} \otimes {}^r\Lambda \tag{4.4}$$

where ${}^r\Lambda$ is the diagonal matrix of order m:

$${}^r\Lambda = \mathrm{diag}\big(\lambda'_{(1,\ldots,1)} = {}^r\lambda_1, \ldots,\ \lambda'_{(f_1,\ldots,f_r,\ldots,f_L)} = {}^r\lambda_{f_r}, \ldots,\ \lambda'_{(m_1,\ldots,m_r,\ldots,m_L)} = {}^r\lambda_{m_r}\big) \tag{4.5}$$

It is obvious that ${}^r\Lambda = \Lambda$ if we have:

$${}^i\lambda_{f_i} = 0 \qquad \forall i \neq r \tag{4.6}$$

Therefore:

$${}^r\Lambda = \underline{0}_{m_1} \oplus \underline{0}_{m_2} \oplus \ldots \oplus \Lambda_r \oplus \ldots \oplus \underline{0}_{m_L} \tag{4.7}$$

where $\underline{0}_s$ is the zero square matrix of order s. The MMPP defined according to (4.4) is stochastically equivalent to the input $\mathrm{MMPP_r}$, hence its event-stationary version is obtained by choosing the initial phase according to the probability measure ${}^r p$:

$${}^r p = (\pi^r\, \Lambda_\iota\, e)^{-1}\, \pi^r\, \Lambda_\iota \tag{4.8}$$

So the component ${}^r p_i$, $i = (\nu, f_1, \ldots, f_L)$, of ${}^r p$ is the stationary probability of an arrival from MMPP$_r$ finding the system in the state i. Therefore the stationary distribution of the number of users in the system at the arrival epochs of MMPP$_r$ is:

$$\left\{ {}^r\gamma'_\nu = \frac{\pi_\nu {}^r\Lambda\, e}{\pi\, {}^r\Lambda_\iota\, e} \;:\; \nu \in \{0, 1, \ldots, N+k\} \right\} \tag{4.9}$$

Consequently the marginal congestion for the input MMPP$_r$ (that is the probability of an arrival from MMPP$_r$, being rejected) is:

$$B_r = \sum_{\nu=0}^{N+k} {}^r\gamma'_\nu \big[1 - \alpha_r(\nu)\big] \tag{4.10}$$

5. OVERFLOW POINT PROCESSES

The characterization of the point processes associated with service requests rejected by this type of systems is of great practical interest in the context of networks of queues. In particular, for circuit–switched communication networks with alternative routing the calls rejected by a first group of servers (overflow traffic) (namely in the case of restricted availability, as is the case of telephone exchanges) may be offered to a second choice group of servers.

5.1. Global Overflow Point Process

The analysis of the global overflow point process (associated with the calls rejected by the system) is very similar to the one presented in [24], based on the conceptual framework given in [14] – Versatile Markovian Point Process model. Only the main points of the analysis, namely some adaptations resulting from the definition of the access function $\alpha(\nu, f)$ need to be obtained.

Let us define the set $L = \big\{(\nu, f, n) : (\nu, f) \in I,\ n \in \mathrm{N} \cup \{0\}\big\}$ and the process $\{O_t : t \geq 0\}$, having L as sample space, where n represents the number of users rejected by the system in $]0, t]$ and (ν, f) is the state of the Markov process $\{X_t\}$ (defined in Section 2) at instant t. $\{O_t\}$ is also a Markov process the infinitesimal generator ${}^oG := [{}^o g_{ij}]$ of which is constructed in the same manner as in [24]. Now we obtain:

$${}^o g_{(\nu,f,n)(\nu,f,n+1)} = \lambda_f\big(1 - \alpha(\nu, f)\big) \tag{5.1a}$$

by taking into account that:

$$P\{O_{t_0+t} = (\nu, f, n+1) \mid O_{t_0} = (\nu, f, n)\} = \lambda_f\big(1 - \alpha(\nu, f)\big)t + o(t)$$

Also:

$$\begin{aligned} {}^o g_{(\nu,f,n)(\nu',f,n)} &= q_{(\nu,f)(\nu',f)} && \text{if } \nu \neq \nu' \\ {}^o g_{(\nu,f,n)(\nu,f',n)} &= q_{(\nu,f)(\nu,f')} && \text{if } f \neq f' \end{aligned} \tag{5.1b}$$

where $q_{(\nu,f)(\nu',f')}$ is the generic element of Q. All the remaining non-diagonal elements of ${}^oG$ are zero. Therefore ${}^oG$ has the following block structure, taking in account that the sum of the elements in each row is zero:

$${}^oG = \begin{bmatrix} Q - \Lambda' & \Lambda' & & \\ & Q - \Lambda' & \Lambda' & \\ & & \cdots & \\ & & & \cdots \end{bmatrix} = I_\infty \otimes (Q - \Lambda') + H_\infty \otimes \Lambda' \tag{5.2}$$

where:

$$\Lambda' = \left[I_{m(N+k+1)} - \text{diag}\big(\underline{\alpha}(0), \ldots, \underline{\alpha}(N+k)\big) \right] (I_{N+k+1} \otimes \Lambda) \tag{5.3}$$

Taking into account that the block Λ' in ${}^{o}G$ is diagonal we may state, according to [14], that the *overflow point process is a $m(N+k+1)$–MMPP with Q as generator of the underlying Markov process and Λ' as diagonal matrix of intensities ((Q, Λ') source).*

The calculations of four important characterizing parameters of the overflow point process (the mean intensity ${}^{o}\lambda$; the squared coefficient of variation of the arbitrary interval of its event-stationary version, ${}^{o}c^2$; the asymptotic value of the dispersion index, ${}^{o}Z(\infty)$ of the counting random variable, ${}^{o}N(t)$, and the asymptotic value of the covariance between the number of arrivals in adjacent time intervals, ${}^{o}C(\infty)$) are perfectly analogous to the ones presented in [24]. The only relevant difference is the fact that matrix Λ' is now given according to (5.3).

Beyond the intrinsic information associated with these four parameters, ${}^{o}\lambda$, ${}^{o}c^2$, ${}^{o}C(\infty)$, ${}^{o}Z(\infty)$, they may be directly used for approximating the overflow point process by a SPP, using the method proposed by Rossiter [19], [20]. The need for approximating arises as a result of the impracticable numerical and storage effort which may be required by the exact model calculations, specially in situations where that process is to be the input process to a second stage system. The SPP approximation using directly those parameters as characterizing parameters is applicable in the domain $\{({}^{o}\lambda, {}^{o}c^2, {}^{o}C(\infty), {}^{o}Z(\infty) : {}^{o}\lambda > 0,\ 1 < {}^{o}c^2 \leq {}^{o}Z(\infty),\ {}^{o}C(\infty) > 0\}$ and enables the approximate modelling of superpositions of independent point processes, taking into account the "commutative" and "associative" properties of the characterizing parameters [19]. This modelling method intends to be a "general" one (see [19]). Alternatively we could use the technique proposed by Heffes [4] for obtaining a SPP approximation for the overflow point process. For further details on these approximations see [12] and [24].

Finally it may be noted that the overflow point process is not renewal in general (in fact it rarely is), unlike the case of the marginal overflow streams appearing in [12], so the SPP approximation is not, in general, an Interrupted Poisson Process.

5.2. Marginal Overflow Processes

Let us define the set $L^{(r)} := \{(\nu, f, n^{(r)}) : (\nu, f) \in I, n^{(r)} \in \mathrm{N} \cup \{0\}\}$ and the process $\{O_t^{(r)} : t \geq 0\}$ having $L^{(r)}$ as sample space, where $n^{(r)}$ represents the number of arrivals from the input process MMPP$_r$ rejected by the system in the interval $]0, t]$ and (ν, f) is the state of the Markov process $\{X_t : t \geq 0\}$ at instant t. The MMPP$_r$ is stochastically equivalent to the MMPP whose underlying Markov process has generator A (2.2) and intensity diagonal matrix ${}^{r}\Lambda$ (4.7). Similarly to Section 5.1, we may conclude that the marginal overflow point process associated with the rth input process is a m(N+k+1)–MMPP characterized by $(Q, \Lambda^{(r)})$:

$$\Lambda^{(r)} = \left(I_{N+k+1} - \text{diag}\,(\underline{\alpha}_r)\right) \otimes {}^{r}\Lambda \tag{5.4}$$

where the row vector $\underline{\alpha}_r$ is given by:

$$\underline{\alpha}_r := \left(\alpha_r(0),\ \alpha_r(1), \ldots,\ \alpha_r(N+k)\right) \tag{5.5}$$

Calculations and formulae similar to those mentioned for the analysis of the global overflow point process could now be obtained in the same manner.

6. ACCEPTED POINT PROCESS

The characterization of the process associated with the accepted customers (accepted process) may be required to support the analysis of the performance of the system being modeled with the queue, namely in the case of teletraffic switching systems, such as in the study [5]. Similarly to the global overflow point process, the stochastic characterization of the global accepted point process (corresponding to the instants of arrival of accepted users) is very similar to the one presented in [24]. Only the main points of analysis will be referred to. The events of this process correspond to the instants of entrance of accepted users into the system. Such instants correspond precisely to the transitions of the type $(\nu, f) \to (\nu+1, f)$, in the process $\{X_t\}$. Therefore this point process is a particular case of the "Versatile Markovian Point Process" of Neuts ([14], [16]) and of the point process analyzed in [22]. Let us define the set $J = \{(\nu, f, n) : (\nu, f) \in I, n \in \mathbf{N} \cup \{0\}\}$ and the process $\{T_t : t \geq 0\}$, having J as sample space, where n represents the number of users accepted by the system in $]0, t]$ and (ν, f) is the state of the Markov process $\{X_t\}$ at instant t. Following an analysis similar to the one of the overflow process we may conclude that the Markov process $\{T_t\}$ has infinitesimal generator ${}^{T}G = [{}^{T}g_{ij}]$, $i, j \in J$:

$$ {}^{T}G = \begin{bmatrix} Q - {}^{T}\Lambda & {}^{T}\Lambda & & \\ & Q - {}^{T}\Lambda & {}^{T}\Lambda & \\ & & \cdots & \\ & & & \cdots \end{bmatrix} = I_\infty \otimes (Q - {}^{T}\Lambda) + H_\infty \otimes {}^{T}\Lambda \quad (6.1) $$

where ${}^{T}\Lambda$ is :

$$ {}^{T}\Lambda_{(\nu,f)(\nu',f')} := {}^{T}g_{(\nu,f,n)(\nu',f',n+1)} = \begin{cases} \lambda_f \, \alpha(\nu, f) & \text{if } f' = f \wedge \nu' = \nu + 1 \\ 0 & \text{otherwise} \end{cases} \quad (6.2) $$

or, in matrix notation:

$$ {}^{T}\Lambda = \mathrm{diag}\big(\underline{\alpha}(0)\,\Lambda, \ldots, \underline{\alpha}(N+k)\,\Lambda\big)\, H^{m}_{N+k+1} \quad (6.3) $$

The whole remaining analysis of this point process is perfectly analogous to the one presented in [24].

7. CONCLUSIONS

An exact analytical model for a finite queue with multiple MMPP inputs with access functions, and negative exponentially distributed service times has been presented, namely regarding the calculation of global and marginal congestions, moments of waiting and sojourn time distributions and characterization of overflow point processes. A basic result of this work is the stochastic equivalence between the queue with multiple MMPP inputs, each with a certain access function, and the queue with a single MMPP input and a "multidimensional", phase-dependent access function.

At a methodological level these models confirm the great potentialities of the matrix analytic methods and the Versatile Markovian Point Process general model, proposed by Neuts ([14], [15], [16]). These works, as well as many application studies in the particular area of teletraffic systems stochastic modelling, namely [12], [24], put in evidence the

necessity of seeking to take advantage of the specific properties of the mathematical entities used in the models (namely with respect to the associated infinitesimal generators and systems of linear equations) in order to enhance their algorithmic tractability.
A key numerical problem of the model is the resolution of the linear system (3.1) enabling to calculate the stationary probability distribution π. In [24] it is shown an iterative method suitable for the matrix Q which corresponds to a block iterative scheme analogous to the one presented in [12]. Beyond the iterative method previously mentioned or the adaptation to this problem of a general sparse technique (see for example [2]), it seems worthwhile to investigate direct methods of resolution because they may give further insight into the properties of the matrix Q and the associated linear system which may be useful both in theoretical and practical terms. Following this line of research a direct method which starts by exploring specific properties of matrices of the type of Q is being developed, for possible application to these type of linear systems, in association with an iterative procedure.

REFERENCES

[1] Arvindson, A., (1991). On the Performance of a Circuit Switched Link with Priorities, *IEEE Journal on Selected Areas in Communications*, **9**, 2.

[2] Barker, V. A., (1989). Numerical Solution of Sparse Singular Systems of Equations Arising from Ergodic Markov Chains, *Stochastic Models*, **5**, 3, 335–381.

[3] Bel, G.; Chemouil, P.; Garcia, J. M.; Le Gall, F.; Bernoussou, J., (1985). Adaptative Traffic Routing in Telephone Networks, *Large Scale Systems Journal*, **8**, 3, 267–282.

[4] Heffes, H., (1980). A Class of Data traffic Processes - Covariance Function Characterization and Related Queueing Results, *Bell Syst. Tech. J.*, **59**, 897–929.

[5] Heffes, H.; Holtzman, J. M., (1973). Peakedness of Traffic Carried by a Finite Trunk Group with Renewal Input, *Bell System Technical Journal*, **52**, 9, 1617–1642.

[6] Heffes, H.; Lucantoni, D. M., (1986). A Markov Modulated Characterization of Packetised Voice and Data Traffic and Related Statistical Multiplexer Performance, *IEEE Select. Areas. Commun.*, SAC-4, 856–868.

[7] Holtzman, J. M., (1986). *Characterization of Superpositions of Traffic Streams*, Brussels Specialist Seminar on ISDN Traffic Issues, Brussels.

[8] Kuczura, A., (1973). The Interrupted Poisson Process as an Overflow Process, *Bell Syst. Tech. Jour.*, **52**, 437–449.

[9] Lucantoni, D. M., (1991). New Results on the Single Server Queue with a Batch Markovian Arrival Process. *Stochastic Models*, **7**, 1, 41–42.

[10] Matsumoto, J.; Watanabe, Y., (1985). Individual Traffic Characteristics of Queueing Systems with Multiple Poisson and Overflow Inputs, *IEEE Trans. Comm.*, COMM-33, 1–9.

[11] Meier, K. S., (1984). *A Statistical Procedure for Fitting Markov-Modulated Poisson Processes*. Ph. D. Dissertation, Univ. Delaware.

[12] Meier, K. S., (1989). The Analysis of a Queue Arising in Overflow Models, *IEEE Trans. Commun.*, **37**, 4, 367–372.

[13] Neuts, M. F., (1971). A Queue Subject to Extraneous Phase Changes, *Adv. Apl. Prob.*, **3**, 78–119.

[14] Neuts, M. F., (1979). A Versatile Markovian Point Process, *J. Appl. Prob.*, **16**, 764–779 .

[15] Neuts, M. F., (1981). *Matrix-Geometric Solutions in Stochastic Models*. Baltimore, MD: The John Hopkins University Press.

[16] NEUTS, M. F., (1989). *Structured Stochastic Matrices of M/G/1 Type and Their Applications.* New York and Basel: Marcel Dekker, Inc.

[17] NEUTS, M. F., (1991). *Modelling Data Traffic Streams.* [In *Proc. 13th International Teletraffic Congress*, Jensen and Iversen (Eds.)]. Amsterdam: Elsevier.

[18] ROSS, S. M., (1983). *Stochastic Processes.* New York: John Wiley and Sons.

[19] ROSSITER, M. H., (1986). *Sojourn Time Theory and the Switched Poisson Process*, Telecom Australia, Report 7835.

[20] ROSSITER, M. H.; HARRIS, R. J., (1989). A Multi-class, Multi-Commodity Flow Model of a Fast Packet Switched Network. *Australian Telecom. Research*, **23**, 2.

[21] RUBINO, G; SERICOLA, B, (1989). Sojourn Times in Finite Markov Processes, *J. Appl. Prob.*, **26**, 744–756.

[22] RUDEMO, M., (1973). Point Processes Generated by Transitions of Markov Chains, *Adv. Appl. Prob.*, **5**, 262-286 .

[23] TRALHÃO, L. M., CRAVEIRINHA, J., PAIXAO, J., (1991). *A Study on a Loss System with MMPP Input and an Access Function.* In *Proc. 5th Int. Symp. Appl. Stoch. Models and Data Analysis, Granada.* Singapore: World Scientific.

[24] TRALHÃO, L; CRAVEIRINHA, J.; PAIXÃO, J., (1991). *Analysis of Key Processes in a Stochastic System with MMPP Input and an Access Function.* (Submitted for publication).

[25] WALLSTROM, B., (1966). Congestion Studies in Telephone Systems with Overflow Facilities, *Ericsson Technics*, 3, 190–351.

Reliability Theory

Recent Adv. in Stat. and Prob., pp. 431-439
J. Pérez Vilaplana and M. L. Puri (Eds)

Re-thinking Reliability Theory: Schur-Concave Survival Functions and Survival Analysis

R.E. BARLOW*[1] and M.B. MENDEL†[2]
[1] Department of Industrial Engineering and Operations Research, College of Engineering, University of California at Berkeley, 4117 Etcheverry Hall, Berkeley, CA 94720, U.S.A.
[2] Department of Industrial Engineering and Operations Research, College of Engineering, University of California at Berkeley, 4117 Etcheverry Hall, Berkeley, CA 94720, U.S.A.

Abstract – This paper presents a Bayesian approach to survival analysis and basic definitions associated with that approach. Relative to this approach, the concept of aging relative to "wear-out" is characterized for finite (as opposed to conceptually infinite) populations of units. Beginning with a finite population of units and the judgment of exchangeability for units with respect to lifetime, we argue that measures of similarity lead to the appropriate probabilistic models for aging (or wear-out). This in turn implies that Schur-concavity of the joint probability function (or more generally, the joint survival distribution) provides the correct probabilistic description of aging. The concept of no aging relative to finite populations is characterized by the Schur-constant joint survival distributions which are then determined up to a mixing distribution.

1. INTRODUCTION

A related research paper [Barlow and Mendel (1992)] has been accepted for publication. In addition, Fabio Spizzichino has given important new research results [Spizzichino (1992)] related to this paper.

In what follows we use the term "aging" to denote the "wear-out" process observed for most units (animate or inanimate) as they age. Aging is an observable physical and biological process and methods for assessing probability models to predict lifetime with this in mind are of great interest. Then what is the appropriate probabilistic description of aging?

1.1. The Approach Based on Limiting Frequency

The usual probabilistic notion of aging for a *single* unit is relative to a conceptually *infinite* population. This has to be so since in this case probability itself is defined as a "limiting

---

*Research partially supported by the U. S. Air Force Office of Scientific Research (AFOSR-90-0087) grant and the Army Research Office (DAAL03-91-G-0046) grant to the University of California at Berkeley.

†Research partially supported by Lawrence Livermore National Laboratories, Livermore, California under contract No. 442446-26701.

frequency". In this approach, properties of life distributions are usually described in terms of the failure rate function. If X is the lifetime of an unit and $F(x) = P(X \leq x)$, then the failure rate function is defined as

$$r(x) = \frac{f(x)}{1 - F(x)}$$

where $f(x)$ is the density when it exists. The probabilistic notion of aging corresponds to this function increasing in x. More generally, the unit is said to have increasing failure rate if $\log[1 - F(x)]$ is concave where finite since this corresponds to the failure rate increasing when the density exists. Properties of this class of life distributions are discussed at length in Barlow and Proschan (1981). From the limiting frequency point of view there is no problem with this definition. However, the limiting frequency approach to probability has been criticized relative to its use in inductive inference [Basu (1988)]. Since probability modeling can itself be considered an act of inductive inference, this criticism led us to a new probabilistic notion of aging using the Bayesian approach. It has connections with the classical notion for conceptually infinite populations of exchangeable units.

1.2. A Bayesian Approach to Modeling Aging

Our argument begins by considering a finite population of units which are exchangeable in our opinion with respect to lifetime. Rather than consider the life distribution of a single unit we instead think about the properties of the joint distribution of a population of N units. That is, in thinking about their joint survival distribution

$$P(X_1 > y_1, X_2 > y_2, ..., X_N > y_N) = \bar{F}_N(\mathbf{y}_N)$$

this function should be invariant under permutation of the coordinates. $\bar{F}_N(\mathbf{y}_N)$ is a *predictive joint survival function* since no parameters have as yet been defined. At this point we are concentrating only on properties of this predictive joint survival function.

From a Bayesian perspective, probability is based on the judgments of a single decision maker. In thinking about a life distribution model for a finite population of exchangeable units, labeled $\{1, 2, \ldots, N\}$, with respect to lifetime, it seems that the notion of wear-out due to aging is really a relative concept. Our concept of aging is derived from past experience; i.e. observing the lifetimes of similar or related units. The question is: "How can we make use of this relativistic concept to develop a probabilistic notion of aging for finite populations?"

We first argue that "similar" lifetimes should be more probable than "diverse" lifetimes since physical and chemical processes leading to aging suggest more similarity with respect to lifetimes. Clearly, the *most similarity* in lifetimes, $\mathbf{x}_N$, corresponds to $x_1 = x_2 = \ldots = x_N$. Let $p_N(\mathbf{x}_N)$ be our probability function or joint probability density for the vector of lifetimes, $\mathbf{x}_N$. Hence in this situation we should have

$$p_N(\mathbf{x}_N) \geq p_N(\mathbf{y}_N) \tag{1}$$

for another vector of lifetimes, $\mathbf{y}_N$, (if *some additional condition is satisfied*). (The analog of this idea in the classical framework corresponds to the notion that a degenerate distribution is the "most" IFR). However (1) does not make sense unless $\mathbf{x}_N$ and $\mathbf{y}_N$ are in some sense *comparable*. In comparing vectors of lifetimes we will argue in section 2 that their sums need to be equal; i. e., $\sum_{i=1}^{N} x_i = \sum_{i=1}^{N} y_i$. This leads to the assumption corresponding to aging that $p_N(\mathbf{x}_N) \geq p_N(\mathbf{y}_N)$ whenever $x_1 = x_2 = \ldots = x_N$ and $\sum_{i=1}^{N} x_i = \sum_{i=1}^{N} y_i$.

1.3. Majorization and Schur-Concavity

To develop the previous ideas further we need some kind of ordering (actually pre-ordering) for all lifetime vectors having the same sum. In Fig.1 the vector $\mathbf{y} = (y_1, y_2)$ is said to *majorize* the vector $\mathbf{x} = (x_1, x_2)$; i. e. $\mathbf{x} <_M \mathbf{y}$. Both vectors lie on the simplex determined by $s = x_1 + x_2 = y_1 + y_2$ but $\mathbf{x}$ is closer than $\mathbf{y}$ to the central line $x_1 = x_2$. Since the ordering should be permutation invariant (units are judged exchangeable) we also require that $(x_2, x_1) <_M (y_2, y_1)$.

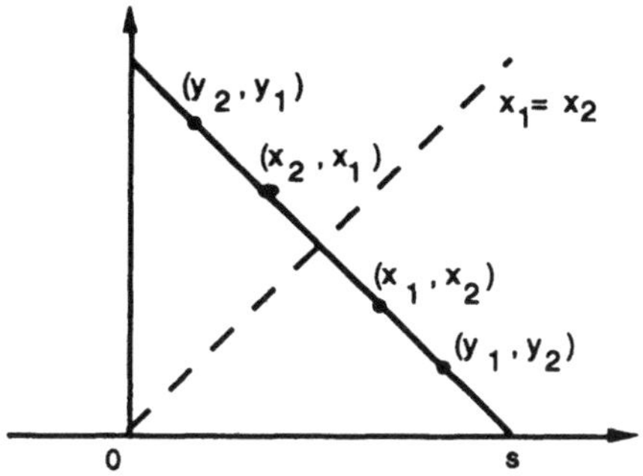

Figure 1. (y_2, y_1) majorizes (x_2, x_1)

Since the ordering is permutation invariant we need only compare vectors whose coordinates are ordered, say decreasing. Therefore let $x_{[1]} \geq x_{[2]} \geq \ldots, \geq x_{[N]}$ denote the reverse ordering of vector coordinates.

Definition 1 [see Marshall and Olkin (1979), pp. 7, 64].

$\mathbf{y}_N$ *majorizes* $\mathbf{x}_N$, *i.e.*, $\mathbf{x}_N \leq_M \mathbf{y}_N$ *iff*

$$1.)\quad \sum_{i=1}^{j} x_{[i]} \leq \sum_{i=1}^{j} y_{[i]} \qquad j = 1, 2, \ldots, N;$$

and

$$2.)\quad \sum_{i=1}^{N} x_i = \sum_{i=1}^{N} y_i .$$

Majorization induces a pre-ordering on vectors of lifetimes. It is *not* a partial ordering since $\mathbf{x}_N \leq_M \mathbf{y}_N$ and $\mathbf{y}_N \leq_M \mathbf{x}_N$ do not imply

$$\mathbf{x}_N \equiv \mathbf{y}_N.$$

The next question is, what real valued functions, f, preserve the majorization ordering?

Definition 2

- $f(\cdot)$ *is* Schur-convex *iff* $\mathbf{x} \leq_M \mathbf{y}$ *implies* $f(\mathbf{x}) \leq f(\mathbf{y})$
- $f(\cdot)$ *is* Schur-concave *iff* $\mathbf{x} \leq_M \mathbf{y}$ *implies* $f(\mathbf{x}) \geq f(\mathbf{y})$

Schur-convex (concave) functions are only meaningful for vector arguments. Hence we assume that $N \geq 2$ from now on.
The name Schur-convex seems to come from the following:

Proposition 1 $\mathbf{x} \leq_M \mathbf{y}$ *holds if and only if*

$$\sum_{i=1}^{N} \psi(x_i) \leq \sum_{i=1}^{N} \psi(y_i)$$

for every convex function ψ

[cf. Marshall & Olkin, pp. 108-109]. (If f is Schur-concave it is not true in general that f is concave.)

2. A NEW PROBABILISTIC NOTION OF AGING

If units {1, 2, ..., N} are aging relative to lifetime in our opinion then one possible probability property capturing this notion is that $p_N(\mathbf{x}_N)$ should be Schur-concave; i.e. $p_N(\mathbf{x}_N)$ reverses the majorization ordering suggested by aging. It is known that $p_N(\mathbf{x}_N)$ Schur-concave implies that

$$P(X_1 > y_1, X_2 > y_2, ..., X_N > y_N) \equiv \bar{F}_N(\mathbf{y}_N)$$

is also Schur-concave [Marshall and Olkin (1979), p. 379]. Spizzichino (1992) has shown that this weaker condition is equivalent to the following: Suppose that exchangeable units { 1, 2, ..., N} have survived to ages $\mathbf{y}_N$. We require that if $y_i < y_j$, then

$$\begin{aligned} &P(X_i > y_i + t \mid X_1 > y_1, ..., X_N > y_N) \geq \\ &\geq P(X_j > y_j + t \mid X_1 > y_1, ..., X_N > y_N) \end{aligned} \tag{2}$$

for all $t > 0$. That is, the younger of the two units, namely unit i, should have the greater survival probability conditional on all units surviving to ages $\mathbf{y}_N$. This can be shown to be true for all age vectors $\mathbf{y}_N$ and $t > 0$ if and only if $\bar{F}_N(\mathbf{y}_N)$ is *Schur-concave*. This motivates the definition:

Definition 3 *Units* {1, 2, ..., N} *age, or "wear-out" with respect to age, in our opinion iff their joint survival distribution* $\bar{F}_N(\mathbf{y}_N)$ *is* Schur-concave.

For completeness we prove the following proposition due to Spizzichino (1992).

Proposition 2 *Inequality* (2) *holds for all* $t > 0$ *and ages* $\mathbf{y}_N$ *if and only if* $\bar{F}_N(\mathbf{y}_N)$ *is Schur-concave.*

PROOF: Note that (2) holds if and only if

$$\begin{aligned} &P(X_1 > y_1, \ldots, X_i > y_i + t, \ldots, X_j > y_j, ..., X_N > y_N) \geq \\ &\geq P(X_1 > y_1, \ldots, X_i > y_i, \ldots, X_j > y_j + t, ..., X_N > y_N) \end{aligned} \tag{3}$$

where $y_i < y_j$ and $t > 0$.
First suppose $\bar{F}_N$ is Schur-concave. Note that

$$(y_1, \ldots, y_i + t, \ldots, y_j, \ldots, y_N) <_M (y_1, \ldots, y_i, \ldots, y_j + t, \ldots, y_N).$$

when $y_i < y_j$ and $t > 0$. (3) follows since $\bar{F}_N$ is Schur-concave.
Now suppose that (2) holds when $y_i < y_j$ and $t > 0$. We need the following Theorem from Marshall and Olkin [(1979), p. 55.]

Theorem 1 *$f(\mathbf{x}_N) \geq f(\mathbf{y}_N)$ for every $x_N >_M y_N$ iff*

1) *f is symmetric*

2) *$f(z, 2\alpha - z, z_3, \ldots, z_N)$ is increasing in $z > \alpha$ for every $z_3, \ldots, z_N$.*

Thus $\bar{F}_N$ is Schur-concave if and only if $-\bar{F}_N$ satisfies these two conditions. In the beginning we assumed exchangeability so that $-\bar{F}_N$ is symmetric. Fix $\alpha > 0$ and choose any $z_1 > z_0 > a$ and $z_3, \ldots, z_N$. Thus, we need to show

$$\bar{F}_N(z_0, 2\alpha - z_0, z_3, \ldots, z_N) \geq \bar{F}_N(z_1, 2\alpha - z_1, z_3, \ldots, z_N),.$$

Let $y_1 = 2\alpha - z_1$, $y_2 = 2\alpha - z_0$, $t = z_1 + z_0 - 2\alpha$, $y_3 = z_3$, $y_4 = z_4, \ldots, y_N = z_N$. Then we have

$$\bar{F}_N(z_0, 2\alpha - z_0, z_3, \ldots, z_N) = \bar{F}_N(y_1 + t, y_2, y_3, \ldots, y_N) \geq \bar{F}_N(y_2 - t, y_1, y_3, \ldots, y_N)$$

by (2). Since $\bar{F}_N(y_2 - t, y_1, y_3, \ldots, y_N) = \bar{F}_N(z_1, 2\alpha - z_1, z_3, \ldots, z_N)$, $-\bar{F}_N$ satisfies 2) of the theorem and is therefore Schur-concave. ∎

Assuming partial differentiability, Schur-concave functions are characterized by

$$\frac{\partial f(\mathbf{x}_N)}{\partial f(x_{[1]})} \leq \frac{\partial f(\mathbf{x}_N)}{\partial f(x_{[2]})} \leq \cdots \frac{\partial f(\mathbf{x}_N)}{\partial f(x_{[N]})}$$

Within the class of Schur-concave survival distributions, the logarithmically concave joint survival distributions play a central role.

Proposition 3 *If a real valued function $f(\mathbf{x}_N)$ is permutation invariant and logarithmically concave, then it is also Schur- concave.*

PROOF: Let $N = 2$ and $\mathbf{x} = (x_1, x_2) <_M (y_1, y_2) = \mathbf{y}$ as in Fig. 1. Since $\mathbf{x}$ and $\mathbf{y}$ lie on the same simplex, there exists $\alpha\,(0 \leq \alpha \leq 1)$ such that $\mathbf{x} = \alpha\,(y_1, y_2) + (1 - \alpha)\,(y_2, y_1)$. Hence

$$\begin{aligned} \log f(\mathbf{x}) = \log f\,[\alpha\,(y_1, y_2) + (1 - \alpha)\,(y_2, y_1)] \geq \\ \geq \alpha \log f(y_1, y_2) + (1 - \alpha) \log f(y_2, y_1) = \log f(\mathbf{y})\,. \end{aligned}$$

The inequality follows from logarithmic concavity and the second equality by permutation invariance. Hence $f(\mathbf{x}) \geq f(\mathbf{y})$.

If $\mathbf{x}_N <_M \mathbf{y}N$, then we can transform $\mathbf{y}_N$ to $\mathbf{x}_N$ by a sequence of pair-wise transpositions as above [cf. Marshall and Olkin, (1979), p. 21] and the above argument can be used for $N > 2$. ∎

EXAMPLE.- Let $X_1, X_2, \ldots, X_N$ be the independent lifetimes of N units, each with specified univariate distribution F. Suppose further that $\log \bar{F}(x)$ is concave where finite. Then

$$\bar{F}_N(\mathbf{y}_N) = \prod_{i=1}^{N} \bar{F}(y_i)$$

is logarithmically concave and therefore Schur-concave.

3. SCHUR-CONSTANT SURVIVAL DISTRIBUTIONS AND INDIFFERENCE RELATIVE TO AGING

According to our probabilistic notion of aging, we would be *indifferent* relative to aging if $\sum_{i=1}^{N} x_i = \sum_{i=1}^{N} y_i$ were to always imply $p_N(\mathbf{x}_N) = p_N(\mathbf{y}_N)$ That is, *indifference relative to aging* means that $p_N(\mathbf{x}_N)$ is *Schur-constant*. Conditioning on $\sum_{j=1}^{N} X_j = N\theta$, we have

$$p\left(\mathbf{x}_{N-1} \mid \sum_{j=1}^{N} X_j = N\theta\right) = \frac{p(\mathbf{x}_N)}{p\left(\sum_{j=1}^{N} x_j = N\theta\right)}.$$

Since $p_N(\mathbf{x}_N)$ is constant on the simplex, $x_N = N\theta - \sum_{j=1}^{N-1} x_j$ is determined and $p\left(\mathbf{x}_{N-1} \mid \sum_{j=1}^{N} x_j = N\theta\right)$ is constant on the "wedge" $\left\{\mathbf{x}_{N-1} \mid 0 \leq \sum_{j=1}^{N-1} x_j \leq \leq N\theta\right\}$. Figure 2 for $N = 3$ illustrates the situation when we are indifferent relative to aging.

Proposition 4 *If $p_N(\mathbf{x}_N)$ is constant on simplices then*

$$p_N\left(\mathbf{x}_N \mid \sum_{j=1}^{N} X_j = N\theta\right) = [(N-1)/N\theta] \ldots [(N-n)/N\theta] \left(1 - \sum_{i=1}^{n} x_i)/N\theta\right)^{N-n-1}$$

for $1 \leq n < N$.

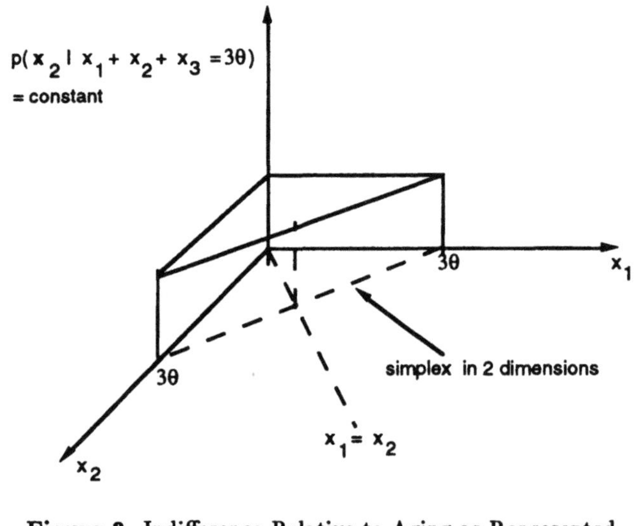

Figure 2. Indifference Relative to Aging as Represented by the Uniform Distribution on the "Wedge"

PROOF.- We first show that if $n = N - 1$, then

$$p\left(\mathbf{x}_{N-1} \mid \sum_{j=1}^{N} X_j = N\theta\right) = \frac{(N-1)!}{(N\theta)^{N-1)}}$$

for $0 \leq \sum_{j=1}^{N-1} x_j \leq N\theta$. Without loss of generality, let $N\theta = 1$ for convenience. Since

$$\int_{0 \leq \sum_{j=1}^{N-1} x_j \leq 1} p\left(\mathbf{x}_{N-1} \mid \sum_{j=1}^{N} X_j = 1\right) dx_1 \ldots dx_{N-1} = 1$$

we need only show

$$\int_{x_1=0}^{1} \cdots \int_{x_{N-1}=0}^{1-\sum_{j=1}^{N-2} x_j} \cdot \; dx_1 \ldots dx_{N-1} = \frac{1}{(N-1)!}.$$

This is the Dirichlet integral. Integrate beginning with x_{N-1}. After one integration, we have

$$\int_{x_1=0}^{1} \cdots \int_{x_{N-2}=0}^{1-\sum_{j=1}^{N-3} x_j} \cdot \; \left[1 - \sum_{j=1}^{N-2} x_j\right] \; dx_1 \ldots dx_{N-2}$$

After two integrations we have

$$\int_{x_1=0}^{1} \cdots \int_{x_{N-3}=0}^{1-\sum_{j=1}^{N-4} x_j} \cdot \; \frac{1}{2}\left[1 - \sum_{j=1}^{N-3} x_j\right]^2 \; dx_1 \ldots dx_{N-3}$$

etc. Finally we have $1/(N-1)!$ as required. The general case follows using the same kind of integrations. ■

Proposition 5 *If $p(\mathbf{x}_N)$ is constant on simplices then*

$$\bar{F}\left(\mathbf{y}_N \mid \sum_{j=1}^{N} X_j = N\theta\right) = \left(1 - \frac{\sum_{j=1}^{N} y_j}{N\theta}\right)^{N-1}$$

for $0 \leq \sum_{j=1}^{N} y_j \leq N\theta$ and is 0 elsewhere.

PROOF: Note that for $1 \leq n < N$

$$\bar{F}\left(\mathbf{y}_N \middle| \sum_{j=1}^{N} X_j = N\theta\right) =$$

$$= \Pr\left(X_1 > y_1, \ldots, X_n > y_n, X_{n+1} > 0, \ldots, X_N > 0 \middle| \sum_{j=1}^{N} X_j = N\theta\right).$$

The n-dimensional density can be calculated as

$$(-1)^n \frac{\partial^n}{\partial y_1 \ldots \partial y_n} \bar{F}\left(\mathbf{y}_n \mid \sum_{j=1}^{n} X_j = N\theta\right)\Bigg]_{y_1=x_1, \ldots, y_n=x_n} =$$

$$= (-1)^n \frac{\partial^n}{\partial y_1 \ldots \partial y_n} \left(1 - \frac{\sum_{j=1}^{n} y_j}{N\theta}\right)^{N-1}\Bigg]_{y_1=x_1, \ldots, y_n=x_n} =$$

$$= [(N-1)/N\theta] \ldots [(N-n)/N\theta] \left(1 - \sum_{i=1}^{n} x_i \,/\, N\theta\right)^{N-n-1}.$$ ■

Such distributions are called *singular* since, although continuous, there is no corresponding probability density in N dimensions. However there is a density corresponding to all marginal distributions of dimension $n < N$.
For $n = 1$, we have the marginal density

$$p_N(x_1 \,|\, \theta) = [(N-1) \,/\, N\theta]\,[1 - x_1 \,/\, N\theta]^{N-2} \qquad 0 \le x_1 \le N\theta$$

and $\bar{F}_N(y_1 \,|\, \theta) = [1 - y_1 \,/\, N\theta]^{N-1}$. The marginal probabilistic failure rate in the classical sense is

$$r_N(x_1 \,|\, \theta) = [(N-1) \,/\, N\theta]\,[1 - x_1 \,/\, N\theta] \tag{4}$$

Note that $r_N(x_1 \,|\, \theta)$ is *strictly increasing* for $0 \le x_1 \le N\theta$.
The expression (4) illustrates the fundamental difference between our notion of no-aging and the classical notion. In the finite population case, our notion of no-aging implies indifference between vector lifetimes having the same sum. Conditional on the sum, $N\theta$, the usual probabilistic notion of failure rate for the first unit must increase as the unit's life approaches $N\theta$. On the other hand as $N \to \infty$, our notion for $n = 1$ coincides with the classical notion of constant failure rate. For $N \to \infty$ and $\sum_{i=1}^{N} x_i/N \to \theta$, we have $p(x_1 \,|\, \theta) = \frac{1}{\theta} e^{-x_1/\theta}$ which of course has constant probabilistic failure rate.
Fix n and $\theta > 0$ and let $N \to \infty$. Then

$$\lim_{N\to\infty} p\left(\mathbf{x}_n \,|\, \sum_{i=1}^{N} X_i = N\theta\right) = \prod_{i=1}^{n} \frac{1}{\theta} e^{-x_i/\theta}.$$

The result follows from a characterization of e, namely

$$\lim_{N\to\infty} \left[1 - \frac{x}{N}\right]^N = e^{-x}.$$

4. PARAMETRIC MODELS

A decision problem begins with a list of relevant alternative decisions or possible acts [Savage (1954)]. If our utility or loss function relative to N units depends only on the total life of these units or the total life of a subset of these units, then our parameter of interest is

$$\Theta = \sum_{i=1}^{N} X_i \,/\, N$$

Our "likelihood" model for analyzing lifetime data relative to this decision problem is

$$\bar{F}\left(\mathbf{y}_N \,|\, \sum_{j=1}^{N} X_j = N\theta\right) = \left(1 - \frac{\sum_{j=1}^{N} y_j}{N\theta}\right)^{N-1}$$

as we saw in section 3.
If our decision problem depends on say the total lifetime operating cost of N units where $\psi(X_i)$ is the cost for unit i, then

$$p\,(\psi(x_1), \ldots, \psi(x_N))$$

would be constant for $\left\{\mathbf{x}_N \mid \sum_{i=1}^{N} \psi(x_i) = N\theta\right\}$. Hence

$$\bar{F}\left(\mathbf{y}_N \mid \sum_{j=1}^{N} \psi(X_j) = N\theta\right) = \left(1 - \frac{\sum_{j=1}^{N} \psi(y_j)}{N\theta}\right)^{N-1}$$

using Proposition 5 and a change of variable. Note that if ψ is convex then

$$\bar{F}\left(\mathbf{y}_N \mid \sum_{j=1}^{N} \psi(X_j) = N\theta\right)$$

is logarithmically concave and therefore Schur-concave by Proposition 2.
If $\Pi_N(\theta) = P(\Theta \leq \theta)$ is our opinion concerning Θ and

$$\bar{F}_N(\mathbf{y}_N) = \int_{\theta \geq \sum_{i=1}^{N} \psi(y_i)/N} \left[1 - \sum_{i=1}^{N} \psi(y_i) \,/\, N\theta\right]^{N-1} d\Pi_N(\theta),$$

then $\bar{F}_N(\mathbf{y}_N)$ is also Schur-concave. This is so since ψ convex implies $\sum_{i=1}^{N} \psi(y_i)$ is Schur-convex by Proposition 1.

REFERENCES

[1] Barlow, R. E., and Proschan, F. (1981), *Statistical Theory of Reliability and Life Testing.* Silver Spring, MD: To Begin With.

[2] Barlow, R. E. and Mendel, M. (1992), "de Finetti-Type Representations for Life Distributions", *Journal of the American Statistical Association*, **87**, 420, 1116-1122.

[3] Basu, D. (1988). *Statistical Information and Likelihood.* (Lecture Notes in Statistics, Number 45). New York: Springer-Verlag.

[4] De Finetti, B. (1937), "Foresight: Its Logical Laws, Its Subjective Sources," *Annales de l'Institute Henri Poincaré*, **7**, 1-68. English translation in *Studies in Subjective Probability.* [H.E. Kyburg, Jr., and H.E. Smokler, (eds), 1980, (2nd ed.)]. Huntington, N.Y.: Robert E. Krieger Pub. Co., pp. 53-118.

[5] Marshall, A. W. and Olkin, I. (1979), *Inequalities: Theory of Majorization and Its Applications.* New York: Academic Press.

[6] Mendel, M. B. (1989), *Development of Bayesian Parametric Theory with Applications to Control.* MIT Ph.D. thesis.

[7] Mendel, M. B. (1991), "Bayesian Parametric Models for Lifetimes", to appear in the *Proceedings of the IV International Meeting on Bayesian Statistics*, Valencia, April 15-20, 1991.

[8] Proschan, F. (1975), "Applications of Majorization and Schur Functions in Reliability and Life Testing". In *Reliability and Fault Tree Analysis*, [R. E. Barlow, J. B. Fussell and N. D. Singpurwalla (eds.)], pp. 237-258.

[9] Savage, L. J. (1954), *The Foundations of Statistics.* New York: John Wiley and Sons.

[10] Spizzichino, F. (1991), "Reliability Decision Problems under Conditions of Aging", to appear in the *Proceedings of the IV International Meeting on Bayesian Statistics*, Valencia, April 15-20, 1991.

Recent Adv. in Stat. and Prob., pp. 441-450
J. Pérez Vilaplana and M. L. Puri (Eds)

de Finetti Representations of Survival Functions Level to a Product Measure *

R.M. COOKE
Department of Mathematics and Informatics, Delft University of Technology, Mekelwes 4, NL-2600GA Delft, The Netherlands

Abstract – de Finetti-type representations of survival functions are considered. Exchangeable continuous strictly monotonic infinite-dimensional survival functions whose finite dimensional marginals have the same level sets as a product survival function, can be represented uniquely as mixtures of positive powers of that product survival function. A functional equation characterizes the survival functions which can be represented in this way, and the product measure is extracted using techniques from functional equations.

1. INTRODUCTION

Two functions, f and g on $\Re^n$, with $n \geq 2$ are called *level* ($f \sim g$) if they have the same level sets, that is, $f(x) = f(y)$ if and only if $g(x) = g(y)$. This concept is meaningful for measures only if the measures are represented as integrals of density functions. It is convenient to study survival functions, that is functions of the form $\bar{F}(x) = \text{Prob}\{X > x\} = \text{Prob}\{X_1 > x_1, \cdots X_n > x_n\}$ for a random vector X. We shall be concerned with continuous survival functions which are strictly monotone in each argument. For univariate survival functions this entails that an inverse exists; for n-dimensional multivariate survival functions, it entails that the level sets are simple $n-1$ dimensional hypersurfaces intersecting all the coordinate axes. Each point of the support is on exactly one such hypersurface.
This paper studies de Finetti-type representations of continuous strictly monotone survival functions which are level to a product measure. Let $\Re_+$ denote the positive reals, and $\bar{\Re}_+$ the non-negative reals. If $\bar{F} : \bar{\Re}_+^\infty \rightarrow [0,1]$ is a survival function and $x \in \bar{\Re}_+^n$, then $\bar{F}(x)$ denotes the n-dimensional marginal $\bar{F}(x_1, \ldots, x_n, 0, 0, \ldots)$, and when the context would be ambiguous, $\bar{F}_n$ will denote the n-dimensional marginal. $\bar{F} \sim \prod \mu$ means that all finite-dimensional marginals of the two functions $\bar{F}$ and $\prod \mu$ are level. Recall that according to the de Finetti theorem, any exchangeable survival function $\bar{F}; \bar{\Re}_+^\infty \rightarrow [0,1]$

*This research was partially supported by a grant to the University of California at Berkeley by the Army Research Office, and was preformed while the author was a visiting research associate at the Department of Industrial Engineering and Operations Research, University of California at Berkeley, on sabbatical leave from the Department of Mathematics and Informatics, Delft University of Technology, The Netherlands. The idea of looking at the de Finetti theorem for survival functions was suggested and motivated by Richard Barlow. Helpful comments were provided by Jolanta Misiewicz, Yu Hayakawa and Max Mendel.

can be written as

$$\forall x \in \Re^n; \bar{F}(x) = \int_{\mu\in\mathcal{M}} \prod_{i=1}^{n} \mu(x_i) d\lambda_F \mu \qquad (1)$$

where $\mathcal{M}$ is the set of survival measures on $[0,\infty)$. The measure λ_F is unique. In some cases the integral can be expressed as an integral over the range of some real parameter. For example, if we require that the density f of $\bar{F}$ is l_p- *isotropic*, that is, has level sets corresponding to the l_p norm, then the k-dimensional marginal density f_k can be uniquely written as ([5], [10], [11])

$$f_k\left(\sum_{i=1}^{k} x_i^p\right) = \int_0^\infty \exp\left[= -t\sum_{i=1}^{k} x_i^p\right]\left[\frac{pt^{1/p}}{\Gamma(1/p)}\right]^k \lambda_F(dt). \qquad (2)$$

Comparing the above two representations, two differences are apparent:
(i) in the second equation the integral runs over values of a "scale parameter" t, and,
(ii) the product measures over which the measure λ_F mixes are level to the density on the left hand side.
These two facts are equivalent for continuous monotone survival functions, in fact:

Theorem 1 *Let $\bar{F} : \bar{\Re}_+^\infty \to [0,1]$ be a continuous strictly monotone survival function and μ be a continuous strictly monotone univariate survival function, then the following are equivalent:*

$$\bar{F} \sim \prod \mu \qquad (3)$$

$$\forall n \in N, \forall x \in \bar{\Re}_+^n;\ \bar{F}(x) = \int_0^\infty \prod_{i=1}^{n} \mu(x_i)^s \lambda_F(ds) \qquad (4)$$

PROOF: It is trivial that (3) follows from (4). To go the other way note first that l_1-isotropic survival functions have l_1-isotropic densities, and converseley (this is not true for l_p densities with $p \neq 1$) [7]. Put $Z_i = -\ln(\mu(X_i))$. Since μ is strictly decreasing, $P(Z_i > y_i) = P(\mu(X_i) < e^{-y_i}) = P(X_i > \mu^{-1}e^{-y_i})$. We have:

$$P(Z_1 > y_1, \dots Z_k > y_k) = P(Z_1 > \tilde{y}_1, \dots Z_k > \tilde{y}_k) \longleftrightarrow$$

$$P(X_1 > \mu^{-1}e^{-y_1} \dots X_k > \mu^{-1}e^{-y_k}) = P(X_1 > \mu^{-1}e^{-\tilde{y}_1} \dots X_k > \mu^{-1}e^{-\tilde{y}_k}) \longleftrightarrow$$

$$\prod \mu\mu^{-1}e^{-y_i} = \prod \mu\mu^{-1}e^{-\tilde{y}_i} \longleftrightarrow \sum y_i = \sum \tilde{y}_i.$$

Hence $Z_1, Z_2, \dots$ is l_1-isotropic. From (2):

$$P(Z_1 > y_1, \dots, Z_k > y_k) = \int_0^\infty \left(e^{\Sigma - y_i}\right)^s \lambda(ds).$$

Substitute $y_i = -\ln \mu x_i$ to obtain (4). ∎

The family of survival functions $e^{-sx}, s \in (0,\infty)$ are said to have proportional hazard functions. There are advantages to looking at the representation of $l-1$-isotropic survival functions from the viewpoint of Theorem 1 In particular it suggests techniques of functional equations for finding equivalent conditions for (3), and for finding μ when these conditions are satisfied. For exchangeable $\bar{F}$, equivalent conditions in terms of a bisymmetry condition on the two-dimensional marginals of $\bar{F}$ are derived in Section 2:. If X is exponentially distributed and $f(\cdot)$ is non-negative, continuous and increasing with $f(0) = 0$, then Mendel and Barlow [3] call the distribution of $f(X)$ "generalized

Weibull". Section 2. shows how to recognize mixtures of generalized Weibulls from their 2-dimensional marginals, and how to extract the transformation $f = -\ln(\mu)$. This is applied in Section 3. to obtain a representation of l_p isotropic survival functions. Section 4. gives two functional equations for survival functions. The last of these could be derived from (2), but a direct proof using techniques of functional equations, enables us to give a short proof of Theorem 1 in the last Section.

2. EQUIVALENT CONDITIONS FOR $\bar{F} \sim \prod \mu$

Throughout this section we assume that $\bar{F}$ is a strictly monotonic exchangeable survival function on $\bar{\Re}_+^\infty$, and that μ is a continuous strictly monotonic univariate survival function. The first theorem collects some obvious properties, the second shows that it suffices to look at 2−dimensional marginals, and the third gives a functional equation which represents all $\bar{F}$ which are level to a product.

Theorem 2 (1) *If $\bar{F} \sim \prod \mu$ and $\mu(x) = \bar{F}(x)$ (i.e. μ is the 1-dimensional marginal of $\bar{F}$), then $\bar{F} = \prod \mu$.*
(2) *If $\bar{F} \sim \prod \mu$ then*

$$\bar{F}(x_{n+1}|x_1, \cdots x_n) = \bar{F}(x_{n+1}|y_1, \cdots y_n) \Longleftrightarrow \bar{F}(x_1, \cdots x_n) = \bar{F}(y_1, \cdots y_n) \tag{5}$$

PROOF: Immediate. ∎

Theorem 3 *Suppose for all $x, x' \in \bar{\Re}_+^2$: $\bar{F_2}(x) = \bar{F_2}(x')$ if and only if $\mu(x_1)\mu(x_2) = \mu(x'_1)\mu(x'_2)$; then $\bar{F} \sim \prod \mu$.*

PROOF: Put $z_i(x_i) = -\ln(\mu(x_i))$ and $\tilde{F}(z_1, \cdots, z_n) = \bar{F}(x_1, \cdots, x_n)$.

Lemma 1 *A function $H : \Re^\infty \to \Re$ which is invariant under finite permutations and l_1 isotropic in the first two coordinates, is l_1 isotropic.*

PROOF OF LEMMA: We must show for any $x_1, \cdots x_n$, $H(x_1, \cdots x_n) = H(\sum_{i=1}^n x_i, \overbrace{0, \cdots, 0}^{n-1})$.

$$\begin{aligned} H(\textstyle\sum_{i=1}^n x_i, \overbrace{0, \cdots, 0}^{n-1}) &= H(x_1, \textstyle\sum_{i=2}^n x_i, \overbrace{0, \cdots, 0}^{n-2}) = \\ = H(0, \textstyle\sum_{i=2}^n x_i, x_1, \overbrace{0, \cdots, 0}^{n-3}) &= H(x_2, \textstyle\sum_{i=3}^n x_i, x_1, \overbrace{0, \cdots, 0}^{n-3}) = \\ &\cdots \\ = H(x_n, \cdots, x_1) &= H(x_1, \cdots, x_n). \end{aligned}$$ ∎

By the hypothesis of the theorem, if $z_1 + z_2 = z'_1 + z'_2$, then $\tilde{F}(z_1, z_2) = \tilde{F}(z'_1, z'_2)$, i.e. $\tilde{F}$ is l_1 isotropic in the first two coordinates. The lemma entails that $\tilde{F}$ is l_1 isotropic, hence for x, $x' \in \bar{\Re}_+^n$, $z(x) = (\ln \mu(x_1), \cdots, \ln \mu(x_n))$; we have: $\bar{F}(x) = \bar{F}(x') \Leftrightarrow \tilde{F}(z) = \tilde{F}(z')$ $\Leftrightarrow \sum_{i=1}^n z_i = \sum_{i=1}^n z'_i \Leftrightarrow \prod_{i=1}^n \mu(x_i) = \prod_{i=1}^n \mu(x'_i)$. It follows that $\bar{F} \sim \prod \mu$. ∎

For the next theorem we define $G : \bar{\Re}_+^2 \to \bar{\Re}_+$ as:

$$G(x, y) = z \text{ if } \bar{F}(z, z) = \bar{F}(x, y).$$

G is called *bisymmetric* if

$$G\big(G(x, y),\, G(z, w)\big) = G\big(G(x, z),\, G(y, w)\big).$$

The problem of identifying those survival functions which are level to a product is essentially a matter of representing bisymmetric forms.

Theorem 4 *$\bar{F}$ is level to a product if and only if G is bisymmetric.*

PROOF: By Theorem 3 it suffices to find a univariate survival function μ such that for all $x, x', y, y' \in \bar{\Re}_+$;

$$\mu(x)\mu(y) = \mu(x')\mu(y') \qquad \text{iff } \bar{F}(x,y) = \bar{F}(x',y').$$

The function G is reflexive and symmetric. There exists a continuous strictly monotonic function $k : \bar{\Re}_+ \to \Re$ such that

$$G(x,y) = k^{-1}\left(\frac{k(x)+k(y)}{2}\right); \qquad x, y \in \bar{\Re}_+. \tag{6}$$

Moreover, k is uniquely determined up to an affine transformation [2, p.287, p 246]. Arranging that $k(0) = 0, k(x) \geq 0$, put $\mu(x) = e^{-k(x)}$, then

$$\mu\Big(G(x,y)\Big)^2 = \mu(x)\mu(y).$$

If $\bar{F}(x,y) = \bar{F}(x',y')$, then $G(x,y) = G(x',y')$ and $\mu(x)\mu(y) = \mu(x)\mu(y)$. ■

From the proof of [2, p.87], one can construct the function k and hence μ. However, if the conditions of Theorem 4 hold it is easier to recover μ in a different way. Let $\bar{F}_1$ denote the one-dimensional marginal of $\bar{F}$, and define

$$g_k(x) = \bar{F}_1^{-1}(\bar{F}(\overbrace{x, x, \cdots, x}^{k})$$

Then $\mu(x)^k = \mu(g_k(x)))$, or more generally,

$$\mu(x)^{k/n} = \mu\Big(g_n^{-1}\big(g_k(x)\big)\Big), \qquad k, n \in N. \tag{7}$$

Hence, if we assign $\mu(x_0) = r$, for some $0 < r < 1$, then (7) determines μ^{-1} on a dense set. Since μ is continuous, μ is determined on a dense set as well. In specific situations more elegant methods may be available, as illustrated in the next section.

3. APPLICATIONS TO LINEAR, INVERSE LINEAR AND l_p ISOTROPIC SURVIVAL FUNCTIONS

For a uniquely defined continuous function $g : \bar{\Re}^+ \to \bar{\Re}^+$ we may write

$$\bar{F}\Big(g(x), g(x)\Big) = \bar{F}(0,x) \tag{8}$$

If $\bar{F} \sim \prod \mu$, then by Theorem 1 μ also satisfies (8), or

$$\log\Big(\mu(g(x)\Big) = \frac{1}{2}\log\Big(\mu(x)\Big) \tag{9}$$

Equation (9) is an example of the Schröder equation, and has been studied extensively [9]. Existence and uniqueness theorems are available for certain g, as illustrated by ([9, pp 68, 69], [8]):

Theorem 5 *Let g be continuous and strictly increasing on $D = [0, a]$, $0 < a \leq \infty$; such that*

(i) $0 < g(x) < x$ in $D\backslash\{0\}$,

(ii) $x \to g(x)/x$ is monotonic in $D\backslash\{0\}$, and

(iii) $\lim_{x\to 0}[g(x)/x] = s$, where $0 < s < 1$.

Then $\phi(g(x)) = s\phi(x)$ has a unique one parameter family of solutions $\phi : D \to \Re$ given by $\phi(x) = c \lim_{n\to\infty} \frac{g^n(x)}{g^n(x_0)}$, where $c \in \Re$ is any constant and x_0 is an arbitrary fixed point in $D\backslash\{0\}$. The function $x \to \frac{\phi(x)}{x}$ is monotonic in $D\backslash\{0\}$.

This result can be applied to represent survival functions level to infinite products of *linear* ($1 - x$; $0 \leq x \leq 1$) and *inverse linear* ($\frac{1}{1+x}$; $0 \leq x \leq \infty$) survival functions respectively. These correspond to uniform and quadratic densities.

Theorem 6 *Let $\bar{F}$: $D^\infty \to [0,1]$ be a continuous monotone exchangeable survival function and suppose that $\bar{F} \sim \prod \mu$.*

(i) If $D = [0,1)^\infty$ and $\bar{F}\left(1 - (1-x)^{1/2}, 1 - (1-x)^{1/2}\right) = \bar{F}(0, x)$, then $\mu(x) = 1 - x$.

(ii) If $D = [0,\infty)^\infty$ and $\bar{F}\left((1+x)^{1/2} - 1, (1+x)^{1/2} - 1\right) = \bar{F}(0, x)$, then $\mu(x) = \frac{1}{1+x}$.

PROOF: We do only (*ii*) as (*i*) is similar. The function $g(x) = (1+x)^{1/2} - 1$ in (9) is easily verified to satisfy conditions (i) and (ii) of Theorem 5, and with L'Hospital's rule $\lim_{x\to 0} \frac{g(x)}{x} = \lim_{x\to 0} \frac{dg(x)}{dx} = 1/2$. It follows that a solution exists and is unique up to a constant. $\mu(x) = \frac{1}{1+x}$ is a solution and is therefore the only solution in the class of univariate survival functions. ∎

Results like Theorem 5 place restrictions on g which are not satisfied for general polynomial and inverse polynomial survival functions. A characterization of survival functions level to products of polynomial and inverse polynomial survival functions is not known. l_p isotropic survival functions violate condition (*iii*). A representation for l_p isotropic survival functions is given in:

Theorem 7 *Let $\bar{F}$: $[0,\infty)^\infty \to [0,1]$ be a continuous monotone exchangeable survival function and suppose that the two-dimensional marginal of $\bar{F}$ satisfies*

$$\forall t \in \Re_+,\ \bar{F}(t,0) = \bar{F}(2c(t/2, t/2)); \tag{10}$$

for some constant c (in fact $c \leq 1$ since $\bar{F}$ is a survival function), and

$$\bar{F} \sim \prod \mu; \tag{11}$$

for some continuous monotone univariate survival function μ; then $\bar{F}$ is l_p-isotropic with $p = -log_2(c)$.

PROOF: By Theorem 1

$$\bar{F} = \int_0^\infty \prod \mu^s \, d\lambda_F(s)$$

It suffices to show that $\mu(t) = K^{-t^p}$ for some constant K, and that $p = -\log_2(c)$. (10) and (11) entail that $\mu(t) = \mu(ct)^2$. After k iterations of the substitution $t \to ct$, this becomes:

$$\mu(t) = \mu(c^k t)^{2^k};$$

Put $t = c^x$; then μ must satisfy the functional equation:

$$\mu(c^x) = \mu(c^{k+x})^{2^k}$$

or

$$U(x) = k + U(k + x) \tag{12}$$

with $U(x) = \log_2 \ln \mu(c^x)$. As U is continuous, the solutions [1] have the form

$$U(x) = -x + \text{constant}$$

so that

$$\mu(c^x) = e^{K2^{-x}} \tag{13}$$

for some constant K. By Theorem 9, all μ satisfying (11) are powers of a given solution, so we may choose $K = 1$. Substitute

$$x = log_c(t) = \frac{\log_2(t)}{\log_2(c)}$$

in (13), then

$$2^{-x} = t^{\frac{-1}{\log_2(c)}}.$$

It remains to show that $p = -\log_2(c)$. For an l_p -symmetric measure satisfying (11):

$$t = \|\, 2c(t/2, t/2) \,\|_p = c\|\, (t,t) \,\|_p = 2^{\frac{1}{p}} tc$$

which is equivalent to

$$c = 2^{-\frac{1}{p}}.$$

■

4. FUNCTIONAL EQUATIONS FOR SURVIVAL FUNCTIONS

Saying that two survival distributions are level entails functional equations from which the following two theorems draw conclusions. If a function f is continuous on $(0,1)$ and satisfies $f(xy) = f(x)f(y)$, then $f(x) = x^q$ for some constant q. ([1, p. 41]). For Theorems 9, 10 the following simple fact will suffice:

Theorem 8 *Let f be a positive function defined on an interval $I \subset \Re_+$, such for all s, y, with $y, y^s \in I$, $f(y^s) = f(y)^s$; then $f(y) = y^q$, for some q.*

PROOF: Pick $y_0 \in I, y_0 \neq 1$, and put $q = \log_{y_0} f(y_0)$. Then $f(y_0) = y_0^q$. Put $y = y_0^{r(y)}$ for $y \in I$. Then $f(y) = f(y_0^{r(y)}) = f(y_0)^{r(y)} = y_0^{qr(y)} = y^q$. ■

Theorem 9 *Let $\bar{F}$ and $\bar{G}$ be continuous monotone univariate survival functions, and suppose for some integer $n \geq 2$,*

$$\prod^n \bar{F} \sim \prod^n \bar{G} \tag{14}$$

then for some non-negative real number s;

$$\forall x \in \Re^n \; \bar{F}(x) = \bar{G}(x)^s. \tag{15}$$

[1] This is an instance of Pexider's equation $f(x+y) = g(x) + h(y)$; having the general solution $f(t) = \phi(t) + a + b$, $g(t) = \phi(t) + a, h(t) = \phi(t) + b$; where a and b are arbitrary constants, and ϕ solves $\phi(x+y) = \phi(x) + \phi(y)$. [1, p. 142]

PROOF: Restricting to arguments of the type $(\overbrace{x_1, x_2, 0, 0\cdots}^{n})$ it suffices to prove the result for $n = 2$. We have

$$\bar{F}(x_1)\bar{F}(x_2) = \bar{F}(t) \iff \bar{G}(x_1)\bar{G}(x_2) = \bar{G}(t).$$

Put $\bar{F}(x) = \bar{H}(\bar{G}(x))$, which is possible because G is invertible. We have

$$\bar{H}\bar{G}(x_1)\bar{H}\bar{G}(x_2) = \bar{H}\bar{G}(t) = \bar{H}(\bar{G}(x_1)\bar{G}(x_2)). \tag{16}$$

This holds for all $\bar{G}(x_1)$, $\bar{G}(x_2) \in (0,1)$. With the multiplicative Cauchy equation [1, p. 41] we conclude for some $s \in \Re$

$$\bar{F}(x) = \bar{H}(\bar{G}(x)) = \bar{G}(x)^s.$$

Since $0 \le \bar{F}(x), \bar{G}(x) \le 1$, we must have $0 < s$. ∎

Theorem 10 *For univariate survival functions* $\bar{F}_i$ *and* $\bar{G}$; *with* $\alpha_i > 0$, $i = 1,\cdots m$, $\sum \alpha_i = 1$; *suppose*

$$\sum_{i=1}^{m} \alpha_i \prod \bar{F}_i \sim \prod \bar{G} \tag{17}$$

where $\sum_{i=1}^{m} \alpha_i \prod \bar{F}_i$, *and* $\bar{G}$ *are continuous and strictly monotone. Then*

$$\bar{F}_i = \bar{G}^{s_i} \tag{18}$$

PROOF: The derivatives of $\bar{F}_i$ exist, though $\bar{F}_i$ are not assumed to be strictly monotone (this follows from the theorem). G^{-1} exists because of strict motonicity. Put $\mathcal{F}_i = \bar{F}_i \bar{G}^{-1}$. $\mathcal{F}_i$ are defined on $(0,1]$, and are continuous ($\bar{G}^{-1}(0)$ might be defined, but in light of continuity, there is no loss of generality in restricting attention to $(0,1]$.). Letting $m!$ denote the set of permutations of $\{1,\cdots m\}$, define:

$$d = \min_{\pi \in m!} \left\{ \inf \{ x | \mathcal{F}_{\pi(1)}(y) \ge \mathcal{F}_{\pi(2)}(y) \ge \cdots \mathcal{F}_{\pi(m)}(y); \quad \forall y \in (x,1) \} \right\} \tag{19}$$

$$D = (d, 1] \tag{20}$$

We note that $d < 1$ since the (one-sided) derivatives of $\mathcal{F}_i$ exist at 1. Also, if $d > 0$, and if S is the subset of indices for which $\mathcal{F}_i(d) = \mathcal{F}_j(d)$, then

$$\text{there exist } y \in D \text{ and } i,j \in S \text{ such that } \mathcal{F}_i(y) \ne \mathcal{F}_j(y) \tag{21}$$

We may assume without loss of generality that

$$\mathcal{F}_1(y) \ge \mathcal{F}_2(y) \ge \cdots \mathcal{F}_m(y); \; y \in D. \tag{22}$$

Suppose $(\prod \bar{G})(\overbrace{x,\cdots x}^{n}) = (\prod \bar{G})(\overbrace{t,\cdots t}^{q}, \overbrace{0,\cdots,0}^{n-q})$, then $\bar{G}(x)^n = \bar{G}(t)^q$. We have from (17)

$$\bar{G}(x)^n = \bar{G}(t)^q \quad \Longrightarrow \quad \sum_{i=1}^{m} \alpha_i \bar{F}_i(x)^n = \sum_{i=1}^{m} \alpha_i \bar{F}_i(t)^q \tag{23}$$

Write $x = \bar{G}^{-1}(y)$; $t = \bar{G}^{-1}(v)$, and (23) is equivalent to:

$$y^n = v^q \quad \Longrightarrow \quad \sum_{i=1}^{m} \alpha_i \mathcal{F}_i(y)^n = \sum_{i=1}^{m} \alpha_i \mathcal{F}_i(v)^q$$

Substitute $v = y^{n/q}$. To study the limit behavior as $n \to \infty$ we restrict the arguments to D: For all y, n, q such that $y, y^{n/q} \in D$:

$$\left[\sum_{i=1}^{m} \alpha_i \mathcal{F}_i(y)^n\right]^{1/n} = \left[\sum_{i=1}^{m} \alpha_i \mathcal{F}_i(y^{n/q})^q\right]^{1/n} \tag{24}$$

We let $n, q \to \infty$, such that $n/q \to k$ with $y^k \in D$. The left hand side of (24) converges to $\mathcal{F}_1(y)$, for all $y, y^k \in D$, because of (22). On the right hand side

$$\left[\sum_{i=1}^{m} \alpha_i \mathcal{F}_i(y^{n/q})^q\right]^{1/n} = \alpha_1^{1/n} \left[\mathcal{F}_1(y^{n/q})^q\right]^{1/n} \left[1 + \sum_{i=2}^{m} \frac{\alpha_i}{\alpha_1} \frac{\mathcal{F}_i(y^{n/q})^q}{\mathcal{F}_1(y^{n/q})^q}\right]^{1/n} \longrightarrow \mathcal{F}_1(y^k)^{1/k}. \tag{25}$$

Comparing (24) and (25) we see that for $y, y^k \in D$; $\mathcal{F}_1(y^k) = \mathcal{F}_1(y)^k$. This means that Theorem 8 can be applied to yield; $\forall y \in D: \ \mathcal{F}_1(y) = y^{s_1}$; or:

$$\forall x \text{ such that } \bar{G}(x) \in D: \ \bar{F}_1(x) = \bar{G}(x)^{s_1}. \tag{26}$$

for some s_1, which must be positive since $\mathcal{F}_1$ ranges over $(0,1]$. We now substitute (26) into (24). For $y, y^{n/q} \in D$ we can eliminate the term in $\mathcal{F}_1$ and conclude:

$$\sum_{i=2}^{m} \alpha_i \mathcal{F}_i(y)^n = \sum_{i=2}^{m} \alpha_i \mathcal{F}_i(y^{n/q})^q. \tag{27}$$

From this we derive, as above

$$\forall x \text{ such that } \bar{G}(x) \in D: \ \bar{F}_2(x) = \bar{G}(x)^{s_2}. \tag{28}$$

Proceeding in this way we show for $i = 1, \cdots m$

$$\forall x \text{ such that } \bar{G}(x) \in D: \ \bar{F}_i(x) = \bar{G}(x)^{s_i}. \tag{29}$$

We show that $d = 0$. Suppose to the contrary that $d > 0$. Then for those i, j such that $\mathcal{F}_i(d) = \mathcal{F}_j(d)$, we have by continuity: $\mathcal{F}_i(d) = d^{s_i}$ and $\mathcal{F}_j(d_j) = d^{s_j}$. This entails that $s_i = s_j$; and that $\mathcal{F}_i(y) = \mathcal{F}_j(y)$ for all $y \in D$. This however contradicts (21). It follows that D is $(0,1]$, and the theorem is proved. ■

In Theorem 10 *all* finite dimensional marginals are level, whereas in Theorem 9, the n-dimensional product measures are level, for a fixed n. It is not known whether the analogue of Theorem 9 holds for mixtures of product measures.

5. A DE FINETTI THEOREM

We recall briefly some facts about the topology of weak convergence. Let S be a metric space, and let $B(S)$ be the Borel field over S generated by the metric. The set $Z(S, B(S))$ of probability measures on $(S, B(S))$ may be endowed with the topology of weak convergence. For $\lambda \in Z(S, B(S))$, a basis neighborhood of λ may be written as:

$$N(\lambda) = \{\eta | \eta(F_i) \leq \lambda(F_i) + \xi; \xi > 0, F_i \in B(S) \text{ closed}; \ i = 1 \cdots k\} \tag{30}$$

Convergence in this topology is equivalent to weak convergence, denoted by $\to_w$. Moreover, the set of $\lambda \in Z(S, B(S))$ with finite support is dense in this topology on $Z(S, B(S))$. Let $\mathcal{M} = \{\mu | \mu \text{ is a univariate survival function}\}$. $\mathcal{M}$ is a metric space with the Prohorov metric, and may be endowed with the topology of weak convergence ([6, p. 236 - 238]). Define: $\Lambda =$ the set of probabilities on $\left(\mathcal{M}, B(\mathcal{M})\right)$. Let $\mu^* \in \mathcal{M}$; define

$$\Lambda^* = \left\{\lambda \in \Lambda \mid \int_{\mathcal{M}} \prod \mu \, d\lambda\mu \sim \prod \mu^*\right\} \tag{31}$$

$$\Lambda_0^* = \{\lambda \in \Lambda^* \mid \lambda \text{ has finite support}\} \tag{32}$$

Let $\mu^* \in \mathcal{M}$; define If $\lambda \in \Lambda_0^*$; then $\lambda = \sum_{i=1}^{n} \alpha_i 1_{\mu_i}$, where 1 denotes the indicator function. We can now prove

Theorem 1 *Let $\bar{F}$: $[0,\infty)^\infty \to [0,1]$ be a continuous monotone symmetric survival function and suppose that*

$$\bar{F} \sim \prod \mu^* \tag{33}$$

for some continuous monotone univariate survival function μ^*. Then $\lambda_{\bar{F}}$ in (1) *is concentrated on* $\{\mu^{*s} | s \in (0,\infty)\}$. PROOF: Take Λ^* as above, with μ^* continuous monotone.

Let $M = \{\mu^{*s} | s \in (0,\infty)\}$. We show that $\lambda_{\bar{F}}(M) = 1$ if $\lambda_{\bar{F}} \in \Lambda^*$. Pick $\lambda_i \to_w \lambda_{\bar{F}}$; $\lambda_i \in \Lambda_0^*$, $i = 1, 2 \cdots$; then $\lambda_i(M) = 1$ by Theorem 10. Let $\bar{M}$ denote the closure of M in $\left(\Lambda^*, B(\Lambda^*)\right)$. We show:

$$\bar{M} = \{\mu^{*s} | s \in (0,\infty]\}$$

where $\mu^{*\infty}$ is unit mass at zero. Suppose $\mu^{*s_i}(x) \to_w \eta$, $i \to \infty$. We show that $\eta \in \{\mu^{*s} | s \in (0,\infty]\}$. Consider two cases:
(1) for all $k \in N$ there are finitely many i with $s_i \in [0,k]$ and
(2) there exists a $k \in N$ such that for infinitely many i, $s_i \in [0,k]$.
In case (1) we have that $\mu^{*s_i}(x)$ converges to unit mass at the origin. In case (2) pick a subsequence s_{i_j} such that $s_{i_j} \to r$, $r \in [0,k]$. For all $x \in [0,\infty)$

$$\mu^{*s_{i_j}}(x) \to \mu^{*r}(x).$$

It follows that $\mu^{*s_{i_j}} \to_w \mu^{*r}$, as $j \to \infty$, and hence that $\mu^{*s_i} \to_w \mu^{*r}$, since μ^{*s_i} converges weakly.
Now $\lambda_i(\bar{M}) = 1$ so that $\lambda_{\bar{F}}(\bar{M}) = 1$, by (30). Moreover, $\lambda_{\bar{F}}$ cannot assign positive mass to $\mu^{*\infty}$, as $\lambda_{\bar{F}}$ is continuous monotone by assumption. ■

REFERENCES

[1] ACZEL, J., (1966). *Lectures on Functional Equations and their Applications.* Academic Press, New York.

[2] ACZEL, J. AND DHOMBRES, J., (1989). *Functional Equations in Several Variables.* Cambridge University Press.

[3] BARLOW, R.E. AND MENDEL, M.B., (1992). " De Finetti - Type Representations for Life Distributions" *JASA*, **87**, 420, 1116-1122. .

[4] BERMAN, S., (1969). "Second Order Random Fields of l_p with Homogeneous and Isotropic Increments" *Z. Wahr. verw. Geb.*, **12**, 107-126.

[5] BERMAN, S., "Stationarity, Isotropy and Sphericity in l_p", *A. Wahr. verw. Geb.*, **54**, 21-23.

[6] BILLINGSLEY, P., (1968). *Convergence of Probability Measures.* John Wiley, New York.

[7] HAYAKAWA, Y., (1992). *Bayesian Parametric Models for Lifetimes from a Subjectivistic Viewpoint: Model Construction and Characterizations of Aging*. PhD dissertation, UC Berkeley.

[8] KUCZMA, M., (1964). "Note on Schröder's Functional Equation. *J. Austral. Math. Soc.*, **4**, 149-151.

[9] KUCZMA, M.; CHOCZEWSKI, B. AND GER, R., (1990). *Iterative Functional Equations*. Cambridge University Press, Cambridge.

[10] MENDEL, M.B., (1989). *Development of Bayesian Parametric Theory with Applications to Control"*. PhD Dissertation, Department of Mechanical Engineering, Massachusetts Institute of Technology.

[11] MISIEWICZ, J. AND COOKE, R., (1988). "l_p-Invariant Probability Measures". Delft University of Technology, Department of Mathematics Report 88-91.

Recent Adv. in Stat. and Prob., pp. 451-458
J. Pérez Vilaplana and M. L. Puri (Eds)

Extendibility of Schur Survival Functions and Aging Properties of Their One–dimensional Marginals

F. SPIZZICHINO

Dipartimento di Matemática "Guido Castelnuovo"', Università "La Sapienza", Piazzale Aldo Moro # 2, I-00185 Roma, Italia

1. INTRODUCTION

For $N \in \mathcal{N}$, let $t' \equiv (t'_1, \ldots, t'_N)$ and $t'' \equiv (t''_1, \ldots, t''_N)$ be two elements of $\mathcal{R}^{\mathcal{N}}$. The vector t' is said to be *majorized* by the vector t'', denoted by $t' \prec t''$, if

$$\sum_1^k {}_j t'_{(j)} \geq \sum_1^k {}_j t''_{(j)} \,, \quad k = 1, 2, \ldots, N-1 \,, \quad \sum_1^N {}_j t'_{(j)} = \sum_1^N {}_j t''_{(j)}$$

where $t'_{(1)} \leq t'_{(2)} \leq \cdots \leq t'_{(n)}$ and $t''_{(1)} \leq t''_{(2)} \leq \cdots \leq t''_{(n)}$ are the *order statistics* of $(t'_1, \ldots, t'_n)$ and $(t''_1, \ldots, t''_n)$, respectively. Symmetric functions $\psi : \mathcal{R}^{\mathcal{N}} \to \mathcal{R}$ for which $t' < t''$ implies $\psi(t') \leq \psi(t'')$ are said to be *Schur–convex*; ψ is *Schur–concave* if $\psi(t') \geq \psi(t'')$.

The concept of Schur–concavity (convexity) has been considered many times in the statistical literature (see e.g. Marshall, Olkin [9], Nevius, Proschan, Sethuraman [11] Marshall, Olkin [10] and references therein). Recently new attention has been payed to non–negative random variables with Schur–concave (convex) survival functions by Barlow and Mendel [2], Spizzichino [13], Barlow and Spizzichino [4]: it has been pointed out that such condition provides a probabilistic notion of aging which fits with a subjectivist interpretation of probability; in particular Schur–concavity models situations of *wear–out* and Schur–convexity models situations of *infant mortality*. These properties of the joint survival functions imply useful properties of conditional survival probabilities, given different kinds of failure and survival data; the latter properties, in turn, can be employed for studing problems of optimal (bayesian) design of life–testing experiments. Such a research is still in progress, some results can be found in [14].

Here we shall consider again non–negative random variables $T_1, \ldots, T_N$ with a Schur–concave (convex) survival function:

$$t' \prec t'' \Rightarrow P\{T_1 > t'_1, \ldots, T_N > t'_N\} \geq (\leq) P\{T_1 > t''_1, \ldots, T_N > t''_N\} \qquad (1)$$

$T_1, \ldots, T_N$ will be interpreted as the lifetimes of N units $U_1, \ldots, U_N$. Since, in any case, the joint survival function is assumed to be symmetric, $T_1, \ldots, T_N$ are exchangeable; the latter can be seen as a condition of *similarity* for $U_1, \ldots, U_N$.

In the following the joint survival function of $T_1, \ldots, T_N$ will be denoted by the symbol $\overline{F}_N(t_1, \ldots, t_N)$; for $n = 1, \ldots, N-1$, we put

$$\overline{F}_n(t_1, \ldots, t_n) \equiv \overline{F}_N(t_1, \ldots, t_n, 0, \ldots, 0) .$$

$\overline{F}_n$ is the marginal survival function of $T_1, \ldots, T_n$; due to exchangeability, any n-tuple of lifetimes $T_{i_1}, \ldots, T_{i_n}$ ($\{i_1, \ldots, i_n\}$ subset of $\{1, \ldots, N\}$) has the same survival function $\overline{F}_n$. We say that $\overline{F}_N$ is M-extendible ($M > N$) if we can find a symmetric M-dimensional survival function $\overline{F}_M$ such that $\overline{F}_N$ is the N-dimensional marginal of $\overline{F}_M$; $\overline{F}_N$ is infinitely-extendible if it is M-extendible for all $M > N$.

In this note we study some aspects of the special case of exchangeability defined by (1); in particular our aim is to illustrate that, in a sense, the motivations for considering (1) are analogous to those for which the concept of exchangeability was originally introduced [7]. Aging properties of one-dimensional marginal distributions have a central role in the discussion. In order to avoid confusion we must keep in mind that, in a subjectivist standpoint, stochastic independence is not synonymous with *physical* independence [8].

In the next Section we shall analyze some relations between aging and extendibility properties; Section 3 will be devoted to the special case of (possibly finitely-extendible) forms of *time-transformed exponential models*.

Consider the lifetimes $T_1, \ldots, T_N$ of N *similar* units $U_1, \ldots, U_N$ which are judged to act *independently* one of the other. Such a situation is of course an extreme and idealized one, but it is necessary to consider it as a starting point; in a probabilistic language it is described by the condition that $T_1, \ldots, T_N$ are independent random variables with a *same* uni-dimensional survival function $\overline{G}(\cdot)$. Now think of a common factor θ (e.g. the environmental temperature, say) which influence the failure mechanism of the units; $\overline{G}(\cdot)$ will depend on θ and we shall rather use the symbol $\overline{G}(\cdot|\theta)$.

In the frequentist setting $T_1, \ldots, T_N$ are independent random variables with an identical unidimensional (possibly unknown) survival function $\overline{G}(\cdot|\theta)$.

In the subjectivist setting, as very well-known, we must formulate this situation by saying that $T_1, \ldots, T_N$ are conditionally independent given θ. When the value of θ is a random quantity (with a distribution Π and values in a measurable space L), the joint (*unconditional*) survival function of $T_1, \ldots, T_N$ is

$$\overline{F}_N(t_1, \ldots, t_N) = \int_L \overline{G}(t_1, \cdot|\theta) \cdot \ldots \cdot \overline{G}(t_N|\theta)\, d\Pi(\theta) \tag{2}$$

and $T_1, \ldots, T_N$ are not independent, even if *the situation of physical independence in their behaviour is maintained.* Of course $\overline{F}_N$ in (2) is symmetric and $T_1, \ldots, T_N$ are exchangeable. Indeed the condition of exchangeability is the common feature in both the two situations of independence and conditional independence (more in particular $\overline{F}_N$ is infinitely extendible).

Exchangeability can arise also in a different way. $T_1, \ldots, T_N$ are again the lifetimes of the similar units $U_1, \ldots, U_N$ which act independently one of the other ($T_1, \ldots, T_n$ are i.i.d). At a certain stage we get a piece of (deterministic or probabilistic) information concerning a statistic of the form $S = \sum_{1}^{N}{}_j H(T_j)$, say (we consider a symmetric statistic in order to maintain a situation of symmetry in our conditional state of information about the units). Conditional on such information, $T_1, \ldots, T_N$ are not independent but, as it is easily seen, their joint distribution is still exchangeable; the latter situation of exchangeability is different from the one above in that the survival function is now, in general, finitely extendible.

We can conclude by stating that exchangeability (both in the finitely or infinitely extendible cases) is the special form of subjective dependence among lifetimes of units which *are similar and act independently*, when different situations of information are considered which may destroy stochastic independence but maintain the condition of similarity.

We must face now a more specific but similar situation. Think of units $U_1, \ldots, U_N$ which are similar, act independently one of the other and *undergo wear-out* during their operative life. This is modelled by saying that their lifetimes $T_1, \ldots, T_N$ are i.i.d random variables with a same uni-dimensional *logconcave survival function* [3]

$$\overline{G}(t) = \exp\{-\rho(t)\} \qquad t \geq 0$$

where ρ is a non-decreasing *convex* function (this is equivalent to the IFR property for absolutely continuous distributions). By replacing the condition of convexity for ρ by the condition of *concavity* (equivalent to the DFR property in the absolutely continuous case), we can model situations of *infant mortality*, instead of wear-out, for the units.

It is easily seen that the joint survival function $\overline{F}_N(t) = \exp\{-\sum_{1}^{N}{}_i \rho(t_i)\}$ is Schur-concave (Schur-convex) if and only if ρ is convex (concave).

As before, think now of an individual who is not sure about the value θ of a factor which influences the failure mechanism of $U_1, \ldots, U_N$:

$$\overline{F}_N(t) = \int_L \exp\{-\sum_{1}^{N}{}_i \rho(t_i|\theta)\}\, d\Pi(\theta) \tag{3}$$

where $\rho(t_i|\theta)$ is a non-decreasing convex function of t for any $\theta \in L$.

As recalled above, $T_1, \ldots, T_N$ are not independent, relative to the unconditional probability distribution (3). But also the property of IFR marginals can be lost, by integrating out θ: in general the convexity property of ρ does not imply that

$$\overline{F}_1(t) = \int_L \exp\{-\rho(t|\theta)\}\, d\Pi(\theta)$$

is log-concave. Indeed it can happen that, at the beginning of the units' life, the individual's conditional survival probabilities of the residual lifetimes increase more and more when the time elapses and no failure has been observed (see the discussion in [1] and [13]). The Schur-concavity property of the unconditional survival function is, on the contrary, maintained: the Schur-concavity property of

$$\overline{G}_N(t|\theta) = \exp\{-\sum_{1}^{N}{}_i \rho(t_i|\theta)\} \qquad \text{for any } \theta \in L$$

implies that $\overline{F}_N(t)$ in (3) is Schur-concave as well.

In what follows we show further instances of change of information which destroy the independence and marginal aging properties of the lifetimes, without modifying the Schur properties of the survival functions. We stress that the change of information that we are going to consider will in general lead to situations of finite extendibility.

Let $T_1, \dots, T_N$ originally be i.i.d. lifetimes with a joint survival function

$$\overline{F}_N(t) = \exp\{-\sum_{1}^{N} {}_i\,\rho(t_i)\}$$

(ρ concave non-decreasing so that $\overline{F}_N$ is Schur-convex) and let $X_1, \dots, X_m$ be observable random variables, conditionally independent on $T_1, \dots, T_N$ given the statistic $S(T) = \sum_1^N {}_i\,\rho(T_i)$. Suppose we observe the values $x_1, \dots, x_m$ taken by $X_1, \dots, X_m$ and consider the conditional distribution of $T_1, \dots, T_N$ given $x_1, \dots, x_m$; the conditional survival function of $T_1, \dots, T_N$ given $x_1, \dots, x_m$ has the form

$$\overline{F}_N(t|x) = \int_0^\infty \left[1 - \frac{\sum_i \rho(t_i)}{\eta}\right]_+^{N-1} \pi(\eta|x)\, d\eta \qquad (4)$$

where $\pi(\cdot|x)$ denotes the conditional density of $S(T)$ given $(X = x)$.

The expression (4) is easily proved by using the conditional independence assumption of T and X, given $S(T)$, and by taking into account that the conditional survival function of $T_1, \dots, T_N$ given $(S(T) = \eta)$ is of the form $\left[1 - \frac{\sum_i \rho(t_i)}{\eta}\right]_+^{N-1}$ (see e.g. [2]).

$\overline{F}_N(t|x)$ in (4) is a symmetric, and in general finitely extendible, survival function; ρ being a concave function does not necessarily imply that the conditional one-dimensional marginal $\overline{F}_1(t|x) = \int_0^\infty \left[1 - \frac{\rho(t)}{\eta}\right]_+^{N-1} \pi(\eta|x)\, d\eta$ is log-convex. But ρ being concave does imply that $\overline{F}_N(t|x)$ in (4) is a Schur-convex function of $t_1, \dots, t_N$, alike the original unconditional survival function $\overline{F}_N(t)$.

As a conclusion of the discussion above we recognize that Schur-concavity (Schur-convexity) describes a special form of *subjective* dependence arising among lifetimes of similar units which act independently one of the other and undergo wear-out (infant mortality) when different situations of information are considered which destroy the stochastic independence but maintain the condition of similarity.

2. RELATIONS BETWEEN DEPENDENCE AND AGING PROPERTIES

As recalled above, in the i.i.d. case Schur-concavity (Schur-convexity) is equivalent to the IFR (DFR) property of the one-dimensional marginal distribution (for simplicity's sake we are thinking here of absolutely continuous distributions) but such an equivalence does not hold in general. However by a simple remark we can expect that other different situations of Schur-concavity (Schur-convexity) imply some constraints on the aging properties of the one-dimensional marginal distribution.

In [4] the following representation has been introduced for joint survival functions

$$\overline{F}_N(t) = \overline{F}_1(h_N(t)) \qquad (5)$$

where $\overline{F}_1$ is the one-dimensional marginal survival function and $h_N(t) = \overline{F}_1{}^{-1}(\overline{F}_N(t))$.

$\overline{F}_N(t)$ is Schur-concave (convex) if and only if $h_N(t)$ is Schur-convex (concave); moreover $\overline{F}_N(t') = \overline{F}_N(t'')$ if and only if $h_N(t') = h_N(t'')$. (5) represents the dependence of $\overline{F}_N$ on t by separating the effects of the one-dimensional marginal survival function from the form of the *level sets*. In principle, by using (5), we might combine an arbitrary one-dimensional survival function $\overline{F}_1$ with a function $h_N(t)$ corresponding to the level sets of an arbitrary Schur-concave (say) survival function in order to build up a Schur-concave survival function with the prescribed one-dimensional survival function $\overline{F}_1$. As a matter of fact such a procedure may lead to a function which is not an N-dimensional survival function (being an N-dimensional survival function requires of course that any N-dimensional interval has a non-negative probability and this condition is not ensured by the construction above). This argument shows that the form of the level sets may imply some constraint on the form of $\overline{F}_1$. In what follows we shall present some simple but interesting results in this direction.

We recall that two random variables T_1, T_2 are *negatively (positively) dependent* if

$$P\{T_1 > y_1, T_2 > y_2\} \le P\{T_1 > y_1\}P\{T_2 > y_2\}$$
$$\left(P\{T_1 > y_1, T_2 > y_2\} \ge P\{T_1 > y_1\}P\{T_2 > y_2\}\right)$$

(for brevity's sake, in the following we shall briefly use the term *dependent* in place of *upper orthant dependent*). Moreover a unidimensional survival function $\overline{G}$ is NBU (NWU) if $\overline{G}(t+s) \le (\ge)\overline{G}(t)\overline{G}(s)$. (Note that IFR property implies NBU and DFR implies NWU).

Proposition 1 [5]. *Let $\overline{F}_N(y)$ be Schur-concave. If $T_i, T_j (i \ne j)$ are negatively dependent then the unidimensional marginal is NBU. If the unidimensional marginal is NWU, then they are positively dependent.*

PROOF: Negative dependence means $\overline{F}_1(y_i) \cdot \overline{F}_1(y_j) \ge \overline{F}_2(y_i, y_j)$. On the other hand Schur-concavity of $\overline{F}_N(y)$ implies Schur-concavity of $\overline{F}_2(y_i, y_j)$ and we can write $\overline{F}_2(y_i, y_j) \ge \overline{F}_2(y_i + y_j, 0) = \overline{F}_1(y_i + y_j)$.

We can then conclude $\overline{F}_1(y_i) . \overline{F}_1(y_j) \ge \overline{F}_1(y_i + y_j)$.

If the unidimensional marginal is NWU, we can write $\overline{F}_1(y_i + y_j) \ge \overline{F}_1(y_i) . \overline{F}_1(y_j)$ and Schur-concavity of $\overline{F}_2$ then yields

$$\overline{F}_2(y_i, y_j) \ge \overline{F}_2(y_i + y_j, 0) = \overline{F}_1(y_i + y_j) \ge \overline{F}_1(y_i) \cdot \overline{F}_1(y_j) . \qquad \blacksquare$$

By using similar arguments, it is easy to prove the following "dual" result for lifetimes with Schur-convex survival function.

Proposition 2 . *Let $\overline{F}_N(y)$ be Schur-convex.*

a) If $T_i, T_j (i \ne j)$ are positively dependent then the uni-dimensional marginal is NWU.

b) If the unidimensional marginal is NBU, then they are negatively dependent.

Since conditional independence with a same DFR (conditional) distribution is a special case of positive dependence with Schur-convex survival function and DFR is a special case of NWU, the statement (a) above can be seen as an extension of the following well-known result.

Proposition 3 [3]. *If $T_1, \dots, T_N$ are conditionally independent with a DFR (conditional) distribution, then their unconditional marginal is DFR as well.*

Negative dependence of exchangeable lifetimes implies a limitation to their extendibility [12]; so, for Schur-convex survival functions, the NBU property is an obstruction to extendibility. In this respect we can give a more precise result; for $\overline{F}_N(y)$ given, denote by M^* the maximum M such that $\overline{F}_N$ is M-extendible and put

$$q = \inf_{y>0} \frac{P\{T_1 > 2y | T_1 > y\} - P\{T_1 > y\}}{P\{T_1 \le y\}} .$$

Proposition 4 *Let $\overline{F}_N(y)$ be Schur-convex. If $q < 0$, then*

$$M^* \le 1 - \frac{1}{q} . \tag{6}$$

PROOF: Fix $\varepsilon \in (0, -q)$ and choose $y > 0$ such that $\frac{P\{T_1 > 2y | T_1 > y\} - P\{T_1 > y\}}{P\{T_1 \le y\}} < q + \varepsilon$. We consider the exchangeable random variables $T_1, \dots, T_N, T_{N+1}, \dots, T_{M^*}$ and denote by Z_i the indicator of the event $\{T_i > y\}$ $(i = 1, 2, \dots, M^*)$.

Note that, by exchangeability, the correlation coefficient of Z_i and $Z_j (i \ne j)$ is

$$\begin{aligned} \rho &= \frac{\overline{F}_2(y,y) - [\overline{F}_1(y)]^2}{\overline{F}_1(y)[1 - \overline{F}_1(y)]} \le \frac{\overline{F}_1(2y) - [\overline{F}_1(y)]^2}{\overline{F}_1(y)[1 - \overline{F}_1(y)]} = \\ &= \frac{P\{T_1 > 2y | T_1 > y\} - P\{T_1 > y\}}{P\{T_1 \le y\}} < 0 . \end{aligned}$$

It results

$$Var\,(\sum_1^{M^*}{}_i Z_i) = M^*\,Var\,(Z_i) + M^*\,(M^* - 1)\rho\,Var\,(Z_i) ,$$

whence $M^* \le 1 - \frac{1}{\rho} < 1 - \frac{1}{q+\varepsilon}$ and the inequality (6) is then achieved by letting ε to go to 0. ∎

3. THE CASE OF TIME-TRANSFORMED EXPONENTIAL MODELS

In the case of i.i.d. life-times the function $h_N(t)$ in (5) has the form

$$h_N(t) = \rho^{-1}[\sum_{i=1}^{N} \rho(t_i)] \tag{7}$$

where ρ is a non-decreasing function such that $\rho(0) = 0$ and $\rho(+\infty) = +\infty (\rho = -\log \overline{F}_1)$. As shown in [4], in general a survival function $\overline{F}_N(t)$ has the level sets specified by $h_N(t)$ of the form (7) (i.e. it has the same level sets of a *product measure*) if and only it corresponds to a *time-transformed exponential model*, namely if and only if it is of the form

$$\overline{F}_N(t) = \int_0^\infty \left[1 - \frac{\sum_i \rho(t_1)}{\eta}\right]_+^{N-1} d\Pi_N(\eta) \tag{8}$$

where Π_N is a probability distribution on $[0, \infty)$.

A joint survival function $\overline{F}_N(t)$ of the form (8) is Schur-concave (convex) if and only if ρ is convex (concave). We have the case of i.i.d. life-times when Π_N has a *gamma* density with parameters N, $\beta(\beta > 0)$; if the density of Π_N is a mixture of gamma densities:

$$\pi(\eta) = \frac{1}{(N-1)!} \int_0^\infty \beta^N \eta^{N-1} \exp\{-\beta\eta\}\, dP(\beta)$$

then we have the case of conditionally i.i.d lifetimes:

$$\overline{F}_N(t) = \int_0^\infty \exp\{-\beta \sum_i \rho(t_i)\}\, dP(\beta).$$

In general, in the other cases for the distribution Π_N, $\overline{F}_N(t)$ in (8) is not infinitely extendible. Relatively to the form of the interdependence among $T_1, \ldots, T_N$ we have the following result

Proposition 5 *$T_i, T_j (i \neq j)$ are negatively (positively) dependent if and only if $\int_0^\infty [1 - \frac{z}{\eta}]_+^{N-1}\, d\Pi_N(\eta)$ is an NBU (NWU) function of z.*

PROOF: Let us consider the variables $Z_1 \equiv \rho(T_i)$ and $Z_z \equiv \rho(T_j)$; of course T_i, T_j are positively (negatively) dependent if and only if Z_1, Z_2 are positively (negatively) dependent. The joint survival function of Z_1, Z_2 is

$$\overline{H}(z_1, z_2) = P\{\rho(T_i) > z_1, \rho(T_j) > z_2\} = \overline{F}_1(\rho^{-1}(z_1 + z_2)).$$

This is Schur-constant and its one-dimensional marginal is given by $\overline{H}_1(z) = \overline{F}_1(\rho^{-1}(z))$.

By (8) we have that

$$\overline{F}_1(t) = \int_0^\infty [1 - \frac{\rho(t)}{\eta}]_+^{N-1}\, d\Pi_N(\eta)$$

and then

$$\overline{H}_1(z) = \int_0^\infty [1 - \frac{z}{\eta}]_+^{N-1}\, d\Pi_N(\eta).$$

Now we can conclude by remarking that in the Schur-constant case the one-dimensional marginal survival function is NBU (NWU) if and only if Z_i, Z_j are positively (negatively) dependent, as it can be immediately shown (see [6]). ■

REFERENCES

[1] R.E. BARLOW. (1985). A Bayesian Explanation for an Apparent Failure Rate Paradox. *IEEE Trans. On Rel.*, R34, **2**, 107–108.

[2] R.E. BARLOW, M. MENDEL. (1992). Similarity as a Characteristic of Wear-out. In *Reliability and Decision Making*, [Barlow, Clarotti, Spizzichino (Eds)] Elsevier Applied Science. London.

[3] R.E. BARLOW, F. PROSCHAN. (1975). *Statistical Theory of Reliability and Life Testing.* Holt, Rinehart and Winston. New York.

[4] R.E. Barlow, F. Spizzichino, Schur-concave Survival Functions and Survival Analysis, *Journal of Computational and Applied Mathematics* (to appear).

[5] L. Caramellino. (1991). *Variabili Aleatorie Non Negative: Proprietà di Schur e di Ageing delle Distribuzioni*, PhD Dissertation in Mathematics, Univ. of Rome "La Sapienza".

[6] L. Caramellino, F. Spizzichino. (1992). *Dependence and Aging Properties of Lifetimes with Schur-constant Survival Functions.* Technical Report 37/92, Department of Mathematics, Univ. of Rome "La Sapienza".

[7] B. De Finetti. (1937). La Prevision: Ses Lois Logiques, Ses Sources Subjectives. *Ann. Inst. H. Poincaré*, **7**, 1–68.

[8] B. De Finetti. (1970). *Teoria delle Probabilità.* Einaudi. Torino. [English translation: *Theory of Probability*, John Wiley and Sons. New York, 1974].

[9] A. Marshall, I. Olkin. (1974). Majorization in Multivariate Distributions, *Ann. of Stat.*, **2**, 1891–1200 .

[10] A. Marshall, I. Olkin. (1979). *Inequalities: Theory of Majorization and its Applications.* Academic Press. New York.

[11] [11] S.E. Nevius, F. Proschan, J. Sethuraman. (1977). Schur Functions in Statistics–II. Stochastic Majorization. *Ann. of Stat.*, **5**, 263–273.

[12] F. Spizzichino. (1982). *Extendibility of Symmetric Probability Measures.* In *Exchangeability in Probability and Statistics* [G. Koch, F. Spizzichino (Eds.)] North–Holland. Amsterdam.

[13] F. Spizzichino. (1992). *Reliability Decision Problems under Conditions of Ageing.* In *Bayesian Statistic 4* [Bernardo, Berger, Dawid, Smith (Eds.)]. Clarendon Press. Oxford.

[14] F. Spizzichino. (1992). *A Unifying Approach to Optimal Design of Life-testing and Burning.* In *Reliability and Decision Making* [Barlow, Clarotti, Spizzichino (Eds)]. Elsevier Applied Science. London.

Bei Fragen zur Produktsicherheit wenden Sie sich bitte an:
If you have any questions regarding product safety,
please contact:

Walter de Gruyter GmbH
Genthiner Straße 13
10785 Berlin
productsafety@degruyterbrill.com